Joachim E. Trümper
Günther Hasinger (Eds.)

The Universe in X-Rays

With 237 Figures, 40 in Color and 19 Tables

Springer

Joachim E. Trümper
Günther Hasinger

Max-Planck-Institut
für extraterrestische Physik
Giessenbachstraße
85748 Garching
Germany
E-mail: jtrumper@mpe.mpg.de
grh@mpe.mpg.de

Cover illustrations: above: ROSAT all-sky survey; from Max-Planck-Institut fuer extraterrestrische Physik; below: 15. Nov. 2006, Credit: NASA/CXC/MIT/UMass Amherst/M.D. Stage et al.

Library of Congress Control Number: 2007933496

ISSN 0941-7834
ISBN 978-3-540-34411-7 Springer Berlin Heidelberg New York

Springer is a part of Springer Science+Business Media
springer.com

Typesetting by SPi using a Springer LATEX macro package

Cover design: eStudio Calamar, Girona/Spain

Printed on acid-free paper SPIN: 11328858 55/SPi 5 4 3 2 1 0

Preface

In the early years of X-ray astronomy, one of us (J. E. T.) had always a book in reach, entitled "X-ray Astronomy" and edited by Riccardo Giacconi and Herbert Gursky about a decade after they had opened the field by the discovery of Scorpius X-1 and the X-ray background. This book summarized all the knowledge at the time, based on the results from the pioneering rocket and balloon experiments and from Uhuru, the first satellite entirely dedicated to X-ray astronomy.

Since those early times X-ray astronomy has evolved with enormous pace. The number of known sources has increased by a factor of thousand, but more important, they now comprise almost all classes of astronomical objects – from planets, moons and comets out to clusters of galaxies and quasars. In the era of multi-wavelength astronomy X-ray observations provide insight into extreme physical conditions prevailing in all these sources – very high temperatures, very strong gravitational fields, super-nuclear densities, extreme concentrations of relativistic particles.

The intent of this book is to summarize the present status of the field, which has become quite challenging, since the number of publications in refereed journals has risen to more than 20 000. Therefore the coverage cannot be complete, but must rather be representative. We apologize for omitting any important ideas, methods or results.

The authors of the various chapters are mainly scientists working at the Max-Planck-Institute for Extraterrestrial Physics, the home of ROSAT, or colleagues who have been working closely with us during the last 20 years. Besides ROSAT, the main sources of information have been the other X-ray satellites of the nineties – ASCA, RXTE and BeppoSAX – and their more recent successors – Chandra, XMM-Newton, INTEGRAL, Swift and Suzaku – which have used novel instrumentation to produce a wealth of knowledge on the universe seen at high energies.

This book addresses mainly scientists who are teaching the subject, and young scientists entering the field, as well as astronomers from neighbouring disciplines and physicists interested in one of the most exciting fields of astrophysics. It is organized in a straightforward way: We start with a discussion of instruments and methods in part I and then continue in parts II and III with the status of galactic and extragalactic X-ray astronomy respectively, ordering the contributions in a geocentric fashion. In Chapter 26 a short summary of the current plans for future missions in X-ray astronomy is given.

We are very thankful to all authors of this book for their contributions. The editorial assistance of Konrad Dennerl who brought the LaTeX manuscript into its final shape is gratefully acknowledged. We thank Birgit Boller and Walburga Frankenhuizen for their dedicated secretarial support, as well as Maria Fürmetz and Barbara Mory for their painstaking work on the index of the book.

Garching, November 2007

Joachim E. Trümper
Günther Hasinger

Contents

Part II Galactic X-Ray Astronomy

List of Contributors

Werner Becker
Max-Planck-Institut für extraterrestrische Physik, Giessenbachstraße,
85748 Garching, Germany
e-mail: web@mpe.mpg.de

Klaus Beuermann
Institut für Astrophysik, Friedrich-Hund-Platz 1, 37077 Göttingen, Germany,
e-mail: beuermann@astro.physik. uni-goettingen.de

Hans Böhringer
Max-Planck-Institut für extraterrestrische Physik, Giessenbachstraße,
85748 Garching, Germany
e-mail: hxb@mpe.mpg.de

Thomas Boller
Max-Planck-Institut für extraterrestrische Physik, Giessenbachstraße,
85748 Garching, Germany
e-mail: bol@mpe.mpg.de

Dieter Breitschwerdt
Universität Wien, Türkenschanzstraße 17, 1180 Wien, Austria,
e-mail: breitschwerdt@univie.ac.at

Konrad Dennerl
Max-Planck-Institut für extraterrestrische Physik, Giessenbachstraße,
85748 Garching, Germany
e-mail: kod@mpe.mpg.de

Michael Freyberg
Max-Planck-Institut für extraterrestrische Physik, Giessenbachstraße,
85748 Garching, Germany
e-mail: mjf@mpe.mpg.de

Peter Friedrich
Max-Planck-Institut für extraterrestrische Physik, Giessenbachstraße,
85748 Garching, Germany
e-mail: pfriedrich@mpe.mpg.de

Jochen Greiner
Max-Planck-Institut für extraterrestrische Physik, Giessenbachstraße,
85748 Garching, Germany
e-mail: jcg@mpe.mpg.de

Frank Haberl
Max-Planck-Institut für extraterrestrische Physik, Giessenbachstraße,
85748 Garching, Germany
e-mail: fwh@mpe.mpg.de

Günther Hasinger
Max-Planck-Institut für extraterrestrische Physik, Giessenbachstraße,
85748 Garching, Germany
e-mail: ghasinger@mpe.mpg.de

Eckhard Kendziorra
Institut für Astronomie und Astrophysik, Universität Tübingen, Sand 1,
72076 Tübingen, Germany
e-mail: kendziorra@astro.uni-tuebingen.de

Stefanie Komossa
Max-Planck-Institut für extraterrestrische Physik, Giessenbachstraße,
85748 Garching, Germany
e-mail: skomossa@mpe.mpg.de

Joachim Krautter
Landessternwarte, Königstuhl, 69117 Heidelberg, Germany
e-mail: jkrautte@lsw-uni-heidelberg.de

Norbert Meidinger
Max-Planck-Institut für extraterrestrische Physik, Giessenbachstraße,
85748 Garching; MPI Halbleiterlabor, Otto-Hahn-Ring 6, 81739 München,
Germany
e-mail: nom@hll.mpg.de

Robert Petre
X-ray Astrophysics Laboratory, NASA / Goddard Space Flight Center,
Greenbelt, MD 20771, USA,
e-mail: robert.petre-1@nasa.gov

Wolfgang Pietsch
Max-Planck-Institut für extraterrestrische Physik, Giessenbachstraße,
85748 Garching, Germany
e-mail: wnp@mpe.mpg.de

Elmar Pfeffermann
Max-Planck-Institut für extraterrestrische Physik, Giessenbachstraße,
85748 Garching, Germany
e-mail: epf@mpe.mpg.de

Peter Predehl
Max-Planck-Institut für extraterrestrische Physik, Giessenbachstraße,
85748 Garching, Germany
e-mail: predehl@mpe.mpg.de

Jürgen H.M.M. Schmitt
Hamburger Sternwarte, Gojenbergsweg 11, 21029 Hamburg, Germany
e-mail: jschmitt@hs.uni-hamburg.de

Rüdiger Staubert
Institut für Astronomie und Astrophysik, Univ. Tübingen, Sand 1, 72076 Tübingen, Germany
e-mail: staubert@astro. uni-tuebingen.de

Beate Stelzer
Osservatorio Astronomico di Palermo, Piazza del Parlamento 1, 90134 Palermo, Italien,
e-mail: stelzer@astropa.unipa.it

Lothar Strüder
Max-Planck-Institut für extraterrestrische Physik, Giessenbachstraße,
85748 Garching; MPI Halbleiterlabor, Otto-Hahn-Ring 6, 81739 München, Germany
e-mail: lts@hll.mpg.de

Yasuo Tanaka
Max-Planck-Institut für extraterrestrische Physik, Giessenbachstraße,
85748 Garching, Germany
e-mail: ytanaka@mpe.mpg.de

Joachim Trümper
Max-Planck-Institut für extraterrestrische Physik, Giessenbachstraße,
85748 Garching, Germany
e-mail: jtrumper@mpe.mpg.de

Klaus Werner
Institut für Astronomie und Astrophysik, Universität Tübingen, Sand 1,
72076 Tübingen, Germany
e-mail: werner@astro.uni-tuebingen.de

Part I

X-Ray Astronomical Instrumentation

1 Overview

R. Staubert and J. Trümper

The advancement of X-ray astronomy since its start about half a century ago has been strongly dependent on the development of instruments and observational techniques. Since the earth's atmosphere is opaque for X- and gamma-rays this field could only develop in parallel to space technology providing the necessary carriers, which can place X-ray astronomy telescopes and detectors near or beyond the boundaries of our atmosphere. In the early days, in the sixties and seventies, stratospheric balloons and rockets played an important role, albeit with severe limitations on altitude ($\sim$40 km, leaving still substantial absorption) and on observing time of a few minutes, respectively. Today, satellites are available allowing X-ray astronomy missions to last for a decade or longer. The principle mode of measurement in X-ray astronomy is to detect individual photons with the aim to determine the complete set of four properties: arrival direction (leading to images), the energy and the time of arrival of the photon, and its polarization angle. The first detectors were proportional counters and scintillation counters, originally developed for detecting charged particles in nuclear physics research. They had effective areas of a few hundred square centimeters and were usually equipped with mechanical collimators providing some indirect imaging capability through the restriction of the field of view (typically to a few square degrees) and the possibility for scanning observations. An important challenge for these detectors was the reduction of the background radiation, both from photons of the diffuse X-ray sky background and from charged particles of the ever present cosmic rays. This was achieved by narrow colimators and the invention of various techniques of anticoincidence and veto schemes, as perfected for example in multiwire proportional counters. The first X-ray satellite Uhuru, launched in December 1970, carried collimated gas proportional counters and was scanning the entire X-ray sky. The detection of $\sim$400 X-ray sources marked a quantum leap in X-ray astronomy. The so called "gas scintillation proportional counter," combined the two physical detector principles and gave an improved energy resolution, but had limited application and scientific impact.

The next major step was the introduction of focusing and imaging X-ray optics, the Wolter telescope, together with imaging detectors in the focal plane providing two-dimensional X-ray images. The first satellite mission, the Einstein Observatory, carrying such a telescope with the imaging proportional counter (IPC) and the high resolution imager (HRI) as focal plane detectors allowed a break through in two areas: extended objects could directly be imaged, and for all sources the sensitivity

J.E. Trümper and G. Hasinger (eds.), *The Universe in X-Rays.*

was greatly improved through the focusing and the corresponding background reduction.

ROSAT performed the first all sky survey with an imaging telescope. Using a greatly improved telescope and detector technology, it provided a large step in the observational capabilities, both in the number of detected X-ray sources ($\sim$125.000), and through the large number of pointings throughout the remaining 8 years of the mission. Today the standard focal plane detector is based on actively cooled pixelized solid state detectors (CCDs), which provide a higher energy resolution and wider energy range than proportional counters. The use of CCDs was pioneered by ASCA and further perfected on Chandra and XMM-Newton.

In parallel to imaging telescopes, high resolution grating spectrometers were developed, first used in the Einstein Observatory and today with great success in the Chandra and XMM-Newton missions. Intensive work has also gone into the development of very deeply cooled bolometers, which have a great potential because of their very high spectral resolution and large throughput. Unfortunately, the first bolometer flown on a satellite exploded with ASTRO-E, and the second attempt on Suzaku failed because the cryogenic coolant was lost before the observations commenced. At higher photon energies ($>$10 keV), focusing becomes difficult and the current technique is imaging by spatial aperture modulation, the so called "coded mask" technique, first used in the Mir-KVANT mission and now on INTEGRAL. Efforts are underway to develop also focusing telescopes for hard X-rays and even gamma-rays by employing multilayer-coded reflecting surfaces or making use of Bragg reflection on crystals. Polarimetry is still in a rudimentary state. Imaging, high resolution spectroscopy, and high time resolution measurements have reached a high level of sophistication with a corresponding wealth of scientific results, but there is still a wide open field for further advances.

2 Proportional Counters

E. Pfeffermann

2.1 Introduction

After the discovery of the first extra-solar X-ray source in 1962 with a gaseous detector, the proportional counter became the workhorse instrument of soft X-ray astronomy for nearly four decades. The origin of gaseous detectors dates back to the early twentieth century, when Rutherford and Geiger published in 1908: "An electrical method of counting the number of α-particles from radioactive substances" [23]. Cosmic rays were discovered in 1912 with a gaseous detector by V. Hess. During the first half of the twentieth century much progress was made in the technology of gaseous detectors in the fields of nuclear and cosmic-ray physics [8]. For instance, the discovery of the effect of quench gases allowed a stable operation of gas detectors [28]. Ionization-dependent output signals of gas counters were observed first by Geiger and Klemperer [7]. About 10 years later, proportional counters were developed [13]. First attempts to operate multiwire detectors were carried out in conjunction with the Manhatten project [22]. In 1968, Charpak and collaborators succeeded in the development and operation of multiwire proportional counters (MWPC) [4]. These detectors combine the advantages of a large sensitive area with multidimensional event parameter sensing. Many innovative ideas arose from the group around Charpak. The development of gaseous detectors is still going on. New detector types like the micro strip gas chamber (MSGC) [15] and micro pattern gas detectors (MPGD) like the gas electron multiplier (GEM) have been described [25].

2.2 Gaseous Detectors

A gaseous detector is basically a capacitor filled with gas. Electrons and ions generated by ionizing radiation in the gas are collected by the corresponding electrodes. The acceleration of the charges in the electrical field extracts energy from the capacitor. Therefore, the electrodes show signals before the charges arrive at the electrodes. Depending on the gas and the electrical field strength in the capacitor, the instrument operates in different modes. At moderate electrical field strengths, electrons and ions are just collected by the electrodes, this is the so called ionization-chamber mode. At a higher field strength, the electrons gain enough energy on a mean free path

J.E. Trümper and G. Hasinger (eds.), *The Universe in X-Rays.*

length to excite the gas atoms (in this case a pure noble gas) resulting in the emission of UV light. The detector operates as a gas scintillation proportional counter. A further increase of the electrical field enables the electrons to ionize gas atoms by collisions and charge multiplication takes place. As long as the charge generated by the multiplication process is proportional to the original number of electrons, the detector is in the proportional counter mode. At higher electrical fields the detector enters the Geiger mode, where the avalanche propagates through the whole detector. The saturated signals are no longer proportional to the original charge. The spark chamber mode is the subsequent mode at the highest field strength. The ionizing event triggers a spark discharge of the counter. It depends on the skill of the detector designer that the instrument stays in the desired operating conditions even in the harsh space environment.

2.3 Operation Principle of a Proportional Counter

The simplest geometry of a proportional counter is a gas-filled cylindrical conductive tube with a coaxial thin wire as shown in Fig. 2.1. The wire is connected to a positive high voltage and coupled via a capacitor to a charge sensitive preamplifier. For the detection of X-rays, the cathode tube has to have a window, transparent to the required energy band. X-rays entering the detector volume through the window interact with the detector gas. At X-ray energies up to 50 keV the predominant interaction process is the photo effect. The photo effect cross section scales as $Z^n E^{-\frac{8}{3}}$, where E is the X-ray energy and Z is the atomic number of the detector gas and $n \approx 4$–5. The number N of electron–ion pairs generated by this event can be written as $N = E/W$, where E is the energy of the absorbed X-ray photon and W the average energy for the creation of one electron–ion pair in the detector gas (usually a noble gas with an additive of a molecular gas like CO_2 or CH_4). The average energy for the creation of an electron–ion pair depends on the detector gas and is about 25–30 eV. A 1 keV X-ray photon creates 30–40 electron–ion pairs. Electrons and ions drift in the electrical field of the detector to the anode wire and the cathode, respectively. If the electrons gain enough energy over a mean free path length to ionize the detector gas, charge multiplication takes place. The actual charge reduplicates on average after each ionizing collision of the electrons. This happens in the vicinity

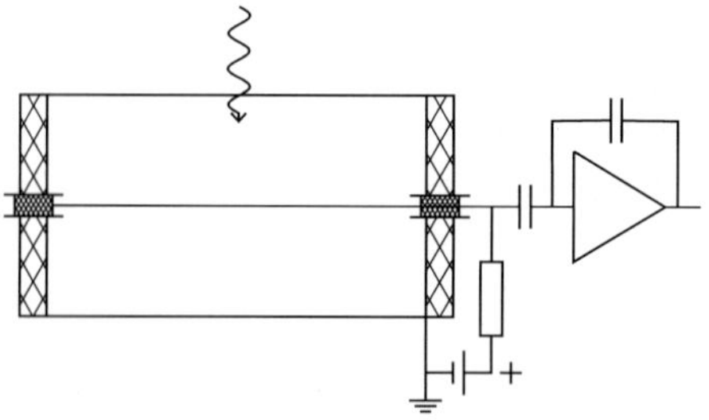

Fig. 2.1 Single wire proportional counter

of the anode wire, where the electrical field is in the order of 10^5 V cm^{-1}. In this cylindrical geometry, the electrical field strength as a function of the radial distance r from the tube center is: $\mathrm{d}U/\mathrm{d}r = U_o/[r(\ln r_c/r_a)]$ with U_o = anode wire voltage, r_a = anode wire radius, r_c = cathode tube radius . The movement of electrons and ions extracts energy from the electrical field generating displacement currents on anode and cathode. The electrons move about three orders of magnitude faster than the ions and the majority of the charge is generated only several mean free path lengths away from the anode wire. Therefore, the waveform of the output signal of the detector has a small fraction with a short rise time, contributed by the electrons. The main portion of the signal, with a rise time of 100 μs or more, is generated by the movement of the ions. Not only charge multiplication takes place in the avalanche, but also the generation of UV photons both by excitation of gas atoms and by the neutralization of positive ions on arrival at the cathode. UV photons hitting the cathode induce the emission of electrons from cathode surface, when the work function of the cathode material is less than the photon energy. These electrons in turn can cause subsequent avalanches possibly leading to a permanent discharge of the counter. The addition of several % of a polyatomic gas (quench gas) to the detector gas prevents this problem. Quench gases absorb UV photons emitted by the noble gas and convert them via radiationless transitions finally into heat. Via charge exchange quench gases reduce also the number of noble gas ions arriving at the cathode. Quench gases can speed up the drift velocity of electrons quite dramatically reducing the influence of gas impurities [24].

2.3.1 Quantum Efficiency of Proportional Counters

The quantum efficiency of a proportional counter for X-rays is determined by the transmission of the window and the absorption of the detector gas. To achieve a high transmission rather thin windows are used. A thin window has to be supported by a grid to withstand the gas pressure. The X-ray transmission of the support structure is usually energy independent and reduces the transmission by a constant factor (T). Proportional counters with a permanent gas filling have to use metallic window materials like beryllium or aluminum. Metallic window materials limit the detectable X-ray band to energies above 1.5 keV. Detectors using plastic window materials like polypropylene (about 1 μm thick) are able to detect X-ray photons down to 0.1 keV. Because of the gas diffusion through the plastic window such detectors have to use a gas supply system. Figure 2.2 shows the X-ray transmission of a 1 μm polypropylene foil and a 25 μm beryllium foil as a function of energy. The absorption of X-rays in the detector gas, usually a mixture of a noble gas with 5–20% quench gas, depends on the atomic numbers of the gas mixture, the gas pressure, and the dimension of the absorption region. For low energies, the basic constituent of the gas mixture is argon, whereas for higher energies increasing admixtures of xenon or pure xenon with a quench gas is used. The quantum efficiency of the detector can be written:

$$Q = Te^{-d\mu_w}(1 - e^{-g\mu_g}) \tag{2.1}$$

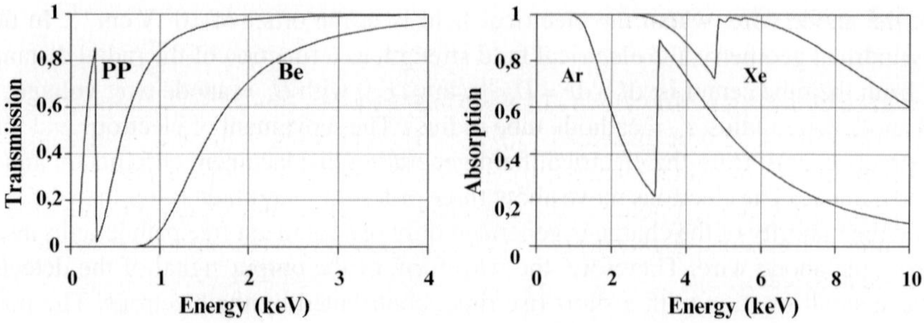

Fig. 2.2 *Left panel* shows the transmission of two window materials 25 μm beryllium (Be) and 1 μm polypropylene (PP). The *right panel* shows the absorption of 1 cm of argon (Ar) and xenon (Xe) at a pressure of 1 bar as a function of X-ray energy [10]

d and g are the window and gas column densities in g cm^{-2}. μ_w, μ_g are the corresponding energy-dependent mass absorption coefficients.

2.3.2 *Energy Resolution*

Proportional counters have a moderate energy resolution. The energy resolution is mainly determined by the statistics of the initial ionization process and the statistics of charge multiplication. The variation of the number N of electron–ion pairs created by the ionizing event is less than that estimated from Poisson statistics, because the collisions of the ionization process are not statistically independent. The Fano factor F is an empirical constant to adapt the experimental observed variance to the predicted one [5].

$$\left(\frac{\sigma_N}{N}\right)^2 = \frac{F}{N}; \qquad F \approx 0.05 - 0.2 \tag{2.2}$$

For large values of multiplication, the variance of the amplification A of a single electron is [12]:

$$\left(\frac{\sigma_A}{\bar{A}}\right)^2 \simeq b; \qquad b \approx 0.5 - 0.6 \tag{2.3}$$

Therefore, the energy resolution of a proportional counter is:

$$\left(\frac{\sigma_E}{E}\right)^2 = \left(\frac{\sigma_N}{N}\right)^2 + \frac{1}{N}\left(\frac{\sigma_A}{\bar{A}}\right)^2 \tag{2.4}$$

$$\frac{\sigma_E}{E} = \sqrt{\frac{F+b}{N}} = \sqrt{\frac{W(F+b)}{E}} \tag{2.5}$$

These estimates give a limit for the best achievable energy resolution. Gas impurities, tolerances of anode wire diameter, and loss of electrons at the entrance window can deteriorate the energy resolution of the proportional counter.

2.3.3 Time Resolution

Because of the high drift velocity of electrons in gases (10^6–10^7 cm s^{-1} at moderate electrical fields, see for instance [24]), proportional counters have a good time resolution. In cylindrical geometry, only the electron drift time between the absorption position of the X-ray photon and the avalanche region contributes substantially to the time uncertainty. Depending on detector geometry time resolution below 1μs can be achieved.

2.3.4 Background Rejection Capability

Detectors for X-ray astronomy operated in space are exposed to the whole spectrum of cosmic rays. The background rate exceeds by far the X-ray event rate of the majority of cosmic X-ray sources. A sophisticated event selection logic is mandatory to distinguish real X-ray events from charged particle events or fluorescent X-rays from surrounding materials. One possibility is to limit the energy band of accepted events. The depth of the detector cell must be chosen large enough so that minimum ionizing particles deposit more energy than the most energetic accepted X-ray event. In this way, minimum ionizing particles can be easily discriminated with an upper event threshold. Another possibility to distinguish particle events from X-ray events is the geometric shape of the related ionization cloud. Particle events leave an ionized track, whereas X-ray events leave a more point-like ionization cloud resulting in different rise times of the detector signal. A further background reduction method is to surround the actual X-ray detector with anticoincidence detectors on three to five sides. Coincident signals in an anticoincidence counter and the X-ray detector indicate with a high probability a non X-ray interaction and the event should be rejected. To eliminate X-rays generated by cosmic rays in the detector housing, the gas column density of the anticoincidence counter should be large enough to absorb X-rays efficiently up to the upper threshold of the accepted energy band. Large area X-ray detectors achieve by these methods background rejection efficiencies of 99.6% [6].

2.3.5 Detector Lifetime

The lifetime of detectors in the harsh space environment is a major concern to the involved experimenters. The radiation environment can damage a proportional counter in two ways. Heavy ionizing particles can deposit energies 3–4 orders of magnitude higher than X-rays within the nominal operating range. These huge amount of charges must not trigger a permanent discharge or destroy the detector by spark discharge. In low earth orbit (ROSAT orbit at 580 km), heavy ionizing events have a trigger rate of about 2×10^{-4} cm^{-2} s^{-1} behind 1.5 cm of aluminum.

Gas detectors suffer a permanent aging because of the cracking of the quench gas molecules during normal operation. Hydrocarbons like CH_4 tend to deposit polymerization products on anode and cathode wires, resulting in gain shifts and or permanent discharge (Malter effect) of the detector after accumulation of a not very well-defined critical charge per millimeter of anode wire. Many parameters like gas purity, gas composition, wire, housing and sealing materials, and last but not least the electrical field contribute to the radiation dose tolerated by the detector [19]. Sealed detectors tolerate a charge dose of 10^{-5} C mm^{-1} anode wire for an Ar–CH_4 gas filling but only 10^{-7} C mm^{-1} for a Xe–CH_4 gas before serious degradation occurs [26]. Therefore, many large area proportional counters for X-ray astronomy use CO_2 as quench gas. CO_2 does not polymerize. Only carbon deposits have been observed [21].

2.4 Large Area Proportional Counters for X-Ray Astronomy

The observation of the weak photon fluxes from cosmic X-ray sources with non-imaging instruments (detectors with mechanical collimators) requires large area detectors with high background rejection capability. Table 2.1 shows the development of collecting area of several proportional counter experiments for X-ray astronomy within the last decades. Multi anode multilayer proportional counters subdivided in cells by cathode grids as shown in Fig. 2.3 were mainly used for such observations in the energy band up to 50 keV. The cell structure of these detectors offers different possibilities for discriminating background events. Using the signals of the detector cells bordering the walls of the detector housing in anticoincidence results in a three-sided anticoincidence. Additional end-veto electrodes of anode or cathode protect the other two sides of the sensitive volume [3, 30]. Requiring that a single event must not show signals in neighboring cells reduces charged particle background from the front side. The loss of real X-ray events, because of photoelectron tracks crossing the border of two cells, is less than 10% in the 1.5–35 keV band [6]. Another approach was used in the RXTE detector by introducing a separate front anticoincidence layer filled with a low Z gas (propane) separated from the lower detector volume by an aluminized mylar foil [2].

Table 2.1 Several large area proportional counters for X-ray astronomy

Experiment	Year	ΔE (ke V)	Area (cm^2)	FOV (FWHM)	Reference
Uhuru	1970	2.0–20	2×840	$5^\circ \times 5^\circ$, $5^\circ \times 0.5^\circ$	[9]
HEAO-1 A1	1977	0.15–20	7×1350–1900	$1^\circ \times 4^\circ - 1^\circ \times 0.5^\circ$	[18]
EXOSAT ME	1983	1.2–50	1800	$0.75^\circ \times 0.75^\circ$	[29]
Ginga LAC	1987	1.5–37	4000	$1.1^\circ \times 2^\circ$ elliptical	[30]
RXTE PCA	1995	2.0–60	6250	1° hexagonal	[2]

Collimator field of view (FOV)

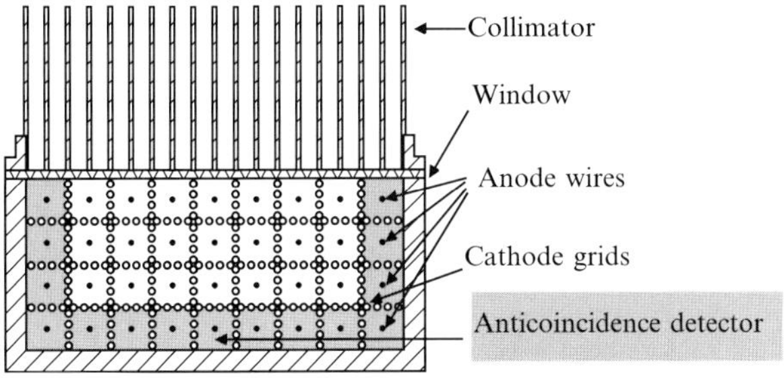

Fig. 2.3 Large area proportional counter for X-ray astronomy

The detection limit of such an instrument for a point source in the presence of a diffuse X-ray background component and a cosmic-ray background can be estimated as follows if the observed quantities are constant.

Q = quantum efficiency of the detector
A_x = geometric detector area for X-rays (cm^2)
A_b = geometric detector area for background (cm^2)
B_c = cosmic-ray background events not rejected by event selection logic (events cm^{-2} s^{-1} keV^{-1})
B_x = diffuse cosmic X-rays background (events cm^{-2} s^{-1} keV^{-1} sr^{-1})
Ω = field of view in (sr)
F_{min} = minimum detectable flux of a point source (Photons cm^{-2} s^{-1} keV^{-1})
ΔE = energy band of detector (keV)
S = desired number of standard deviations
t = observing time (s)

$$F_{min} = \frac{S}{QA_x}\sqrt{\frac{B_c A_b + Q\Omega B_x A_x}{t\Delta E}} \tag{2.6}$$

2.5 Gas Scintillation Proportional Counters

Gas scintillation proportional counters (GSPC), developed in 1972 [20], offer the advantage of an enhanced energy resolution when compared with proportional counters. In conventional proportional counters, the charge generated by an ionizing event is multiplied in a high electrical field. The amplification grows exponentially with the number of ionizing collisions of an electron in the high field region. Only 14 ionizing collisions result in a charge multiplication of four orders of magnitude. The variation in the number of ionizing collisions during multiplication degrades the Fano limited energy resolution of proportional counters by almost a factor of two (see Chap. 2.3.2 energy resolution). In gas scintillation proportional counters (see Fig. 2.4), the charge released by an ionizing event is not amplified. Similar to multiwire proportional counters, X-rays are absorbed by the detector gas (usually a

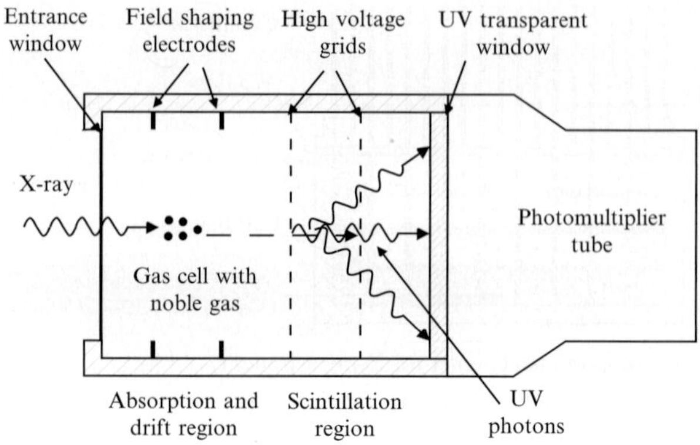

Fig. 2.4 Gas scintillation proportional counter

noble gas or a mixture of noble gases) in an absorption and drift region. The electrons drift from this low field region into the high field scintillation region where they acquire sufficient energy to excite the scintillation of the detector gas, but not to ionize it. In case of xenon, diatomic molecules, formed by the collision of excited atoms, deexcite by the emission of UV photons in the wavelength band of 150–195 nm [14]. The number of scintillation photons increase linearly with the number of exciting collisions of the electrons with the gas atoms. These collisions are independent events. Therefore, the variation of the light output generated during the scintillation process depends on the statistics of the final number of photons registered by the photomultiplier. The integral intensity of the light flash is proportional to the energy of the ionizing event. GSPCs can reach an energy resolution nearly at the Fano limit because of the large amount of scintillation photons.

GSPCs are rather intolerant to gas impurities because of the low electron mobility in xenon. Therefore, the gas cell of GSPCs has to be manufactured with ultrahigh vacuum technology to avoid contamination of the detector gas. In addition, gas purification systems like getter pumps are used. The slow velocity of the electrons in the absorption and drift region with a low electrical field strength intensifies the susceptibility of the GSPC to gas impurities. This effect can be reduced in the so called “driftless” GSPC. The “driftless” GSPC has a common high field absorption–scintillation region located directly below the detector window. The high electrical field mandatory for the scintillation excitation of the gas by the electrons results in a high drift velocity of the electrons from the beginning. The high field reduces in addition the loss of electrons from the ionization cloud of X-ray events absorbed near to the entrance window. But nothing is for free and the light output of an event in this configuration depends on the absorption depth of the X-ray event in the absorption–scintillation region. To recover the original energy of the event, the signal has to be corrected with a burst length factor.

X-ray astronomy with nonimaging detectors requires large apertures and a good background rejection efficiency. Rise time discrimination, burst length discrimination, and limitation of the energy band are the main background suppression

Table 2.2 Characteristics of GSPC instruments for X-ray astronomy

Satellite	BeppoSAX	EXOSAT	Tenma
Year	1996	1983	1983
Experiment	HPGSPC	GS	SPC-A, SPC-B, SPC-C
Effective area	240 cm^2 *	~100 cm^2	320 cm^2, 320 cm^2, 80 cm^2
FOV (FWHM)	1.1°	0.75°	3.1°, 2.5°, 3.8° mod. collimator
Energy range	4–120 keV	2–40 keV	2–60 keV
$\Delta E/E$ (% FWHM)	$31 \times (E(keV))^{-0.5}$	$27 \times (E(keV))^{-0.5}$	$23 \times (E(keV))^{-0.5}$
Reference	[1]	[16, 17]	[27]

* @ 30 keV

methods for GSPCs to distinguish background events from real X-ray signals. The background rejection efficiency by the burst length discrimination can be improved substantially by using a gas mixture of xenon with helium. Addition of helium increases the electron drift velocity considerably compared with pure xenon [11]. The full field energy resolution of large area GSPCs with a single photomultiplier readout is degraded because of solid angle variations of the light emission regions for events distributed over the whole sensitive area. A focusing electrical field in a conical absorption and drift region concentrating the electrons on a small scintillation region reduce the effect of solid angle variation. Large aperture detectors with such focusing geometries were operated on the X-ray satellites Tenma and EXOSAT [11, 16]. Another approach to overcome the problem of solid angle variations in large area detectors is to view the scintillation region of the GSPC by an Anger camera arrangement of several photomultipliers. The event position derived from the ratio of the photomultiplier signals is used to correct the event energy. This method was used in the HPGSPC experiment aboard BeppoSAX. Table 2.2 gives the characteristics of GSPCs with mechanical collimators operated on several X-ray astronomy satellite missions.

References

1. Boella, G., Butler, R. C., Perola, G. C., et al. 1997, Astron. Astrophys. Suppl., 122, 299
2. Bradt, H. V., Rothschild, R. E. and Swank, J. H. 1993, Astron. Astrophys. Suppl., 97, 355
3. Brunner, A. N., Kraushaar, W. L., McCammon, D., et al. 1973, Rev. Sci. Instrum., 44, 418
4. Charpak, G., Bouclier, R., Bressani, T., et al. 1968, Nucl. Instr. Meth., 62, 235
5. Fano, U. 1947, Phys. Rev., 72, 26
6. Fraser, G.W. 1989, X-Ray Detectors in Astronomy, Cambridge University Press, Cambridge, pp 35–95
7. Geiger, H., Klemperer, O. 1928, Z. Phys., 49, 753
8. Geiger, H., Müller, W. 1928, Phys. Z., 29, 839
9. Giacconi, R., Kellogg, E., Gorenstein, P., et al. 1971, Astrophys. J., 165, L27
10. Henke, B. L., Gullikson, E. M., Davis, J. C. 1993, Atomic Data and Nuclear Data Tables, 54(2), 181

11. Inoue, H., Koyama, K., Mae, T., et al. 1980, Nucl. Instr. Meth., 174, 301
12. Knoll, G.F. 1979, Radiation Detection an Measurement, Wiley, New York, pp 196–200
13. Korff, S.A., Danford, W.E. 1939, Phys. Rev., 55, 980
14. Manzo, G., Peacock, A., Andresen, R.D., Taylor, B.G. 1980, Nucl. Instr. Meth., 174, 301
15. Oed, A. 1988, Nucl. Instr. Meth., A 263, 351
16. Peacock, A., Andresen, R.D., Leimann, E.A., et al. 1980, Nucl. Instr. Meth., 174, 301
17. Peacock, A., Andresen, R.D., Manzo, G., et al. 1981, Space Sci. Rev., 30, 525
18. Peterson, L.E. 1975, Ann. Rev. Astron. Astrophys. Suppl., 13, 423
19. Pfeffermann, E., Briel, U.G., Freyberg, M.J. 2003, Nucl. Instr. Meth., 515(1–2), 65
20. Policarpo, A.J.P.L., Alves, M.A.F., Santos, M.C.M., Carvalho, M.J.T. 1972, Nucl. Instr. Meth., 102, 337
21. Ramsey, B.D., Austin, R.A., Decher, R. 1994, Space Sci. Rev., 69, 139
22. Rossi, B., Staub, H. 1949, Ionization Chambers and Counters – Experimental Techniques, McGraw-Hill Book Company, Inc., New York, pp 97-100
23. Rutherford, E., Geiger, H. 1908, Proc. Roy. Soc. London, A 81, 141
24. Sauli, F. 1977, CERN REPORT, 77-09, 531
25. Sauli, F. 1997, Nucl. Instr. Meth., A 386, 531
26. Smith, A., Turner, M.J.L. 1982, Nucl. Instr. Meth., 192, 475
27. Tanaka, Y., Fujii, M., Inoue, H., et al. 1984, Publ. Astron. Soc. Jpn., 36, 641
28. Trost, A. 1937, Z. Phys., 105, 399
29. Turner, M.J.L., Smith, A., Zimmermann, H.U. 1981, Space Sci. Rev., 30, 513
30. Turner, M.J.L., Thomas, H.D., Patchett, B.E., et al. 1989, Publ. Astron. Soc. Jpn., 41, 345

3 Scintillation Counters

E. Kendziorra

3.1 Introduction

For the observation of hard X-rays above 15 keV inorganic scintillation counters have been commonly used right from the early days of X-ray astronomy. The device consists of a scintillating crystal optically coupled to a phototube or (more recently) to a photodiode. Incident radiation energy is converted into optical photons, which are then measured as an electrical pulse by the light sensor. An ideal scintillation material should have a high linear scintillation efficiency and transparency to its own light. The emission spectrum should fit to the spectral response function of the light sensor. Materials used as photon detectors should have a high photon absorption cross section. In addition, the decay time of the induced luminescence should be short. In scintillation detectors for X-ray astronomy, thallium activated sodium iodide NaI(Tl) and caesium iodide, either activated with sodium, CsI(Na), or thallium, CsI(Tl), are the most commonly used detector materials. In Table 3.1, we summarize the main properties of these three alkali halide crystals together with those of bismuth germinate ($Bi_4Ge_3O_{12}$), commonly named BGO, and gadolinium silicate Gd_2SiO_5, abbreviated GSO. The latter two scintillators have a lower photon yield than alkali halide crystals, but much higher density, and are therefore well suited as shielding material or for phoswich scintillators for higher energies. CsI(Tl) has the highest integrated light yield of all crystals and is also easy to handle and to machine, because it is not hygroscopic like NaI or CsI(Na). However, if coupled to a phototube with S-11 or bialkali photocathode, the resulting pulse height is a factor of two smaller than for NaI(Tl), because the emission spectrum of NaI(Tl) (and CsI(Na)) matches the response function of typical phototubes much better. To make full use of the high light yield of CsI(Tl), it has to be coupled to photodiodes that show an extended sensitivity in the red. Up to now NaI(Tl) is the most widely used scintillator material for hard X-ray detectors. For application in space, it has to be packed in a vacuum tight housing, to protect the hygroscopic material from degradation during ground handling.

The energy resolution of a scintillation counter is determined by the statistical fluctuations of the number of photoelectrons generated at the photocathode.

Advanced photocathodes like Bialkali (K–Cs) achieve a maximum quantum efficiency of $(25-30)\%$ resulting in a maximum energy resolution of 6 keV (FWHM) at 60 keV. However, this figure can only be achieved for small crystals, where

J.E. Trümper and G. Hasinger (eds.), *The Universe in X-Rays.*

Table 3.1 Properties of crystals used in hard X-ray phoswich detectors, data taken from Knoll [5] and references therein

Crystal	Density ($g\,cm^{-3}$)	Decay time (ns)	Light yield (photons/100 keV)	Wavelength of max. emission (nm)
NaI(Tl)	3.67	230	3 800	415
CsI(Tl)	4.51	680 (64%) 3 340 (36%)	6 500	540
CsI(Na)	4.51	460	3 900	420
GSO	6.71	56 (90%) 400 (10%)	900	440
BGO	7.13	300	820	480

nearly all photons are collected by the cathode. For larger crystals of $>100\,cm^2$ area, typically used in hard X-ray detectors, even a careful packing of the crystal in white Teflon foil or micropor paper will finally only scatter about half of the emitted photons to the cathode. A relative energy resolution ΔE(FWHM)/E of 15% at 60 keV can be achieved. The reduction of internal background, mainly produced by charged particles, is an important issue for the design of scintillation counters for X-ray astronomy. Two different approaches have been followed in the past. A well-type mounting of the detector was used for early satellite and balloon instruments for small detectors up to $100\,cm^2$. The detector crystal and the collimator are surrounded on five sides by anticoincidence crystals, leaving only the field of view open. For larger detectors of several $100\,cm^2$, a different approach has been used since mid 1975, the so-called phoswich detector configuration. Here, the main detector crystal is optically coupled to a thick shielding crystal of different fluorescence decay time. The light of both crystals is viewed by the same phototube and events from the two crystals are distinguished by means of pulse shape discrimination. This technique works well for NaI(Tl) ($\tau = 230$ ns) as the principle detector and CsI(Tl) ($\tau = 680$ ns) or CsI(Na) ($\tau = 460$ ns) as shielding detectors. The background of these phoswich counters can further be reduced by plastic anticoincidence shields and passive-graded shields.

3.2 Scintillation Counters for X-Ray Astronomy

In 1964, the first observation of hard X-rays from the Crab was performed by Clark [2] with a NaI(Tl) scintillation counter on a balloon payload. Subsequent balloon and satellite payloads used the well-type configuration. In balloon instruments, large ($100\,cm^2$) crystals were coupled to standard 5 in. phototubes. In the mid 1970s, the first phoswich detectors were built, first by the UCSD and MIT groups [8] followed soon by the MPE/AIT collaboration [9]. Because of this improved shielding technique, larger detectors with several hundred cm^2 collecting area and low instrumental background could be developed. The MPE/AIT balloon payload HEXE carried an array of 3×4 NaI(Tl)/CsI(Tl) phoswich detectors with a total collecting area of $2\,300\,\text{cm}^2$. The series of Orbiting Solar Observatories OSO-3, OSO-5, OSO-7, and OSO-8, launched between 1967 and 1975, carried the first scintillation counters into earth orbit. Theses instruments with effective areas from $10\,\text{cm}^2$ on OSO-3 to $65\,\text{cm}^2$ on OSO-7 were mainly devoted to the exploration of the diffuse X- and Gamma-ray background [1]. The UCSD/MIT hard X-ray and low Gamma-ray experiment A4 on HEAO-1 (1977–1978) , which consisted of 7 phoswich detectors of three different types, performed the first high energy all sky survey. In this instrument, a combination of well-type and phoswich shielding was used resulting in a remarkably low instrumental background of $(1\text{–}2)10^{-4}$ counts $\text{cm}^{-2}\,\text{s}^{-1}\,\text{keV}^{-1}$ in the $220\,\text{cm}^2$ low energy detector [4].

Ten years later a space qualified version of the MPE/AIT HEXE detector [10] was launched with the Kvant module and docked to the Soviet space station Mir. Figure 3.1 shows a cross section of this instrument. The Mir HEXE consisted of 4 phoswich detectors surrounded on five sides by a graded shield and covered on all six sides by Ne 110 plastic scintillator (5-mm thick at the top, 10-mm thick elsewhere). NaI(Tl) crystals (3.2-mm thick) were coupled to 50-mm thick CsI(Tl) crystals and sealed by a window made of an aluminium honeycomb structure covered on both sides by 0.1-mm thick aluminium foil. The field of view (FOV) was defined to $1.6^\circ \times 1.6^\circ$ (FWHM) by two collimators made of hexagonally shaped tungsten tubes. The collimators could separately be tilted by 2.5° to provide simultaneous source and background measurements. The tungsten collimators reduced the geometric area by less than 10%. For the gain control of the phoswich detectors there were four calibration sources mounted inside the passive shield. Am^{241} was

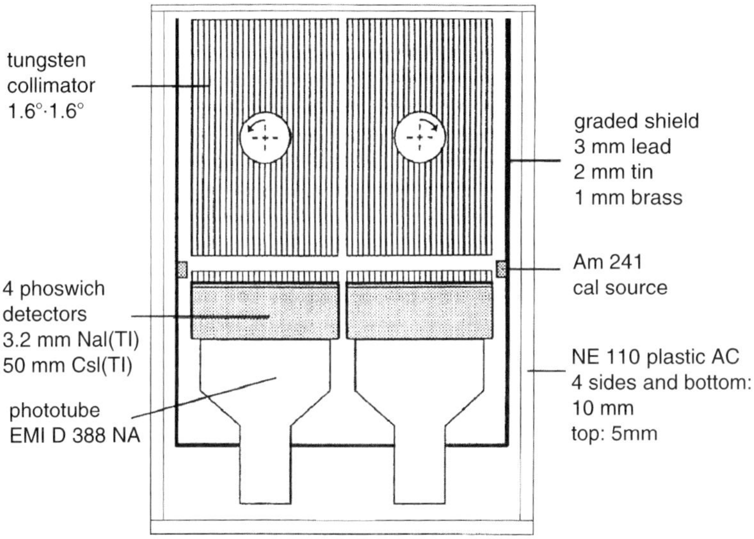

Fig. 3.1 Cross section of the Mir HEXE phoswich detector assembly

embedded in plastic scintillators providing tagging of the calibration photons by the light pulses from the simultaneous alpha particles with better than 95% efficiency. Although the orbit of the Mir station ($i = 56°$) regularly passed through the South Atlantic Anomaly and through the polar radiation belts, the detector background was quite low on average, see Table 3.2. The quadratic shape of the phoswich crystals allowed for a dense packing of the detector assembly, providing mutual shielding. However, the transition from the quadratic crystal to the circular phototube was not ideal in terms of light collection, resulting in a reduced energy resolution of only 25% at 60 keV. Until January 1993, about 1 800 observation sessions with a total observing time of $1.9\,10^6$ s were performed with the Mir HEXE detector.

About 9 years after Mir Kvant, two further hard X-ray phoswich detectors were launched into low earth orbit, first the high energy X-ray timing experiment (HEXTE) on RXTE (December 1995) [11] and then the Phoswich detection system (PDS) on BeppoSAX (April 1996) [3]. Similar to Mir HEXE, both instruments use rocking collimators for continuous background monitoring. Both detectors reach about 15% relative energy resolution at 60 keV. This was achieved by a careful shaping of the shielding crystal and in the case of RXTE by the use of circular detectors. In Table 3.2, we summarize the main properties of both instruments in comparison to Mir HEXE and the Suzaku HXD detector.

While in the past all balloon and satellite phoswich detectors used alkali halide crystals, the hard X-ray detector (HXD) on board of the Japanese X-ray astronomy satellite Suzaku (launched July 2005 into an 31° inclined low earth orbit) utilizes an array of 5-mm thick GSO crystals as the main detector for the 40–600 keV energy band [7]. Low energy (10–70 keV) photons are measured by 2-mm thick silicon PIN type diodes in front of the GSO crystals with very good energy resolution of

Table 3.2 Hard X-ray phoswich detector instruments

Instrument	Mir HEXE	RXTE HEXTE	SAX PDS	Suzaku HXD
Type	3.2 mm NaI(Tl) 50 mm CsI(Tl)	3 mm NaI(Tl) 57 mm CsI(Na)	3 mm NaI(Tl) 50 mm CsI(Na)	2 mm Si PIN 5 mm GSO
Energy band	15–200 keV	15–250 keV	15–300 keV	10–600 keV
Rel. energy res. at 60 keV	25%	15.4%	15%	30%
Net open area	750 cm^2	1 600 cm^2	640 cm^2	404 cm^2
FOV (FWHM)	$1.6° \times 1.6°$	$1° \times 1°$	$1.3° \times 1.3°$	$0.5° \times 0.5°$
Background[a]	4	4.4^b	4^c	1^d
in units of 10^{-4}	3	2.8	2	3.8
cts cm^{-2} s^{-1} keV^{-1}	1.5	1.0	1.4	1.6

[a] Typical background values (for Mir HEXE and RXTE HEXTE for part of orbit with high magnetic rigidity) are quoted for three energy bands: 20–40, 40–80, and 80–200 keV (top to bottom)
[b] R. Rothschild, priv. comm.
[c] From BeppoSAX Observer's Handbook
[d] PIN and GSO background 220 days after launch [6]

3 keV (FWHM). The HXD is composed of a matrix of 4×4 well-type units surrounded at four sides by active BGO shielding units. Each well unit contains four stacks of a PIN diode and a GSO crystal deep inside an active collimator made of 3-mm thick BGO plates for the four side walls and a central cross. The 320-mm long collimator plates and the four GSO crystals are optically glued to a 60-mm thick BGO shielding crystal, coupled to one phototube. The FOV of the active collimator is $4.5^\circ \times 4.5^\circ$ (FWHM). For photons below 100 keV, it is further confined to $0.57^\circ \times 0.57^\circ$ (FWHM) by fine collimators made of 50 μm phosphor bronze. Because of the combination of well-type and phoswich-type shielding, the background between 20 and 50 keV is much lower than in previously flown missions, while above 50 keV the background is similar to standard phoswich detectors [6].

References

1. Bradt, H.V.D., Ohasi, T., Pounds, K.A. 1992, Annu. Rev. Astron. Astrophys. 30, 391
2. Clark, G. 1965, Phys. Rev. Letters 14, 91
3. Frontera, F., Costa, E., Dal Fiume, D., et al. 1997, Astron. Astrophys. Suppl. 122, 357
4. Gruber, D. E., Matteson, J.L., Peterson, L. E., et al. 1999, ApJ 520, 124
5. Knoll, G.H. 2000, Radiation Detection and Measurement, 3rd edn, Wiley, New York, pp 219–245
6. Kokubun, M., Makishima, K., Takahashi, T., et al. 2007, Publ. Astron. Soc. Jpn. 59, S53
7. Takahashi, T., Abe, K., Endo, M., et al. 2007, Publ. Astron. Soc. Jpn. 59, S35
8. Peterson, L.E. 1975, Annu. Rev. Astron. Astophys. 13, 423
9. Reppin, C., Trümper, J., Pietsch, W., et al. 1978, Proc. Esrange Symp., Ajaccio, ESA SP-135, 296
10. Reppin, C., Pietsch, W., Trümper, J., et al. 1985, in G. C. Perola and M. Salvati (eds.), Nonthermal and Very High Temperature Phenomena in X-Ray Astronomy (Istituto Astronomico, Universita La Sapenzia, Rome 1985), pp 279–282
11. Rothschild, R.E., Blanco, P.R., Gruber, D.E., et al. 1998, Astrophys. J. 496, 538

4 Imaging Proportional Counters

E. Pfeffermann

4.1 Introduction

The sensitivity of astronomical observations in the soft X-ray band has been increased dramatically by the use of imaging optics and position sensitive focal plane detectors. The tremendous sensitivity increase of imaging telescopes compared with detectors with mechanical collimators has two reasons. The effective collecting area of the telescope is determined by the sensitive area of the mirror system and not any more by the area of the detector. For ROSAT (ROentgen SATellit) [23] the sensitive detector area is a factor of ten smaller than the effective collecting area of the mirror. Because of the imaging capability of the instrument, the relevant background for the detection of a point source is limited to an angular resolution element of the mirror-detector system (compare NXB$\times$RE in Tables 4.1 and 4.2). For these geometrical reasons, the background noise of an imaging system is several orders of magnitude less for the detection of a point source compared with a collimated detector with the same sensitive area. Three astronomy satellites with imaging X-ray telescopes had position sensitive proportional counters. The Einstein satellite [8] and ROSAT used multiwire proportional counters, while EXOSAT (European X-ray Observatory SATellite) [22] had a proportional counter with parallel plate geometry. JEM-X [11], the X- ray monitor aboard INTEGRAL (INTErnational Gamma-Ray Astrophysics Laboratory) [25] has a position sensitive micro strip gas chamber in a coded aperture mask telescope.

4.2 Geometry of Multiwire Proportional Counters

Figure 4.1 shows a schematic cross section of a multiwire proportional counter for X-ray astronomy. An anode grid with thin wires of the order of 10 μm and a wire pitch of 1–2 mm sandwiched between 2 cathode wire grids is the position sensing X-ray detector. A second anode grid below the X-ray detector operates as anticoincidence detector for background rejection. The wire grid assembly is mounted in a gas tight detector housing. X-rays entering the detector through the window interact predominantly by the photo effect with the detector gas in the absorption and drift region. Electrons generated by this interaction drift in the electrical field through the

J.E. Trümper and G. Hasinger (eds.), *The Universe in X-Rays.*

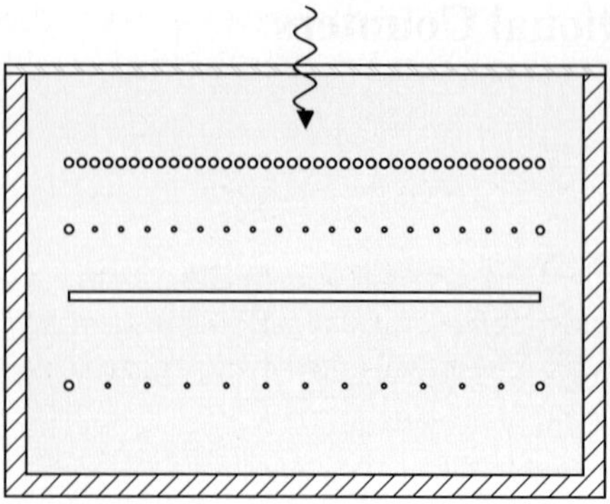

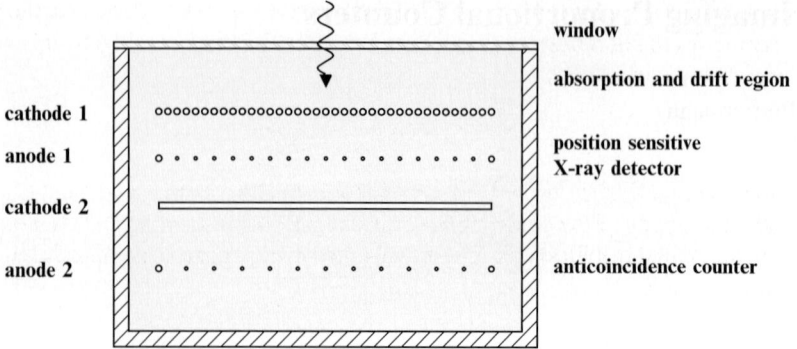

Fig. 4.1 Multiwire proportional counter for X-ray astronomy

cathode grid into the high field region of the anode grid, where they are amplified in an avalanche by a factor of about 10^4–10^5. The acceleration of electrons and ions in the electrical field generates a negative signal on the anode and positive signals on the upper and lower cathode of the proportional counter. The anode signal is normally used to determine the energy and the cathode signals to derive position of the event. Multiwire detectors have the same energy resolution capability like single wire detectors with one exception. The individual anode wires of the anode grid act as separate amplifiers and the amplification is very sensitive to the pitch tolerances of the anode grid. Pitch tolerances in the order of micrometer have to be realized to achieve a homogeneous gain in the order of several percent.

4.3 Position Resolution of Multiwire Proportional Counters

The position resolution of such detectors is limited by several factors. Nonvertical incidence of the X-rays from the mirror system in conjunction with the absorption depth of the X-rays degrades the position resolution (σ_a). Displacing the focus of the mirror about 1 mm (depending on incidence angle from the mirror and the accepted energy range) below the window surface inside the detector can reduce the broad wings of the point spread function for the higher energies. Because of window bulging by the gas pressure, the focus position varies as a function of the offset angle relative to the window surface.

A further limitation of the position resolution is the finite size of the charge cloud generated by the interaction of X-rays with the detector gas. This size is determined by the range of the photo electron, the range of the fluorescent photon or Auger electron (σ_r), and the lateral diffusion of the electron cloud during the drift into the amplification region (σ_d). The position accuracy of such an electron cloud is given by the geometrical size arriving at the amplification region divided by the square root of the number of electrons before amplification. Amplifier noise reduces the position resolution a bit further, especially at low energies (σ_e). Therefore, the spacial accuracy (σ_x) of multiwire proportional counters is mainly limited by following independent contributions.

$$\sigma_x{}^2 = \sigma_a{}^2 + \sigma_r{}^2 + \sigma_d{}^2 + \sigma_e{}^2 \tag{4.1}$$

For low energy X-rays, the largest contribution for the spatial uncertainty is the statistical accuracy of the center of the electron cloud before amplification. For this reason, the spatial resolution for X-rays in the low energy band scales like the energy resolution with $E^{-0.5}$.

4.4 Position Readout Methods

Several methods can be employed to measure the position of X-ray events in multiwire proportional counters. Only readout methods used in satellite borne imaging X-ray detectors are addressed here. In general, the signals of the two cathode grids provide the event position as shown in Figs. 4.2 and 4.3. The wire grids of the two cathodes are oriented orthogonal to each other to get both coordinates. One method is to use the wires of the continuous Z-wound cathode grid as distributed R-C delay line as shown in Fig. 4.2. Measuring the rise time of the signals arriving at both ends of each cathode permits the determination of both coordinates of the event [2]. This method was used in the IPC (imaging proportional counter) detector aboard the Einstein satellite. The IPC had a position resolution of about 1 mm FWHM [9].

Another method is to measure the charge signals induced on the individual stripes of the segmented cathodes as shown in Fig. 4.3. The charge distribution on the cathode has a (FWHM) width of about twice the distance between anode and cathode grid. Cathode strips with a width equal to the grid distance of anode to cathode reproduce this charge distribution by their signal magnitudes on several (2–5 depending on the event energy) strips. By a centroid computation of the signals,

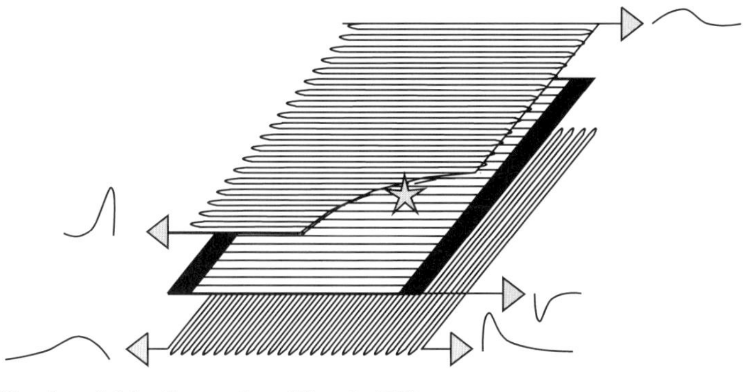

Fig. 4.2 Distributed delay line readout (Einstein IPC)

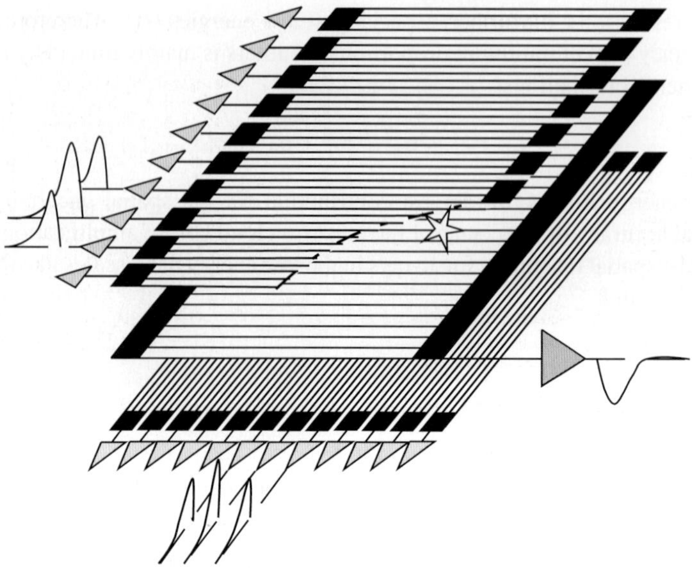

Fig. 4.3 Center of gravity readout via a segmented cathode (ROSAT PSPC)

the event position can be determined with a resolution small compared with the strip width [4, 10]. This was the readout method of the ROSAT PSPC (position sensitive proportional counter) with a spatial resolution of 250 μm (FWHM) at 0.93 keV [18].

The PSD of EXOSAT used a parallel plate proportional counter. In this detector type, the charge amplification takes place in a homogeneous high electrical field between a wire grid and a resistive plate acting as anode. By signal rise time measurements on four pick-up electrodes of the resistive anode, the event position is derived [20]. This detector, although very attractive due to its compact set up, suffered shortly after launch radiation damage by heavy ionizing particles [13].

A new detector concept was used in the JEM-X instrument aboard INTEGRAL [11]. The position sensitive proportional counter in the coded aperture mask telescope is a microstrip gas chamber [14]. This detector type uses, instead of wire grids, conductive micro structures on a partially isolating glass substrate. The micro structures can be manufactured by photolithographic processes with high accuracy resulting in high gain uniformity and fast response time. Alternating thin anode and broader cathode structures as shown in Fig. 4.4 form the intense electrical field around the anode, where gas amplification takes place. A second set of cathode stripes is placed on the rear side of the glass substrate orthogonal to the cathodes on the front side. The event position is derived from both sets of cathode stripes by capacitive charge division. In space environment this detector suffered by heavy ionizing cosmic rays. During the first week of operation about one anode per day was lost because of discharges triggered by heavy ionizing events. After lowering the gain by a factor of three, the radiation damage by heavy ionizing events was

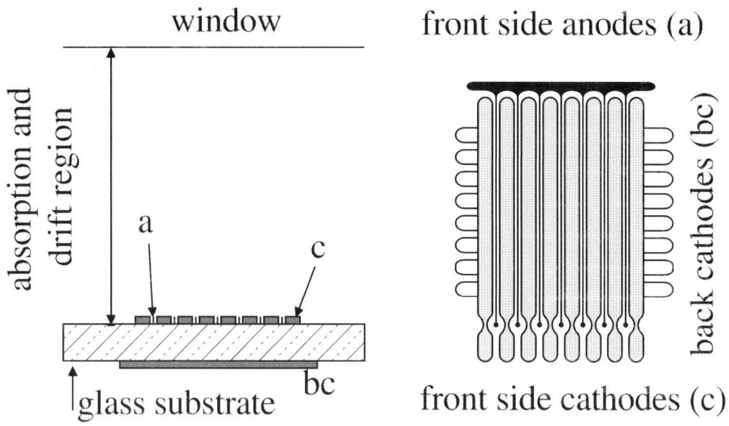

Fig. 4.4 Electrode structure of a microstrip gas chamber

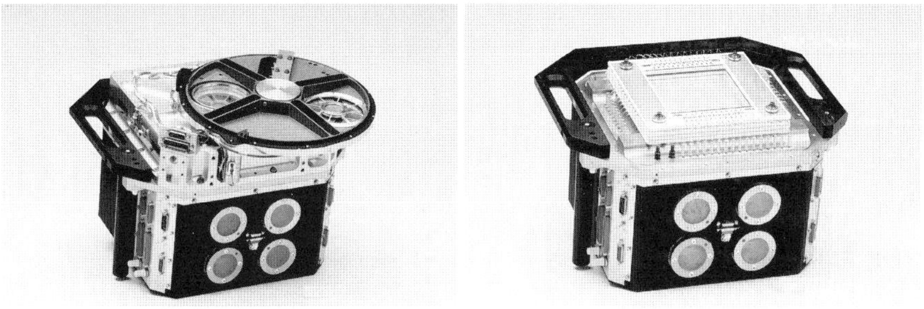

Fig. 4.5 The position sensitive proportional counter of the ROSAT telescope with closed and open detector housing [18]

reduced to a level of one anode loss in 2 months assuring a life time of more than 5 years [11]. Characteristics of these four IPC are summarized in Table 4.1.

4.5 The ROSAT PSPC

With a continuous operational life time of 4 years and an additional life of 4 years with alternating hibernation and operation (due to the nearly exhausted gas supply), the ROSAT PSPC (Fig. 4.5) was one of the most successful imaging X-ray detectors [18, 19]. Because of a high mechanical accuracy of the wire grids of the order of micrometer, the detector gain was uniform within 3% over the sensitive area. The position resolution, mainly limited by the size of the primary electron cloud arriving at the anode, was 250 µm at 930 eV Cu–L_α and about 400 µm at 280 eV C–K_α. A five-sided anticoincidence counter system in connection with an event

Table 4.1 Performance of imaging proportional counters aboard several X-ray astronomy satellites

Satellite	Einstein	EXOSAT	ROSAT	INTEGRAL
Telescope	Imaging	Imaging	Imaging	Coded mask
Focal length	340 cm	110 cm	240 cm	–
Detector	IPC	PSD	PSPC	JEM-X
Active area	58 cm^2	6.2 cm^2	50 cm^2	500 cm^2
Energy range	0.2–4 keV	0.1–2 keV	0.1–2.4 keV	3–35 keV
Window	Lexan 1.8 μm PP C	Lexan 0.5 μm PP 0.8 μm PP+ C	Lexan 1 μm PP C	Be 250 μm
Gas	84% Ar 6% Xe	80% Ar	65% Ar 20% Xe	90% Xe
Quench gas	10% CO_2	20% CH_4	15% CH_4	10% CH_4
Pressure	1.05 bar	1.1 bar	1.46 bar	1.5 bar
Sealed/Flow	F	F	F	S
ΔE/E % (FWHM)	$\frac{122}{\sqrt{E(\text{keV})}}$	$\frac{42}{\sqrt{E(\text{keV})}}$	$\frac{41}{\sqrt{E(\text{keV})}}$	$\frac{40}{\sqrt{E(\text{keV})}}$
ΔX mm (FWHM)	$\frac{1.0}{\sqrt{E(\text{keV})}}$	$\frac{0.25}{\sqrt{E(\text{keV})}}$	$\frac{0.25}{\sqrt{E(\text{keV})}}$	$\frac{1.94}{\sqrt{E(\text{keV})}}$
Ang. Res. (HPD) Det. + Mirror	1 arcmin	0.75 arcmin	0.42 arcmin	3.4 arcmin
NXB $\text{arcmin}^{-2}\,\text{s}^{-1}\,\text{keV}^{-1}$	3×10^{-5}	Not available	5×10^{-6}	1×10^{-5} *
NXB×RE $s^{-1}keV^{-1}$	2.4×10^{-5}	Not available	1×10^{-6}	-
Reference	[8,9]	[7,13]	[18,21]	[3,11]

PP, polypropylene; C, carbon; NXB, non X-ray background;
HPD, half power diameter; RE, angular resolution element *arcmin*^{2}
* = $\text{mm}^{-2}\,\text{s}^{-1}\,\text{keV}^{-1}$

selection logic, based on the cathode pattern signature of the events, resulted in a background rejection efficiency of 99.8%. The excellent background rejection efficiency of the PSPC allowed deep observations with high contrast. Radiation hardness of the detector against heavy ionizing particle events was tested on a heavy ion tandem accelerator. A careful prelaunch calibration was carried out concerning quantum efficiency, energy resolution, and spatial resolution. [5,6]

4.6 Imaging Gas Scintillation Proportional Counters

In conventional imaging multiwire proportional counters, the event position is usually derived from the avalanche charge distribution, which has a rather limited spatial extent. The statistical spatial accuracy of the avalanche charge distribution is not the limiting factor for the spatial resolution of a proportional counter. In gas scintillation proportional counters (GSPC), X-ray photons trigger UV light bursts, which carry information about the energy and the location of the events (see Chap. 2.5). The energy resolution of a GSPC is about a factor two better compared with proportional counters. In contrast to proportional counters, where the charge distribution is confined to a small region, the UV photons in the GSPC are distributed over the whole sensitive area of the UV photon read out device. Solid angle and UV detection efficiency reduce the fraction of the registered photons. The large distribution area of the UV photons in combination with the reduced number of photons limit the statistical accuracy of the center of gravity determination of the UV light burst to several 100 μm. The spatial resolution of a GSPC can be expressed as follows:

$$\sigma_{\mathrm{X}}{}^{2} = \frac{\sigma_{\mathrm{P}}{}^{2}}{N_{\mathrm{P}}} + \frac{\sigma_{\mathrm{e}}{}^{2}}{N_{\mathrm{e}}} \tag{4.2}$$

σ_{X} = spatial resolution
σ_{P} = rms width of UV photon distribution at the photon detector
σ_{e} = rms width of the electron cloud
N_{P} = number of detected UV photons
N_{e} = number of electrons generated by the absorption of the X-ray photon

Several methods can be employed in GSPCs to localize the event position via a center of gravity determination of the UV light burst. By measuring the UV light burst with an array of photomultipliers the relative signal amplitudes can be used to extract the event position. Further methods are to image the light output of GSPCs

Table 4.2 Characteristics of gas scintillation proportional counters for imaging X-ray telescopes

Satellite	ASCA	BeppoSAX	BeppoSAX
Focal length	350 cm	185 cm	185 cm
Detector	GIS	LEGSPC	MEGSPC
Active area (diameter)	50 mm	20 mm	30 mm
Energy range	0.7–20 keV	0.1–10 keV	1.3–10 keV
Window	Be 10 μm	PI 1.25 μm, AlN 35 nm Al 44 nm	Be 30 μm
Gas cell			
Gas	96% Xe 4% He	Xe	Xe
Pressure	1.2 bar	1.2 bar	1 bar
Drift region	10 mm	Driftless	20 mm
Scintillation reg.	15 mm	49 mm	17.5 mm
$\Delta E/E$ %(FWHM)	$19.4(E(\mathrm{keV}))^{-0.5}$	$18.8(E(\mathrm{keV}))^{-0.42}$	$19.6(E(\mathrm{keV}))^{-0.5}$
ΔX mm (FWHM)	$0.5\times\sqrt{\frac{5.9}{E(\mathrm{keV})}}$	$0.86\times\sqrt{\frac{6}{E(\mathrm{keV})}}$	$0.7\times\sqrt{\frac{6}{E(\mathrm{keV})}}$
Ang. res. (HPD*)			
Det. + Mirror	3 arcmin	2.1 arcmin	2.5 arcmin
NXB			
$\mathrm{arcmin}^{-2}\,\mathrm{s}^{-1}\,\mathrm{keV}^{-1}$	$5–9\times10^{-6}$	5.2×10^{-6}	$2–4\times10^{-6}$
NXB×RE			
$\mathrm{s}^{-1}\,\mathrm{keV}^{-1}$	3.5×10^{-5}	1.8×10^{-5}	1.0×10^{-5}
Reference	[12, 15]	[16, 17]	[1, 24]

PI, polyimide; AlN, aluminiumnitride
* = half power diameter at 6 keV

with continuous UV sensitive instruments like micro channel plates or position sensitive proportional counters. The imaging GSPCs aboard the X-ray satellites ASCA (GIS) and BeppoSAX (GSPCs of LECS and MECS) had multianode photomultiplier for the readout of the detectors. Table 4.2 gives an overview of the characteristics of these detectors.

References

1. Boella, G., Chiappetti, L., Conti, G., et al. 1997, A&AS, 122, 327
2. Borkowski, C.J., Kopp, M.K. 1972, IEEE Trans. Nucl. Sci. NS-19, 161
3. Brandt, S., Budtz-Jørgensen, C., Lund, N., et al. 2003, A&A, 411, L243
4. Breskin, A., Charpak, G., Demierre, D., et al. 1977, Nucl. Instr. Meth., A 143, 29
5. Briel, U.G., Pfeffermann, E., Hartner, G., Hasinger, G. 1988, Proc. SPIE, 982, 401
6. Briel, U.G., Burkert, W., Pfeffermann, E. 1989, Proc. SPIE, 1159, 263
7. Fraser, G.W. 1989, X-Ray Detectors in Astronomy, Cambridge University Press, Cambridge, New York, New Rochelle, Melbourne, Sydney, 92
8. Giacconi, R., Branduardi, G., Briel, U., et al. 1979, ApJ, 230, 540
9. Gorenstein, P., Harnden Jr., F.R., Fabricant, D.G. 1981, IEEE Trans. Nucl. Sci. NS-28, 869
10. Jeavons, A.P., Ford, N., Lindberg, B., Parkmann, C., Hajduk, Z. 1976, IEEE Trans. Nucl. Sci. NS-23, 259
11. Lund, N., Budtz-Jørgensen, C., Westergaard, N.J., et al. 2003, A&A, 411, L231
12. Makishima, K., Tashiro, M., Ebisawa, K., et al. 1996, PASJ, 48, 157
13. Mason, I., Branduardi-Raymont, G., Culhane, J.L., et al. 1984, IEEE Trans. Nucl. Sci. NS-31, 795
14. Oed, A. 1988, Nucl. Instr. Meth., A 263, 351
15. Ohashi, T., Ebisawa, K., Fukazawa, Y., et al. 1996, PASJ, 48, 157
16. Parmar, A.N., Martin, D. D. E., Bavdaz, M., et al. 1997, A&AS, 122, 309
17. Parmar, A.N., Oosterbroek, T., Orr, A., et al. 1999, A&AS, 136, 407
18. Pfeffermann, E., Briel, U.G., Hippmann, H., et al. 1986, Proc. SPIE 733, 519
19. Pfeffermann, E., Briel, U.G., Freyberg, M. 2003, Nucl. Instr. Meth., A 515, 65
20. Sanford, P.W., Mason, I., Dimmock, K., Ives, J.C. 1979, IEEE Trans. Nucl. Sci. NS-26, 169
21. Snowden, S. L., Plucinsky, P.P., Briel, U., et al. 1992, ApJ, 393, 819
22. Taylor, B. G., Andresen, R. D., Peacock, A., Zobl, R. 1981, Space Sci. Rev., 30, 479
23. Trümper, J. 1981, J., of the Washington Academy of Science, 71 Nr.2, 114
24. Vecchi, A., Molendi, S., Guainazzi, M., Parmar, A.N. 1999, A&A, 349, L73
25. Winkler, C., Courvoisier, T.J.-L., Di Cocco, G., et al. 2003, A&A, 411, L1

5 Aperture Modulation Telescopes

R. Staubert

5.1 Principle of Aperture Modulation

The standard way of imaging in astronomy is by making use of refraction and reflection in employing lenses and mirrors. In the X-ray range, traditional glass lenses cannot be used because the X-rays are absorbed. Reflection under small incidence angles is possible at low ($<20\,\text{keV}$) X-ray energies and imaging is realized today through Wolter telescopes, which achieve imaging through double reflection on parabolic and hyperbolic surfaces. Developments are underway to extend the useful energy range to several tens of kiloelectronvolt by multilayer coated reflecting surfaces. The first detector systems for X-ray astronomy were just flat X-ray sensitive detectors (gas proportional counters or anorganic scintillation counters). They could hardly be called *telescopes*. In placing a mechanical, X-ray absorbing *collimator* in front of a flat X-ray detector (e.g., a "slat collimator" consisting of parallel metallic plates) the field of view (FOV) is restricted and an *indirect imaging* capability is achieved: when such a system scans the sky (in the direction perpendicular to the orientation of the plates) the flux of any existing X-ray source will be modulated according to the triangular collimator response function. This is the basic principle of *imaging through aperture modulation*. Aperture modulation telescopes ("shadow cameras") for high energy X-ray and gamma-ray astronomy make use of *temporal* and/or *spatial aperture modulation*.

5.1.1 Temporal Aperture Modulation

Through movement of the "aperture" a temporal modulation of the signal from the X-ray source of interest is introduced. For this method, a spatial resolution of the detector surface is not required.

5.1.1.1 Moon and Earth Occultations

A historically interesting way to "modulate the aperture" is to make use of the occultation of an X-ray source by another celestial body, e.g., the moon. As seen from the

J.E. Trümper and G. Hasinger (eds.), *The Universe in X-Rays.*

earth (or from a satellite), the moon scans certain parts of the sky and occults sources that happen to lie in the half degree wide strip of the moons path. By observing an X-ray source while the occultation is taking place, it is possible to measure the source position and in the case of extended sources their angular extent. Historically, important events were the moon occultations of the Crab nebula [23]. The first European X-ray satellite EXOSAT was planned as a moon-occultation satellite and was put into a highly excentric earth orbit to increase the total area of the sky, which was occulted at one time or the other. Ironically, EXOSAT did not perform a single observation of a moon occultation. The 4-day long, highly excentric orbit, however, turned out to be highly useful, as it provided long uninterrupted observations and thereby allowed the performance of long-term timing studies of X-ray sources, especially of active galactic nuclei (AGN). Occultation by the earth has extensively been used by the BATSE experiment onboard the Compton Gamma Ray Observatory (CGRO) to monitor sufficiently strong persistent X-ray sources [11].

5.1.1.2 Scanning with Slat Collimators

Historically, an important method to "image" the sky by "nonimaging instruments" was to perform scanning observations with flat detectors equipped with slat collimators. Of course, in one linear scanning measurement, the position is determined only in one coordinate (the sanning direction), such that at least a second scanning measurement in a different direction is needed, most usefully perpendicular to the first direction. The first All Sky Survey in the X-ray range was performed by Uhuru between 1970 and 1972, using two gas proportional counters with metal collimators defining fields of view of $5^\circ \times 10^\circ$ and $2^\circ \times 10^\circ$ (FWHM), respectively. Positions, flux levels, limited information on source extent, and rough spectra in the 2–6 keV range were measured for 339 sources (4th Uhuru catalog [7]). Several further satellites repeated such observations with higher sensitivity and extended energy ranges (most successfully HEAO-1, [13]), before the first truly imaging X-ray telescope was launched in 1978 on board HEAO–2, later called the "Einstein Observatory." The accuracy reachable through such scanning observations is a fraction of the instrumental angular resolution given by the opening angle of the collimator. The exact fraction is determined by the statistical accuracy by which the collimator response is reproduced by the observation.

There is an interesting proposal for a new hard X-ray (15–200 keV) All Sky Survey with greatly increased sensitivity and imaging capability: the Hard X-ray Modulation Telescope (HXMT) (planned in China) is supposed to carry 18 phoswich detectors (286 cm^2 each) equipped with rectangular FOV collimators ($0.5^\circ \times 5^\circ$ FWHM), the orientation of which cover the entire circle in steps of 10° [14]. The imaging capability of this multiangle scanning collimator instrument will be utilized by the direct demodulation method.

5.1.1.3 ON/OFF Observations

Even though it is not directly connected to the problem of "imaging," we mention here the technique of *ON/OFF observations*, because of its basic importance for all nonfocussing observational techniques. Assuming the position of the source is known, a series of observations (for a few minutes integration time each) are performed alternating between an orientation directly to the source ("ON") and an orientation to a background position ("OFF") (where the source is clearly outside the collimator response). The source flux and spectrum is found by taking the difference between "ON" and "OFF." The individual ON/OFF pointings are kept so short in time such that the background flux is not appreciably changing between successive pointings. This technique is used by the high energy X-ray timing experiment (HEXTE) onboard the Rossi X-ray Timing Explorer (RXTE) [18]. For weak sources, the optimum strategy is to spend equal time in "ON" and "OFF." Then the sensitivity of such observations can be expressed through the minimum detectable flux, given by

$$S_{\min}\,[\mathrm{cts/cm^2\,s}] = k\left(\frac{2B\,[\mathrm{cts/cm^2\,s}]}{T\,[\mathrm{s}]\,F\,[\mathrm{cm^2}]}\right)^{1/2} \tag{5.1}$$

with k being the number of standard deviations by which the source is required to be detected above the background. An alternative method is to use a model of the background (found from repeated observations under different conditions) for subtraction. This method is used for the proportional counter array (PCA) on RXTE.

5.1.1.4 Scanning Grid Collimators

For measurements with higher angular resolution, collimators have been constructed, which consist of two or more planes (parallel to the detector plane) of parallel rods of absorbing material. The optical axis of such a system is perpendicular to the plane of the detector and the grids. A modulation of the incident flux occurs when the optical axis performs a linear scanning movement in the direction perpendicular to the orientation of the absorbing rods. For the case of two grids (see Fig. 5.1), it is easy to see that the overall transmission of the double grid varies regularly between zero (when the light that travels through the open slits of the upper grid is completely blocked by the lower grid) and one-half (when the two grids are aligned such that they form a common shadow). The transmission function (for small scanning angles) is a repeated triangle with wave length given by d/D (the instrumental angular resolution), where d is the width of the rods (and the slits between them) and D is the distance between the two grids. A higher resolution is achieved, when three or more grids are used (e.g., 4 in the A-3 experiment on board HEAO-1, see below). With a special choice of spacing of the grids, a response function can be produced, which avoids the ambiguity of a regularly repeated pattern and achieves a fine angular

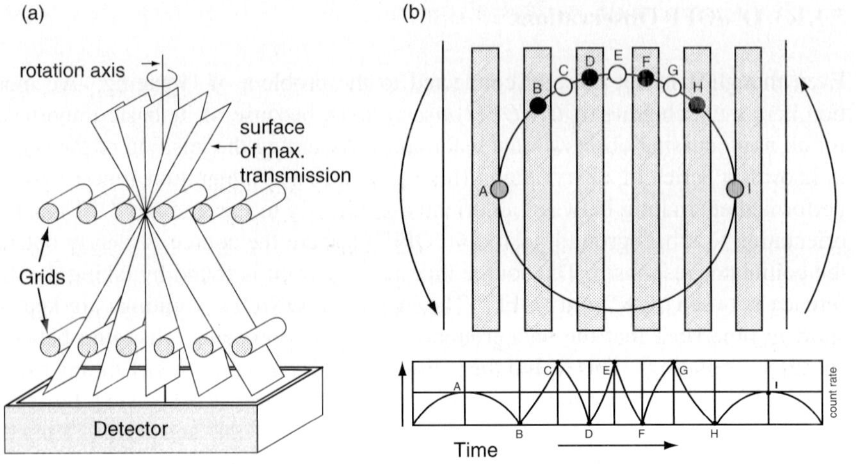

Fig. 5.1 (**a**) Principle of the aperture modulation by a double grid. (**b**) A rotation modulation collimator (RMC) measures a unique modulation curve (lower right) from which a sky image can be reconstructed [19]

resolution (e.g., 22 arcsec [16]). Again, for two-dimensional measurements, scans in two or more directions are necessary.

5.1.1.5 Rotation Modulation Collimators

A combined measurement of both coordinates and in fact imaging of the complete field of view (FOV) is possible if a double grid collimator as described in the previous section is placed in front of a detector and rotated (with constant angular velocity) around its optical axis. The flux reaching the detector is modulated in time by the variable transmission of the grid collimator. Such a device is called *rotation modulation collimator*. It was first proposed by Mertz [15] (see also [19]). Depending on the position of the source, a unique modulation curve is produced. For more than one source in the FOV, the resulting modulation curve is the superposition of the curves of all individual sources. The *image* is generated by a cross correlation procedure: The FOV is thought to consist of small image cells (sky pixels), each cell is then filled by the value of the cross correlation integral of the theoretical modulation curve (expected if the source is at the position of this pixel) and the actually observed modulation curve. Existing sources are represented in the image by a central peak and a system of concentric rings. The amplitude of the peak/rings is a measure of the source flux. Multiple sources can easily be imaged simultaneously and the position and flux of each individual source can be determined. Image imperfections like ghost peaks can be avoided by shifting the grid patterns against each other (by $d/2$) and by placing the rotation axis at the edge of the desired FOV [24].

5.1.2 Spatial Aperture Modulation

The alternative to the temporal aperture modulation is the spatial aperture modulation. It requires two-dimensional position sensitive detectors. The spatial modulation of the aperture is achieved by a pattern of holes in an otherwise X-ray absorbing plate, providing a unique spatial code. The combination of such a mask with a detector constitutes a *coded-aperture-* or *coded-mask telescope*. The use of coded apertures for X-ray astronomy was proposed in 1968 (independently of one another) by Ables [1] and by Dicke [5]. The most popular masks are those with a transmission of one-half, that is, the total open and blocked areas are equal. For an overview of coded-mask techniques see [4] and *Imaging in High Energy Astronomy*, edited by Bassani and di Cocco in 1995 [2].

5.1.2.1 Principle and Image Reconstruction

The basic principle of coded-mask imaging is that the mask pattern (in the form of the shadow produced by the parallel beam of an X-ray source) is recognized by a two-dimensional position sensitive detector (Fig. 5.2). Any shift of the pattern on the detector surface is directly related to a shift in source position. There are different types of mask patterns and mask/detector configurations, which will be discussed in the next paragraph. The "fully-coded field of view" (FCFOV) is defined as follows: photons from any source within this area of the sky cannot reach the detector without passing through the mask (that is the complete detector surface is "coded").

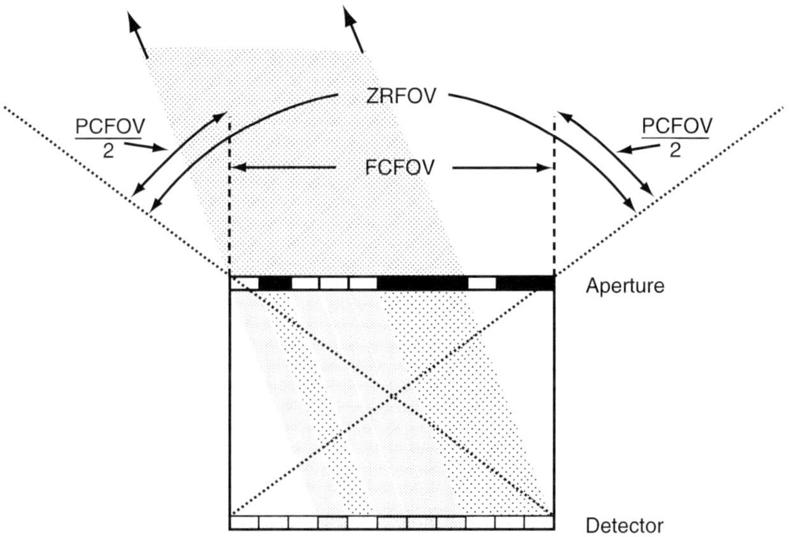

Fig. 5.2 Coded-mask box camera. Full coding is only achieved for on-axis sources

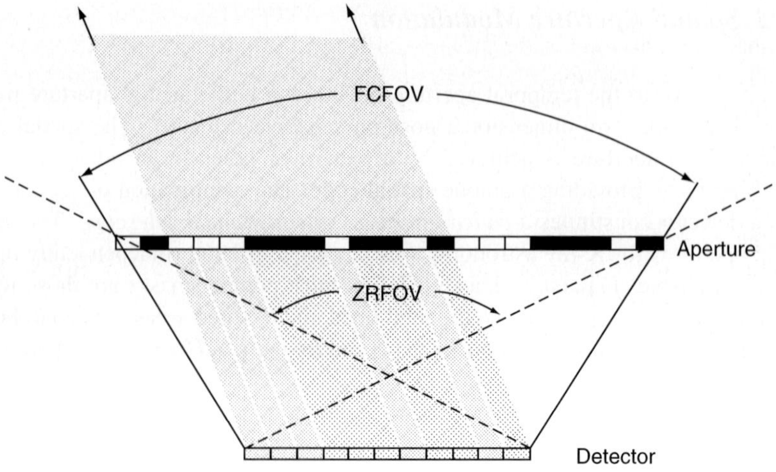

Fig. 5.3 Coded-mask mosaic camera. The FCFOV is increased by a larger mask (e.g., with a repeated coding pattern)

The boundaries of the "partially-coded field of view" (PCFOV) are reached when the fraction of the detector area which is coded reaches zero. In real systems, the effective FOV may be limited by an additional conventional collimator (generally to a FOV smaller than the total PCFOV). The instrumental angular resolution of such a telescope is given by d/D, with d being the characteristic length scale of the mask pattern (the width of the holes) and D being the distance between the mask and the detector (here it is assumed that the spatial resolution of the detector is equal or better than d). Note, however, that the accuracy with which the position of a point source is determined can be substantially better than d/D (depending on photon statistics).

The ultimate goal is to produce an image of the sky, represented by a two-dimensional array of pixels filled with intensity values. To construct this image, the observed intensity distribution on the detector surface must be interpreted ("unfolded") using the *coding function* provided by the mask pattern. This is a classical "inversion problem" in which an unknown cause (the sky distribution) is to be reconstructed from a measurement (the detector distribution). The basic *coding equation* is given by

$$D(x) = M(x) \times S(x) \tag{5.2}$$

with $D(x)$ being the observed detector distribution as the result of the "folding" of the sky distribution $S(x)$ by the aperture modulation function $M(x)$ (x is the vector describing the two-dimensional coordinate in the respective plane). For the inversion of this equation, several techniques are available. It is, however, important to understand that – even with "optimum patterns" (see below) – the resulting sky image is not unique, but rather subject to uncertainties that can be quite large. The main reason for this is the uncertainty from counting (Poisson) statistics, which is generally substantial because of the presence of background: the sky distribution $S(x)$ is actually the superposition of the uniform diffuse X-ray background radiation and all existing sources (point-like or extended) in the total FOV ($B_{\text{sky}}(x) + \text{sum}(S_i(x))$), both coded by $M(x)$). In addition there is the so called "detector background," counts in the detector, indistinguishable from photon events, which are actually caused by charged particles and locally produced secondary photons. Also, for each individual source, all the other sources in the FOV constitute background. In general, the counting rate in the detector due to any single source is small compared with the combined background. For sources in the partially-coded field of view (PCFOV), the incomplete coding adds a "coding noise."

The reconstruction of the image (the best estimate of the sky distribution $S(x)$) can be achieved by several different methods. Only for special codes, it is possible to invert the coding matrix $M(x)$ directly. The most widely used techniques are based on correlation procedures, that is, by cross-correlating the aperture code with the (suitably binned) observed intensity distribution. Alternative techniques are "mismatched filtering" (employing the inverse of the Fourier transform of the point spread function) or "Backprojection." In backprojection, one starts with the position of each detected photon and projects the mask pattern back onto the sky, marking all areas from which this photon could have originated. The superposition of all backprojections leads to the repeated marking of positions from which the photons could have come and produces an image of the source distribution in the sky. It can be shown that for optimum masks all methods are equivalent [21]. Figure 5.4 shows an image of a point source in the center of the FOV plus uniform background, reconstructed from simulated data by the cross-correlation technique. For the simulation, a four times replicated 11×13 URA mask (see below) was used. The optimum-coded FCFOV in the center is a flat top of a pyramid, which falls off to the edges of the PCFOV. By subtracting a "flat field" the image can be "balanced" over the complete FOV. In the PCFOV, so called "ghost images" (of the central point source) are visible, which are due to the incomplete coding.

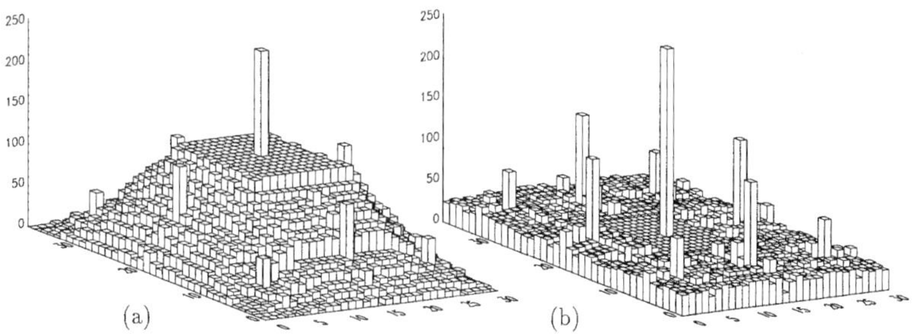

Fig. 5.4 Coded-mask images (Fig. 5.5 in [10])

5.1.2.2 Telescope Configurations and Types of Masks

The geometry of coded-mask telescopes can be quite different: for a given detector, the mask can be (a) smaller, (b) equal, or (c) larger than the detector. Configurations (b) and (c) are shown in Figs. 5.2 and 5.3. For configuration (b), the so called "box camera," full coding is only for on-axis sources (FCFOV equals zero). Here, it is necessary to have a closed (X-ray opaque) telescope tube (an effective collimator, defining the zero response field of view - ZRFOV), such that no photons can reach the detector without passing through the mask. A wide FOV is achieved by configuration (c). Often the mask is chosen to be twice as wide as the detector. Also here an additional collimator limiting the FOV is generally employed. While configuration (a) has some advantage over (b) (e.g., nonzero FCFOV), it suffers from a smaller effective area and therefore reduced sensitivity.

A coded aperture is defined by four parameters: the dimension of the mask elements (pixels), the number of pixels, the fraction of open pixels, and the coding pattern. The *pixel size* is dictated by the desired angular resolution, the spatial resolution element of the detector (which should be smaller than the mask pixel), and the desired or allowed geometry (the size of the mask and its distance D to the detector). The *number of pixels* is constrained by the overall dimension of the mask and the pixel size as well as by the design of the coding pattern (e.g., if certain prime numbers are to be used). In the case of background dominated measurements, the optimum for the *open fraction* is 50% (this is equivalent to the above discussed ON/OFF observation where the optimum on-source time is 50%). For observations with negligible background, e.g., in observing gamma-ray bursts, the optimum open fraction is close to 33%, as was chosen for the BeppoSAX WFC [12]. For the *coding pattern* of the mask, numerous configurations have been proposed (for a compendium of coded-mask designs see [4] and [22] (and references therein)). Although the original proposal by [1] and [5] was to use either regular grids or truly random hole positions, it has become clear that "optimum masks" (those which allow to construct images with a flat background and well-defined source peaks) can be found following certain mathematical construction procedures, which ensure that the reconstructed images carry the desired characteristics. In one-dimension sequences of n (with n being a prime number) randomly selected binary values (0 or 1 for "open" and "closed" mask pixels) in cyclic repetition are the basis for many coding patterns. For two-dimensional aperture codes often *twin-primes* (certain combinations of number of elements in x and y) are chosen. When such basic patterns are repeated (except for one element), so called "uniform redundant arrays" (URA) are formed [6]. A more general procedure, allowing to use other combinations of prime numbers (and to construct quadratic masks) was introduced by [8] under the name of "modified uniform redundant arrays"(MURA). Summaries of the mathematical foundation of mask generation techniques can be found in [4, 10, 17] and in references given in *Imaging in High Energy Astronomy* [2]. Figure 5.5 shows the mask of the imager on INTEGRAL: it is a four times replicated 53×53 MURA (with 11 rows and 11 columns cut off).

Fig. 5.5 Mask of IBIS – the Imager on board of INTEGRAL

The performance of an actual coded-mask telescope can be much worse than theoretically expected, e.g., if there is a variable background: a nonuniform background across the detector surface (caused by a nonregular distribution of matter in the spacecraft next to the detector) and/or a time variable background, which is usually present (certainly in satellites with nonequatorial orbits). These effects lead to false features in the reconstructed images and can severely limit the sensitivity of the instrument. The artifacts can in part be removed by special analysis software (as an example see [9]). An alternative is to perform the observation with a "variable aperture," such that a large fraction of the total detector area is alternately illuminated and not illuminated. One way to do this is to not stare at the sky in a fixed orientation, but rather step the optical axis of the telescope through a pattern of positions. This *dithering* technique is quite usefully applied even with truly imaging telescopes. An elegant (but not easy to realize) technique is to alternately observe through a mask and an "antimask" (where the open and closed pixels are inverted with respect to the mask). This assures that those areas of the detector, which are illuminated observing through the mask, are not illuminated when observing through the antimask. This technique is equivalent to the ON/OFF observations and allows direct background subtraction [3].

5.2 Various Coded-Mask Telescope Missions

In Table 5.1, an overview is given about the main aperture modulation instruments and missions. The table is not nearly complete, giving only examples for those instruments that have either produced important scientific results or constituted a step

Table 5.1 Important aperture modulation telescopes

Year	Carrier	Instr.	No.of instr. & aperture	Mode of operation	Energy [keV]
1970	Uhuru		2 Coll[a]	Scan[b]	2–20
1971	OSO-7		4 Coll	Scan	1–1 000
1974	Ariel V		5 Coll, 1 RMC[c]	Spin,scan	0.3–40
1975	SAS-3		6 Coll, 2 RMC	Spin, point[d]	0.1–60
1976	Rocket	SL1501	CM	Point	2–10
1977	HEAO-1	A1-A4	4 Coll(1 grid)[e]	Scan	0.2–10 000
1983	Tenma	HXT	1 Coll, 1 CM[f]	Spin[h]	2–10
1985	Balloon	GRIP	CM	Point	30–600
1985	SL2	XRT	CM	Point	2–25
1987	Mir	TTM	CM	Point	1.8–30
1989	Granat	SIGMA	CM	Point	30–1 300
		ART-P	CM		4–60
		ART-S	CM		3–100
		WATCH	RMC		6–120
1991	CGRO	BATSE	Coll	Point, EaOc[g]	20–10 000
1995	RXTE	ASM	CM	Scan	2–10
1996	BeppoSAX	WFC	CM	Point	2–30
2000	HETE-2	SXT	2 CM	Point	0.5–14
		WXM	2 CM	Point	2–25
2002	INTEGRAL	JEM-X	CM	Point	3–35
		IBIS	CM		15–10 000
		SPI	CM		20–8 000
2004	Swift	BAT	CM	Point	15–150

For nonscanning missions simple collimated instruments are not mentioned
No individual references are given here. General information about X-ray astronomy satellites can be found in http://heasarc.gsfc.nasa.gov/docs, followed by the name of the satellite (e.g., "/heao1/heao1.html"). A useful web page for coded aperture is: http://lheawww.gsfc.nasa.gov/docs/cai/coded.html
[a]Slat or tube collimators
[b]Scanning
[c]Rotation modulation collimator (RMC)
[d]Pointing
[e]Grid collimators
[f]Coded mask (CM)
[g]Earth occultations
[h]Spinning satellite

forward in technology. The reader is referred to the web addresses given in the footnote of Table 5.1.

The two most advanced satellites carrying Coded Mask instruments are INTEGRAL and Swift. While INTEGRAL, operating since November 2002, is devoted to pointed observations with combined high spatial, spectral and temporal resolution, the main scientific objective of Swift is to detect and accurately localize gamma-ray bursts and to perform fast response follow-up pointed observations at X-ray and UV/optical wavelengths.

References

1. Ables, J.G. 1968, Proc. Astron. Soc. Aust., 1, 172
2. Bassani, L., di Cocco, G. (eds.) 1995, Imaging in High Energy Astronomy, Kluwer, Dordecht, ISBN 0-7923-3788-3
3. Busboom, A., et al. 1997, J. Opt. Soc. Am., 14, 1058
4. Caroli, E., et al. 1987, Space Sci. Rev., 45, 349
5. Dicke, R.H. 1968, Astrophys. J., 153, L101
6. Fenimore, E.E., Cannon, T.M. 1978, Appl. Opt., 17, 337
7. Forman, W., et al. 1978, Astrophys. J. Suppl., 38, 357
8. Gottesmann, S.R., Fenimore, E. E. 1989, Appl. Opt., 28, 4344
9. Grebenev, S.A., et al. 1995, in L. Bassani, G. di Cocco (eds.), Imaging in High Energy Astronomy, Kluwer, Dordecht, 155
10. Groeneveld, H.A. 1998, Design, Simulation und Optmierung kodierter Aperturen, PhD Thesis, Univ. of Tübingen
11. Harmon, B.A., et al. 2002, Astrophys. J. Suppl., 138, 149
12. in 't Zand, J.J.M. , et al., 1994, Astron. Astrophys., 288, 665
13. Levine, A.M., et al. 1984, Astrophys. J. Suppl., 54, 581
14. Li, T.P., et al. 2003, Proc. 28th Int. Cosmic Ray Conf., Tsukuba, Japan, 2775
15. Mertz, L. 1967, in R. Fox (ed.), Proc. Symp. Modern Optics, Polytechnic Press Brooklyn, New York, 787
16. Ogawara, Y., et al. 1984, in M. Oda, R. Giacconi (eds.), X-Ray Astronomy '84, Institute of Space and Astronautical Science, Tokyo, 313
17. Rideout, R. 1995, Coded Imaging Systems for X-ray Astronomy, PhD Thesis, Univ. of Birmingham
18. Rothschild, R., et al. 1996, Astrophys. J., 496, 538
19. Schnopper, H., et al. 1968, Space Sci. Rev., 8, 534
20. Schnopper, H., Delvaille, J.P., 1972, Scientific American, 27
21. Skinner, G.K. 1995, in L. Bassani, G. di Cocco (eds.), Imaging in High Energy Astronomy, Kluwer, Dordecht, 1
22. Skinner, G.K., Rideout, R.M. 1995, in L. Bassani, G. di Cocco (eds.), Imaging in High Energy Astronomy, Kluwer, Dordecht, 177
23. Staubert, R., et al. 1975, Astrophys. J., 201, 15
24. Theinhardt, J., et al. 1984, Nucl. Instr. Meth., 221, 288

6 Wolter Optics

P. Friedrich

6.1 Principle

Optical elements for X-rays are based on the principle of grazing incidence reflection. Reflection, absorption, and transmission are expressed through the complex index of refraction, which can be written as

$$n = 1 - \delta - i\beta$$

where δ describes the phase change and β accounts for the absorption. The optical constants δ and β are functions of the wavelength or the photon energy (Fig. 6.1). For X-rays, the real part of n, $1 - \delta$, is slightly less than unity for matter whereas it is exactly unity in vacuum. X-rays propagating in vacuum, therefore, undergo total external reflection when incident below the critical grazing angle α_t with

$$\cos \alpha_t = 1 - \delta$$

according to Snell's law. Because of the nonvanishing value of β the reflection is actually not total for $\alpha \leq \alpha_t$, but is less than unity, and X-rays are reflected at angles even larger than the critical angle α_t [1]. There are also characteristic absorption edges resulting from the specific atomic structure of each element.

Generally, δ is proportional to the atomic number Z and proportional to the squared wavelength λ. In the case of heavy elements, the critical angle for $\delta \ll 1$ can be estimated as

$$\alpha_t = 56 \times \sqrt{\rho} \times \lambda$$

for a material with density ρ and $\alpha_t \approx \sqrt{2\delta}$, with α_t in arcminutes, ρ in g/cm^3, and λ in nanometer [1]. Converting the wavelength λ into a photon energy E – which is more appropriate for X-rays – one gets

$$\alpha_t = 69 \times \sqrt{\rho}/E$$

with E in kiloelectronvolts.

Imaging X-ray optical systems were introduced by Hans Wolter when he published his paper (in German) on "Grazing Incidence Mirror Systems as Imaging Optics for X-rays" in 1952 [12]. Originally meant for X-ray microscopy, they were

J.E. Trümper and G. Hasinger (eds.), *The Universe in X-Rays.*

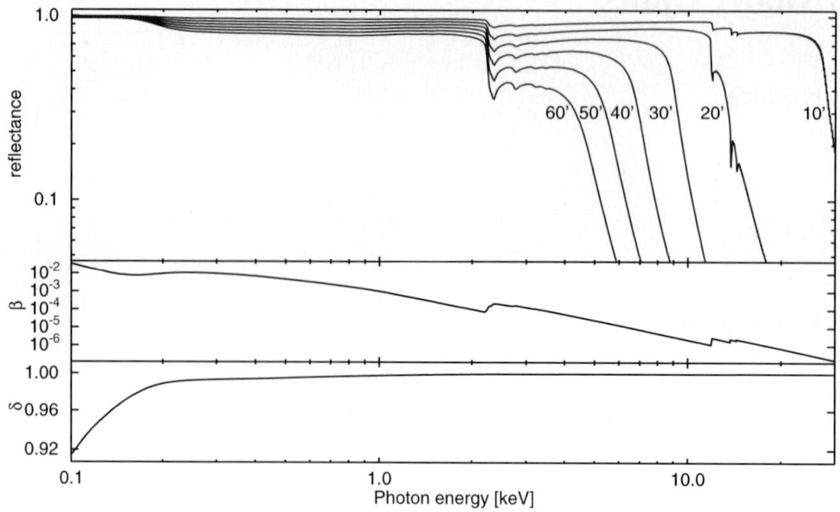

Fig. 6.1 Optical constants δ and β and reflectance of gold in the energy range 0.1–10 keV for different grazing angles

recognized as the appropriate optics for X-ray telescopes when the new branch of X-ray astronomy developed, after the detection of the first X-ray source in the sky in 1962 [8].

The type of optics proposed by Wolter – now called Wolter optics – makes use of the fact that X-rays are reflected by smooth surfaces under small angles of incidence. However, a single mirror optics like a paraboloid is not able to focus X-rays properly because it cannot satisfy Abbe's sine condition (Fig. 6.2).

$$d/\sin\vartheta = f$$

The sine condition means that the principal surface has to be a sphere with radius f. For single mirrors, the principal surface is always identical to the mirror surface itself. Therefore, the sine condition is approximately satisfied only if the reflection is almost perpendicular to the mirror's surface. Unlike optical mirrors, grazing incidence X-ray mirrors obviously do not fulfil this condition. The solution for X-ray optics is – as Wolter elaborated – an optical system of at least two mirrors. Wolter could also show that mirror systems with even numbers of mirror elements can satisfy the sine condition whereas systems with odd numbers of mirror elements cannot (as long as reflections are less than 90° to the optical axis). In practice, two-mirror systems are strongly favored because any system with four and more mirror elements would increase the losses due to scattering and reflection significantly; furthermore, the alignment procedure for such systems would be complicated.

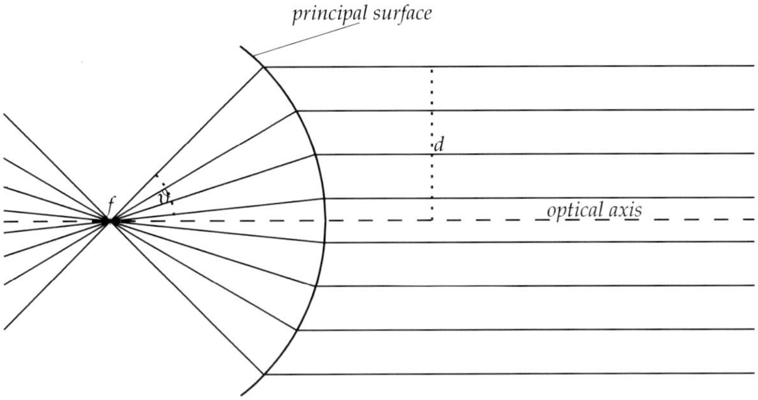

Fig. 6.2 Abbe's sine condition: The optic's principal surface has to be a sphere making the distance to the focus f the same for all paraxial rays

6.2 Wolter-Type Telescopes

Wolter describes three configurations of two-mirror systems each with a pair of different figures of revolution in a confocal arrangement. They are mostly called Wolter-1, Wolter-2, and Wolter-3. Both Wolter-1 and Wolter-2 configurations are composed of a paraboloid and a hyperboloid (see Fig. 6.3, 6.4). A Wolter-3 system consists of a paraboloid plus an ellipsoid.

Wolter-1 systems have become most important in X-ray astronomy because – unlike the other types – Wolter-1 mirrors can be nested to achieve large effective aperture areas. Furthermore, manufacturing and alignment is less difficult.

The more compact Wolter-2 systems are sometimes preferred where nesting can be renounced, e.g., for solar X-ray telescopes. Wolter-3 systems have never found application in X-ray astronomy.

Already Wolter pointed out that it is possible to design two-mirror systems, which fulfill the sine condition exactly [4, 13]. They require more complicated figures of revolution. Such systems are called Wolter-Schwarzschild optics because they base on an idea from Karl Schwarzschild [9]; like normal Wolter optics Wolter-Schwarzschild optics can be divided into three different configurations. In practice, such systems have not yet gained much importance because they are difficult to fabricate and to measure. Furthermore, such optics would be perfect only on the optical axis. One Wolter-Schwarzschild-2 optics has been realized for the coronal diagnostic spectrometer (CDS) on *SOHO*.

Special types of Wolter optics are proposed for wide field X-ray imaging [3]. These are designed to optimize the imaging quality over a large field-of-view. Here, higher order terms are added to the figures of revolution to correct for the off-axis

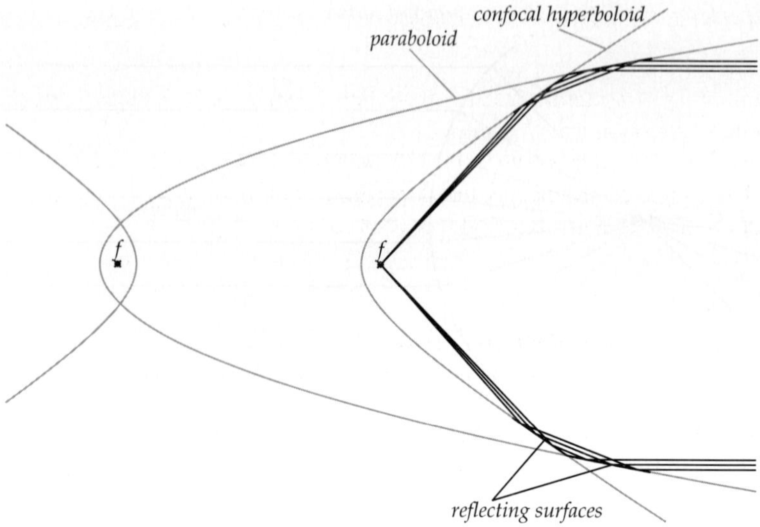

Fig. 6.3 Scheme of a Wolter-1 optics

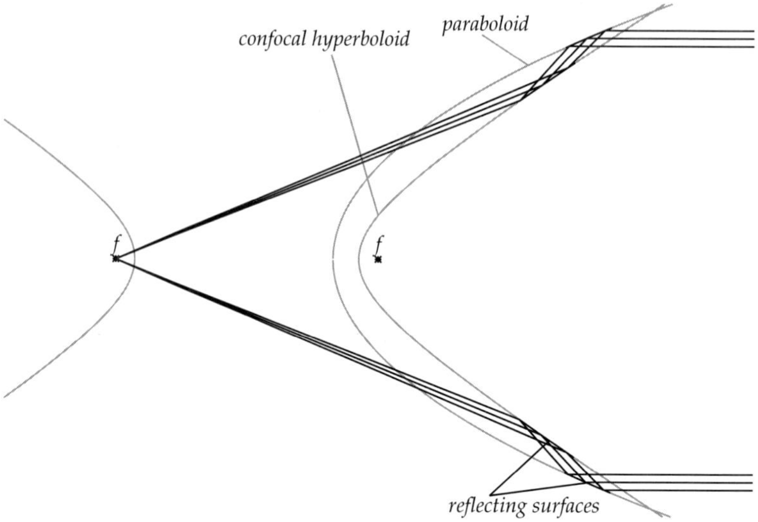

Fig. 6.4 Scheme of a Wolter-2 optics

bluring of a normal Wolter optics (see below). This implies of course a less perfect on-axis imaging.

In general, all modifications of Wolter-1 optics are more complicated and are worth to realize only if other effects such as figure errors, scattering, detector resolution no longer dominate the imaging quality.

6.3 General Imaging Properties

The following considerations refer to Wolter-I systems, which incorporate the vast majority of astronomical X-ray optics.

Small grazing angles result in large focal lengths. The ratio of aperture size $2r$ to focal length f is determined by the slope angle α of the first mirror element, which is approximately the grazing angle for paraxial rays:

$$r/f = \sin 4\alpha$$

The ratio $2r/f$ is typically around 1/10 if the energy band ranges up to 10 keV because the grazing angles have to be less than 1°; soft X-ray telescopes, e.g., ROSAT, can work with a ratio of about 1/3.

The full diameter of the field-of-view is limited both by geometrical vignetting, because of the aperture stops, and shape of the reflecting surfaces and by the requirement on the grazing angle of reflection [7]. Also nested mirror shells act as aperture stops. Practically, the usable field-of-view is also limited by the off-axis blurring (see later). Detector sizes are chosen such that the central part of the field-of-view with an acceptable vignetting and blurring is covered. In general, Wolter optics with a lower ratio $2r/f$ have smaller fields-of-view than those with large ratios; the ROSAT PSPC, for example, corresponds to field-of-view diameter of 2°, the CCD cameras on XMM-Newton cover about $30'$.

Beside the optical axis, the imaging quality degrades continuously with increasing off-axis angles. The off-axis blurring is shown in Fig. 6.5. An empirical formula for the rms blur circle σ for a flat detector plane is given by Van Speybroeck and Chase [11]:

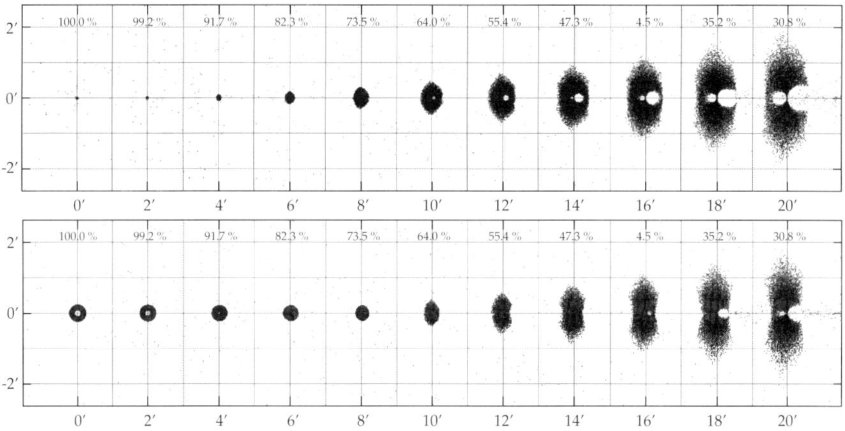

Fig. 6.5 Ray tracing calculated point spread function for off-axis angles from $0'$ to $20'$ for a nested mirror system with $2r_{\max}/f = 1/10$ and $l = 1.8r_{\max}$ with an unshifted detector plane (*upper panel*) and with a shifted one – the shift is $0.00125f$ (*lower panel*); the percentage numbers denote the vignetting

$$\sigma = \frac{\zeta+1}{10} \times \frac{l}{f} \times \frac{2\tan^2\theta}{\tan\alpha} + 4\tan\theta\tan^2\alpha$$

where α again is the slope angle of the first mirror element (the paraboloid), θ the off-axis angle of the incoming rays, l the length of a mirror element, f the focal length, and ζ the ratio of the slope angles of the paraboloid and that of the hyperboloid. Usually, the grazing angles of both optical elements are chosen so that both reflections of an incoming paraxial ray occur under the same angle of incidence thus minimizing the losses due to reflection. The slope angle of the second element, defined with respect to the optical axis, is therefore three times larger than that of the first so that ζ becomes 3.

A hyperbolically curved detector surface would be better adapted to the actual focal plane and thus reduce the blur circle by up to 50% because the factor 2 in the third term is then omitted. In practice, a much simpler but less efficient possibility to reduce blurring is a slight shift of the detector toward the mirror system (see Figs. 6.5 and 6.6) [1,7]. The amount of this shift is chosen according to the actually achieved

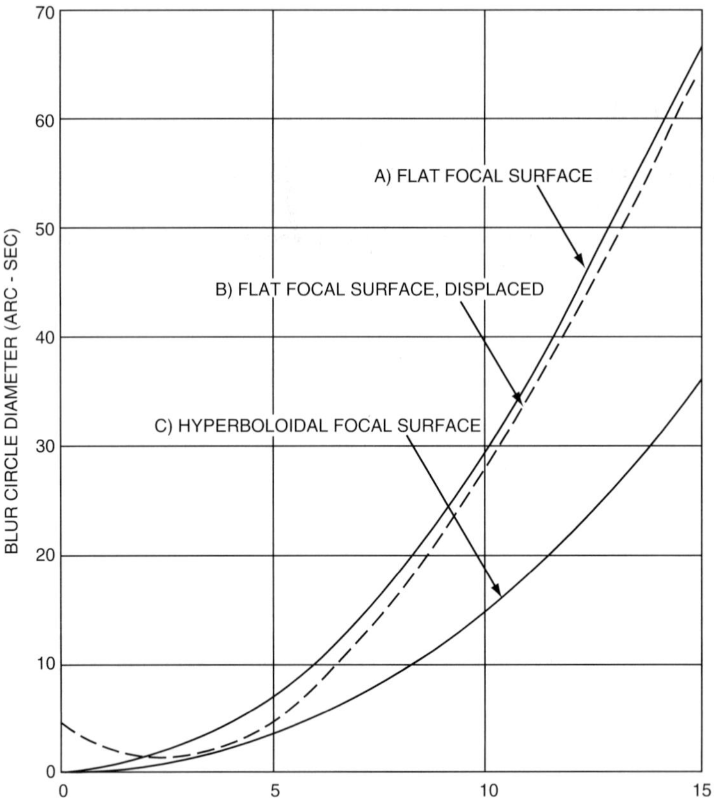

Fig. 6.6 Total blur circle diameter vs. off-axis angle in arcminutes, from a ray tracing done on a telescope system with $2r/f = 1/10$ and $l = 2r$; the detector shift is $0.00018f$ (from Giacconi et al. [7])

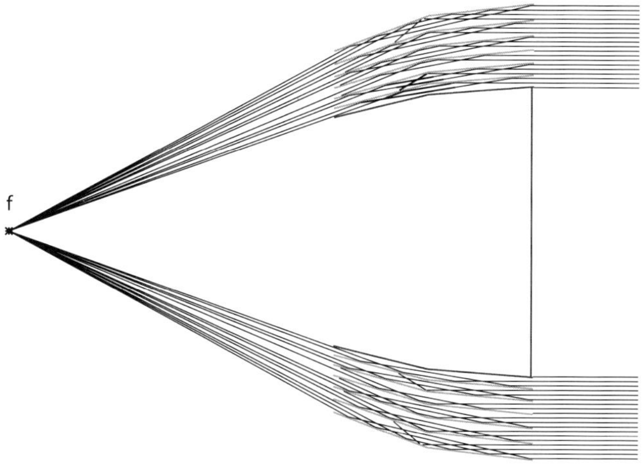

Fig. 6.7 Nested Wolter-1 mirror shells provide large effective apertures; as nested shells act as aperture stops, they limit the field-of-view but can also prevent direct light and singly reflected rays from penetrating the mirror system

Fig. 6.8 ROSAT mirror (*left*) and an integrated mirror module for XMM-Newton with 58 replicated gold-coated nickel shells (*right*)

on-axis performance of an optical system and with respect to the field-of-view to be optimized.

6.4 Nesting of Mirror Shells

For small slope angles α (see above) the geometrical collecting area of a mirror shell is a thin circle with an projected area a of

$$a \approx 2\pi r \times l \sin\alpha$$

where r is the mirror radius and l the length of a mirror element. For small angles α it can also be expressed with the focal length f:

$$a \approx 0.5\pi r^2 \times l/f$$

Table 6.1 Satellite missions with Wolter telescopes

Mission	Year of launch	Upper energy limit (keV)	Focal length (m)	Mirror modules	Degree of nesting	Effective area @ 1 keV (cm^2)	On-axis resolution (HPD)
S-054/Skylab	1973	4	2.13	1	2	15	48″
S-056/Skylab	1973	1.3	1.90	1	1	9	3″
Einstein (HEAO-2)	1978	4	3.44	1	4	100	4″
EXOSAT	1983	2.5	1.09	2	2	70	24″
ROSAT	1990	2.5	2.40	1	4	420	3″
BBXRT	1990	12	3.77	2	118	450	5″
Yohkoh SXT	1991	4.0	1.54	1	1	23	<5″
ASCA (Astro-D)	1993	10	3.50	4	120	1 200	180″
Soho CDS	1995	0.5	2.58	1	1	23	<5″
BeppoSAX	1996	10	1.85	4	30	344	60″
ABRIXAS	1999	10	1.60	7	27	560	25″
Chandra (AXAF)	1999	10	10.00	1	4	780	<1″
XMM-Newton	1999	15	7.50	3	58	4 260	16″
Swift	2004	10	3.50	1	12	130	18″
Suzaku (Astro-E2)	2005						
XRT-I		12	4.75	4	175	2250	120″
XRT-S		12	4.50	1	168	2250	120″

This is much less than the polished mirror surface. Nesting several mirror shells with the same focal length and the same optical axis into each other is an appropriate solution for an enlarged effective collecting area while keeping the telescope as compact as possible (Fig. 6.7). Nesting was first practiced for the solar X-ray telescope *S-054* mounted on the Apollo platform on *Skylab* in 1973. It consisted of two small Wolter-1 mirror shells. The nesting of many thin-walled mirror shells allows good filling factors for a given area. Examples for such highly nested optics are ASCA, XMM-Newton, and Suzaku.

6.5 Fabrication Techniques for Wolter Telescopes

There are two main challenges in the manufacturing of Wolter optics: to achieve a perfect figure and to keep the micro-roughness of the optical surface low. The accuracy of the figure described in terms of roundness and slope errors governs the overall point spread function and thus the angular resolution of the optics. A particular difficulty is the proper figuring of the break between the parabolic and hyperbolic part or the alignment of separate parboloids and hyperboloids. Surface errors range from long-wave deformations to high spatial frequencies where interference effects dominate the imaging quality rather than geometric optics. In the regime of interference, the surface errors are called micro-roughness. As X-rays have wavelengths about a factor of hundred shorter than optical light, the requirement on the micro-roughness is correspondingly high. There is a distinct relationship between the X-ray scattering and the surface topography described by vector electromagnetic scattering theory [1, 5]. The reducing of micro-roughness required extreme efforts in the early days of X-ray optics, whereas super-polishing down to a few tenth of a nanometer micro-roughness is now standard.

Most of the first Wolter telescopes have been manufactured in the classical way adopted from optical mirror production. They were made from thick glass blanks, which were ground, polished, and finally coated. For example, fused quartz glass with nickel coating was used for Einstein and Zerodur glass with gold coating for ROSAT (Fig. 6.8). Chandra, the most precise Wolter telescope so far, is made of iridium-coated Zerodur and has an angular resolution of better than $1''$ half power diameter (HPD). High-Z corrosion-resistant metals like gold, iridium, and nickel are well suited for the coating of the reflecting mirror sufaces. Multilayer coatings combined of different materials can enhance the reflectance beyond the limit of total external reflection by using constructive interference from the thin layers. Multilayer Wolter optics are under study for future X-ray missions.

After the pioneering astronomical X-ray missions the demand for high throughput X-ray optics initiated the development of light-weight, thin-walled X-ray mirrors. One result of this development, mainly driven by the XMM-Newton mission, was the technique of coated electroformed nickel shells replicated from super-polished mandrels (Fig. 6.8). Another result, e.g., applied for ASCA, was the technique of replicated foil mirror segments (aluminum foil plus epoxy) [10]. Both techniques allow a high degree of nesting. The nickel mirrors are heavier but are suited for an angular resolution up to $16''$ HPD (XMM-Newton) whereas the very light foil mirrors achieve a resolution in the arcminute range only. Replication techniques have the advantage that once a mandrel is polished several integration-ready replications can be drawn from it. Even the coating is part of the replication process.

Scientific goals of today require X-ray telescopes with even larger collecting areas on the order of several square meters and angular resolution in the range of arcseconds to avoid source confusion. At the same time, however, limited launch masses are a severe constraint driving the development of new technical solutions for extremely light weight and rigid Wolter optics. Two approaches are currently under study: thin glass mirror segments can be formed in appropriate moulds when the glass has a certain viscosity through heating [6, 14]. Still lighter are pore optics made from stacked silicon sheets with small grooves as "light channels" [2]. These techniques are favored for the Wolter optics of the planned X-ray missions: "slumped glass" for *Constellation-X* and pore optics for *XEUS*.

6.6 Missions with Wolter Telescopes

The first X-ray instruments for astronomical purposes were equipped with collimators and large area detectors rather than with imaging capabilities. First X-ray mirrors designed as Wolter optics were flown in 1963 on a sounding rocket for solar imaging. The first long term X-ray observations were done with two small Wolter telescopes for solar imaging operated on the space station *Skylab* from 1973 to 1975. The era of satellite-based Wolter telescopes begun in 1978 with HEAO-2, which was then renamed to Einstein Observatory. Currently, two major X-ray observatories with Wolter telescopes are operating: Chandra and XMM-Newton. They complement each other because Chandra has an unprecedented angular resolution of better than $1''$ whereas XMM-Newton by far has the largest collecting area. Since 2005 they are joined by Suzaku with five highly nested foil telescopes (Table 6.1). Missions with much larger collecting areas like *Constellation-X* and *XEUS* are planned for the second decade of the twenty-first century.

References

1. Aschenbach, B. 1985, X-ray telescopes, Reports on Progress in Physics, 48, 579
2. Bavdaz, M., Lumb, D., Gondoin, P. et al. 2006, The XEUS X-ray telescope, SPIE 6266, 62661S
3. Burrows, C.J., Burg, R., Giacconi, R. 1992, Astrophys. J., 392, 760
4. Chase, R. C., Van Speybroeck, L. P. 1973, Appl. Opt., 12, 1042
5. Church, E. L. 1979, SPIE 184, 196
6. Friedrich, P., Aschenbach, B., Braig, C. et al., 2006, SPIE 6266, 62661G
7. Giacconi, R., Reidy, W.P., Vaiana, G.S. et al. 1969, Space Sci. Rev., 9, 3
8. Giacconi, R. 1974, in R. Giacconi, H. Gurski (eds.), X-Ray Astronomy, Astrophysics and Space Science Library, Vol. 43, 1
9. Schwarzschild, K., 1905, Untersuchungen zur geometrischen Optik I–III, Astronomische Mittheilungen der Königlichen Sternwarte zu Göttingen
10. Serlemitsos, P.J., Soong, Y. 1996, Astrophys. Space Sci., 239, 177
11. Van Speybroeck, L.P., Chase, R.C. 1972, Appl. Opt., 11, 440
12. Wolter, H. 1952, Annalen der Physik (6. Folge) 10, 94
13. Wolter, H. 1952, Annalen der Physik (6. Folge) 10, 286
14. Zhang, W.W., Chan, K.-W., Content, D.A., 2006, SPIE 6266, 62661V

7 CCD Detectors

L. Strüder and N. Meidinger

7.1 Introduction

Since several years, charge coupled devices (CCD) type detectors are used in X-ray satellite missions in the focus of imaging optics (e.g., ASCA, Chandra, XMM-Newton, Swift, Suzaku). They measure position, energy, and arrival time of individual X-rays from 100 eV to 15 keV energy. The Japanese ASCA satellite was the first X-ray mission to employ CCDs as a focal plane detector in 1993. The US American Chandra and the European XMM-Newton missions followed with larger detector arrays and improved performance in 1999.

Since the launch of the European XMM-Newton satellite on December 10, 1999, reliably operating X-ray CCDs are delivering extraordinary images, recorded in a single photon counting mode, imaged through the largest X-ray telescope ever built. Behind two of the three X-ray mirror systems reflecting grating spectrometers are measuring high resolution X-ray spectra, recorded with 9 CCDs along the dispersion direction [26]. About 40% of the incident X-ray flux is transferred to a focal plane with a mosaic of 7 CCDs covering a field of view of approximately 30 arcmin [5]. All of those CCDs are of the conventional MOS type.

The telescope without gratings getting 100% of the X-rays directed onto the focal plane is equipped with the novel pn-type CCD. The fully depleted, back side illuminated pn-junction CCD (pnCCD), operated in full frame mode, has been developed for applications in X-ray astronomy. A monolithic $6 \times 6\ \mathrm{cm}^2$ large pnCCD is working since more than 7 years in orbit aboard the European X-ray satellite XMM-Newton.

Similar devices are working today in many different fields like hadron physics, synchrotron radiation research, quantum optics, X-ray microscopy, material analysis, and others. In the last years, the capabilities of those devices were largely extended for their use in high speed photography in the visible and near infrared.

For future missions, such as eROSITA, we have developed a new generation of frame store pnCCDs: As the depleted, sensitive thickness of the pnCCDs is increased up to 500 μm, their high energy response to X-rays is extended to 30 keV with a quantum efficiency of still 20%. Frame store pnCCDs have been fabricated with up to 512 readout nodes, funneled into eight parallel output chains, enabling the detector with a format of 256×512 pixels and pixel sizes of $51 \times 51\ \mu\mathrm{m}^2$ and $75 \times 75\ \mu\mathrm{m}^2$ to be read out in 1 ms with a read noise of two electrons (rms).

J.E. Trümper and G. Hasinger (eds.), *The Universe in X-Rays.*

A typical operation temperature is −60° C, where Fano-limited X-ray spectroscopy is reached.

The basic concept of CCDs, their charge transfer mechanisms, their limitations in energy resolution, their long-term stability, and the achieved results of measurements are presented in the following. The main focus is set on the concept of pnCCDs as they exhibit excellent energy and position resolution at high quantum efficiency from the near infrared up to 30 keV, high readout speed (resp. high time resolution), low noise, and high radiation hardness – key parameters for X-ray missions to date and in the future.

X-ray astronomy is pushing since several years the instrumentation for broad-band imaging nondispersive X-ray spectrometers. A brief outlook will, therefore, be finally given on new concepts for imaging X-ray spectrometers based on the concept of the DEPFET active pixel sensors.

7.2 MOS CCDs

CCDs have originally been developed as electronic imagers for optical purposes [1,16]. Until now this is still the main commercial application. This use implies that only a few micrometers of surface depletion are necessary to cover the bandwidth of light from 350 to 650 nm. The depletion is achieved through a conductor–oxide–semiconductor structure (MOS) in nonequilibrium condition. The devices can be modified to make them more sensitive to X-rays by increasing the depleted thickness up to almost 50 μm. A substantial effort was made to manufacture back-illuminated devices, some of which are operating on Chandra and XMM-Newton in the dispersive spectrometers with a thickness of about 30 μm. The operating temperature of the MOS CCDs in orbit is about −120° C and the frame rate a few seconds. The energy resolution of the MOS devices is Fano-limited above 1 keV. The susceptibility to radiation is consistent with previous tests of MOS-type devices. As the pnCCDs were tailored to the requirements of X-ray astronomy, we will concentrate on this concept in the following in more detail.

7.3 Fully Depleted Back-Illuminated pnCCDs

Silicon is used as a detector material since the early fifties of the last century. The fact that an energy of only 3.7 eV is needed to create an electron–hole pair was soon exploited to fabricate detectors for photons and particles with high sensitivity. In addition, the band gap of silicon ($E_{\mathrm{gap}} = 1.1$ eV) is large enough to operate detectors at moderate temperatures avoiding thermally generated leakage currents. The atomic number of silicon ($Z = 14$) is still high enough to obtain a high detection probability for e.g., X-rays up to an energy of 30 keV for a device thickness of the order of 1 mm.

A major breakthrough in the development of silicon detectors occurred when detector grade silicon was processed in planar technology [6]. It opened the opportunity to directly implement signal processing electronics monolithically into the detector silicon. In the eighties and nineties, more and more complex structures were developed making use of the possibility to shape the potential inside the detectors and of the on-chip electronics for optimum sensor performance (sensor means detector in combination with on-chip electronics).

All experimental results shown here are from devices that have been designed, fabricated, and tested at the MPI Halbleiterlabor.

Conceptually, the pnCCD is a derivative of the silicon drift detector [3]. The development of the pnCCDs started in 1985. In the following years, the basic concept was simulated, modified, and designed in detail [19]. N-channel JFET electronics was integrated in 1992 [14, 15] and the first reasonably fine working devices were produced in 1993. Up to then, all presented devices were "small" devices, i.e., 3 cm^2 in sensitive area [17]. The flight type large area (36 cm^2) detectors were produced from 1995 to 1997, with a sufficiently high yield to equip the X-ray satellite missions ABRIXAS and XMM-Newton [9, 18, 21] with defect-free focal plane pnCCDs.

7.3.1 The Concept of Fully Depleted, Back-Illuminated, Radiation Hard pnCCDs

The pnCCD concept for XMM-Newton and the associated fabrication technology allow for an optimum adaption of the pixel size to the X-ray optics, varying from 30 to 500 μm pixel size. Up to now systems with 36–300 μm have been produced. The XMM-Newton mirror performance of 13 arcsec half energy width (HEW) translates to 470 μm position resolution in the focal plane. The FWHM of the point spread function (PSF) is about 7 arcsec. Therefore, a pixel size of 150×150 μm was chosen, giving a position resolution of $\leq$120 μm, resulting in an equivalent spatial resolving capability of $\leq$3.3 arcsec. This is sufficient to fully conserve the positional information of the X-rays from the XMM-Newton mirrors. The quantum efficiency is about 90% at 10 keV because of the sensitive thickness of 300 μm defined by the wafer thickness. The low energy response is given by the very shallow implant of the p^+ back contact; the effective "dead" layer is shallower than 200 Å [4]. The excellent time resolution is achieved by the parallel readout of 64 channels per subunit, altogether 768 channels for the entire camera. High radiation hardness is built in by avoiding active MOS structures and by the fast transfer of the charge in a depth of about 10 μm. The spatially uniform detector quality over the entire field of view is realized by the monolithic fabrication of the pnCCD on a single wafer. For redundancy reasons, 12 individually operated 3×1 cm^2 large pnCCDs subunits were defined. Inhomogeneities were not observed over the entire sensitive area in the calibration energy band from 0.5 to 8 keV, within the precision of the measurements limited by Poisson statistics. The insensitive gap in the vertical separation of the

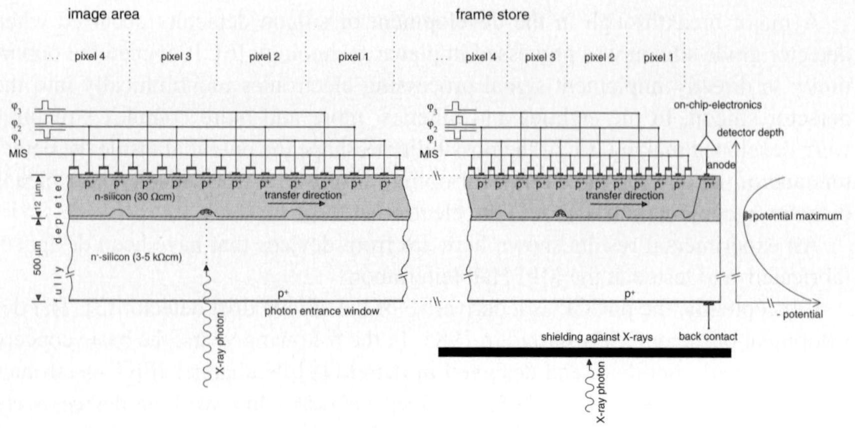

Fig. 7.1 Cross section through the pnCCD along a transfer channel. The device is back-illuminated and fully depleted over the wafer thickness. The electron potential perpendicular to the wafer surface is shown on the *right hand side* [8]

pnCCDs is about 40 µm, neighboring CCDs in horizontal direction have insensitive border regions of 190 μm (see Fig. 7.1).

The basic concept of the pnCCD is shown in Fig. 7.1 and is closely related to the functional principle of silicon drift detectors (SDDs). A double-sided polished high resistivity *n*-type silicon wafer has both surfaces covered with a rectifying p^+-boron implant. On the edge of the schematic device structure (see Fig. 7.1) an n^+-phosphorus implant (readout anode) still keeps an ohmic connection to the non-depleted bulk of the silicon. A reverse bias is now applied to both p^+ junctions, i.e., a negative voltage is applied with respect to the n^+ anode. For simplicity let us assume that the silicon bulk is homogeneously doped with phosphorus with a concentration of 1×10^{12} cm^{-3}. Depletion zones in the high resistivity substrate, with a resistivity of about 4 kΩ cm, develop from both surfaces, until they touch in the middle of the wafer. The potential minimum for electrons is now located in the middle of the wafer. An additional negative voltage on the p^+ back diode shifts the potential minimum for electrons out from the center toward the surface containing the pixel structure. Typical depletion voltages on the backside are -150V. To make a CCD-type detector, the upper p^+ implant must be divided in p^+ strips as shown in Figs. 7.1, 7.2, and 7.6. Adequate voltages should now be applied to the three shift registers such that they form local potential minima for electrons. Three p^+ strips (shift registers) with the potentials Φ_1, Φ_2, and Φ_3 define one pixel. Charges are collected under Φ_1, the potential minimum for electrons. A reasonable change with time of the applied voltages transfers the charges to the new local e^- potential minimum in a discrete way toward the n^+ readout node. In reality, the side having the p^+ shift registers has an additional phosphorus-doped epitaxial layer, 12 µm thick, with a concentration of approximately 10^{14} donors per cm^3. The interface of the epilayer and the high resistivity bulk silicon fixes the electron potential minimum to a distance of about 10 µm below the surface.

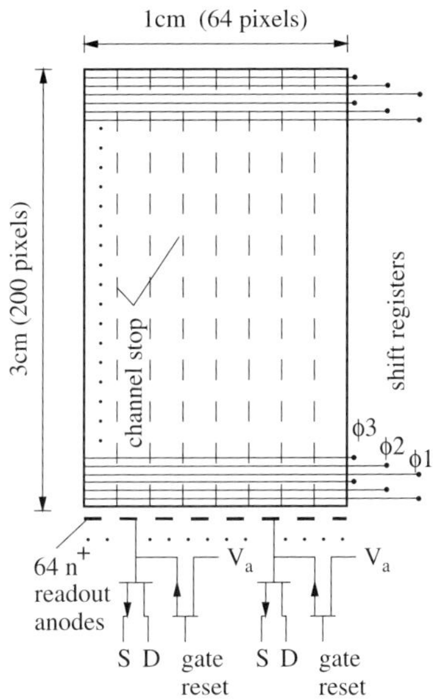

Fig. 7.2 One pn-CCD subunit with 64 on-chip amplifiers and a size of $3 \times 1\ \text{cm}^2$. Each of the 64 columns is terminated by an on-chip JFET amplifier [22]

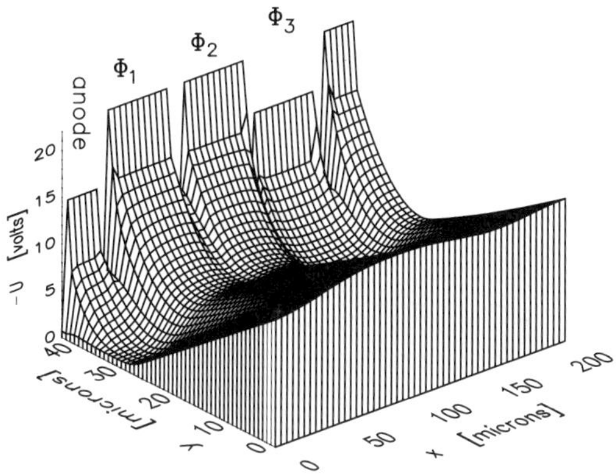

Fig. 7.3 Negative potential distribution inside the pnCCD pixel. In this operating condition, the signal charges are stored under the register Φ_3 only. The p^+ back side potential is only shown up to the depth of 40 μm for clarity. It expands to −150 V at 300 μm distance from the pixel surface [22]

As can be seen in Fig. 7.2, one pnCCD subunit consists of 64 individual transfer channels each terminated by an on-chip JFET amplifier. Figures 7.3–7.5 show the charge transfer mechanism. The p^+ backside contact is not shown: it expands quite uniformly an additional 260 μm toward a negative potential of −150 V. The sequence of changing potentials shows nicely the controlled transfer from register Φ_3 to register Φ_2, one third of a pixel.

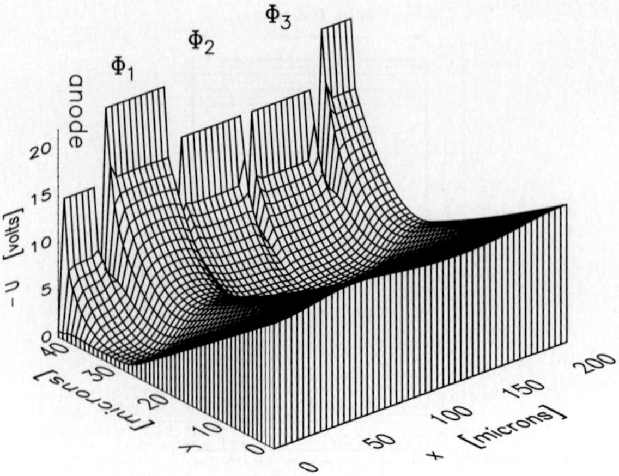

Fig. 7.4 Negative potential distribution inside the pnCCD pixel according to charge transfer sequence. In this operating condition, the signal charges are stored under the registers Φ_2 and Φ_3. The electrons now share a larger volume for a short time. Note that the electrons are still nicely confined in the potential well [22]

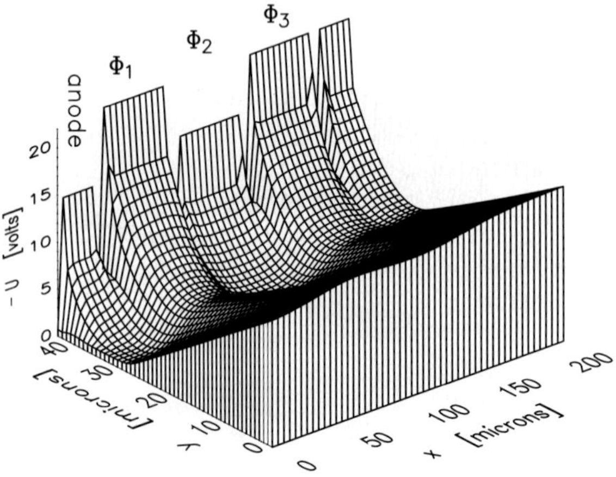

Fig. 7.5 Negative potential distribution inside the pnCCD pixel at next transfer step. In this operating condition, the signal charges are stored under the register Φ_2 only. The charge was transferred by one third of the pixel length in approximately 150 ns [22]

This concept is seen from a different point of view in Fig. 7.6, seen from the inside of a pnCCD: X-rays hit the detector from the back side (indicated as back contact in Fig. 7.6). The positively charged holes move to the negatively biased back side and the electrons to their local potential minimum in the transfer channel, located about 10 μm below the surface having the pixel structure. The electrons are fully collected in the pixels after 5 ns at most, the collection of holes is completed

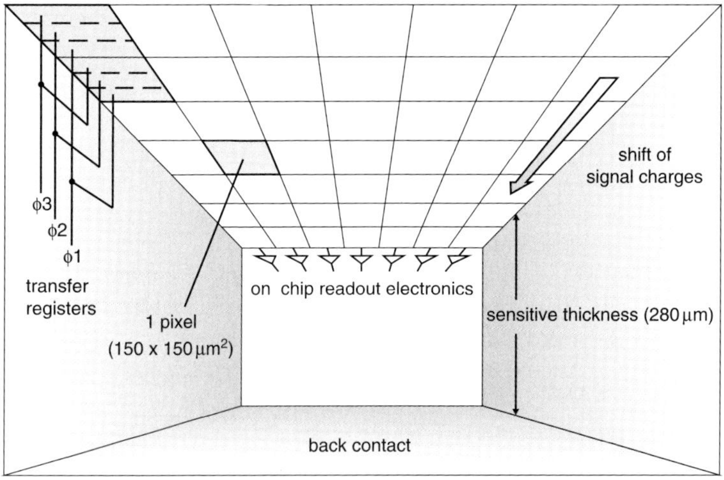

Fig. 7.6 Inside the pnCCD, the X-rays hit the device from the back side (*bottom*). The charges are collected in the pixel well close to the surface having the pixel structure. After integration, they are transferred to the on-chip amplifier [22]

in no more than 15 ns because of their reduced mobility. As can be seen in Figs. 7.2 and 7.6, each CCD column is terminated by a readout amplifier. The on-chip single-sided JFET is described in [15].

The focal plane layout of the XMM-Newton pnCCD camera is depicted in Fig. 7.7. Four individual quadrants each having three pnCCD subunits are operated in parallel. The camera housing and its mechanical, thermal, and electrical properties are described in [13].

7.3.2 Limitations of the CCD Performance

The performance of the CCDs is subject to several limitations: physical and technical. We will treat some of the limitations:

7.3.2.1 Quantum Efficiency

The quantum efficiency at the lowest energies from several tens of electron volts to slightly above the Si–K edge at 2 keV is determined by the transmission through insensitive or partially insensitive layers. This directly leads to incomplete charge collection resulting in an asymmetric signal peak with a "shoulder" on the low energetic side of the spectrum and a shift of the peak position. The remedy for this effect

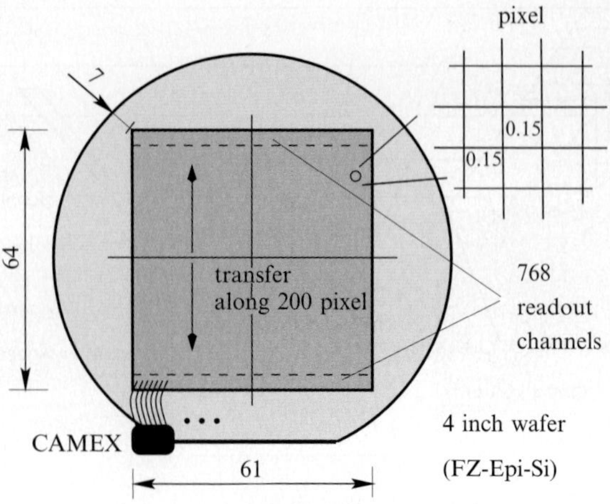

Fig. 7.7 The focal plane of the pnCCD camera on XMM-Newton and ABRIXAS consists of 12 independent, monolithically integrated pnCCDs with a total image area of approximately $60 \times 60\,\text{mm}^2$. In total 768 on-chip amplifiers process the signals and transfer them to a VLSI JFET-CMOS amplifier array. The 12 output nodes of the CAMEX arrays are fed into 4 ADCs, i.e., one ADC is dedicated to a quadrant [22]

is to avoid lattice perturbations, which could lead to charge trapping and recombination. Additionally, electrons from the ionization process in the insensitive layers may reach the partially or totally sensitive regions and leave their remaining energy there. Both effects reduce the proper detection of the X-ray energy and worsen the peak-to-valley ratio of an energy spectrum.

The detector response to high X-ray energies is simply given by the depleted and thus sensitive thickness d:

$$d = \sqrt{\frac{2\,\varepsilon_0\,\varepsilon_\mathrm{r}\,(N_\mathrm{A}+N_\mathrm{D})}{q N_\mathrm{D} N_\mathrm{A}} \cdot V_\mathrm{bias}} \approx \sqrt{\frac{2\,\varepsilon_0\,\varepsilon_r}{q}\,\frac{V_\mathrm{bias}}{N_\mathrm{D}}} \quad \text{for } N_\mathrm{A} \gg N_\mathrm{D} \tag{7.1}$$

The depleted device thickness d can be increased by using silicon with higher resistivity, i.e., with small donor concentration N_D in the case of n-type material. Or the reverse bias V_bias must be increased. But for long term stable operation V_bias should be kept below e.g., 500 V. with given resistivity that limits the achievable sensitive thickness. With 4.5 kΩcm n-type silicon ($N_\mathrm{D} = 10^{12}$ donors per cm^3) a depletion width of 800 µm can be achieved at a reverse bias of 500 V. In the above equation, $\varepsilon_\mathrm{o}\,\varepsilon_\mathrm{r}$ is the dielectric constant of silicon, q the charge of one electron and N_A the high acceptor concentration of the boron implant.

7.3.2.2 Ionization Statistics

Once above 50 eV of photon energy, UV and X-rays need approximately $w = 3.7$ eV (average) of energy to generate an electron–hole pair, despite of the fact that the band gap of silicon is only 1.1 eV. Most of the incident energy is converted into phonons, only about 30% goes into the creation of electron–hole pairs. In addition the ionization cascade is not an uncorrelated process; therefore, straightforward Poisson statistics does not apply. The energy response of incident monochromatic X-rays is broadened by the competing relaxation processes in silicon. The resultant equivalent noise charge contribution ENC_{fano} is given by:

$$ENC_{\text{fano}}^2 = \frac{FE}{w} \,. \tag{7.2}$$

E is the photon energy, F the Fano factor, and w the electron–hole pair creation energy. Assuming a material dependent Fano factor F of 0.12 for silicon and a pair creation energy of 3.7 eV, the intrinsic line width of a photon of 5.9 keV cannot be better than 121 eV (FWHM). According to Fig. 7.8, the Fano noise is dominant for energies above 1 keV if the electronic noise contribution (ENC_{el}) amounts to 5 electrons. For 1 electron noise (ENC_{el}) this threshold is lowered to 50 eV only.

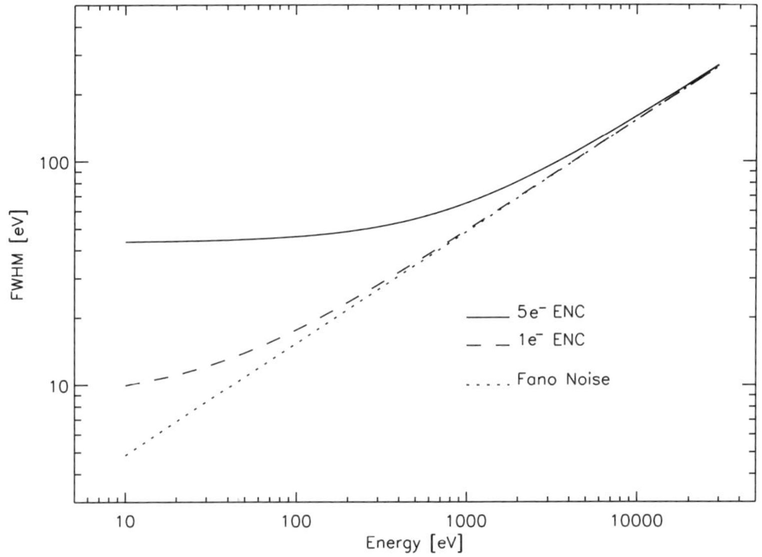

Fig. 7.8 Energy resolution as a function of the photon energy. The Fano noise is taken into account as well as a 5 e^- and 1 e^- electronic equivalent noise charge (ENC_{el}), respectively [22]

7.3.2.3 Charge Transfer Noise

In CCDs, where the signal electrons are transferred over many pixels, the charge shifting mechanism from pixel to pixel must be excellent. Leaving charges behind during the transfer means a reduction in signal amplitude. This loss can be corrected, but adds noise to the signal amplitude measurement.

$$ENC_{\text{trans}}^2 \approx \frac{E}{w}(1-\text{CTE})\,N_{\text{trans}} \qquad (7.3)$$

In a simple model the lost charges can be parametrized according to (3), where the left behinds can be considered as a backward flow loss of electrons. As the loss process is of statistical nature, it is treated similar to a signal leakage current. N_{trans} denotes the number of pixel transfers of the signal in the CCD to the anode and CTE is the charge transfer efficiency–a number close to one, so that (1-CTE) is in the order of 10^{-5}, depending on CCD type, radiation damage, temperature, etc.

7.3.2.4 Electronic Noise

The origin of electronic noise and the processing of signals is described in ref. [14] in detail. We simply summarize the result:

$$\begin{aligned}
ENC_{\text{el}}^2 &= \left(\alpha \frac{2kT}{g_{\text{m}}} C_{\text{tot}}^2 A_1\right)\frac{1}{\tau} && \text{series noise} \\
&+ \left[\left(2\pi a_f C_{\text{tot}}^2 + \frac{b_f}{2\pi}\right) A_2\right] && \text{low frequency noise} \\
&+ \left[\left(qI_l + \frac{2kT}{R_f}\right) A_3\right]\tau && \text{parallel noise}
\end{aligned} \qquad (7.4)$$

C_{tot} is the total input capacitance, kT Boltzmann's constant times absolute temperature, τ is the shaping time constant, I_l the leakage current and R_f the feedback resistor of the charge sensitive amplifier, a_f and b_f parameterize the low frequency $1/f$ noise. A_1, A_2, and A_3 are constants depending on the shaper's filter function.

For the pnCCD (as any other silicon drift detector type system) the electronic noise can be reduced by reducing the read node capacitance, by lowering the leakage current and the $1/f$ noise constants and by optimizing the shaping time constant τ. The total read noise, if not correlated, can be added quadratically and delivers the total equivalent noise charge ENC_{tot}:

$$ENC_{\text{tot}}^2 = ENC_{\text{el}}^2 + ENC_{\text{fano}}^2 + ENC_{\text{trans}}^2 + \ldots \qquad (7.5)$$

State of the art pnCCD systems of today exhibit electronic noise figures around two electrons and a readout speed of several pixels per microsecond and operating temperatures warmer than $-60°$ C.

In pixellated detectors, the signal charges of one single X-ray photon will not always be collected in a single pixel, so the electronic content of several pixels must be added. This increases for the so called split events the electronic noise floor by the factor $\sqrt{N}$, with N as the number of pixels involved.

7.3.3 Detector Performance (On Ground and In Orbit)

For the most recent devices, the best value for the readout (or electronic) noise of the on-chip electronics is 2 e^- rms at $-60°$ C, whereas typical values for the previous devices scatter around 5 e^- rms, e.g., for the XMM-Newton pnCCD camera system. This includes all noise contributions described in (4). The charge transfer properties of the pnCCDs on XMM-Newton are reasonably good, in the order of a several per cent signal loss from the last to the first pixel over a distance of 3 cm charge transfer. As the charge transfer losses describe the position-dependent energy resolution, it is one of the key parameters for the spectroscopic performance, especially after radiation damage may have occurred. Figure 7.12 shows an ^{55}Fe spectrum of a pnCCD in a flat field measurement resulting in a typical energy resolution of 125 eV at an operating temperature of $-120°$ C [17]. The XMM-Newton flight camera was operated at $-90°$ C during calibration on ground with a resolution of about 145 eV (FWHM) over the entire area of 36 cm^2. The degradation of energy resolution was mainly caused by the reduction of the charge transfer efficiency. Leakage currents and on-chip JFET properties only played a minor role. The impact of the material properties of silicon and related impurities and their consequences for the operation of scientific grade X-ray pnCCDs including the effects of radiation damage are treated in detail in the references [7, 10]. The radiation damage accumulated over the expected life time of XMM-Newton was estimated to be equivalent to a 10 MeV proton fluence of $4-5 \times 10^8$ p cm^{-2}. The Figs. 7.9 and 7.10 show the results of the irradiation tests with 10 MeV protons: the expected decrease of energy resolution over the 10-year dose is from 146 to 164 eV FWHM at an operating temperature of $-100°$ C. At the actual operating temperature of $-90°$ C, the expected effect of trapping and detrapping at A-centers [19], generated by the radiation, is even more reduced.

In a single photon counting mode, the quantum efficiency was measured with respect to a calibrated solid state detector. Figure 7.11 shows measurements from the synchrotron radiation facilities in Berlin and Orsay. At 525 eV a 5% dip can be seen because of the absorption at the oxygen edge in the SiO_2 layers. The same happens at the Si–K edge at 1840 eV showing the fine structure of a typical XAFS spectrum (see insert of Fig. 7.11). For all energies, the quantum efficiency is nicely represented by a model using the photo absorption coefficients from the atomic data tables. The quantum efficiency on the low energy side can be further improved with respect to the measurements shown in Fig. 7.11, by increasing the drift field at the p$^+$n - junction entrance window [4] and by using $\langle 100 \rangle$ silicon instead of $\langle 111 \rangle$

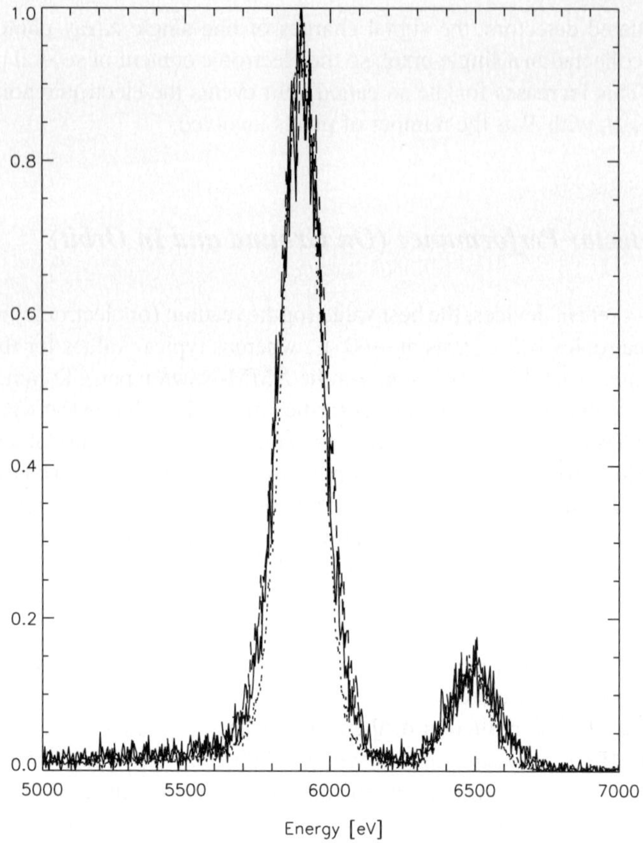

Fig. 7.9 Energy spectrum from an ^{55}Fe source after different 10 MeV proton fluences of 0 p cm^{-2} (*dotted line*), 4.1×10^8 p cm^{-2} (*solid line*), 6.1×10^8 p cm^{-2} (*dashed line*), measured at the low (and after irradiation unfavorable) temperature of 142 K. The expected dose over a life time of 10 years is equivalent to approximately 5×10^8 10-MeV p cm^{-2} for the pnCCD on XMM-Newton [22]

silicon. The useful dynamic range of the pnCCD camera on XMM-Newton was adjusted to the energy band from 100 eV to 15 keV (see Fig. 7.11).

Split events, i.e., events with electrons spread over more than one pixel, originating from one single photon, were reconstructed and summed to one photon event. In total, about 70% of all events are single pixel events, 28% are two pixel events, and 2% are events with three and four pixels involved. In the case of the XMM-Newton pnCCDs, a single X-ray photon spreads the generated signal charge never over more than four pixels.

The readout electronics of the pnCCD system is described in the references [2, 20]. A charge sensing amplifier followed by a fourfold double-correlated sampling stage, multiplexer, and output amplifier (CAMEX64B JFET/CMOS ASIC chip) guide the pnCCD pixel content as a voltage signal to a 10 MHz 12 bit flash ADC system. The whole system, i.e., CCD and CAMEX64B amplifier array

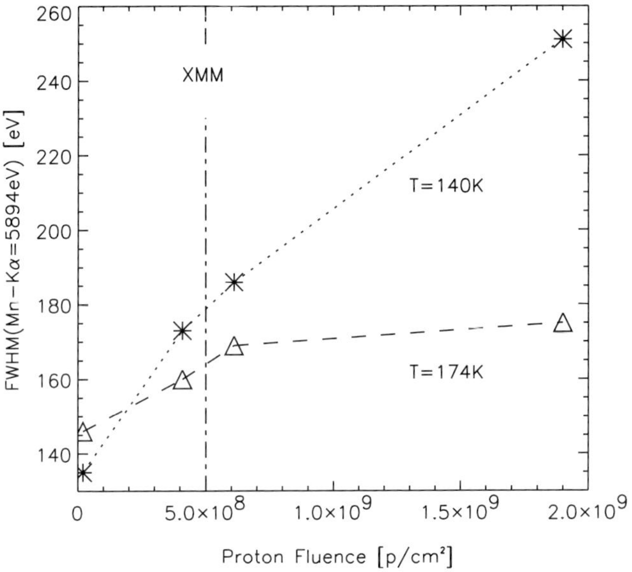

Fig. 7.10 FWHM of the Mn–K_α-spectrum (5 894 eV) in dependence on proton fluence and temperature. Before proton exposure, the lower operating temperature of 140 K gains better results. After a 10 MeV proton fluence of more than $2 \times 10^8/\mathrm{cm}^2$ the higher temperature of 174 K results in a better enrgy resolution. The FWHM is degraded from 135 eV (140 K) to 160 eV and 175 eV (174 K) after 4.1×10^8 p/cm^2 and 1.9×10^9 p/cm^2, respectively. A FWHM of 164 eV is expected after the 10-year XMM-Newton mission [22]

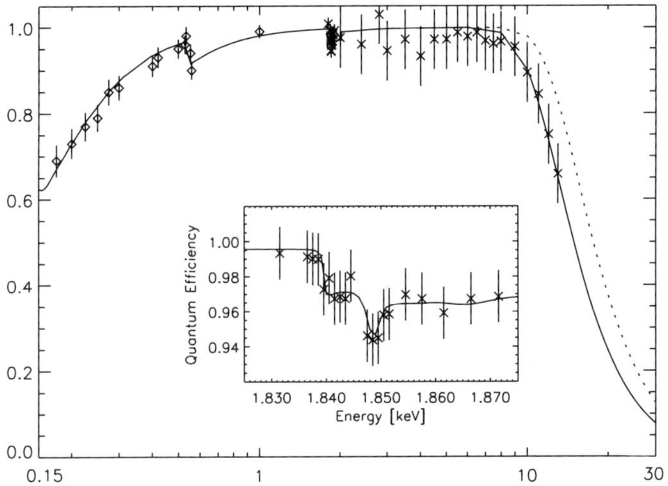

Fig. 7.11 Quantum efficiency of the pnCCD as a function of the incident photon energy. The energy scale ranges from 0.15 to 30 keV. The *solid line* represents a 300 µm thick sensitive volume, the *dotted line* 500 µm [22]

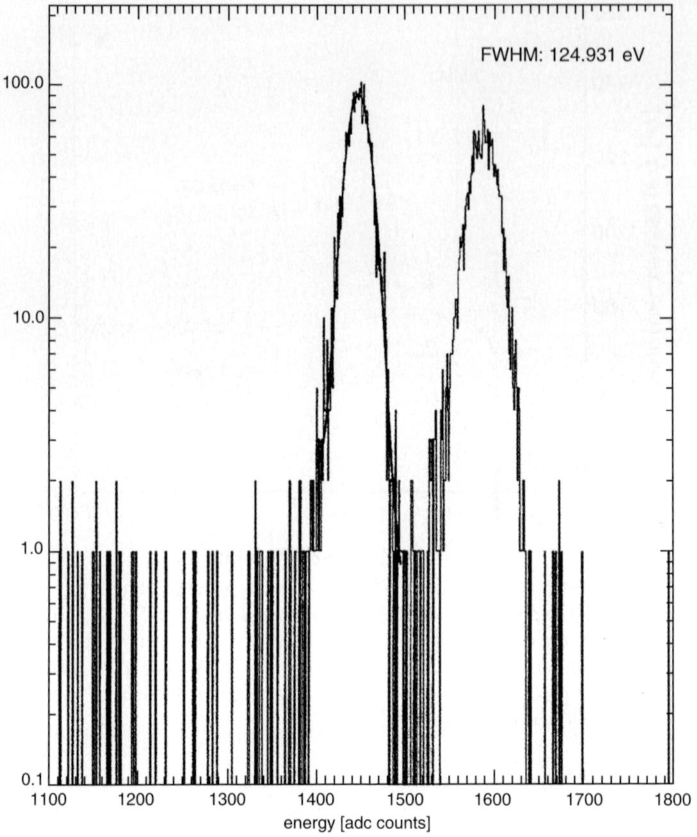

Fig. 7.12 Mn–K_α and Mn–K_β spectrum of an ^{55}Fe source. The measured FWHM is 125 eV at −120° C. Note the log scale

dissipates a power of 0.7 W for the entire camera (768 readout channels), a value that is acceptable in terms of thermal budget on XMM-Newton realized through passive cooling. A further increase of the readout speed can be made only at the expense of further increase of power, or a degradation of the noise performance.

The charge handling capacity of the individual pixels was tested with 5.5 MeV alpha particles from a radioactive ^{241}Am source. Around 10^6 electrons can be properly transferred in every pixel [22]. The spatial resolution was intensively tested in the PANTER facility with the flight mirror module in front of the focal plane. The first light image of the Large Magellanic Cloud in orbit (see Fig. 7.13), as well as the quantitative analysis of the point spread function have shown a perfect alignment of the telescope system as on ground.

Until now, more than seven years after launch, no instrumental surprise occurred: The energy resolution is almost equal to the ground measurements as is the case for the charge transfer efficiency [23]. To date, the electrical stability of the instrument

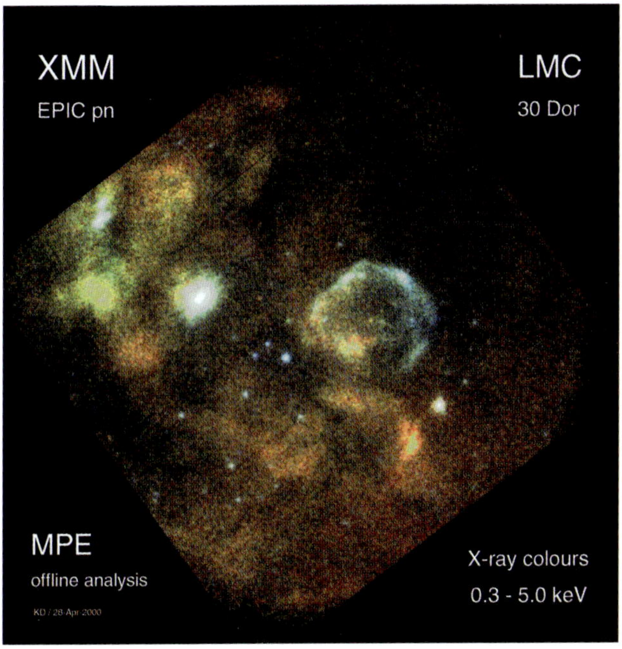

Fig. 7.13 The Large Magellanic Cloud in X-ray colors. The figure shows the first light image of the pnCCD camera in orbit. The field of view of 30 arcmin corresponds approximately to the angular size of the moon. The image shows the area of 30 Doradus, a supernova remnant as an extended source of X-rays. The "north-west" of 30 Dor shows an emission of X-rays up to 5 keV (*blue*), while the "south-east" rim appears much softer in X-rays (*yellow* and *red*). The supernova 1987A is the bright source "south west" of 30 Doradus. About 40 new X-ray objects have been found in this exposure [22]

is perfect. The first light image in Fig. 7.13 qualitatively summarizes the above enthusiastic statements.

7.3.4 Frame Store pnCCDs for Basic and Applied Science

Future missions and other applications require pnCCDs (see Fig. 7.14) with smaller pixels and even faster readout. Two potential applications are the German/Russian eROSITA mission and ESA's XEUS mission. The eROSITA mission shall be launched at the beginning of the next decade, and the XEUS satellite around 2020.

As in conventional CCDs, pnCCDs equally can be designed in a frame store format. This optimizes the ratio of exposure to transfer time, but requires more space on a chip because the store area does not serve as active area but as an analog storage region (Fig. 7.15).

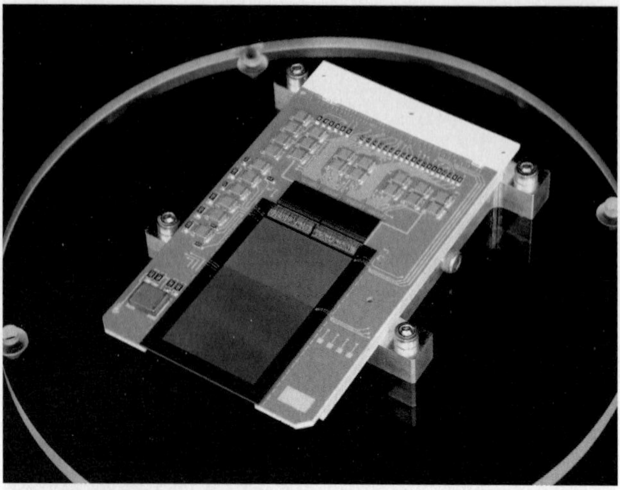

Fig. 7.14 The tested prototype of frame store pnCCD has a format of 256 × 256 pixels with a size of 75 × 75 μm^2 in the image area and 256 × 256 pixels with a size of 75 × 51 μm^2 in the frame store. The number of pixels will be increased for the eROSITA flight devices to 384 × 384 in image area and frame store each

Table 7.1 Properties of CCD detectors used in current and future X-ray missions

Property	XMM-Newton	eROSITA	XEUS
Status	Operating	Prototyping	Research
Type	Full frame pnCCD	Frame store pnCCD	Frame store pnCCD or Active Pixel Sensor (DEPFET APS)
Format	400 × 384	384 × 384	1024 × 1024
Pixel size (μm^2)	150×150	75 × 75	50 × 50 or 75 × 75
Readout noise	5 electrons	2 electrons	2 electrons
Sensitive thickness (μm)	295	450	450
Frame rate (fr. s^{-1})	14	20–1 000	200–1 000
Output nodes per CCD	12	3	32
FWHM(Mn–K$_\alpha$) (eV)	150	130	125
FWHM(O–K) (eV)	90	60	50
Energy range (keV)	0.3–15	0.2–20	0.1–20

The energy resolution (FWHM) refers to incident X-rays of the Mn–K$_\alpha$ line at 5.9 keV and O–K$_\alpha$ at 525 eV measured at temperatures around −60° C

For the pnCCD aboard the XMM-Newton satellite, the signals of one row (64 pixels) are processed in parallel in 23 μs. The extension to 128 channels on the CAMEX amplifiers, to match the smaller pixel pitch, was already realized for first prototype devices. In addition, the signal processing time must be shortened by a factor of two to obtain the same readout time per row. The increased readout speed will certainly have an impact on the power consumption, which is actually below 1 W for the 36 cm^2 large image area.

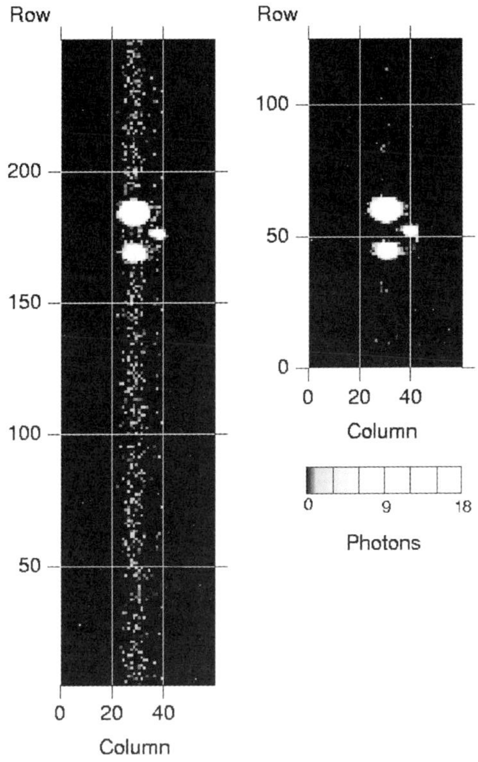

Fig. 7.15 Operation of a frame store pnCCD in fullframe mode (*left figure*) and in frame store mode with 125 rows for the image area and the other 125 rows as frame store (*right figure*). A mask with three different pinhole sizes was mounted in front of the detector. The same frame rate (20 images per second) and the same Al–K X-ray flux were used for both measurements. The out-of-time events appear outside the illuminated spots distributed all over the transfer channels. The fast transfer of the image into the frame store (frame store mode) instead of transfer and read out row by row as performed in the full frame mode reduces the out-of-time event occurrence by a factor of 40 to a value of 0.35% [12]

If the CCD is read out with 12.8 MHz, 10 μs would be required for the parallel readout of one pixel line comprising 128 channels. For the parallel transfer from the image to the storage area, 100 ns are needed per line transfer. A device of $1\,000 \times 1\,000$ pixels would be divided (as in the XMM-EPIC case) in two identical halves of the image area, i.e., $500 \times 1\,000$ pixels each. For the parallel 500 shifts thus 50 μs would be needed for the entire transfer from the image to the shielded storage area. The readout time for the storage area while integrating X-rays in the image part would then be 500×10 μs = 5 ms. That means, within 5 ms the whole focal plane would be read out. The out-of-time probability for X-ray events will then be 1:100. In this operation mode, 200 image frames can be taken in 1 s with a full frame time resolution of 5 ms. The key parameters of future X-ray detector systems are summarized in Table 7.1.

7.3.5 New Devices

Recently, the first prototypes of the pnCCDs for the eROSITA mission have been tested with excellent results [11, 12]. The charge transfer efficiency was improved at least by a factor of 10 at the critical lower energies. In addition, the low energy response was significantly improved (see Fig. 7.16). The trigger threshold was as low as 30 eV with a 4 σ cut, the FWHM for C–K (277 eV) is around 50 eV and the peak-to-background ratio is 50:1. This width is still larger than the theoretical limit (around 40 eV) and reveals some additional improvements to be done in the near future. At Al–K_α the FWHM is about 75 eV. These improvements enable us to get closer to the limits given by silicon as a detector material.

7.4 New Detector Developments: Active Pixel Sensors for X-Rays

The conceptual weakness of CCDs for X-ray detection is the long subsequent transfer of signal charges parallel to the detector surface, in large sensors up to several centimeter. In every individual pixel, the signal charge may undergo a charge loss,

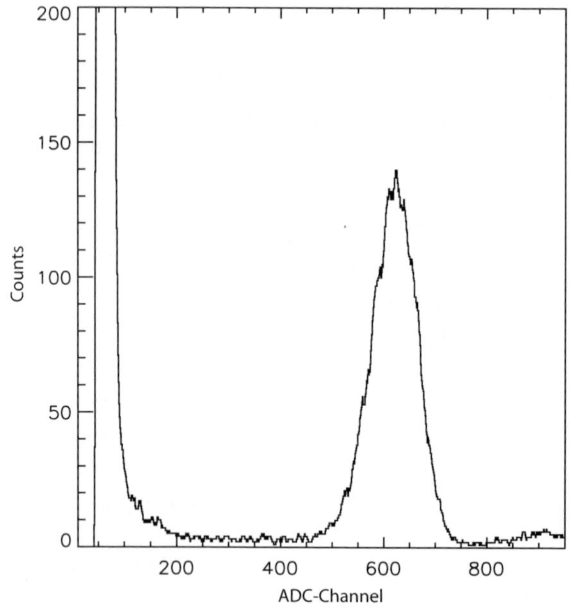

Fig. 7.16 Carbon spectrum recorded with a frame store pnCCD. The C-line energy is at 277 eV corresponding to 75 electrons generated by the incoming low energy X-ray. The measured FWHM is around 50 eV. Because of partial absorption of signal carriers, the peak is shifted by 20 eV toward lower energies, if compared with the peak position of the Mn–K_α line at 5.9 keV. The peak to valley ratio is approximately 50:1

expressed in the charge transfer inefficiency CTI $= 1 - $ CTE. As the number of pixels concerned may be as large as several thousands, the charge transfer efficiency must be almost perfect. In good devices, the CTE >0.99999, i.e., a CTI of $<10^{-5}$. This is achieved in many cases with nonirradiated CCDs. But once in orbit, the CTE can drop substantially after a few years of operation because of radiation damage. This degradation can be avoided if an amplifier is integrated in every pixel, such that charge transfer is not needed any more. This is realized in the DEPFET concept of active pixel sensors (APS) [24]. They are detector and first amplifying element at the same time (Fig. 7.17). The signal charges drift to the internal gate of the DEPFET where they induce a positive charge in the transistor's p-channel and thus increasing the source–drain current. At temperatures as warm as -30° C we have obtained superb X-ray spectra with the system shown in Fig. 7.18. The energy resolution was 130 eV (FWHM) with 6 keV X-rays.

We are currently developing back-illuminated APS detectors in a format of $1\,024 \times 1\,024$ pixels with a pixel size of $75 \times 75\ \mu m^2$. Their collecting fill factor is 100%, no signal charges get lost from any conversion position in the bulk volume of the detector. The readout rate will be 1 000 frames per second in the standard mode and of the order of a few microseconds in dedicated window modes. Operated as a multipurpose focal plane detector DEPFET-type APS systems show clear advantages with respect to other solid-state imagers.

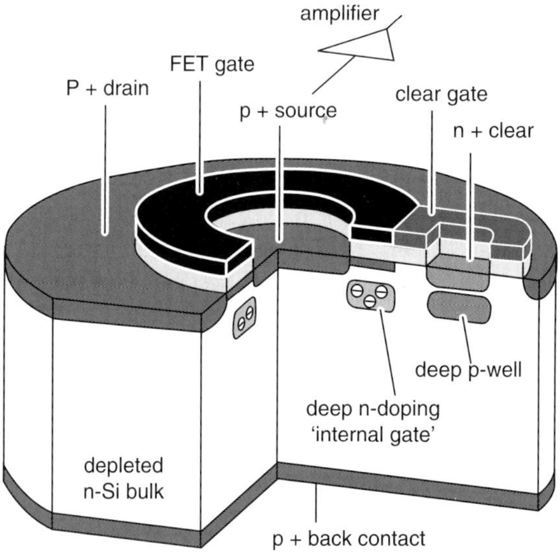

Fig. 7.17 Schematic view on a depleted p-channel FET (DEPFET). A drain contact in the center is surrounded by an MOS gate structure, again surrounded by a source contact. Underneath the gate, a potential minimum for electrons is located, collecting all charges generated within the detector volume. Every electron in the internal gate gives rise to an increase of the FET current by typically 1 nA. This increase in current can be easily measured. The contact on the right-hand side serves as a clear contact of the charges in the internal gate [24] ©2005 IEEE. [25]

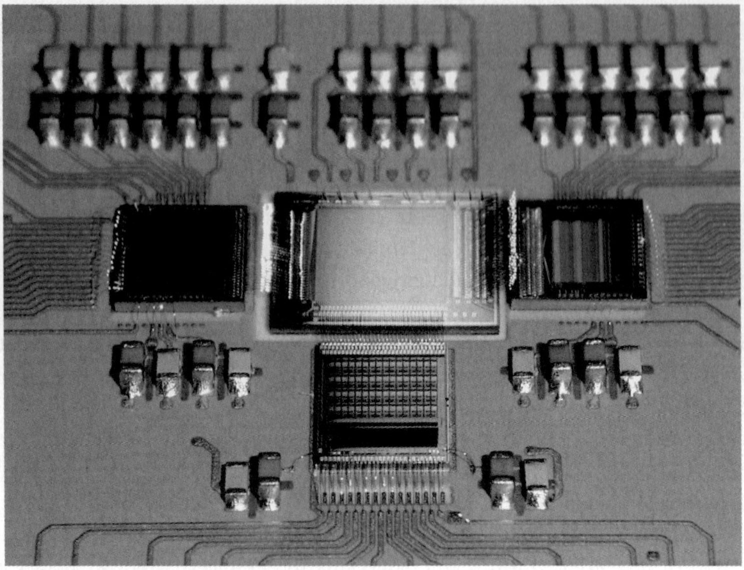

Fig. 7.18 Photograph of the active pixel sensor system. The DEPFET matrix is in the center (64×64 pixels of $75 \times 75\,\mu m^2$ size). The ASIC on the left side selects the rows to be read out, the chip on the right side addresses the clear voltages to the selected rows. The ASIC below is a 64 channel signal processor for the low noise amplification of the on-chip preamplified signals. It comprises a selectable gain within a range of 1.000. The highest gain stage operates from 100 eV to 10 keV, the lowest gain from 100 keV to 10 MeV [24] ©2005 IEEE. [25]

BUT: That does not mean the end of CCD-type detectors. Up to now, the best and cleanest spectra are still recorded with CCDs. In addition, the integration of avalanche-type amplifiers enables them to become a fast detector for single photon counting in optical wavelengths. This is conceptually not possible in DEPFET-based systems. Both concepts CCDs and APSs will establish their niches in science and applied research.

7.5 Conclusion

Since the invention of the silicon drift detector by Gatti and Rehak in 1983, a large variety of new detector structures basing on the principle of sideward depletion was developed. Those detectors have expanded their initial fields of applications in high energy physics, astrophysics, and synchrotron radiation research. They are now a mature technology and open many new industrial applications. Experiments in basic research have driven the performance parameters toward the optimum for the specific applications: high quantum efficiency, excellent energy resolution, high radiation tolerance, good position resolution, high speed, large and cosmetic defect free devices, homogeneous response to the full bandwidth of radiation, and high background rejection efficiency. It will be the aim of future developments to reach the physical limits in radiation detection and to breathe in additional intelligence into the local detector systems to face the steadily increasing amount of data and power dissipation.

References

1. Boyle, W.S., Smith, G.E. 1970, Bell Syst. Techn. J., 49, 587
2. Buttler, W., Lutz, G., Cesura, G., et al. 1993, Nucl. Instrum. Methods A, 326(1–2), 63
3. Gatti, E., Rehak, P. 1984, Nucl. Instrum. Methods A, 225, 608
4. Hartmann, R., Strüder, L., Kemmer, J., et al. 1997, Nucl. Instrum. Methods A, 387(1–2), 250
5. Holland, A.D. 1996, Nucl. Instrum. Methods A, 377, 334
6. Kemmer, J. 1984, Nucl. Instrum. Methods A, 226, 89
7. Krause, N., Soltau, H., Hauff, D., et al. 2000, Nucl. Instrum. Methods A, 439, 228
8. Meidinger, N., Andritschke, R., Hartmann, R., 2006, NIM A, 565, 251
9. Meidinger, N., Bräuninger, H., Briel, U., et al. 1999, SPIE, 3765, 192
10. Meidinger, N., Schmalhofer, B., Strüder, L., 2000, Nucl. Instrum. Methods A, 439, 319
11. Meidinger, N., Bonerz, S., Bräuninger, H., et al. 2002, SPIE, 4851, 1040
12. Meidinger, N., Bonerz, S., Eckhardt, R., 2003, NIM A, 512, 341
13. Pfeffermann, E., Bräuninger, H., Bihler, et al. 1999, SPIE, 3765, 184
14. Pinotti, E., Bräuninger, H., Findeis, N., et al. 1993, Nucl. Instrum. Methods A, 326(1–2), 85
15. Radeka, V., Rehak, P., Rescia, S., et al. 1989, IEEE Electr. Dev. Lett., 10(2), 91
16. Sangster, F.L.J., US patent 3546490, filed Oct. 25, 1966, granted Dec. 8, 1970
17. Soltau, H., Holl, P., Krisch, S., et al. 1996, Nucl. Instrum. Methods A, 377, 340
18. Soltau, H., Kemmer, J., Meidinger, N., et al. 2000, Nucl. Instrum. Methods A, 439(2–3), 547
19. Strüder, L., Holl, P., Lutz, G., Kemmer, J. 1987, Nucl. Instrum. Methods A, 257, 594
20. Strüder, L., Bräuninger, H., Meier, M., et al. 1990, Nucl. Instrum. Methods A, 288, 227
21. Strüder, L., Bräuninger, H., Briel, et al. 1997, Rev. Sci. Ins., 68(11), 4271
22. Strüder, L., 2000, NIM A, 454, 73
23. Strüder, L., Englhauser, J., Hartmann, R., et al. 2003, Nucl. Instrum. Methods A, 512, 386
24. Treis, J., Fischer, P., Hälker, O., et al. 2004, SPIE, 5501, 89
25. Treis, J., Fischer, P., Hälker, O., 2005, IEEE-TNS, 52, 4, 1083
26. Van den Berg, M.L., den Boggende, A.J.F., Bootsma, T.M.V., et al. 1996, Nucl. Instrum. Methods A, 377, 312

8 High Resolution Spectroscopy

P. Predehl

8.1 Introduction

On most instruments in the past, spectroscopy was performed with the same instruments as used for photometry. Imaging proportional counters have an energy resolution of about 400 eV at 1 keV and CCDs less than 100 eV. While the resolution slightly degrades with increasing energy, *the resolving power* $\Delta E/E$ of those instruments severely degrades toward lower energies. However, the X-ray spectroscopic data provide the most direct insight into the physical phenomena, which drive the X-ray energy release. Furthermore, most of the spectral features, both in emission and absorption occur at energies up to a few kiloelectronvolt, requiring a much better spectral resolution than provided by detectors alone. At low energies this can be only accomplished by means of dispersive elements, i.e., diffractive elements in the beam path of a telescope. Diffraction spectrometers as transmission or reflection gratings have an almost constant resolution $\Delta\lambda$, their resolving power $\Delta\lambda/\lambda$ degrades, in contrast to detectors, with decreasing wavelength or increasing energy, respectively. This gives a "break even point" between both instrumentations (Fig. 8.1) at intermediate energies, which is now shifting more and more toward lower energies with the development of better energy resolving X-ray cameras, like bolometers and tunnel junction detectors.

8.2 Transmission Gratings

Transmission gratings have been flown on the Einstein [8], EXOSAT [16], and Chandra observatories [2, 4]. They generally consist of a grating ring filled with little grating elements. Those rings can be placed into the beam path between the Wolter telescope and the focal plane detector preferably close to the mirror exit, thereby dispersing the focussed light into spectral orders. The ring has to follow a special curvature, called Rowland Torus, to guarantee equal beam paths for all rays traversing any of the grating elements [1]. This avoids third order optical aberrations and resulting degradation of the spectral resolution (Fig. 8.2).

The resolution of a transmission grating spectrometer is given by the grating line density and the angular resolution of the telescope: the dispersion angle corresponding to a given wavelength λ is

$$\sin\alpha = m\lambda/d, \tag{8.1}$$

J.E. Trümper and G. Hasinger (eds.), *The Universe in X-Rays.*

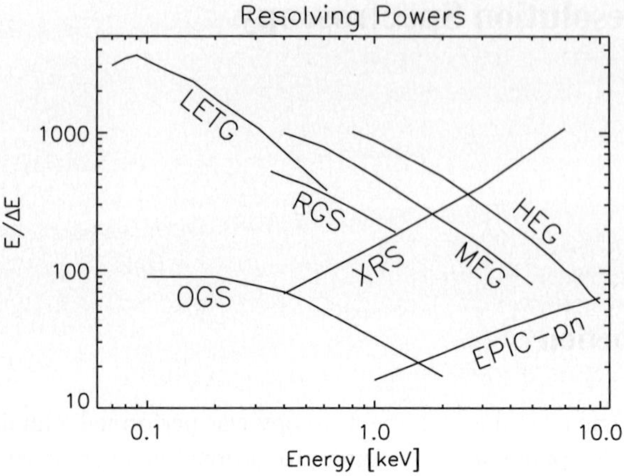

Fig. 8.1 Resolving power of various spectroscopic instruments (Einstein Observatory-OGS, Chandra Observatory-LETG, Chandra Observatory-HETG, XMM-Newton RGS, XMM-Newton EPIC pn-CCDs, Suzaku XRS)

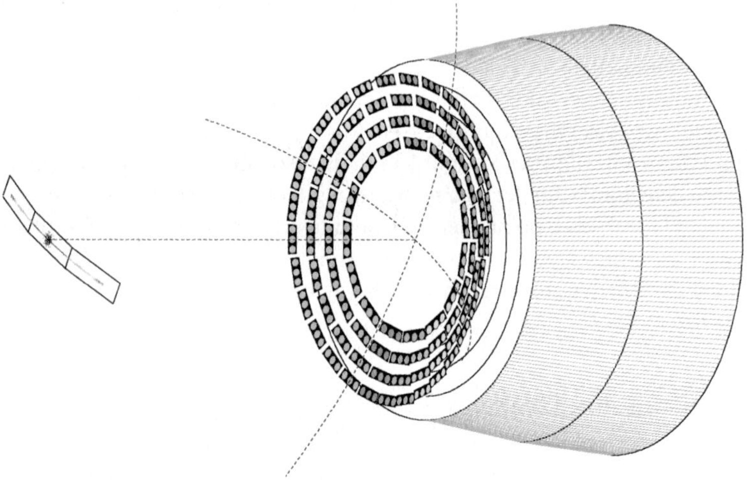

Fig. 8.2 Chandra Transmission grating spectrometer. Individual grating elements are mounted onto a ring following the "Rowland Torus," a surface with two different curvatures: within the dispersion plane, the radius is half of the distance between grating and focal plane, perpendicular to the dispersion plane, the radius is equal to the distance between grating and focal plane. The image of any (point-) source is "stretched" on both sides of the central order image according to the photon energies, thereby producing a spectrum

with d is the grating spacing and m is diffraction order. Typically, only the first diffraction order is used. Line densities vary between 500 l mm^{-1} (Einstein Observatory) and 5 000 l mm^{-1} (Chandra Observatory). Resolving powers more than 2 000 have been reached. Since gratings perform best at low energies, they are mostly freestanding because any substrate would absorb the soft X-rays. Then the grating bars are held only by a coarse support grid.

Transmission gratings are often produced by using an electroforming process, mostly combined with a replication technique [13]. In a first step, a "master" is produced either by a mechanical (diamond-tip) ruling or by laser interference. From this master, the individual gratings are replicated by a photochemical process. Subsequently, the support grid is added by a similar technique (Fig. 8.3). The typical grating material is gold, because it is well suited for electroforming processes and provides, because of its high-Z, sufficient opacity even with grating bar thicknesses below 1 μm.

The efficiency of a classical amplitude grating (bars completely opaque, slits completely transparent) is given by

$$\frac{I}{I_0} = \frac{(\sin m \frac{a}{d} \pi)^2}{(m\pi)^2}, \tag{8.2}$$

with a = slit width between grating bars. In the ideal case with $a/d = 0.5$, the maximum efficiency for each of the symmetric first orders reaches $1/\pi^2 = 10.1\%$, and all even orders vanish, a desired effect to prevent spectral line confusing. In reality,

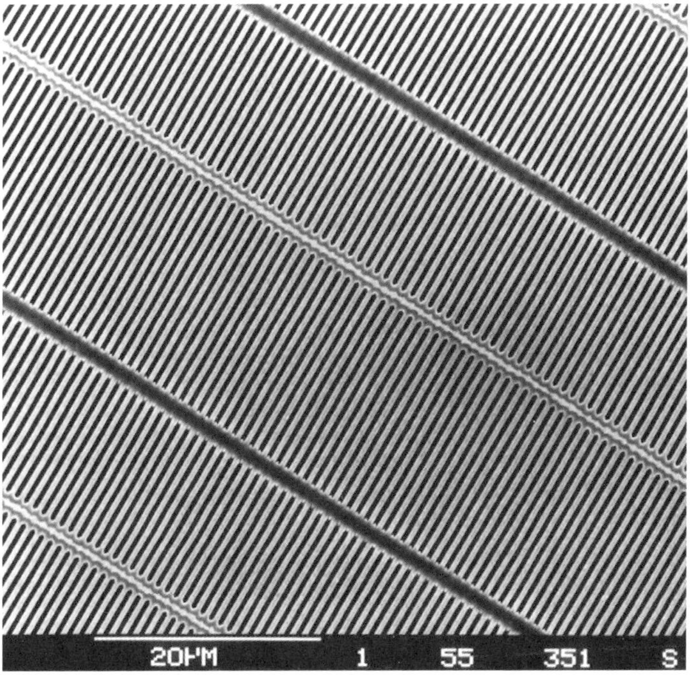

Fig. 8.3 Electron microscope image of a freestanding transmission grating with 1 000 lines per millimeter. The grating bars are held by support grid having a pitch of 25 μm (courtesy DR. JOHANNES HEIDENHAIN GmbH)

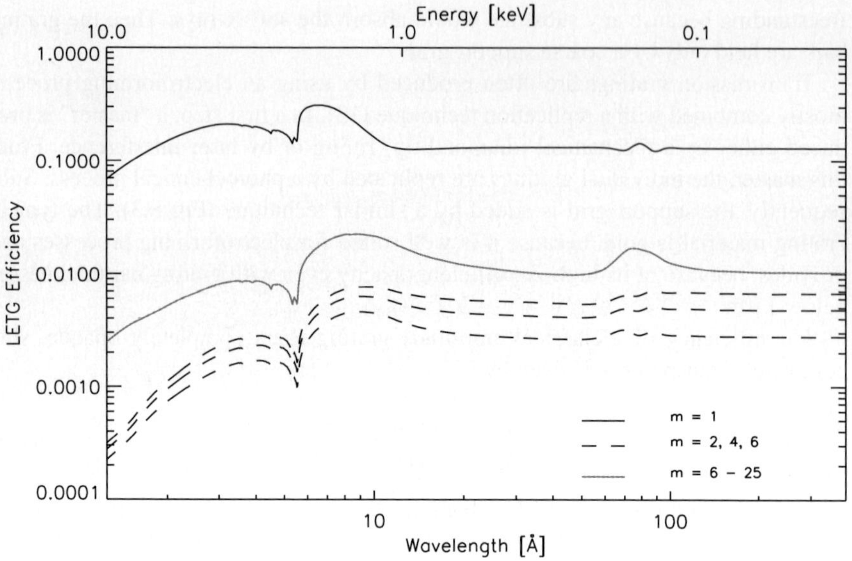

Fig. 8.4 Diffraction efficiencies of the Chandra Observatory LETG for first to sixth and sum of all higher diffraction orders. The design of the grating (bar/slit ratio = 0.5) suppresses the even (2nd and 4th) orders. With a thickness of ≈0.4 μm, the grating bars are partially transparent, and constructive interference enhance the efficiency at the Au–M edge below 10 Å. Source: Chandra Proposer's Guide [14]

the grating bars are not completely opaque. Then, waves passing through the bars are absorbed and phase-shifted and interfere with waves coming through the grating slits. Whether this interference is constructive or destructive depends on the bar thickness and the photon energy. For gold wires with a thickness around 0.4 μm, a maximum of constructive interference around 1.5 keV can enhance the first order efficiency for each of both orders up to 25% (Fig. 8.4).

8.2.1 Einstein OGS

The objective transmission grating (OGS) onboard the Einstein Observatory, launched in 1978, consisted of two transmission gratings (500 and 1 00 1 mm^{-1}), either of which could be placed in the X-ray optical path at the exit from the mirror [8]. The mirror had a focal length of 3.4 m. The gratings consisted of gold bars nominally 0.2 μm thick and with a bar width equal to half of the grating period. The OGS was supplied by SRON in Utrecht/Netherlands. The high resolution imager (HRI) was used to detect the diffracted spectra. The grating resolution was limited by the telescope/detector angular resolution at short wavelengths and by optical aberrations at long wavelengths because the OGS did not follow the Rowland curvature. Therefore, the resolving power was limited to less than 50. The maximum effective area including telescope and detector efficiency was 3 cm^2 at 44 Å [15].

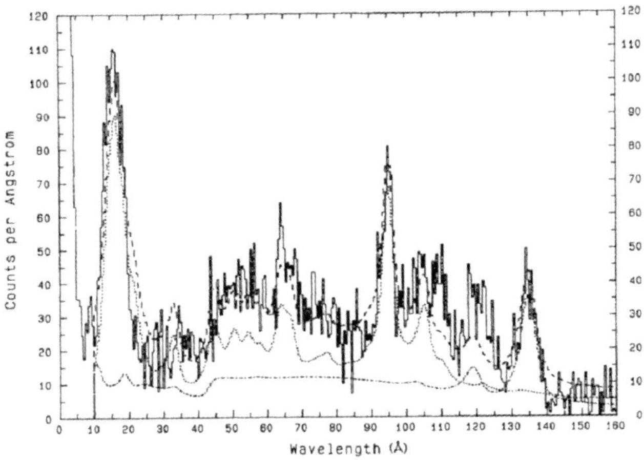

Fig. 8.5 EXOSAT TGS observation of Capella compared with a best fit spectrum (*dashed*) of two temperature components, 3.3×10^6 K (*dotted*) and 1.8×10^7 K (*dashed dotted*) [9]

The instrument was used between 1978 and 1983 to investigate about 40 targets for a total of 500 ks observing time.

8.2.2 EXOSAT TG

The EXOSAT transmission grating spectrometer (TGS) consisted of two gratings, a 1 000 $l\,mm^{-1}$ grating behind one of the two LE (low energy) telescopes, and a 500 $l\,mm^{-1}$ behind the other telescope. Also this instrument was developed by SRON. Channel multiplier arrays (CMA), detectors similar to the Einstein HRI, were used to read out the spectra. Additionally, different photometric filters could be placed in front of the detectors. The spectrometers had a maximum effective area of about 10 cm^2 at 100 Å [12]. With the TGS, 23 objects were observed in an estimated total observing time of 600 ks starting in 1983 (Fig. 8.5). Because of a malfunction of the grating insertion mechanism, most of these 23 objects could be observed only with insufficient observing time, apart from the very bright X-ray/EUV sources HZ43 and Sirius B, which have been extensively studied.

8.3 Chandra

NASA's Great Observatory Chandra, launched 1999, has the highest resolution X-ray mirror system built so far. With less than 0.5 arcsec angular resolution, it is much better suited also for high resolution grating spectroscopy than any other instrument before. Consequently, with the two different gratings, the low energy transmission grating (LETG) and the high energy transmission grating (HETG), a resolving power of more than 2 000 at low energies are achieved.

8.3.1 Chandra HETG

The HETG consists of two sets of gratings, each with a different period [4]. One set, the medium energy grating (MEG), intercepts rays from the outer mirror shells and is optimized for medium energies (0.4–5 keV). Its ruling density is 2 500 l mm^{-1}. The second set, the high energy grating (HEG), intercepts rays from the two inner shells and is optimized for high energies (0.8–10 keV). The ruling density of the HEG is 5 000 l mm^{-1}. Both sets are mounted onto a single support structure and are therefore used simultaneously. HEG and MEG are mounted with their rulings at different angles so that the dispersed spectra form a shallow X, centered on the undispersed (zeroth) diffraction order position. The HETG has been provided by the MIT in Cambridge/Massachusetts (USA). The HETG comprises 336 grating elements in total. The gratings themselves consist of fine gold bars deposited onto a polyimide foil. HETG spectra are recorded with ACIS-S, a CCD detector. Its intrinsic energy resolution provides the separation between different diffraction orders, thereby avoiding line confusing. This allows to some extent also the spectroscopy on extended objects like supernova remnants (Fig. 8.6).

8.3.2 Chandra LETG

The low energy transmssion grating (LETG, Fig. 8.7) is optimized for the energy range between 0.07 and 7 keV having its highest resolving power exceeding 2 000 at 150 Å [2]. It comprises 540 facets with 1 000 l mm^{-1} gratings. Unlike the HETG, these gratings are freestanding, held only by a coarse support grid because any substrate would absorb the soft X-ray radiation. The LETG has been developed in a collaborative effort between SRON and the Max-Planck-Institut für extraterrestrische Physik in Garching/Germany. The grating elements have been fabricated by Dr. Johannes Heidenhain GmbH in Traunreut/Germany. The spectrum readout

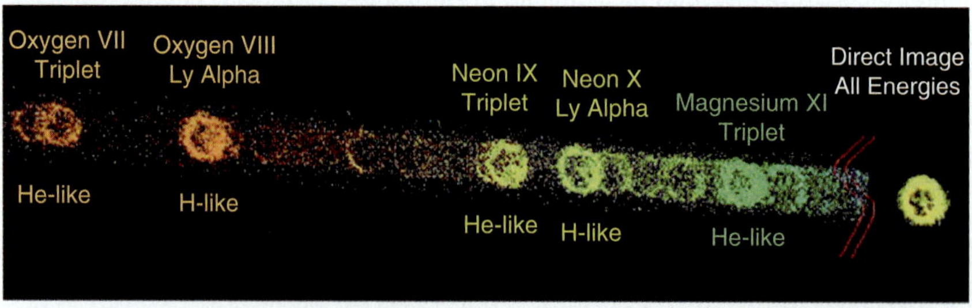

Fig. 8.6 Dispersed high-resolution spectrum of 1E 0102-7219 [7]. Shown here is a portion of the MEG -1 order, color coded to suggest the ACIS energy resolution. At right in the figure (with different intensity scaling) is the zeroth order, which combines all energies in an undispersed image. Images formed in the light of strong X-ray emission lines are labeled

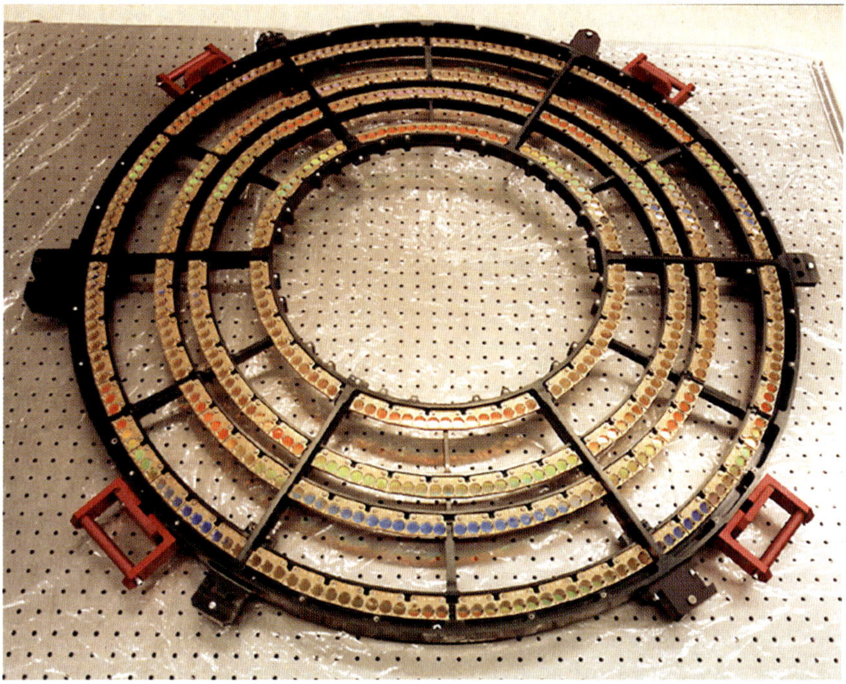

Fig. 8.7 The Chandra LETG consists of four rings corresponding to the apertures of the four mirror shells. In total, 540 single grating facets are mounted on the toroidally shaped ring, each aligned with less than 20 arcsec accuracy (in rotation)

is performed by the HRC-S camera, a microchannelplate detector. This kind of detector is sensitive for very low energetic photons but does not provide any energy resolution. Therefore, emphasis was given to an accurate bar/period ration $a/d = 0.5$ to suppress at least all even diffraction orders. With this optimization, the LETG is the most efficient transmission grating ever been built (Fig. 8.7).

As an example for the spectroscopic capabilities of high resolution spectrometers, plasma density diagnostics based on the study of helium-like triplets is shown in Fig. 8.8. The intensity ratio of the three lines (forbidden, intercombination, and resonance) is a sensitive measure for both temperature and density of a hot thin plasma.

8.4 Reflection Gratings: XMM-Newton RGS

ESA's XMM-Newton, launched in 1999, carries three X-ray telescopes with the largest effective area up to now. Each of the telescopes has about twice the area of

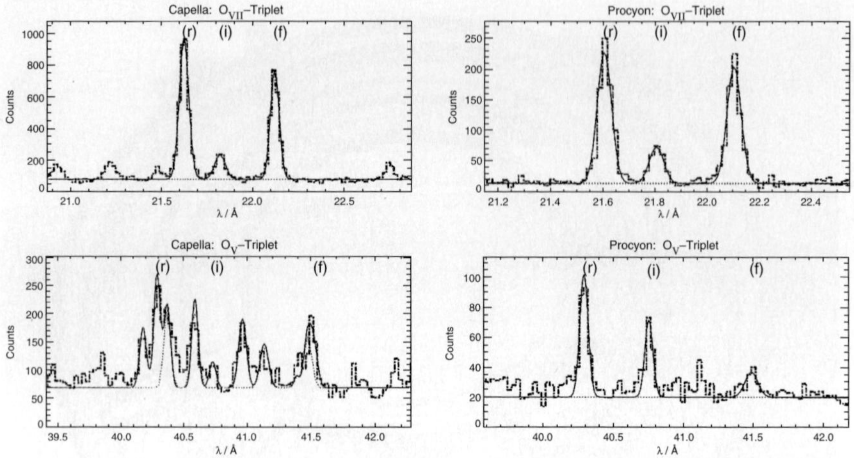

Fig. 8.8 With Chandra-LETG measured spectrum (*line-dotted*) and fit (*solid*) for the O VII and C V triplets of Capella (*left*) and Procyon (*right*). The *dotted lines* represent the total background. The binsize is 0.02 Å for O VII and 0.03Å for C V [11]

the Chandra Observatory telescope but, on the other hand, with an angular resolution 20 times worse. To achieve an appreciable spectral resolution, transmission gratings are needed to have an unfeasible ruling density, reflection gratings are, therefore, the first choice. Their ruling density can be much lower because they are operated under shallow angles.

The reflection grating spectrometer (RGS) consists of two identical instruments behind two of the three X-ray mirrors [3]. The design incorporates an array of 182 reflection gratings placed in the beam behind the mirror in the in-plane classical

configuration. Since the beam is converging, the gratings are not parallel, and they all lie on a Rowland circle. The gratings pick off roughly half of the X-ray light and diffract it into an array of CCD detectors offset from the telescope focal plane (Fig. 8.9).

The spectral bandwidth is limited to 5–35 Åby the size of the detector and vignetting within the stacked gratings. The RGS provides a spectral resolution of $\lambda \approx 0.04$ Å or a resolving power $\lambda/\Delta\lambda$ between 150 and 800 in first order diffraction. With up to 250 cm^2 effective area (both instruments including mirror and detector), the RGS is the most efficient spectrometer so far [6]. To give an example for the spectral capabilities of the RGS, absorption by oxygen in the Interstellar Medium has been studied in detail in the spectral region around 0.53 keV using the (Fig. 8.10). The complexity of this oxygen region led to the conclusion that the absorption feature is produced by atomic as well as compound oxygen.

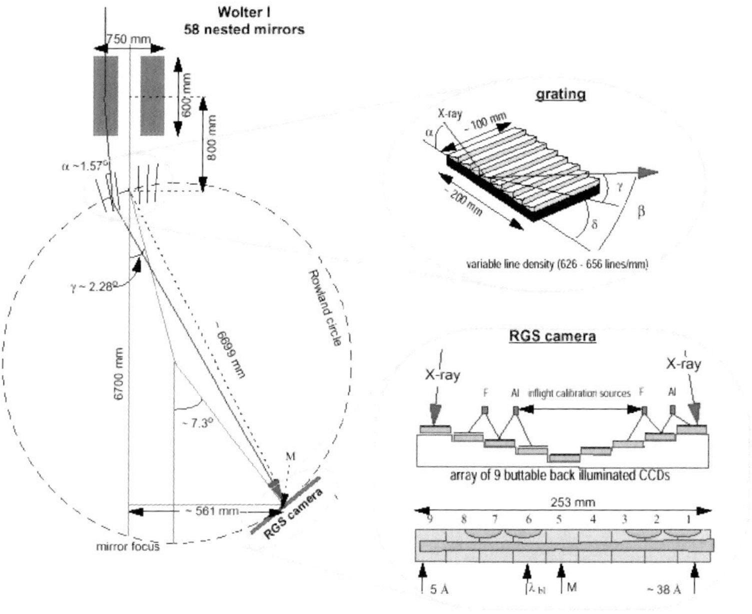

Fig. 8.9 RGS instrument [3]

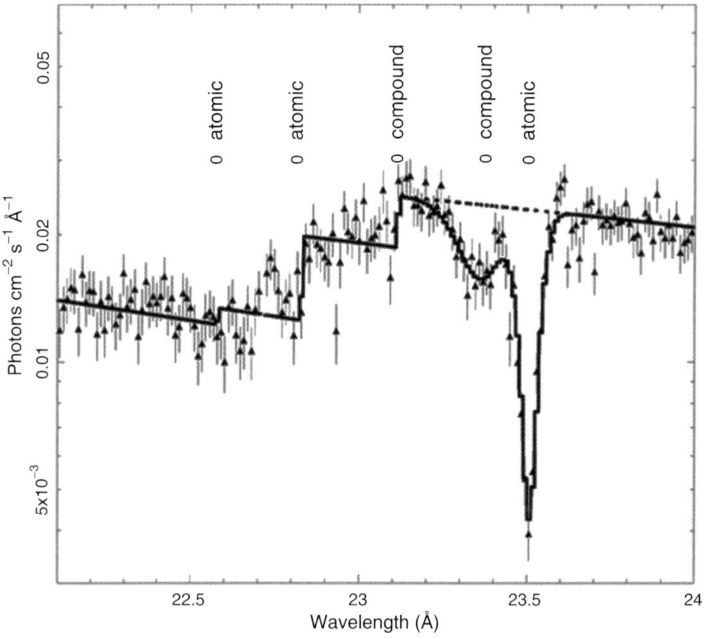

Fig. 8.10 Oxygen absorption due to the interstellar medium in the spectrum of Cyg X-2 [5]

8.5 Bolometers

At very low temperatures below 0.1 K, it is possible to construct devices with high responsitivity, low thermal noise and an energy resolution of a few electonvolt. such that heat pulses from individual X-ray photons can be sensed with a signal-to-noise ratio of up to one part in several thousand.

The Japanese X-ray observatory Suzaku, launched in June 2005, is the first satellite carrying a cryogenic X-ray detector (X-ray Spectrometer, XRS). XRS utilizes a microcalorimeter array of 32 pixels operating at 60 mK [10]. The basis of the XRS array is micro-machined silicon with ion-implanted thermistors and semimetallic crystal mercury telluride (HgTe) as X-ray absorbers. With an energy resolution of 10 eV at 6 keV, it provides a spectral resolving power $E/\Delta E \sim 600$, a value that was reserved in the past to dispersive elements only. The effort required for cooling such an instrument is huge and comprises three stages consisting of solid neon, liquid helium, and an adiabatic demagnetization refrigerator. Unfortunately, the instrument failed shortly after launch because of a leakage in the helium dewar.

References

1. Beuermann, K.P., Bräuninger, H., Trümper, J. 1978, App. Opt. 17, 15
2. Brinkman, A.C., van Baren, C., Gunsing, C.J.T., Kaastra, J.S., Kamperman, T.M., van der Meij, Z.N., Mewe, R., Valkenburg, C., Bräuninger, H., Kettenring, G., Lochbihler, H., Predehl, P. 1996, MPE-Report 263, 677
3. Brinkman, A.C., Aarts, H.J.M., den Boggenden, A.J.F., Dubbeldam, L., den Herder, J.W., Kaastra, J.S., de Korte, P.A.J., Hailey, C.J., Kahn, S.M., Paerels, F., Branduardi-Raymont, G., Bixler, J., Thomsen, K., Zehnder, A. 1996, MPE-Report 263, 675
4. Canizares, C.R., Dewey, D., Galton, E.B., Markert, T.H., Smith, H.I., Schattenburg, M.L., Woodgate, B.E., Jordan, S. 1992, Proc. of AIAA, 11
5. Costantini, E., Freyberg, M., Predehl, P. 2005, Astron. Astrophys., 444, 187
6. den Herder, J.W. 2002, SPIE 4851, 196
7. Flanagan, K.A., Canizares, C.R., Dewey, D., Houck, J.C., Fredericks, A.C., Schattenburg, M.L., Markert, T.H., Davis, D.S. 2004, Astrophys. J., 605, 230
8. Giacconi, R. et. al. 1979, Astrophys. J., 230, 540
9. Heise, J. 1988, Memorie della Societa' Astronomical Italiana (MmSAI) 59, 53
10. Kelley, R.L., Audley, M.D., Boyce, K.R., Breon, S.R., Fujimoto, R., Gendreau, K.C., Holt, S.S., Ishisaki, Y., McCammon, D., Mihara, T., Mitsuda, K., Moseley, S.H., Mott, D.B., Porter, F.S., Stahle, C.K., Szymkowiak, A.E. 1999, SPIE 3765, 114
11. Ness, J.-U., Mewe, R., Schmitt, J.H.M.M., Raassen, A.J.J., Porquet, D., Kaastra, J.S., van der Meer, R.L.J., Burwitz, V., Predehl, P. 2001, Astron. Astrophys., 367, 282
12. Paerels, F.B.S., Brinkman, A.C., den Boggenden, A.J.F., de Korte, P.A.J., Dijkstra, J. 1990, Astron. Astrophys. Suppl. Ser. 85, 1021
13. Predehl, P., Bräuninger, H., Trümper, J., Kraus, H. 1982, SPIE 316,128
14. http://cxc.harvard.edu/proposer/POG/index.html
15. Seward, F.D., Chlebowski, T., Delvaille, J.P., Henry, J.P., Kahn, S.M., Van Speybroeck, L., Dijkstra, J., Brinkman, A.C., Heise, J., Mewe, R., Schrijver, J. 1982, App. Opt. 21, 11
16. Taylor, B.G. 1985, Adv. Space Res. 5, 35

Part II

Galactic X-Ray Astronomy

9 Solar System Objects

K. Dennerl

9.1 Introduction

The history of X-ray astronomy is closely related to solar system research: the first attempts ever to detect X-rays from a celestial object concentrated onto the Sun [21], and the (unsuccessful) attempt to detect X-rays from the Moon, in 1962, is generally considered as the birth of X-ray astronomy [22]. In the recent decade, our knowledge about the X-ray properties of solar system objects was considerably enhanced. We know today that not only the Sun and Moon radiate in X-rays, but that our solar system is full of X-ray sources, including all planets from Venus to Saturn, and even comets.

Before 1996, the only solar system objects that were known to be X-ray sources were the Sun, the Moon, the Earth, and Jupiter. Weak X-ray emission from the solar corona was detected in September 1949 with small Geiger counters aboard a V-2 rocket. Evidence for auroral X-rays from the Earth (below the auroral zones) was found in 1957 [41], and fluorescent X-rays from the Earth were first indicated in 1967, when unexpectedly high background radiation was observed during a daytime stellar X-ray survey by a rocket [24]. X-ray fluorescence from the Moon was studied during the Apollo flights from lunar orbit (e.g., [34]). It took more than one decade until the first X-ray image of the Moon was obtained with ROSAT (Fig. 9.1, left) [35]. For the Earth, the gap between getting first evidence of X-ray radiation and obtaining a first X-ray image was almost four decades: it was not until 1996 that the first X-ray image of the Earth was taken, with the Polar satellite (Fig. 9.5).The only other solar system object that was known to be an X-ray source before 1996 was Jupiter, which was detected in X-rays in 1979 with the Einstein satellite [28].

In 1996, comets were discovered as a new class of X-ray sources [16,27,30]. This ROSAT discovery also opened a conceptual breakthrough for the understanding of unexplained properties of the soft X-ray background, which could be the result of X-ray emission from the geocorona [16, 20] and heliosphere [9, 11, 12]. In 2002, X-ray radiation from Venus (Fig. 9.3, left) [15] and Mars (Fig. 9.4, left) [13] was detected. In the same year, additional X-ray sources in the Jupiter system were found: the Galilean satellites Io and Europa and the Io plasma torus [18]. In 2004, X-ray emission from Saturn was detected (Fig. 9.6, left) [32], and it was finally proven that the geocorona is emitting X-rays [39] which, together with the X-rays from the heliosphere, do contribute to the soft X-ray background [36].

J.E. Trümper and G. Hasinger (eds.), *The Universe in X-Rays.*

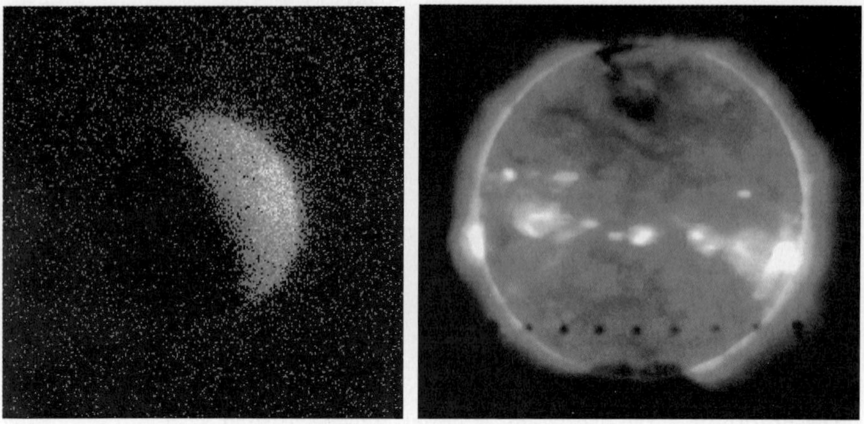

Fig. 9.1 *Left*: First X-ray image of the Moon, taken with ROSAT [35]. This image shows three important features: (i) scattering of solar X-rays on the lunar surface at the sunlit side, (ii) shadowing of the diffuse cosmic X-ray background on the dark side, (iii) excess X-rays at the dark side due to X-ray emission from the geocorona in the foreground. *Right*: X-ray image of the Sun during the transit of Venus on June 8, 2004, obtained with NOAA GOES-12 in the energy range 0.2–2 keV (from `http://www.noaanews.noaa.gov/stories2004/s2240.htm`)

The Sun is in fact responsible for practically all of the X-ray radiation found in the solar system: the interaction of the solar X-ray emission and the solar wind with the conditions found at the individual objects gives rise to an astonishing variety of observable phenomena. In the following sections, the solar system objects are sorted according to the X-ray generation processes, which are known to occur there and which can be studied by observing these objects. More detailed information about the X-ray emission from the solar system (excluding the Sun) is found in [7].

9.2 Solar X-Rays

The solar corona is emitting soft X-rays because of its high temperature (Fig. 9.1, right; see also Chap. 10.1). These X-rays irradiate the environment in a similar way as does the optical light. When this radiation hits a target of sufficient density, some of the incident photons will be scattered, either elastically or by fluorescence. Thus, the X-ray appearance of objects that radiate due to this process generally resembles their optical appearance. However, as the X-ray luminosity of the Sun is considerably lower than its optical luminosity, and as the X-ray albedo of the irradiated objects is also considerably smaller than the optical albedo, their X-ray brightness is extremely small compared with their optical brightness. Another difference to the optical appearance is pronounced variability on a wide variety of time scales, ranging from minutes, due to solar flares, to decades, due to the solar cycle. Depending on the solar X-ray emission, the X-ray albedo, and the solid angle as seen from the Sun, their X-ray luminosities due to this process range from $<10^5$ W to $\sim 10^8$ W.

Until now, X-ray radiation due to scattering of solar X-rays has been detected from the Moon (Fig. 9.1, left), from Venus (Fig. 9.3, left), Earth (Fig. 9.5, right), Mars (Fig. 9.4, left), Jupiter (Fig. 9.4, right), and Saturn (Fig. 9.6, left). Also the other planets Mercury, Uranus, and Neptune, are expected to be X-ray sources because of scattering of solar X-rays, as well as the planetary moons and rings, the cometary nuclei, and the asteroids. An X-ray detection of small objects, however, will be very challenging, because of their extremely low luminosity. Nevertheless, X-rays have already been detected from the asteroid 433 Eros, though not from a satellite in Earth orbit, but from a satellite in a 35–50 km orbit around the asteroid. The scattered solar X-ray radiation could be utilized for determining the elemental composition of the surface [37].

Interestingly, while the list above contains practically all major solar system objects, it does not include comets. The reason is that the gas and dust in the cometary coma (except for the tiny cometary nucleus and its immediate environment) is so tenuous that it does not provide enough targets for scattering solar X-rays[1]. Cometary comae, on the other hand, are huge. They can even exceed the size of the Sun. Thus, they are excellent targets for interactions with solar wind particles, which are characterized by large cross sections, as will be shown in the next section.

9.3 Solar Wind

9.3.1 Comets

The Sun is emitting a wind that contains mainly H and He ions and electrons, plus a minor fraction ($\sim$0.1%) of heavier, highly charged elements. This wind is ejected at some hundred kilometers per second and expands so quickly that the ions and electrons usually have no chance to recombine. The ions remain in their highly charged state until they encounter the gas in a cometary coma, where electrons are available in large numbers. When the ions capture such electrons, they attain highly excited states and radiate a large fraction of the excitation energy in the extreme ultraviolet and X-ray range. As the process of charge exchange is characterized by a large cross section ($\sim 3 \times 10^{-15}$ cm^2), it works very efficiently in the tenuous cometary coma. Although this process and all its constituents were well known for a long time, the importance of the minor fraction of heavy, highly charged ions for the generation of X-rays was overlooked. So the discovery of comets as a new class of X-ray sources came as a big surprise to many scientists.

Prior to the discovery, there were two ideas about a possible generation of X-rays in comets, but charge exchange was not considered. One idea was that the interaction

[1] An exception might be very small dust particles having a size similar to the X-ray wavelengths, as this dust would scatter X-rays very efficiently. However, it is a controversial question whether X-ray scattering on such particles has been observed.

of the magnetized solar wind with the cometary ionosphere could create an extended magnetosphere, which might be capable of accelerating electrons under favorable conditions. These energetic electrons might then produce bremsstrahlung X-ray emission when they decelerated in the cometary coma. There were, however, great uncertainties in this model and its parameters, and the X-ray emission would have occurred only during episodes of cometary auroral substorms. The other idea considered collisions of cometary dust particles with solar system dust. If such particles collided with high velocity, they might evaporate, briefly forming a high-temperature plasma, which might emit an X-ray flash. These ideas were the motivations for searching for X-ray emission from comets, but so far no firm observational evidence has been found that X-rays are produced by these effects.

At the end of 2005, at least 20 comets were detected in X-rays, with the satellites ROSAT, EUVE, BeppoSAX, Chandra, XMM-Newton, FUSE, and Swift [26]. The first record of cometary X-ray emission dates back to July 1990, when Comet 45P/Honda-Mrkos-Pajdušáková happened to cross the field of view of ROSAT during the all-sky survey [16]. The most famous comet is C/1996 B2 (Hyakutake), where X-ray and extreme UV emission was discovered for the first time, with ROSAT [27] and EUVE [30], during its close encounter with the Earth in March 1996. The first high-resolution X-ray spectrum was obtained on Comet C/1999 S4 (LINEAR) with Chandra. It exhibited pronounced emission lines, which could be attributed to the charge exchange process [25].

Probably the best cometary X-ray data obtained so far are from comet C/2000 WM1 (LINEAR), observed in December 2001 with XMM-Newton [14]. Figure 9.2 shows the X-ray contours superimposed on an optical image. It is apparent that

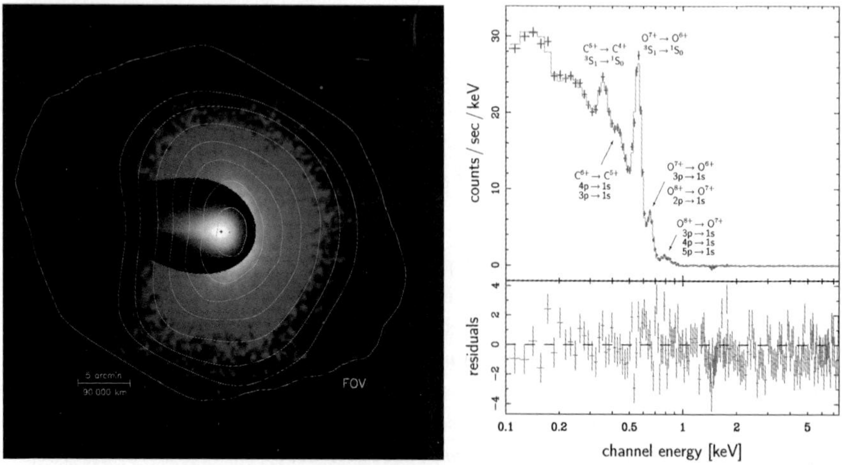

Fig. 9.2 X-ray properties of Comet C/2000 WM1, observed with XMM-Newton [14]. *Left*: X-ray image with contours superimposed on an optical image, illustrating the different morphology and the large extent of the X-ray emission. The Sun is to the right. *Right*: X-ray spectrum, showing pronounced line emission due to solar wind charge exchange. The most likely transitions are labeled. Note that there is no X-ray emission above 1 keV

the X-ray morphology is different from the optical appearance: the maximum of the X-ray emission is shifted toward the Sun and the X-ray coma is elongated and considerably extended perpendicularly to this direction, while the cometary tail is completely invisible. This morphology, which is characteristic for all sufficiently active comets in the undisturbed solar wind, can be well understood as a direct consequence of charge exchange [10, 40]: when the highly charged ions (coming from the right in Fig. 9.2) first encounter the tenuous outer coma, the probability for charge exchange processes is low, as is the X-ray emission. Further inward, the density of available electrons is increasing. It eventually becomes so high that the ionization state of the ions is reduced so much that they are loosing their ability to emit X-rays. This explains why there is no X-ray emission from the cometary tail. The X-ray flux, which we see in the "shadow region" (left of the nucleus in Fig. 9.2), comes mainly from the outer coma and appears there due to projection along the line of sight.

Details of the charge exchange processes are revealed in the X-ray spectrum (Fig. 9.2, right). The most prominent emission line, at 0.561 eV, results when a H-like O^{7+} solar wind ion captures an electron from the cometary gas (usually an outer electron from a H_2O or OH molecule) to become a He-like O^{6+} ion in an excited state, from which it deexcites to the ground state. Thus, the X-ray spectra of comets contain direct information about the heavy ion content of the solar wind, which would otherwise be only accessible by in-situ measurements.

A recent comparative analysis of the spectra of eight comets observed with Chandra showed that spectral differences can indeed be ascribed to different solar wind states, like (i) the fast, cold wind, (ii) the slow, warm wind, and (iii) disturbed, fast, hot winds associated with interplanetary coronal mass ejections [6]. Comets can thus be utilized as natural spaceprobes, which sample the heavy ion content of the solar wind at various states, various heliographic latitudes, and at various phases in the solar cycle. Because of its high cross section, the charge exchange process is so efficient in comets that it may power X-ray luminosities of several 10^9 W.

9.3.2 Geocorona, Mars Exosphere, and Heliosphere

While comets provide the cleanest case for studying charge exchange interactions, this is a general process that occurs whenever highly charged ions interact with gas. One such place is the geocorona, an extended, tenuous cloud of hydrogen around the Earth, which is well visible in the EUV because of fluorescent L_α scattering of solar EUV radiation (Fig. 9.3, right).

Although the existence of the geocorona was known for a long time, its relevance to X-ray astronomy was realized only after the discovery of cometary X-ray emission, which demonstrated the presence of an efficient process capable of converting a tenuous cold cloud into an X-ray emitter. Once it was realized that many X-ray observations are made with satellites from inside an X-ray emitting geocorona, a straightforward explanation was offered for peculiarities in the soft X-ray

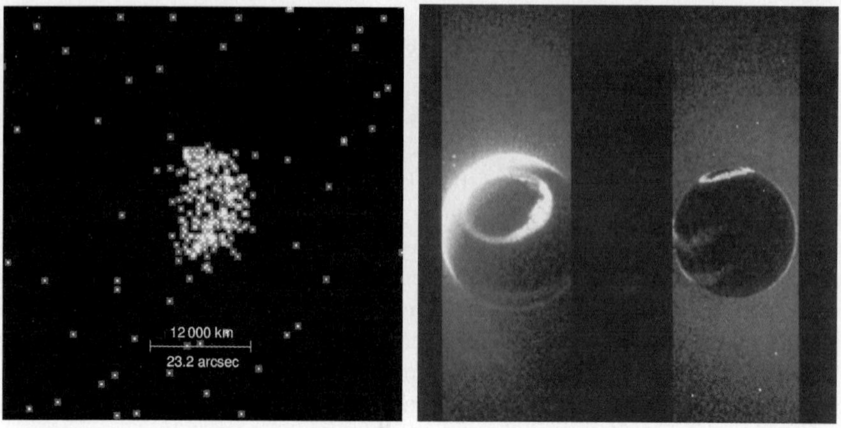

Fig. 9.3 *Left*: First X-ray image of Venus, obtained with Chandra [15]. The X-rays result mainly from fluorescent scattering of solar X-rays on C and O in the upper Venus atmosphere, at heights of 120–140 km. In contrast to the Moon, the X-ray image of Venus shows evidence for brightening on the sunward limb. This is caused by the fact that the scattering takes place on an atmosphere and not on a solid surface. *Right*: EUV images of the Earth, taken with DE-1 SAI [19]. They show the three basic components, which constitute the X-ray properties of Earth: the auroral oval, the sunlit crescent, and the geocorona. The image at right was taken when the Sun was behind the Earth; the two bands straddling the magnetic equator in the premidnight section are caused by airglow

background, which were not understood before. Occasional brightenings of the X-ray sky observed during the ROSAT all-sky survey could be explained by variations of the geocoronal X-ray brightness in response to a temporally variable solar wind, and the presence of X-ray emission, which appeared at the dark side of the Moon, a long-standing puzzle, finally found a straightforward answer (Fig. 9.1, left). Although compelling evidence had accumulated over the last years that the geocorona does influence soft X-ray observations, the direct proof was obtained only recently, when high-resolution spectra of the dark Moon, taken with Chandra [39], revealed the presence of charge-exchange signatures. The same lines were also found in spectra of the diffuse X-ray sky [36]. X-ray observations of the geocorona are complicated by the fact that they cannot be performed from a sufficiently large distance. The narrow field of view of the satellites that are capable of sensitive imaging at low X-ray energies introduces a mixture of temporal and spatial effects, and does not provide full coverage of the geocorona. This complication, however, does not apply to other planets.

In the first X-ray observation of Mars, with Chandra (Fig. 9.4, left), evidence for a faint, extended X-ray halo produced by charge exchange was detected in addition to the bright X-ray fluorescence [13]. A subsequent observation with XMM-Newton confirmed the presence of the halo and made it possible to study its unique charge exchange signatures with an unprecedented spectral resolution of $\sim$4 eV. The O^{6+} multiplet was found to be dominated by the spin-forbidden $2\,^3S_1 \rightarrow 1\,^1S_0$ transition, proving that charge exchange is the origin of the emission. Several X-ray flares

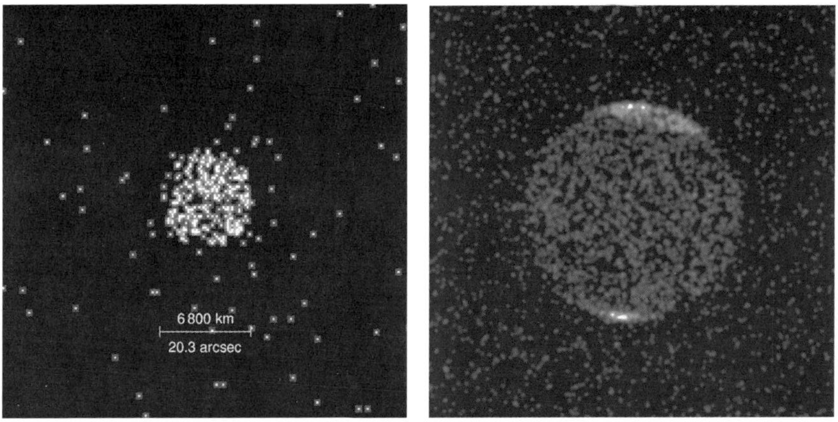

Fig. 9.4 *Left*: First X-ray image of Mars, obtained with Chandra [13]. The X-rays result mainly from fluorescent scattering of solar X-rays on C and O in the upper Mars atmosphere, at heights of 110–130 km, similar to Venus. The X-ray glow of the Martian exosphere is too faint to be directly visible in this image. *Right*: Chandra X-ray image of Jupiter [23]. It shows pronounced auroral emission at the poles superimposed on a uniform brightness distribution. According to the current understanding, the auroral emission is caused by the precipitation of highly ionized oxygen and sulfur into the polar regions, while the nonauroral emission is the combined result of predominantly elastic scattering of solar X-rays and energetic heavy ion precipitation

of the halo were observed, obviously caused by the solar wind, because they were not related to the solar X-ray flux. X-ray emission from the exosphere could be traced in individual emission lines out to at least 8 Mars radii, revealing a highly structured emission with morphological differences between individual ions and ionization states [17]. This first X-ray detection of an exosphere around another planet should also lead to a better understanding of the X-ray properties of our geocorona.

On a larger scale, also the interstellar gas, which is streaming through the solar system, provides electrons for charge exchange with the solar wind ions. Model calculations show that the X-ray surface brightness of the heliosphere due to this process should be roughly comparable with that of the geocorona [12]. However, in contrast to the geocorona, which responds almost instantaneously to variations in the solar wind, the heliosphere is expected to react more slowly and smoothly, because of its large extent.

9.3.3 Magnetized Planets

The process of charge exchange does not require the presence of a magnetic field. However, if a magnetic field of considerable strength is present, like at Earth, Jupiter, and Saturn, then it provides an additional mechanism for creating X-rays:

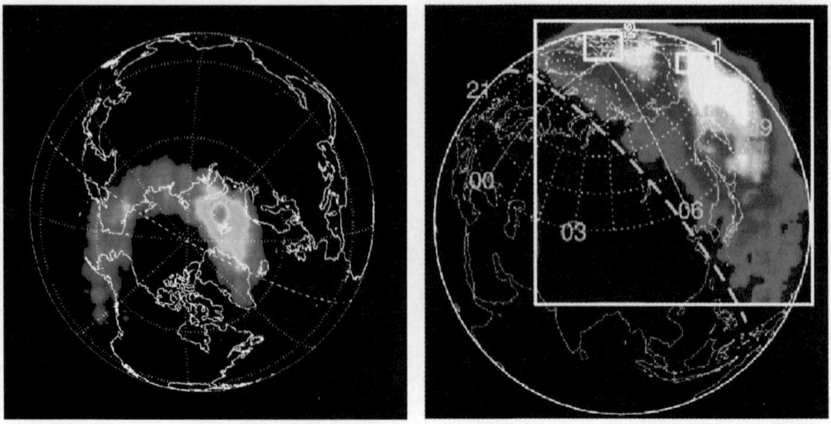

Fig. 9.5 X-ray images of the Earth, taken with Polar PIXIE in the energy range 2–10 keV. *Left*: First global X-ray image of the Earth's aurora, taken on March 20, 1996 during the onset of a small magnetic disturbance [http://apod.nasa.gov/apod/ap000916.html]. *Right*: X-ray image of the Earth during a solar flare [adapted from 33]. The PIXIE instrument sees the emission from the sunlit Earth only during solar flares, because only then the solar flux above 2 keV is high enough to light up the Earth atmosphere in the PIXIE energy range by elastic scattering and by fluorescence of atmospheric Ar at 2.96 keV

energetic electrons, accelerated in the magnetosphere, can produce X-ray emission by bremsstrahlung, when they are decelerated in the planetary atmosphere.

For the Earth, this process has been known for a long time. The auroral oval above the Earth's magnetic north pole is prominent in EUV (Fig. 9.3, right) as well as in X-rays (Fig. 9.5, left). As the X-rays are emitted when energetic electrons from the Earth's magnetosphere strike the upper atmosphere, their intensity is directly related to the intensity of the precipitated electron flux. Dedicated searches for auroral X-ray emission from Earth at energies <2 keV have been made only recently. The first Chandra 0.1–2 keV X-ray observations of Earth's aurora (in the northern polar region) show that it is highly variable in morphology and intensity, exhibiting sometimes intense single or multiple arcs or diffuse patches, while being almost absent at other times. The spectrum is consistent with bremsstrahlung and K-shell line emissions of nitrogen and oxygen produced by electron precipitation [5].

Jupiter also displays pronounced auroral X-ray emission (Fig. 9.4, right). The interpretation of this emission has changed with time; it is currently understood to be caused by precipitation of highly ionized oxygen and sulfur from the outer magnetosphere into the polar regions, and their interaction with Jupiter's upper atmosphere. Evidence for this interpretation came from the Chandra observation of a hot spot of X-ray emission, which was fixed in magnetic latitude and longitude, and which was pulsating with an approximately 45-min period. This indicated that the processes that are operating in the outer magnetosphere are capable of producing highly localized and highly variable emission [23]. The 45-min oscillations did not appear in subsequent observations. Chandra also discovered faint X-ray emission from the Galilean moons Io, Europa, and possibly Ganymede, as well as from the Io plasma torus [18]. Although the exact origin of the X-ray emission from the torus is not clear, the X-ray emission from the Galilean moons is almost certainly due to bombardment of their surfaces by energetic hydrogen, oxygen, and sulfur atoms and ions with subsequent fluorescent emission.

In addition to its X-ray aurora, Jupiter exhibits also low-latitude disk emission (Fig. 9.4, right). This emission was first recognized in ROSAT observations [38] and initially thought to be result of precipitation of energetic sulfur and oxygen ions from Jupiter's inner radiation belts into its atmosphere. High-resolution spectra obtained with XMM-Newton [8] and Chandra [4], however, provided clear evidence of spectral differences between the low-latitude disk emission and the aurorae and supported the view that the low-latitude disk emission is mainly caused by scattered solar X-rays. Further support for this interpretation came from the fact that a large solar X-ray flare was found to have a corresponding feature in the X-ray lightcurve of Jupiter's disk [1].

By analogy with Jupiter, it was expected that Saturn would also display prominent auroral X-ray activity. The first Chandra image of this planet (Fig. 9.6, left), however, did not show any sign of such an activity, at least not at the south pole (the north pole was not visible) [32], while in a subsequent Chandra observation, significant X-ray emission from the south polar cap was detected [2]. However, it was considered as unlikely that this emission was auroral in nature, because the X-ray spectrum was different from that of Jupiter's aurora. Similar to Jupiter, a solar X-ray flare was observed to be mirrored in the X-ray lightcurve of Saturn's disk, supporting the idea that the X-ray radiation of Saturn is primarily due to scattering of solar

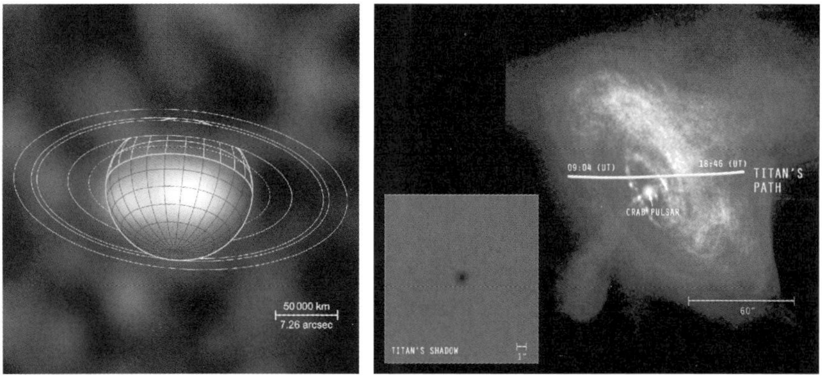

Fig. 9.6 *Left*: First X-ray image of Saturn, obtained with Chandra [32]. The emission is found to be concentrated to regions of low latitude. Its properties resemble that of the nonauroral emission of Jupiter, indicating a similar origin. In contrast to Jupiter, no auroral emission is present at the south pole; the north pole was not visible. *Right*: First X-ray image of Saturn's moon Titan, seen in absorption in front of the Crab nebula, taken with Chandra [29]. This observation made it possible to measure Titan's atmospheric extent at X-ray wavelengths

X-rays [2]. Also the rings of Saturn were detected in X-rays. This radiation was attributed to fluorescent scattering of solar X-rays on oxygen atoms in the tenuous oxygen atmosphere and ionosphere over the rings [3]. Because of the large distance from the Sun, the observed X-ray flux from Saturn is extremely low: Chandra detected on average only one photon every 10 min. For the outer planets Uranus and Neptune, the X-ray flux is expected to be even considerably lower, and only upper limits are available at present [31].

Amazingly, Chandra succeeded in obtaining an X-ray image of Saturn's largest moon Titan, despite its large distance and small size. This was done by utilizing a method of observing solar system objects, which does not even require that these objects are emitting X-rays: by observing them in absorption when they pass across bright, extended X-ray sources. On January 5, 2003, the extremely rare situation occurred that Titan transited the Crab Nebula. Titan was clearly seen in absorption (Fig. 9.6, right), and its atmospheric extent could be measured at X-ray energies for the first time [29]. The transit of Venus across the disk of the Sun, on June 8, 2004, provided another opportunity of applying this special observing method (Fig. 9.1, right).

9.4 What do We Learn from the X-Ray Observations?

X-ray observations of solar system objects provide a novel, complementary view of these objects and the interaction with their environment. For Venus and Mars, X-ray observations make it possible to study the chemical composition and density structure of the upper atmospheric layers above 100 km. This opens up the possibility of using X-ray observations for remotely monitoring the properties of these regions, which are difficult to investigate otherwise, and their response to solar activity. For Earth, Jupiter, and Saturn, the X-rays contain valuable information about the interaction of these bodies with the magnetized plasma medium in which they are embedded, and offer important clues as to the nature of this interaction. Comets can be used as natural space probes for investigating the chemical composition and ionization structure in the solar wind, which would otherwise only be acccessible by in-situ measurements, at various heliographic latitudes and at various phases in the solar cycle.

Moreover, the discovery of cometary X-ray emission has drawn our attention to the relevance of charge exchange processes for X-ray astronomy in general. This has already stimulated considerations at which places outside the solar system such processes might also occur. We now know that the glow of the geocorona is present in many X-ray data, and that the emission of the heliosphere affects *any* X-ray observation made from within the solar system. The recent discovery of a charge exchange induced X-ray halo around Mars, makes it now possible to study the global properties of exospheric X-ray emission in unprecedented spectral and spatial detail.

Having accompanied X-ray astronomy from the very beginning, solar system X-ray research has experienced a boost during the last decade. Successful observations, many of them challenging, have considerably enriched our knowledge with new insights. They have revealed that even our immediate astronomical environment is full of surprising phenomena that are awaiting discovery.

References

1. Bhardwaj, A., Branduardi-Raymont, G., Elsner, R.F., et al. 2005, Geophys. Res. Lett., 32, L03S08
2. Bhardwaj, A., Elsner, R. F., Waite Jr., et al. 2005, Astrophys. J., 624, L121
3. Bhardwaj, A., Elsner, R. F., Waite Jr., et al. 2005, Astrophys. J., 627, L73
4. Bhardwaj, A., Elsner, R. F., Gladstone, G. R., et al. 2006, J. Geophys. Res., 111, A11, 225
5. Bhardwaj, A., Gladstone, G.R., Elsner, R.F., et al. 2007, J. Atm. Solar Terr. Phys., 69, 179
6. Bodewits, D., Christian, D.J., Torney, M., et al. 2007, Astron. Astrophys., accepted
7. Bhardwaj, A., Elsner, R.F., Gladstone, G.R., et al. 2007, Planet. Space Sci., 55, 1135
8. Branduardi-Raymont, G., Elsner, R. F, Gladstone, G. R., et al. 2004, Astron. Astrophys., 424, 331
9. Cox, D.P. 1998, in D. Breitschwerdt, M.J. Freyberg, J. Trümper (eds.), LNP, Vol. 506: IAU Colloq. 166: The Local Bubble and Beyond, 121, Springer-Verlag, Berlin, Heidelberg
10. Cravens, T.E. 1997, Geophys. Res. Lett., 24, 105
11. Cravens, T.E. 2000, Astrophys. J., 532, L153
12. Cravens, T.E., Robertson, I.P., Snowden, S.L. 2001, J. Geophys. Res., 106, 24, 883
13. Dennerl, K. 2002, Astron. Astrophys., 394, 1119
14. Dennerl, K., Aschenbach, B., Burwitz, V., et al. 2003, in J.E. Trümper, H.D. Tananbaum (eds.), X-Ray and Gamma-Ray Telescopes and Instruments for Astronomy, Vol. 4851, SPIE, Bellingham, WA, 2003, ISSN 0277-786X, ISBN 0-8194-4630-0
15. Dennerl, K., Burwitz, V., Englhauser, J., Lisse, C., & Wolk, S. 2002, Astron. Astrophys., 386, 319
16. Dennerl, K., Englhauser, J., & Trümper, J. 1997, Science, 277, 1625
17. Dennerl, K., Lisse, C. M., Bhardwaj, A., et al. 2006, Astron. Astrophys., 451, 709
18. Elsner, R. F., Gladstone, G. R., Waite, J. H., et al. 2002, Astrophys. J., 572, 1077
19. Frank, L.A., Craven, J.D., Rairden, R.L. 1985, Am. Soc. Rev., 5, 53
20. Freyberg, M.J. 1998, in D. Breitschwerdt, M.J. Freyberg, J. Trümper (eds.), LNP, Vol. 506: IAU Colloq. 166: The Local Bubble and Beyond, 113
21. Friedman, H., Lichtman, S.W., Byram, E.T. 1951, Phys. Rev., 83, 1025
22. Giacconi, R., Gursky, H., Paolini, F.R. 1962, Phys. Rev. Lett., 9, 439
23. Gladstone, G.R., Waite, J.H., Grodent, D., et al. 2002, Nature, 415, 1000
24. Grader, R.J., Hill, R.W., Seward, F.D. 1968, J. Geophys. Res., 73, 7149
25. Lisse, C.M., Christian, D.J., Dennerl, K., et al. 2001, Science, 292, 1343
26. Lisse, C.M., Cravens, T.E., Dennerl, K. 2005, in M.C. Festou, H.U. Keller, H.A. Weaver (eds.), Comets II, ISBN 0-8165-2450-5, The University of Arizona Press, Arizona, 631
27. Lisse, C.M., Dennerl, K., Englhauser, J., et al. 1996, Science, 274, 205
28. Metzger, A.E., Gilman, D.A., Luthey, J.L., et al. 1983, J. Geophys. Res., 88, 7731
29. Mori, K., Tsunemi, H., Katayama, H., et al. 2004, Astrophys. J., 607, 1065
30. Mumma, M.J., Krasnopolsky, V.A., Abbott, M.J. 1997, Astrophys. J., 491, L125
31. Ness, J.-U., Schmitt, J.H.M.M., 2000, Astron. Astrophys., 355, 394
32. Ness, J.-U., Schmitt, J.H.M.M., Wolk, S.J., Dennerl, K., Burwitz, V. 2004, Astron. Astrophys., 418, 337
33. Petrinec, S.M., McKenzie, D.L., Imhof, W.L., Mobilia, J., Chenette, D.L. 2000, J. Atm. Solar Terr. Phys., 62, 875

34. Podwysocki, M.H., Weidner, J.R., Andre, C.G., et al. 1974, Lunar and Planetary Institute Conference Abstracts, 611
35. Schmitt, J.H.M.M., Snowden, S.L., Aschenbach, B., et al. 1991, Nature, 349, 583
36. Snowden, S.L., Collier, M.R., Kuntz, K.D. 2004, Astrophys. J., 610, 1182
37. Trombka, J.I., Squyres, S.W., Brückner, J., et al. 2000, Science, 289, 2101
38. Waite Jr., Gladstone, J.H., Lewis, G.R., et al., 1997, Science, 276, 104
39. Wargelin, B.J., Markevitch, M., Juda, M., et al. 2004, Astrophys. J., 607, 596
40. Wegmann, R., Dennerl, K., Lisse, C.M. 2004, Astron. Astrophys., 428, 647
41. Winckler, J.R., Peterson, L., Arnoldy, R., Hoffman, R. 1958, Phys. Rev., 110, 1221

10 Nuclear Burning Stars

J.H.M.M. Schmitt and B. Stelzer

10.1 The Sun, Stars, and Stellar X-Ray Astronomy

10.1.1 Advances of Stellar X-Ray Astronomy

The solar corona has been known to mankind for a very long time, ever since humans consciously observed the first total solar eclipse, when the corona appears as a somewhat irregularly formed "crown" around the Sun. However, the true nature of the solar corona was only recognized in the fourties of the last century, when optically observed coronal emission lines, previously erroneously attributed to a readily introduced new element "coronium," were correctly identified by Grotrian and Edlen. These emission lines turned out to be forbidden transitions of rather highly ionized iron atoms, the production of which requires temperatures in excess of 1 MK. Thus, rather unexpectedly, the very outer layers of the Sun were found to be much hotter than its photosphere, and ever since the search for the dearly required heating mechanism(s) of the solar corona has become the holy grail of solar coronal physics.

Hot plasmas with temperatures in excess of 1 MK radiate most of their energy at soft X-ray wavelengths, and soon after World War II, X-ray emission from the Sun was first detected using Geiger counters onboard rockets originally developed for warfare. Yet, the overall X-ray losses of the Sun are rather weak and less than 1 part in a million of its whole energy budget is emitted at X-ray wavelengths. Thus an extrapolation of the observed solar X-ray properties to stars at large led to rather pessimistic expectations as to the detectability of stellar coronae, and indeed, none of the first couple of hundreds of extrasolar X-ray sources detected in the sixties and seventies of the last century were "normal" stars. However, the introduction of soft X-ray imaging into X-ray astronomy, first with the Einstein Observatory (operated between 1978 and 1981) and later with EXOSAT (1983–1985) and ROSAT (1990–1998), has led to the detection of X-ray emission from many thousands of stars similar to the Sun. Stellar X-ray sources were also intensely studied at extreme ultraviolet wavelengths with the EUVE satellite operated between 1992 and 2000. In particular, the first high spectral resolution observations of coronal emissions from a larger sample of stars were obtained with the spectrometers onboard EUVE. Stars were also observed with other imaging X-ray satellites such as ASCA

J.E. Trümper and G. Hasinger (eds.), *The Universe in X-Rays.*

and BeppoSAX, but the contributions from those missions were geared mostly to other astrophysical topics. At present, the large observatories XMM-Newton and Chandra allow X-ray observations of coronal sources with unprecedented sensitivity and spectral resolution. The gratings onboard XMM-Newton and Chandra utilize the full power of spectroscopy for X-ray astronomy for the first time and have revolutionized this field of research. Since this field is so rapidly developing, we purposely leave out all detailed discussions of spectroscopy and focus on well established results mostly obtained from imaging and lightcurve analyses.

The precise physics of the mechanism(s) responsible for heating the solar corona beyond 1 MK have remained rather elusive up to the very present. However, while originally nonmagnetic processes ("acoustic heating") were thought to be important, the magnetic character of solar (and stellar) coronal heating is now universally accepted, and X-ray emission from cool stars is generally considered to be a key indicator of the so-called "magnetic activity" of these objects. We note, however, that there is no general consensus or generally applied definition of solar and stellar activity. Usually one associates spots, plage, flares, spicules and related phenomena with magnetic activity on the Sun, and similar definitions apply for (cool) stars. Linsky [25] defines solar-like (activity) phenomena as "nonradiative in character, of fundamentally magnetic origin and almost certainly due to a magnetic dynamo operating in or at the base of a convection zone." In magnetically active regions of the Sun and the stars, one finds departures from pure radiative equilibrium caused by some kind of heating and probably momentum deposition processes.

Linsky's definition is very useful because it provides a recipe for identifying activity through searching for evidence of nonradiative heating and showing its magnetic nature. Direct measurements of coronal magnetic fields are very difficult for the Sun, and the situation is worse for stars. The observations of Zeeman broadening go along with large uncertainties because in the optical other line broadening mechanisms (most notably rotational line broadening) dominate over Zeeman splitting. However, it is straightforward to search for the heating effects associated with magnetic activity. Evidence for nonradiative heating can be obtained by observations of the thermal plasma in the UV or X-ray domain or by observations of nonthermal emission from highly energetic particles often accompanying and possibly intimately linked with the heating process(es). Further, nonradiative heating is – usually – confined in space and time, and can be diagnosed by studies of time variability and spatial structure of the emitting regions.

Therefore, X-ray observations of stars are considered a key diagnostics for magnetic activity of stars throughout the Hertzsprung-Russell (HR) diagram. At least for cool stars, i.e., stars with outer convection zones like the Sun, all magnetic activity is thought to be ultimately due to the action of an hypothesized dynamo process operating at the interface between the outer convection zones and radiative interiors of these stars. Such dynamo processes are probably fundamental for astrophysics as far as the generation of magnetic fields is concerned. Most astrophysical objects such as planets, stars, compact objects, accretion disks, jets, galaxies, and clusters of galaxies are associated with magnetic fields, and magnetic dynamos apparently produce these fields on quite different spatial scales, from planet-sized objects to large-scale galactic structures.

10.1.2 The X-Ray Sun

Let us consider the Sun as a starting point of our discussion of X-ray stars. A whole armada of satellites has observed and is observing the corona of the Sun. With only a few hours delay, the obtained images are made available to the general scientific and nonscientific public through the internet (at the web site http://www.sec.noaa.gov/sxi). In Fig. 10.1 we show the X-ray morphology of the Sun as seen by the Japanese *Yohkoh* satellite through the solar cycle; note that the spectral sensitivity and angular resolution of *Yohkoh* is very similar to that of ROSAT or XMM-Newton. The solar X-ray corona appears far from being spherically symmetric, rather it shows an extremely high degree of spatial inhomogeneity and small-scale structure with vast brightness differences ($>10^4$) between adjacent features. Plasma confinement in numerous loop-like structures is readily apparent. Coronae of stars appear as point-like sources, and the solar example as shown in Fig. 10.1 leads us to expect lots of substructure below the instrumental resolution. For the X-ray Sun temporal variability is observed on essentially all time scales covered by the *Yohkoh* observations, starting from seconds/minutes during eruptions up to decades (cf., Fig. 10.1) manifesting the solar cycle. At solar maximum the appearance of the solar corona is dominated by so-called "active" regions, at minimum few or none of those regions are on the solar disk and the X-ray output is produced by

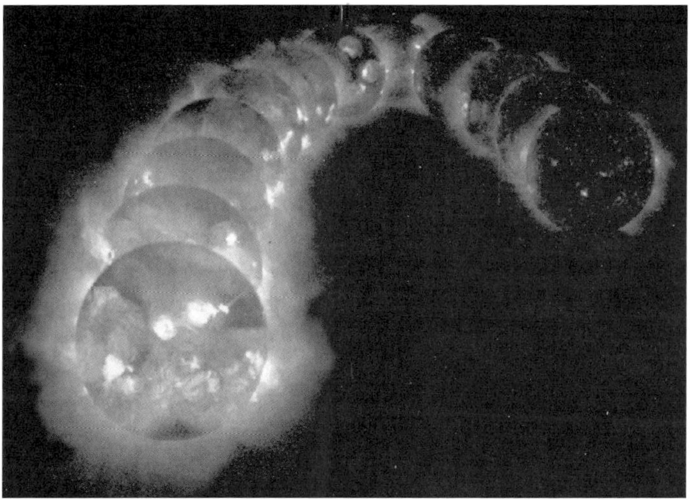

Fig. 10.1 The appearance of the X-ray Sun throughout the solar cycle as seen by *Yohkoh* (from the web site at http://www.lmsal.com/SXT/index.html)

more diffuse, much fainter emission regions. As a consequence, in the course of its 11-year cycle the solar X-ray luminosity varies by up to two orders of magnitude, depending on the considered spectral band pass.

10.1.3 Spatial Structure of Stellar Coronae

For a stellar astronomer, X-ray images of the Sun are a cause of substantial frustration. Stellar coronae cannot be angularly resolved and hence no direct information on their sizes and no direct comparisons to the Sun are available. The only remedy in this situation are eclipsing binaries, where size information can be inferred from light curve analysis. Particularly useful for coronal studies are those eclipsing binaries, in which one component is X-ray dark, e.g., a late B or early A type star (see Sect. 10.2.3). One such example is the system α CrB, composed of a G5 V star in orbit around an A0 star. A total X-ray eclipse was detected during secondary optical minimum, i.e., at the time when the X-ray dark A star is positioned in front of the X-ray bright G star, allowing to deduce the size of the corona around the latter star [42]. Another such system is the prototypical eclipsing binary Algol, which consists of a B8 V primary and a K2 III secondary with an orbital period of 2.87 days. On two occasions, eclipses were observed in the decay of the lightcurves of large flares on Algol. At one instance, a flare rise on Algol B and subsequent eclipse was observed in the middle of an X-ray observation at optical secondary minimum, when the X-ray dark B star is in front of the X-ray bright K star [44]. Both eclipse ingress and egress were observed, allowing the reconstruction of the flare location and size. The resulting image shows the flare on Algol as a limb flare with surprising similarity to limb flares observed on the Sun with *Yohkoh* (shown in Fig. 10.2).

10.1.4 X-Ray Flaring

Flares are a very common type of variability observed on stars. During flares intensity outbursts with timescales of minutes up to several days go along with substantial increases of the temperature of the emitting plasma. Stellar flares are thought to be signatures of magnetic field reconnection. However, the picture for the origin of flares contains many unknowns. In one of the most favored scenarios, magnetic energy builds up as subphotospheric convective motions shuffle the footpoints of the magnetic loops around, thereby exerting stresses on the field lines. The energy stored in the magnetic field is then transported up into the corona and finally liberated in reconnection events [15, see e.g., for a review on solar and stellar flares.]

By their very nature, flares are restricted to magnetic stars, and therefore they are a characteristic feature of stars harboring a dynamo. Indeed, X-ray flaring is observed for almost all types of cool stars. Although flares on F and G stars are relatively rare,they are frequently observed on M dwarfs, RS CVn systems, and similar

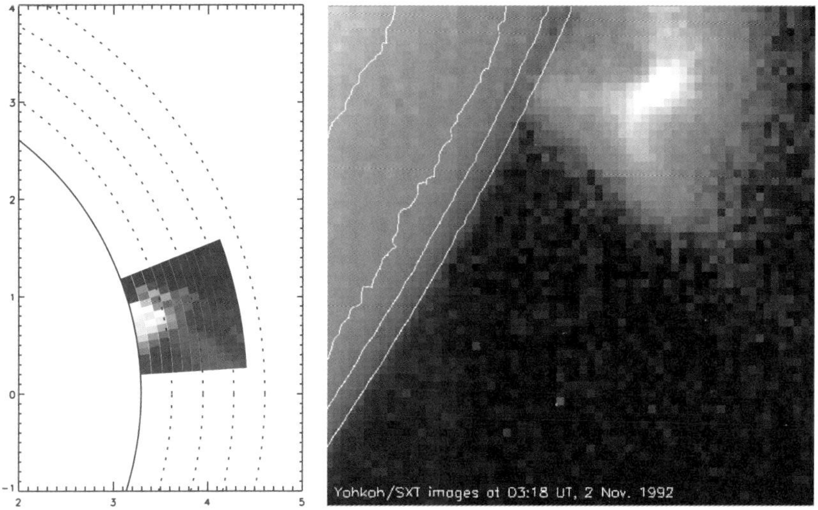

Fig. 10.2 *left*: Reconstructed spatial distribution of the flaring region on Algol B. All of the flaring plasma was assumed to be located at a fixed longitude of $\phi = 70°$. The *solid circle* represents the limb of the K star, units are in solar radius; *dashed circles* indicate heights in steps of 0.1 stellar radii. *Fig. 4 from [44]*; *right*: *Yohkoh* image of a solar limb flare on November 2, 1992 available under http://www.lmsal.com/SXT/. Limb flares on the Sun allow to study the geometry of the X-ray emitting plasma; the loop footpoints and the hot loop top are particularly well visible

active stars. The range of observed stellar flare energetics extends over at least eight orders of magnitude with the largest reported events exceeding 10^{37} erg in released X-ray energy, i.e, larger than the strongest solar flares by five orders of magnitude.

One of the largest stellar flares ever observed was studied with the BeppoSAX satellite on the eclipsing binary Algol [8]. In Fig. 10.3, we show the BeppoSAX lightcurve of Algol in terms of the orbital phase of the binary. The 2–10 keV lightcurve recorded with the MECS instrument on board BeppoSAX is dominated by a huge flare starting at phase $\phi \sim 1.0$. This phase corresponds to the instance when the cool secondary is seen in front of the primary B star obscuring it, i.e., at secondary eclipse. The flare rise took about 8.3 h, then the lightcurve first decayed rather rapidly until, at $\phi \sim 1.25$, a basically exponential decay began. This light curve morphology is in fact quite typical for the decay of solar long duration flares (such as the flare on November 2, 1992), which, however, do not extend over 2 days. Around $\phi = 1.5$ the presumably X-ray dark primary is in front of the X-ray emitting secondary and thus the dip in the lightcurve observed at the same phase must clearly be interpreted as an eclipse of the flaring plasma by the early-type primary. A clear flare-related signal is also seen in the PDS detectors and spectral analysis shows evidence for X-ray emission out to 80 keV. This is a record for the highest energy photons received from stellar X-ray sources. Interestingly, spectral analysis shows full consistency of the recorded spectrum with a thermal source and no evidence for nonthermal X-ray emission. Contrary, large solar flares are always accompanied by

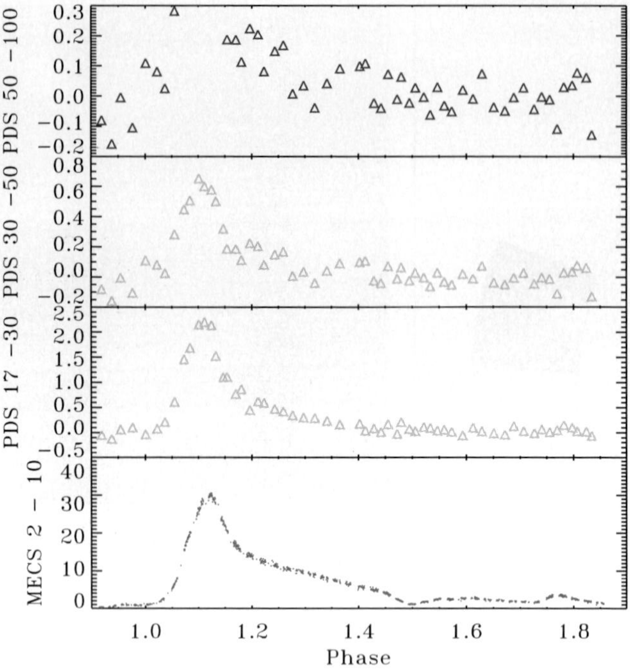

Fig. 10.3 BeppoSAX lightcurves of Algol in different energy bands; MECS (2–10 keV) lowest panel, PDS (17–30 keV), PDS (30–50 keV), PDS (50–100 keV)

nonthermal hard X-ray and γ-ray emission. By analogy, one expects the same for stellar flares. However, so far no convincing examples of nonthermal emission from stellar flares have been presented at X-ray wavelengths, although nonthermal radio emission has been observed.

Flares come in a wide variety of sizes and shapes [see, e.g., 13 for a summary]. Light curves as the one of EQ Peg shown in Fig. 10.4 are typical for flares on dMe stars. This event was observed simultaneously in soft X-rays and in the optical *B* band. The observed optical emission in such flares is considered to be a proxy of the nonthermal particles presumably generated in the first phases of the flare. These accelerated particles penetrate into the deeper photospheric layers, leading to the observed impulsive phase emission. This is followed by thermal soft X-ray emission produced by photospheric and chromospheric material heated and evaporated into the corona. Typical delays between optical and soft X-ray peaks are of the order of a few minutes.

10.1.5 ROSAT All-Sky Survey: Which Stars are X-Ray Emitters?

Within the framework of the ROSAT All-Sky Survey (RASS), it was possible to carry out a sensitive and unbiased survey of X-ray emission from all types of stars.

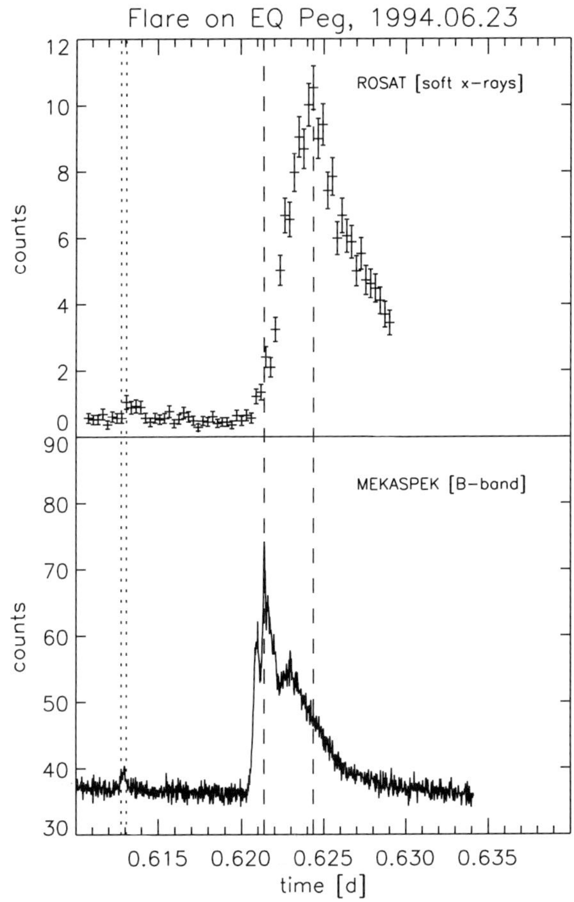

Fig. 10.4 Simultaneous optical and ROSAT X-ray observation of the flare star EQ Peg during a flare

Quite a number of stellar surveys had previously been carried out with the Einstein Observatory [52], and many of the key results of stellar X-ray astronomy had already been obtained with such data. The big surprise of the first data obtained with the Einstein Observatory was the fact that X-ray emission was produced by many different types of stars at levels exceeding the X-ray output of the Sun by many orders of magnitude. If stars emitted X-rays at the level observed from the Sun, only stars in the immediate neighborhood of the Sun would be detectable. The characteristic value for the sensitivity limit of the RASS is a limiting X-ray flux of $\approx 2\times 10^{-13}$ erg cm^{-2} s^{-1}, which implies that X-ray emission at solar-like levels of $L_X \approx 2\times 10^{27}$ erg s^{-1} can be detected only out to distances of 10 pc in the RASS.

As a starting point of our investigations of the X-ray properties of the solar-like stars, let us consider the brightest stars contained in the *Bright Star Catalog*

(BSC) and the nearest stars as contained in the Gliese catalog ([12, 18]). The BSC comprises all stars down to a visual magnitude of 6.5 mag, and has around 10 000 entries. Being a magnitude-limited catalog, its composition in terms of spectral type is biased toward intrinsically bright stars, and consequently contains many stars of spectral type A and F as well as giant stars, with a deficit of intrinsically faint but nearby G, K, and M dwarfs. The Gliese catalog on the other hand is a volume-limited catalog of all stars (≈3 200) known within a distance of 25 pc around the Sun. Note that the latter catalog is incomplete at the very faintest magnitudes and especially among ultracool dwarfs. By its very construction, the Gliese catalog is composed mostly of late-type dwarf stars of spectral type K and M, while its content of earlier type stars overlaps with the BSC.

In the left panel of Fig. 10.5, we plot a color–magnitude diagram of all RASS detected stars contained in the BSC and/or Gliese catalogs. As is apparent, all types of stars commonly placed in the color–magnitude diagram – with the exception of white dwarfs – are found to be X-ray emitters. For most stars shown in Fig. 10.5 (*left panel*) trigonometric distances are known and thus reliable X-ray luminosities can be computed. Figure 10.5 (*right panel*) shows X-ray luminosity vs. $B-V$ color, which is used as an indicator of the stars' effective temperature, for the BSC stars (green) and Gliese stars (red). A huge spread of X-ray luminosities of up to four

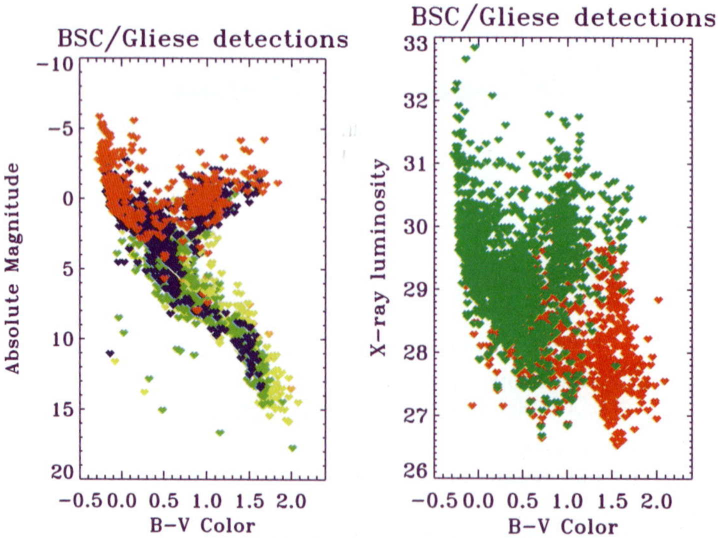

Fig. 10.5 *Left*: Color-magnitude diagram of RASS detected BSC and/or Gliese catalog stars from [14, 19]. For main-sequence stars, the B-V color is a measure of spectral type. Color coding refers to the X-ray luminosity: *yellow* denotes $L_X < 10^{28}$ erg s^{-1}, *green* $L_X = 10^{28} - 10^{29}$ erg s^{-1}, *blue* $L_X = 10^{29} - 10^{30}$ erg s^{-1}, and *red* $L_X > 10^{30}$ erg s^{-1}. *Right*: X-ray luminosity for RASS detected BSC and Gliese stars vs. $B-V$ color

orders of magnitude from stars with given $B-V$ color is apparent. Since most of the stars shown in Fig. 10.5 are main-sequence stars, it is also clear that the "fractional" X-ray luminosity, i.e., the ratio L_X/L_{bol}, also varies over many orders of magnitude from star to star. An understanding of the cause of these variations is one of the central themes of stellar X-ray astronomy.

10.1.6 Connection of X-Ray Emission with Other Stellar Parameters

Given the observed spread of X-ray luminosity for stars of given spectral type, a natural question to ask is which other stellar parameters except those determining the position of a star in the HR diagram determine the X-ray output level. For late-type main-sequence stars the X-ray emission scales with rotation velocity [34]. This relation is believed to trace directly back to the stellar dynamo. On theoretical grounds, the most adequate measure for the dynamo efficiency should be the Rossby number R_0, defined as the ratio between the rotation period and the convective turnover time. However, in observational studies R_0 often is replaced by the rotation period because the convection time scale is difficult to access observationally.

The current observational situation is summarized in Fig. 10.6, which displays rotation-activity relations for a sample of dwarf stars from the field and various young open clusters (Hyades, Pleiades, α Per, IC 2391, IC 2602). Since stellar rotation is thought to be braked by magnetic winds, the younger stars in clusters rotate much faster than the field stars, and they cover the left part of the diagram.

Stars rotating with periods shorter than about 5 days are in the so-called "saturation limit" of $L_X/L_{bol} \sim 10^{-3}$ with no obvious dependence on rotation rate [53]. If one computes for a solar-like star the mean X-ray surface flux corresponding to $L_X \approx 10^{-3} L_{bol}$, one finds $F_X = 6.5\times 10^7$ erg cm^{-2} s^{-1}, which agrees quite well with the maximally observed surface flux values (cf., Fig. 10.7), if one assumes filling factors at the $\sim$10% level.

For slowly rotating stars (periods longer than $\approx$5 days), the observed X-ray luminosity scales inversely with the period, albeit with an unexplained scatter of at least one order of magnitude. Possible explanations for this scatter include long-term variability. Many of the data points used in Fig. 10.6 are derived from snapshot X-ray exposures, and may not be representative for the "mean" X-ray activity. Little information exists on the long-term variability of stellar coronae and, in particular, on the issue if stellar coronal activity cycles exist as observed for the Sun. The solar X-ray emission varies – depending on the X-ray band considered – at least by an order of magnitude during a cycle, while the rotation rate stays clearly constant. Therefore, cycles or other kinds of long-term variability would provide a natural explanation for the observed scatter (Fig. 10.6)

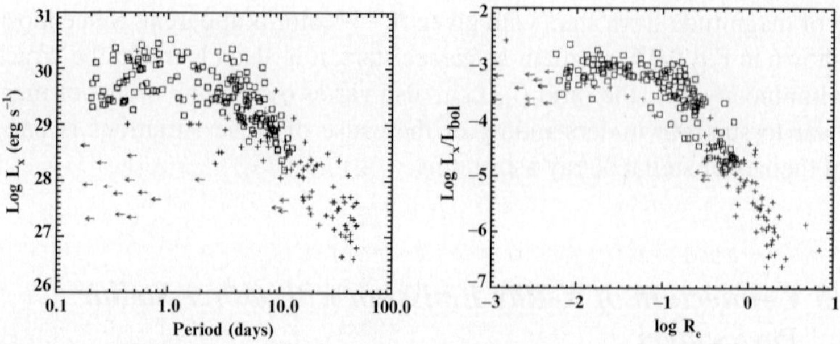

Fig. 10.6 Rotation-activity relation for a sample of ~250 field stars (crosses) and members of young open clusters (squares). *Left*: X-ray luminosity vs. rotation period; *Right*: fractional X-ray luminosity vs. Rossby number. *Leftward arrows* indicate field stars with periods derived from measurements of the rotation velocity $v_{\rm rot} \sin i$, which yields upper limits to the rotation period because of the generally unknown inclination angle. *Figs. 3 and 9 from [36]*

10.2 Cool Stars On and Off the Main-Sequence

10.2.1 Stellar Interiors and Magnetic Dynamos

Stars on the main-sequence derive their energy from nuclear burning in their hot and dense cores. Convection and radiation are the main processes transporting the energy produced in the stars' cores to the photosphere, where it is radiated into space. According to stellar structure theory early-type stars (spectral type O and B) have convective cores and radiative exteriors, A-type stars are purely radiative, and cool main-sequence stars with spectral types F to mid-M have radiative cores and outer convection zones. These stars are – presumably – the nearest analogs to the Sun in terms of their interior structure and in terms of their activity. According to theory, solar-like dynamos ($\alpha\Omega$-dynamos) are localized at the interface region connecting the outer convection zones with the radiative interiors [35]. Therefore, solar-type dynamo action and related manifestations of solar-type magnetic activity ought to be restricted to moderately cool main-sequence stars.

Stars of spectral type ≈M3 or later are expected by theory to be fully convective and should, therefore, lack any boundary layer between convective and radiative zones. Consequently, dynamos operating in their interiors should be different from the "classical" $\alpha\Omega$-dynamo thought to be operating in the Sun. Nuclear burning is expected to persist for stars down to a mass of approximately $0.075\,M_\odot$. Stars with lower masses do not reach central temperatures large enough to start main-sequence hydrogen burning and can, therefore, by definition not be on the "main-sequence." Such "substellar" objects are called *brown dwarfs*. The X-ray properties of fully convective stars and brown dwarfs are discussed in Sect. 10.3. In this section, we focus on solar-like stars, by which we mean stars with an outer convection zone and an interior radiative core, i.e., stars, where presumably the same dynamo mechanism can operate as on the Sun.

10.2.2 Cool Field Stars in the Solar Neighborhood

Cool stars with spectral types F to M are abundantly found in the solar neighborhood and in nearby open clusters. A comparison of these populations allows the study of solar-like stars with a range of masses, ages, and rotation rates, parameters that seem to be crucial for magnetic activity. In this fashion, one also hopes to improve our understanding of solar activity. While, obviously, the Sun can only be observed, but not changed in its physical parameters, the study of the stars enables us to study how X-ray emission – and activity in general – depend on fundamental stellar parameters. This is the concept of the solar-stellar connection.

To study the X-ray properties of solar-like stars in an unbiased manner and to compare the Sun to the stars in a fair way, one, therefore, has to consider *volume-limited* samples. The RASS was a spatially complete but *flux-limited* all-sky survey. Flux-limited surveys tend to find the intrinsically brightest objects of any given population since these are sampled over the largest spatial volumes, while low-luminosity objects are sampled only over small spatial volumes and are therefore quite sparse in the RASS data.

Volume-limited samples with very large detection rates were constructed for F, G, K, and M dwarfs by combining RASS data with ROSAT pointed observations of stars not detected in the RASS. All ROSAT observations of nearby cool stars are summarized in the NEXXUS data base [43]. The NEXXUS catalog shows that stellar X-ray emission is detected down to an absolute magnitude of $M_V = 20$ mag, i.e., down to the very bottom of the main sequence.

What about the completeness of the X-ray detections in the NEXXUS database? Within a volume of 12 pc all F/G stars have been detected. Out of 51 K stars within 12 pc only two stars have not been detected (detection rate of 96%), while out of 65 M stars within 6 pc 6 stars have not been detected (detection rate of 91%), and most of the nondetected M stars are brown dwarfs or very low-mass stars (see discussion in Sect. 10.3). The obvious conclusion is that the detection rate in these volume-limited samples of the nearest stars is very large and in fact, there is no reason to expect that the remaining few undetected stars will not be detected with more sensitive X-ray observations. Therefore, the formation of X-ray emitting coronae appears to be universal for solar-like main-sequence stars, and intrinsically X-ray dark solar-like stars do not exist (at least within the immediate solar environment). The observed mean X-ray surface flux distribution of the NEXXUS sample displays a rather well-defined lower envelope of $F_{\mathrm{x,lim}} \approx 10^4\,\mathrm{erg\,cm^{-2}\,s^{-1}}$ (Fig. 10.7).

This lower limit to the X-ray surface flux actually compares well with the observed X-ray surface flux from solar coronal holes and it is suggestive to interpret the stars observed at their minimum flux levels as stars surrounded by coronal holes without any active regions.

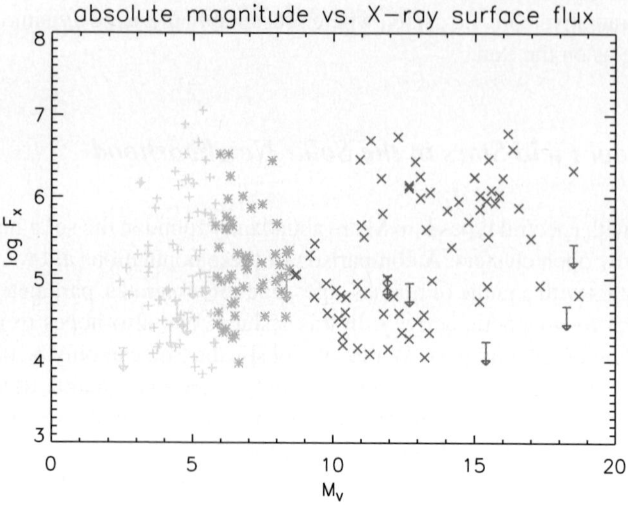

Fig. 10.7 X-ray surface flux vs. absolute *V* band magnitude for a volume-limited sample of nearby cool stars; $+$, F/G stars ($d < 14$ pc); $*$, K stars ($d < 12$ pc); $\times$, M stars ($d < 6$ pc)

10.2.3 Intermediate-Mass Stars

The above-demonstrated cool star paradigm states that all stars with outer convection zones possess hot X-ray emitting coronae. Main-sequence stars of spectral type A are devoid of outer convective zones and should hence not display any coronal emission. A stars and late-B stars also have weak stellar winds or no winds at all, such that the X-ray emission mechanism thought to operate in hot stars (see Sect. 10.5) does not work either.

Indeed, the prototypical nearby A star Vega, one of the brightest stars in the sky, could not be detected as an X-ray source [41]. The upper limit to its X-ray luminosity of $L_X \sim 5.5 \times 10^{25}$ erg s^{-1} places Vega at the very bottom of the distribution for cool stars in terms of activity. However, contrary to expectations, the detection rate of A stars in the RASS data is not zero, but about 15% (cf., Fig. 10.8). The standard hypothesis to explain the observed but unexpected X-ray emission from A stars is to attribute it to (unknown) optically faint, late-type companions. Obviously, such a cool companion star can rather easily escape detection because of the much brighter optical emission of the A star.

In a few cases, the companion hypothesis can be tested by an analysis of the X-ray lightcurve. For example, in the eclipsing binary α CrB the late-type secondary (spectral type G5 V) is occulted by the early-type primary (spectral type A0 V), and in X-rays a total eclipse is seen at the time of optical secondary minimum (see Fig. 10.9 and Sect. 10.1.3).

Further support for the companion hypothesis comes from imaging observations in the infrared using the technique of adaptive optics. Such observations have

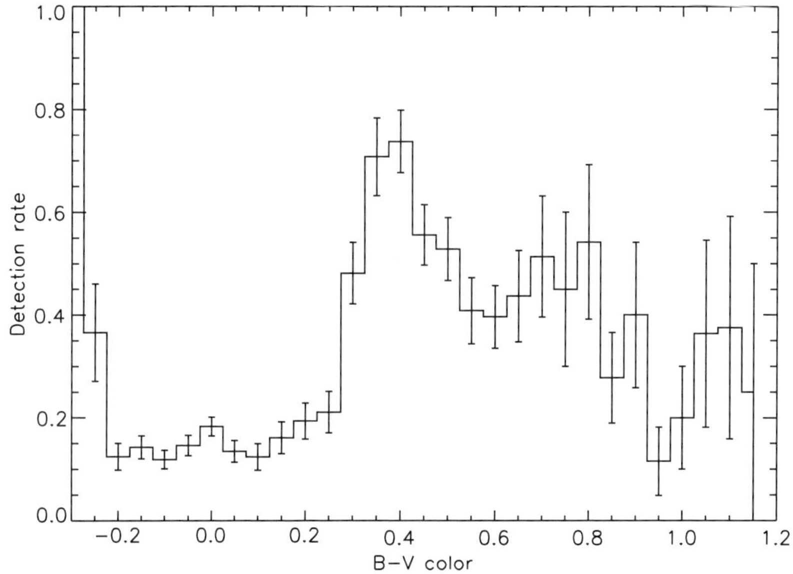

Fig. 10.8 Detection rate of BSC stars in the RASS shown as a function of $B-V$ color, a proxy for spectral type. The transition from A to F spectral types is located at $B-V \sim 0.3$ mag, and coincident with a sharp increase of the X-ray detection rate

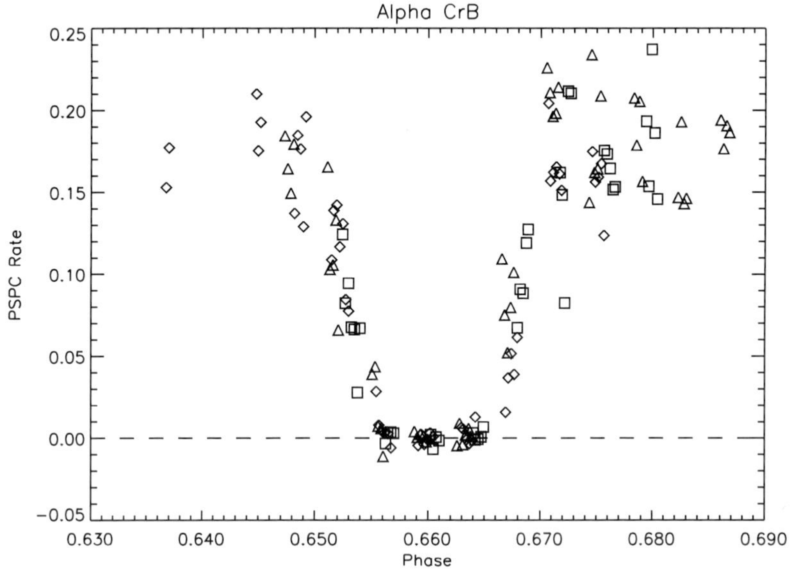

Fig. 10.9 ROSAT PSPC lightcurve of the eclipsing binary α CrB at optical secondary eclipse (A star in front of G star); note the totality of the eclipse

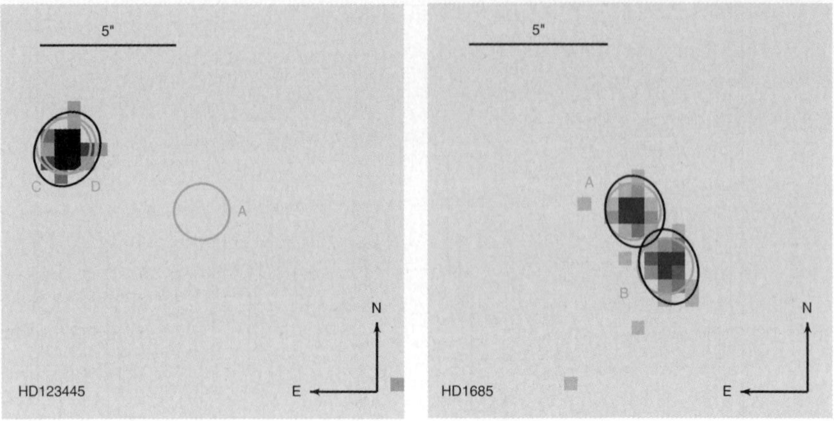

Fig. 10.10 Chandra images of main-sequence B stars (labeled "A" because they represent the primary in the stellar multiple) and their faint companions (labeled "B," "C," and "D") discovered in high angular resolution infrared observations. For the case shown on the left the Chandra image shows that all X-ray emission must be attributed to the companions; in the example on the right both the companion and the B star are detected with Chandra. *Figure 1 from [48]*

recently identified a substantial number of faint objects at separations between $\sim 1''$ and $15''$ from apparently X-ray emitting B stars. The imaging instruments onboard the Chandra satellite provide a spatial resolution comparable to adaptive optics in the infrared ($\leq 1''$), such that the newly identified systems can now be studied in X-rays. However, investigations with Chandra did not unambiguously solve the question of whether the X-ray flux is to be attributed to the early-type primary or the late-type secondary component, as X-ray emission is detected from more than half of the B stars even after resolving them spatially from the new faint infrared objects [48]. An example of this is shown in Fig. 10.10 (right panel), where X-ray emission is clearly detected from both the optical primary and from the secondary identified on recent infrared images. Whether these (primary) stars have additional even closer secondary companions or whether they are intrinsic X-ray emitters by themselves has remained unknown to date. Thus, the old saying "Absence of Evidence is not Evidence of Absence" still holds. Clearly the physics of coronal formation in those stars – if coronae exist at all – must be very different.

10.2.4 Open Clusters

Given the dependence of magnetic activity on rotation and the dependence of rotation on age, one expects a dependence of activity on age. Already more than 30 years ago, prior to the advent of stellar X-ray astronomy, the so-called "Skumanich law" has been formulated, describing the decay of chromospheric Ca II emission in the form $\sim t^{-1/2}$ [45].

Individual stellar ages are rather hard to come by, and thus X-ray studies of open clusters, where the age is known from main-sequence fitting, are the main source of information on the coronal age-activity relations. Already the Einstein Observatory observations of the Hyades cluster ($d = 45$ pc, age $= 600$ Myr), Pleiades cluster (130 pc, 70 Myr) and Ursa Major stream (40 – 100 pc, 400 Myr) showed the dramatic increase of X-ray luminosity for young stars compared to the Sun. In particular, they demonstrated that solar-like stars at an age of the Hyades emit X-rays at levels two orders of magnitude above solar levels. With ROSAT the numbers and especially the age range of the observed clusters has been tremendously expanded.

The very young clusters observed with ROSAT, XMM-Newton, and Chandra include IC 2391 (160 pc, 40 Myr), IC 2602 (160 pc, 30 Myr), α Per (160 pc, 50 Myr) and IC 4665 (350 pc, 40 Myrs), NGC 2516 (380 pc, 60 Myr), and the intermediate age clusters Coma (80 pc, 500 Myr), Praesepe (160 pc, 700 Myrs), NGC 6475 (240 pc, 130 Myrs) , and IC 4756 (480 pc, 600 Myrs). Old clusters containing stars with ages comparable to that of the Sun such as M 67 (800 pc, 4500 Myrs) and NGC 752 (380 pc, 2500 Myr) were also studied, but because of their large distances, solar-like X-ray emission levels are not detectable even with XMM-Newton or Chandra. The stellar X-ray sources detected in such old clusters are mostly active binaries (see section on Close Binaries) with X-ray luminosities of up to $L_X \sim 10^{31}$ erg s^{-1}.

Figure 10.11 shows a plot of the median of the X-ray luminosity of G dwarfs derived for a number of star forming regions and clusters, the Sun, and field stars. During the first 100 million years in a star's life, its activity changes only little, afterwards the activity decays by a factor of 10 in a little less than a decade in logarithmic age. Figure 10.11 strongly suggests that the early evolution of the solar system occurred under conditions very different from the present conditions. The exploration of the consequences of the enhanced UV, X-ray, and particles fluxes in the early solar system has become vital for an understanding of planet formation [9].

10.2.5 Evolved Stars

Stellar structure theory predicts cool giants to have deep convection zones, one of the requirements for a magnetic dynamo to work. On the other hand, single giants tend to be slow rotators, which are not expected to show high levels of activity. Furthermore, because of the complicated evolutionary paths of giants in the HR diagram, without an at least approximate knowledge of mass, the age and hence the main-sequence progenitors of the giant stars and their state of activity during the main-sequence phase are unknown.

Figure 10.5 shows that X-ray emission is frequently found among single giants. An X-ray dividing line (XDL) separates the X-ray emitting G and K giants from apparently X-ray dark M giants [27]. In a complete volume-limited sample of giants within 25 pc around the Sun all giants (to the left of the XDL) were observed and detected in the ROSAT pointing program. This finding led to the conclusion that

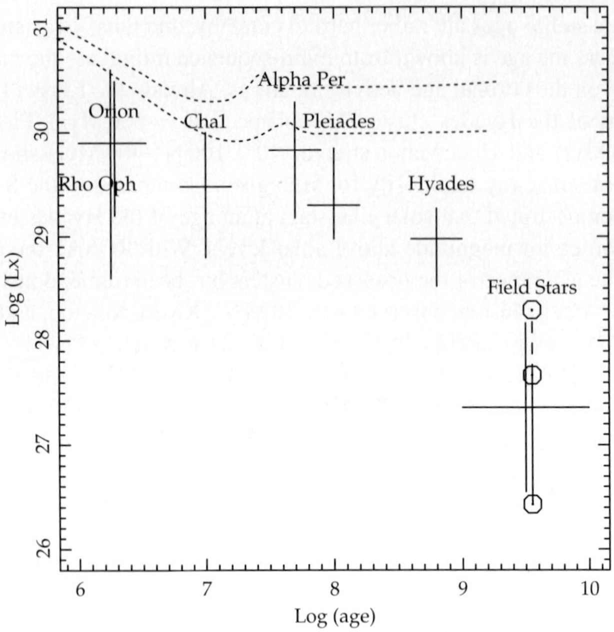

Fig. 10.11 X-ray luminosity of dG stars from the Orion, ρ Oph, and Cha I star forming regions, the α Per, Pleiades, and Hyades open clusters, the Sun and field stars together with the 1 σ equivalent data spread as a function of age; the *lines* show the X-ray luminosities corresponding to the saturation limit of a 0.8 $M_\odot$ and 1.0 $M_\odot$ star; *Fig. 21 from [7]*

giants to the left of the XDL, which all have outer convection zones, are ubiquitous X-ray emitters just like cool main-sequence stars.

The XDL occurs across a very narrow region in the HR diagram. Thus, for example, a G giant (to the left of the XDL) can have a rather high X-ray luminosity of up to $L_X \sim 3 \times 10^{30}$ ergs s^{-1}, while K giants to the right of the XDL can be almost five orders of magnitude fainter. The most sensitive upper limit in terms of X-ray luminosity has been obtained for the nearby K giant Arcturus ($L_X < 3 \times 10^{25}$ ergs s^{-1}). Expressing this upper limit in terms of mean X-ray surface flux, this X-ray nondetection corresponds to a flux value more than 1 000 times fainter than a solar coronal hole. Therefore, coronae of giants to the right of the XDL must be quite different from the solar corona – if at all they exist. Using the Chandra HRC-I camera, T. Ayres and collaborators have recently obtained a tentative X-ray detection of Arcturus. Around the known position of Arcturus 3 photons have been detected with 0.2 counts expected from background [1].

The concept of a dividing line appears to disappear among the brighter giants and supergiants. Among those stars there is a group exhibiting both signatures of transition region material (as inferred from C IV line detections) as well as cool winds (inferred from UV line profiles), the so-called hybrid stars. As a result of

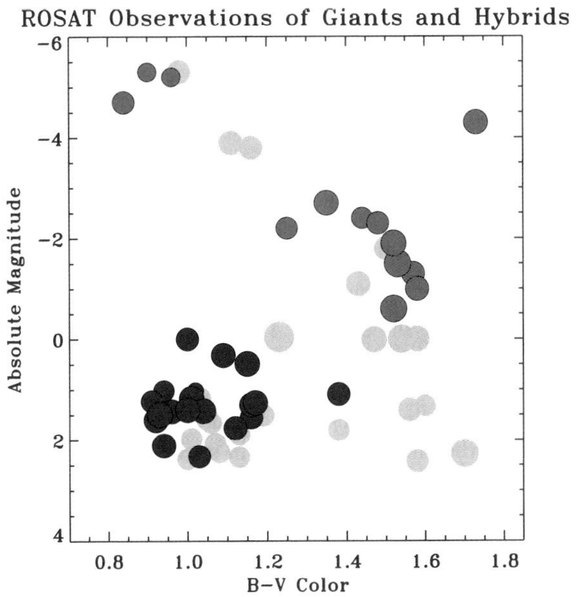

Fig. 10.12 X-ray "bubblegram" of a complete sample of giants within 25 pc around the Sun and hybrid stars. Plotted are the color–magnitude diagram positions of nearby giants detected as X-ray sources (*black circles*), the positions of nearby giants not detected as X-ray sources (*dark circles*), and the positions of hybrid stars detected as X-ray sources (*grey circles*)

extensive ROSAT observations, hybrid stars are now known to possess hot coronal plasma, presumably quite similar to cool main-sequence stars.

The current observational situation is summarized in Fig. 10.12, where giant stars within 25 pc around the Sun and the hybrid stars are shown in a color–magnitude diagram. As is obvious from Fig. 10.12, the XDL shows very clearly up for giants of luminosity class III, while most of the hybrid stars have been detected as X-ray sources, some of them having a spectral type that puts them well beyond the XDL for luminosity class II giants. The general concept of a dividing line has been questioned when it was recognized that all X-ray detections lie to the left of an evolutionary track with $M = 1.25\ M_{\odot}$ in the HR diagram. Low-mass stars ascending the giant branch are restricted to a rather narrow mass range and must therefore be rare. In this light the XDL may be interpreted as an effect of stellar evolution.

X-ray emission from M giants is extremely rare. Only about a dozen candidates for X-ray emitting M giants have been identified with ROSAT. Utilizing the high angular resolution of the Chandra telescope, M. Hünsch and collaborators [20] confirmed the previously detected X-ray emission from these objects, and in particular showed that the X-ray emission must indeed be attributed to the M giant stars, rather than to a coincidental nearby object. Since most of these M giants must be old, their extremely large X-ray luminosities in excess of 10^{30} erg s^{-1} are difficult to explain.

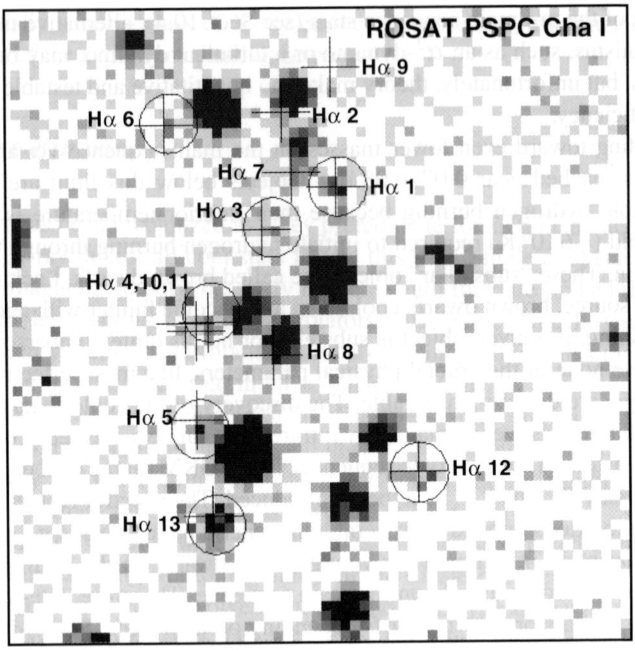

Fig. 10.13 ROSAT PSPC image of the central part of the Cha I dark cloud with the objects ChaHα 1 to 13 marked as crosses. X-ray sources identified with any of those objects are marked by *circles* (centered on the X-ray positions). ChaHα 1 was the first X-ray detected bona fide *brown dwarf* (*Fig. 1 from [32]*)

dwarfs with ROSAT have proven little successful, yielding mostly rather high upper limits to their X-ray luminosity.

The new and far more sensitive X-ray observatories Chandra and XMM-Newton have begun to provide meaningful constraints for X-ray emission from VLM stars and brown dwarfs. However, the bulk of bona-fide brown dwarfs in star forming regions has still remained below the detection threshold of currently available X-ray observations because of the considerable distance of all star forming clouds (e.g., $\sim$140 pc for the nearby Taurus and ρ Ophiuchus, and $\sim$450 pc for the Orion Nebula), imposing a limiting sensitivity of $\log L_{\rm X}\,[\rm erg\,s^{-1}]\sim 28$ even for the most advanced X-ray instruments. However, the sensitivity of the instruments onboard Chandra and XMM-Newton has proven sufficient for first studies of the X-ray spectra and lightcurves of the brighter ones among the young brown dwarfs. Comparing their coronal temperature, flare frequency, and fractional X-ray luminosity to those of higher-mass premain sequence stars indicates no dramatic differences, providing evidence that their emission originates from the same mechanism.

Recent investigations have also addressed the issue of coronal emission from evolved brown dwarfs with their far lower bolometric luminosities and thus presumably also far lower X-ray luminosities. Quite a number of evolved brown dwarfs are located in the solar neighborhood at distances on the order of 10 pc. In Fig. 10.14,

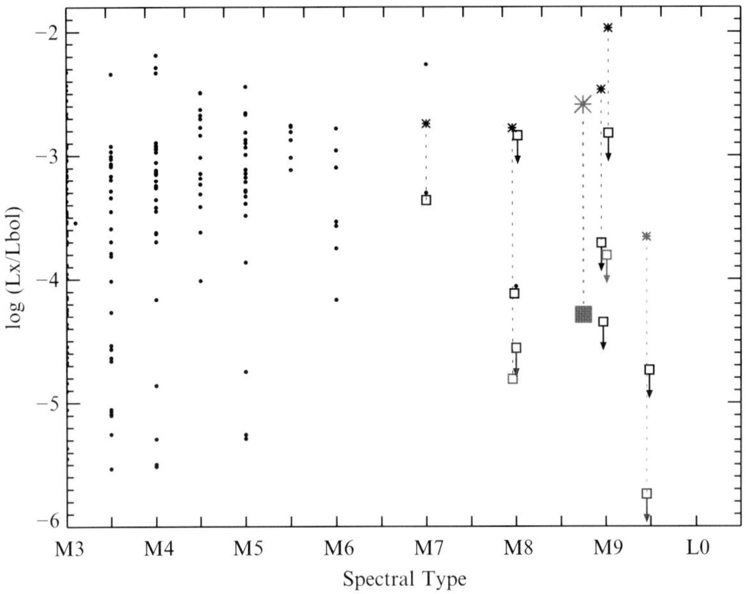

Fig. 10.14 Fractional X-ray luminosity vs. spectral type for M and L field dwarfs. *filled circles*, M dwarfs detected during the RASS; *asterisks*, flares on ultracool dwarfs; *squares*, quiescent emission or upper limits for ultracool dwarfs; *large asterisk and square*, the most recent X-ray detection of a brown dwarf (Gl 569 Bab). For clarity VLM dwarfs with the same spectral type have a small offset with respect to each other on the horizontal axis (*Fig. 4 from [47]*)

we summarize our current knowledge about the activity level in ultracool dwarfs, expressed in terms of the fractional X-ray luminosity vs. spectral type. The RASS has revealed a large number of X-ray sources among the earlier M stars, but beyond spectral type ~M6 the coronae have remained largely inaccessible to ROSAT. The coolest dwarf detected so far in X-rays is LP 944-20, an intermediate age M9.5 brown dwarf ($t \sim 500$ Myrs). LP 944-20 was seen by Chandra during a flare, and XMM-Newton has provided the lowest upper limit for the quiescent flux of any field dwarf so far: $L_X < 3.1 \times 10^{23}\,\mathrm{erg\,s^{-1}}$ [29, 40]. As yet, no X-ray detection of an L dwarf was reported.

One of the striking features of Fig. 10.14 is that almost all detections of evolved ultracool dwarfs are attributed to flares, and it has remained unclear whether VLM stars and brown dwarfs are capable of producing quiescent, persistent X-ray emission. Only very recently, quiescent X-ray emission was for the first time reported from a confirmed brown dwarf [47]. Gl 569 Bab is composed of two brown dwarfs in a very tight orbit around a main sequence star with which it forms a hierarchical triple system. In X-rays only Chandra can resolve the brown dwarf binary (at a separation of only 5″) from the primary. Gl 569 Bab showed both a huge flare and persistent emission after the outburst (see Fig. 10.15).

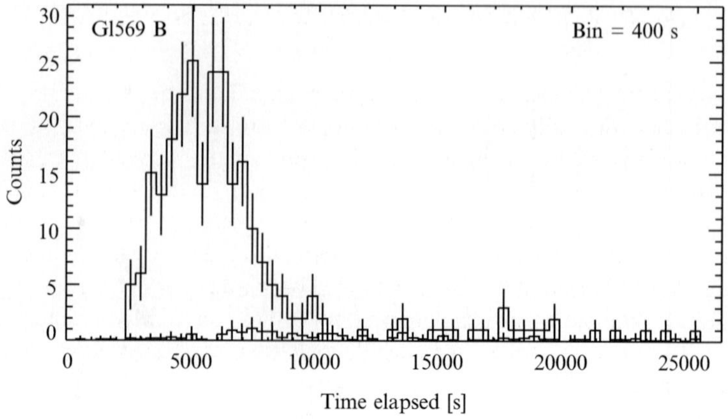

Fig. 10.15 X-ray lightcurve of Gl 569 Bab observed with Chandra during a large flare. The subsequent weak emission represents the first detection of quiescent X-rays from an evolved brown dwarf (*Fig. 2 from [47]*)

10.4 Premain Sequence Stars

During the process of star formation from the start of the gravitational collapse of a molecular cloud core to the arrival of a star on the zero-age main-sequence, young stars pass through several phases. Low-mass objects ($M \leq 2 M_{\odot}$) in evolutionary stages before the quasi-steady hydrogen burning phase are collectively referred to as *Young stellar objects* (YSOs). They are classified by their spectral energy distribution from infrared to millimeter wavelengths, which are less affected by extinction than the optical passband.

In their youngest phases, YSOs are completely hidden behind their circumstellar material in the optical (*Class I protostars*; age $\sim 10^5$ yr), and even earlier also in the near-infrared (*Class 0 protostars*; age $\sim 10^4$ yr). Once the forming star emerges from its gaseous envelope, it enters the T Tauri phase. In *classical T Tauri stars* (*cTTS*; age $\sim 10^{6...7}$ yr) the remaining circumstellar material has settled into a disk, from which it is accreted onto the star. In the infrared classification scheme, cTTS correspond roughly to *Class II sources*, dominated by emission from the disk (see below for differences between cTTS and Class II objects). Later on, when all circumstellar material has been dispersed and/or accreted, the stars' spectral energy distribution resembles that of main-sequence stars, although these *weak-line T Tauri stars (wTTS) or Class III sources* are still in their contraction phase to the main sequence.

Higher-mass premain sequence stars ($M \sim 2$–$10 M_{\odot}$) are called *Herbig Ae/Be stars (HAeBe stars)*. The evolutionary timescale of HAeBe stars is comparable to the dissipation timescale for their circumstellar matter. As a consequence – contrary to the T Tauri stars – *all* HAeBe stars show signatures for circumstellar material that in some cases has been shown to form a disk-like geometry.

10.4.1 The Role of Magnetic Fields on Premain Sequence Stars

Low-mass premain sequence stars, i.e., T Tauri stars, begin their evolution along the Hayashi tracks with fully convective interiors. As for ultracool dwarfs this should prevent solar-type $\alpha\Omega$-dynamo action, which presumably depends on the presence of a transition between a radiative and a convective zone. However, magnetic fields may be generated by alternative types of dynamos operating in the convection zone. The most popular versions of nonsolar-like dynamos are α^2-dynamos and turbulent dynamos [6, 39]. Different dynamos may coexist in all types of magnetically active stars, and the dominating mechanism may depend on the interior structure and thus on mass. Turbulent dynamos create small-scale fields and are thought to be independent of rotation. Therefore, a study of the rotation-activity connection on the pre-MS might give insight into the nature of the operating dynamo(s).

HAeBe stars are fully radiative and should not drive any dynamo. Similarly to low-mass protostars, their magnetic fields, as evidenced by their outflows, may be fossil in nature, i.e., remnants of the magnetic fields of the parent molecular cloud.

In cTTS accretion is believed to occur along the magnetic field lines onto distinct regions on the stellar surface, forming so-called hot accretion spots [22]. In approaching the star, the accreting material reaches nearly free-fall velocities. Upon hitting the photospheric material, the infalling gas is shocked and heated. Temperatures in the postshock region reach a few Mega Kelvin with ensuing production of X-rays. Recent high resolution spectroscopic observations of a few cTTS with XMM-Newton provide in fact evidence for shocked X-ray emitting material, but it is unclear at present how much accretion processes contribute to the overall energy budget of cTTS. Activity phenomena on accreting stars (protostars, cTTS and HAeBe stars) may also include magnetic reconnection of field lines connecting star and disk.

10.4.2 X-Ray Emission from T Tauri Stars

Contrary to the cTTS, which are easily identified in the optical by their conspicuous spectral features related to the accretion process, wTTS are difficult to distinguish from more evolved main-sequence stars based on their optical spectra. Indeed, they were discovered and established as a class of premain sequence stars only, thanks to their strong X-ray emission. Einstein Observatory observations first revealed numerous unknown X-ray sources in nearby star forming regions, many of which were later confirmed to belong to the wTTS class by optical spectroscopy.

The X-ray emission from both cTTS and wTTS can be described as thermal emission from an optically thin plasma with temperatures up to 20 MK and X-ray luminosities of $10^{29...30}$ erg s^{-1}. Their activity is often termed *scaled-up solar-like*, because it seems to share the same emission process but the energetics involved are 3–4 orders of magnitude above solar levels. Similar to cool main-sequence stars, the X-ray luminosity of T Tauri stars comprises between $10^{-3\ldots-5}$ of the bolometric luminosity, with the spread probably being due to the influences of various stellar parameters such as mass, age, and rotation.

10.4.2.1 Mass and Age-Dependence of T Tauri Star X-Ray Emission

Generally, the average X-ray luminosities of young stars increase continuously with mass for $M \leq 3M_{\odot}$. The same holds also for $L_{\rm X}/L_{\rm bol}$ as activity indicator. Here, the upper envelope corresponds roughly to the canonical saturation limit of 10^{-3}, although some observations indicate a saturation limit somewhat higher than this value. Age may also play a role in determining the amount of X-rays liberated from magnetically active stars. Clearly in the course of stellar evolution (from the pre-main sequence to open clusters and field stars) the intensity of stellar X-ray emission declines (see Sect. 10.2.4). A similar effect is seen within a given star forming region if individual ages are assigned to the stars based on their position in the HR diagram in comparison with evolutionary isochrones. Figure 10.16 summarizes the mass and age dependence of T Tauri star X-ray emission derived from a recent Chandra observation of the Orion Nebula Cluster. For stars with $M \leq 2M_{\odot}$ $L_{\rm X}$ decreases with age in each mass bin, but the $L_{\rm X}/L_{\rm bol}$ ratio remains roughly constant, because of the simultaneous decrease of the bolometric luminosity as the stars approach the main sequence. A drastic drop in the fractional X-ray luminosity is seen for stars near $2M_{\odot}$. The exact location of this kink in terms of mass depends on the stellar age, and coincides roughly with the transition where the convective envelope disappears and the stars become fully radiative according to models of premain sequence evolution.

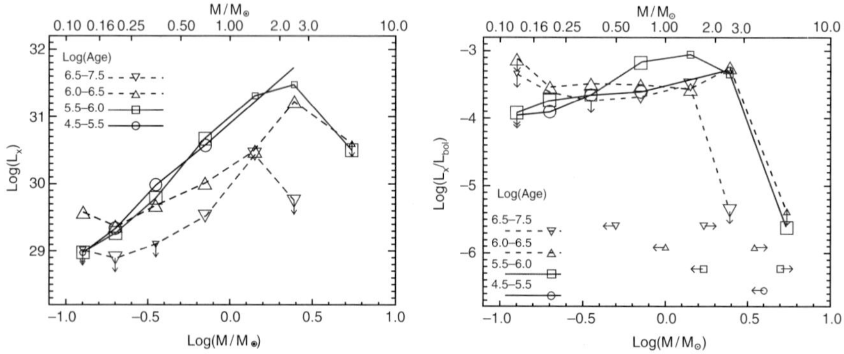

Fig. 10.16 The dependence of X-ray luminosity and $L_{\rm X}/L_{\rm bol}$ ratio on stellar mass and age for a sample of premain sequence stars in Orion. On the *right panel*, masses at which the interior structure of stars is expected to become fully convective and fully radiative are indicated by leftward and rightward pointing *arrows* for each age bin, respectively (*Figs. 10 and 11 from [10]*)

10.4.2.2 Rotation-Activity Connection in T Tauri Stars

For main-sequence stars, the most direct observational signature that X-ray emission is related to dynamo activity is the anticorrelation between X-ray luminosity and rotation period or Rossby number (see Sect. 10.1.6). For premain sequence stars, the situation is less clear. On the theoretical side, the importance of rotation in regulating the dynamo efficiency of fully convective objects is unknown. On the observational side, good samples for studies of the rotation-activity connection exist for only a few star forming regions, e.g., Taurus (which is little obscured) and Orion (which is compact). An anticorrelation between L_x and P_{rot} like in main-sequence stars was found for premain sequence stars in Taurus, but derived on a relatively small subsample of X-ray detected T Tauri stars with known rotation periods. Recent studies in Orion, where about two fifths of the X-ray detected premain sequence sample have measured rotation periods, show no indication for any rotation-activity relation. A possible explanation is the large convective turnover time (~100–1 000 days) of premain sequence stars, which leads to a concentration of all premain sequence stars in the left side of the diagram of $\log(L_x/L_{bol})$ vs. R_0 (see Fig. 10.17), far off the linear regime.

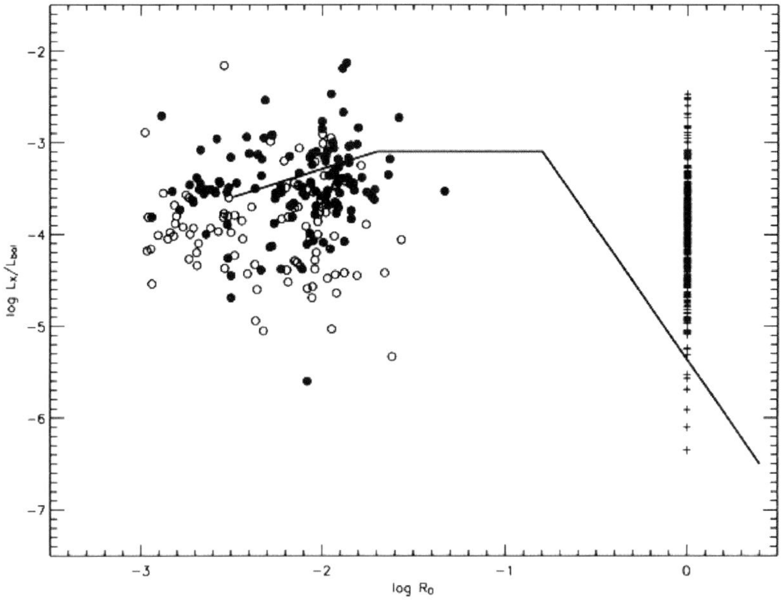

Fig. 10.17 X-ray-rotation relationship for premain sequence stars in the Orion Nebula Cluster as a function of the Rossby number (R_0). The main-sequence relationship from Pizzolato et al. (2003) is indicated by the *solid line*. *Crosses* denote the Orion stars without known rotation period (plotted arbitrarily at $\log R_0 = 0$) (*Fig. 9 from [46]*)

10.4.2.3 X-Ray Emission from Classical and Weak-Line T Tauri Stars

Comparisons of the X-ray properties of cTTS and wTTS have led to controversial results. While in some star forming regions, cTTS and wTTS seem indistinguishable on basis of their X-ray luminosities, other studies – mainly those of the Taurus-Auriga complex – found that wTTS are on average brighter in X-rays than cTTS (see Fig. 10.18a). Incompleteness of the wTTS samples in some areas of star formation was put forth among the possible explanations. Since most wTTS are discovered through their X-ray emission, the known wTTS population is biased toward the brightest X-ray emitters if it is incomplete. This effect could produce artificial differences between the distributions of X-ray luminosities for the two samples.

Recent observations with Chandra and XMM-Newton suggest a different solution to the problem. They have shown that the outcome of such comparisons depends on the diagnostic used to distinguish the two groups of stars. Traditionally, cTTS and wTTS are distinguished on basis of the equivalent width of their H_α emission, stars with $W_{H_\alpha} > 10$ Å being classified as cTTS and those with $W_{H_\alpha} < 10$ Å being considered wTTS; in more sophisticated studies, the spectral type dependence of the boundary between the two groups is also taken into account. This criterion is efficient in separating accreting from nonaccreting stars. Another possibility is to define T Tauri star subgroups based on the presence or absence of IR excess emission, separating disk-bearing from diskless stars. This method corresponds roughly to the IR classification of Class II (stars with disk) and Class III (stars without disk).

Concerning the X-ray luminosity, [38] found in a sample of premain sequence stars in the IC 348 cluster that IR excess and nonexcess stars are equally X-ray luminous (Fig. 10.18b). To summarize the available observations, pre-main sequence star

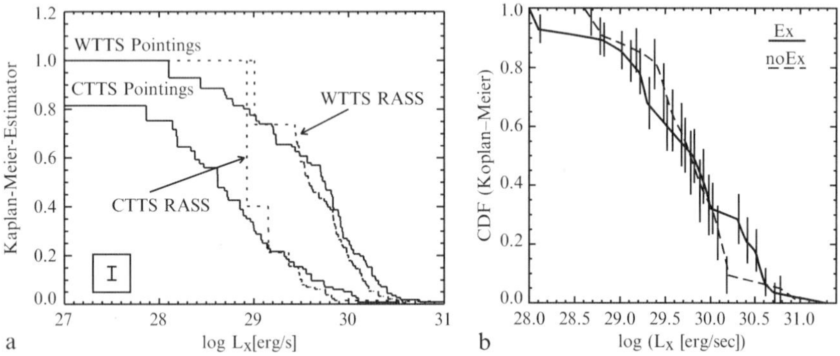

Fig. 10.18 (**a**) X-ray luminosity functions for cTTS and wTTS in the Taurus-Auriga star forming region observed with ROSAT. The pointed observations (*solid lines*) yield better sensitivity than the RASS (*dotted lines*). Both pointed and survey data demonstrate clear differences between the X-ray emission level of the two samples, with wTTS being brighter than cTTS (*Fig. 3 from [49]*). (**b**) X-ray luminosity functions for premain sequence stars in the IC 348 star forming region. The X-ray luminosities for the two subgroups of IR excess and nonexcess stars are indistinguishable (*Fig. 12 from [38]*)

X-ray luminosities appear different if the stars are separated according to accretion signatures such as H_α or Ca II equivalent width, but they appear similar, if distinguished on basis of the presence or absence of a circumstellar disk. This suggests that the disk itself does not play a major role in regulating young star X-ray emission, but the accretion process influences the amount of observable X-rays, e.g., by changing the coronal geometry such as to decrease the fraction of the surface available for the allocation of closed magnetic structures.

10.4.2.4 X-Ray Variability on T Tauri Stars

Strong variability of the X-ray emission is characteristic of T Tauri stars, and has been used as an indicator of youth. Flares on T Tauri stars were long thought to be short (hour-long) events, reminiscent of solar impulsive flares and characterized by a fast rise of the intensity and subsequent exponential decay representing the cooling phase. However, continuous X-ray observations over many hours with the new generation of X-ray satellites have shown that T Tauri star flares comprise also long events that extend up to days. Prolonged flare decays may be interpreted as evidence for continued heating, similar to the case of main-sequence stars.

10.4.3 X-Ray Emission from HAeBe Stars

According to evolutionary models, HAeBe stars are on fully radiative tracks, i.e., they are lacking a convection zone. Therefore, they are not expected to drive a solar-like dynamo and exhibit the ensuing phenomena of magnetic activity. Nevertheless, even the relatively low-sensitivity instruments onboard ROSAT and ASCA achieved detection fractions between 30–50 % [17,54]. Indeed, HAeBe stars turned out to be comparatively X-ray luminous objects, with typical emission levels of $\log L_\mathrm{x}\,[\mathrm{erg/s}] \sim 30\ldots31$. Figure 10.19 shows the position of a sample of HAeBe stars in the HR diagram. The individual X-ray luminosities are indicated by the size of the plotting symbols. The X-ray brightest stars are also the most massive ones, and there is some trend toward lower L_x with increasing age.

The plasma temperatures inferred from modeling of low-resolution CCD spectra reach up to several kiloelectronVolts (~50 MK), similar to the X-ray temperatures of young low-mass stars, and significantly higher than found in high- and intermediate-mass stars on the main sequence. In some cases, strong Fe Kα emission at 6.7 keV is seen, a tracer of extremely high temperature.

Two cases of X-ray flares from HAeBe stars have been reported in the literature [11,16]. Flares arise from confined plasmas heated by magnetic reconnection events. Whether the magnetic structures are anchored on the stellar surface or connecting star and disk is difficult to settle on basis of X-ray observations. Furthermore, in both cases mentioned earlier the identification of the flaring source with the HAeBe star is not unambiguous.

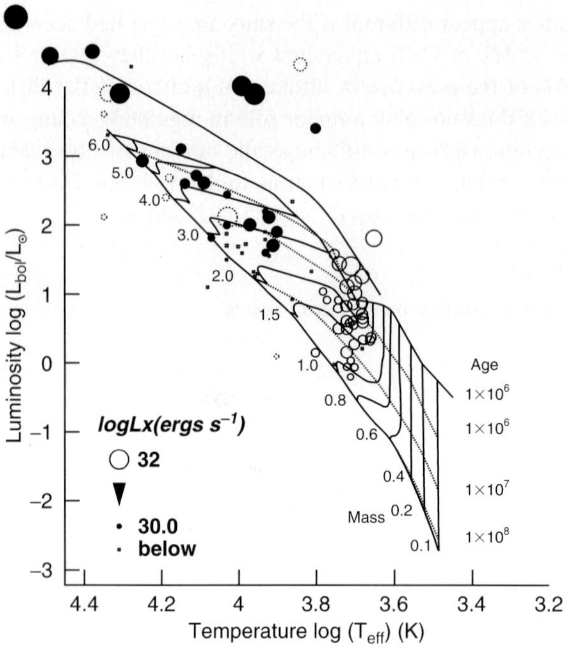

Fig. 10.19 X-ray luminosity of young intermediate-mass stars in the HR diagram. The size of the plotting symbols is scaled to X-ray luminosity. *Black* and *grey* denote ASCA and ROSAT observations, respectively. *Open circles* represent sources in the Orion cloud. The diagram is overlaid by evolutionary models of [33] (*Fig. 16 from [17]*)

Similar to the B and A stars on the main sequence, late-type T Tauri like companions could be the cause for the observed X-ray emission from HAeBe stars. The high X-ray luminosities of many HAeBe stars have often been cited against the companion hypothesis, but recent studies showed that their range of X-ray luminosities is compatible with the typical emission level of a late-type premain sequence star. The brightness ratios of known HAeBe binaries suggest that most companions are of significantly lower mass than the primaries, i.e., they are T Tauri stars, and must be strong X-ray emitters by their nature. So far no systematic studies have attempted to resolve the HAeBe stars from their companions in the X-ray regime.

Other proposals to explain intrinsic X-ray emission from HAeBe stars include a nonsolar dynamo operating in the absence of convection [50]. In this model, the energy related to differential rotation between the stars' center and its outer edge is converted into magnetic energy. The timescale on which the shear energy is exhausted according to the theory is on the order of 10^6 yrs, and is in line with observational hints for a decrease of X-ray emission with age in HAeBe stars. However, the predicted X-ray luminosities tend to underestimate the observed luminosities.

Finally, a number of X-ray production scenarios involve stellar winds. Winds of A and B stars on the main sequence are too weak to generate shocks that reach X-ray emitting temperatures, but calculations show that in the lower gravity HAeBe stars strong winds exist. Shocks, which provide the heating mechanism for X-ray production, may be related to instabilities in the winds themselves or to collision of the wind with the circumstellar material of the HAeBe star. Furthermore, the magnetically confined wind shock model, developed to explain the X-ray emission from magnetic Ap and Bp stars on the main-sequence (see Sect. 10.6), may be applicable in the case of HAeBe stars as well.

10.4.4 X-Ray Emission from Low-Mass Protostars

X-ray observations are particularly successful in peering through dense molecular clouds, because of the large penetrating power of X-ray photons above $\sim$1 keV with respect to optical light, which is more strongly affected by extinction. X-rays from Class I objects of the nearby ρ Oph cloud have been reported in deep ASCA and ROSAT images [23]. More recent observations with Chandra and XMM-Newton have yielded detection fractions between 20 and 70% for Class I sources in the ρ Oph, Orion, IC 348, and Serpens star forming regions, indicating that YSOs are ubiquitous X-ray emitters. The detection of protostars in the Class I stage seems to be mainly a question of sensitivity, and is determined by distance and absorption. In contrast, no convincing evidence for X-ray emission from Class 0 sources has been presented. In this early evolutionary phase the extinction may be too high even for X-ray photons – if any are produced – to penetrate the circumstellar envelope.

The X-ray properties of Class I sources are not vastly different from those of the more evolved T Tauri Stars. They tend to show somewhat higher X-ray temperatures and absorbing columns than these latter ones. Their X-ray luminosites range between $10^{29...31}$ erg s^{-1}, with fractional luminosities $L_{\mathrm{x}}/L_{\mathrm{bol}}$ between $10^{-2...-5}$.

Variability studies indicate that the occurrence of X-ray flares may be higher in protostars when compared with T Tauri stars [21]. The elevated flare rates imply a different emission mechanism. Magnetic energy release in a star-disk magnetosphere has been proposed by [31] as explanation for quasi-periodic flares observed on the Class I protostar YLW 15. In this scenario, magnetic field lines that are anchored with one footpoint on the star and with the other footpoint on a circumstellar disk become twisted due to the differential rotation of star and disk. This leads to buildup of magnetic energy, which is ultimately released and converted into thermal energy by reconnection events. The ensuing heating of the enclosed plasma gives rise to X-ray emission. The phenomenon is intrinsically noncontinuous as a result of the finite time needed to stress the field lines beyond a critical point, and therefore predicts periodically recurring outbursts. Despite its intriguing properties, this model has not been confirmed by further observations of periodic flaring.

10.4.5 Other Types of X-Ray Sources Related to Star Formation

High-mass star forming regions are located at typical distances of a few kpc, about a factor 10 further than the closest regions of low-mass star formation. Therefore, low spatial resolution and low sensitivity X-ray instruments were not able to establish whether the related X-ray emission is produced by a superposition of point sources or whether it is of extended nature. The extinction in the H II nebulae surrounding newly forming massive stars assumes extreme values ($A_V \sim 100 \ldots 1000$ mag), such that soft X-ray emission is completely absorbed.

Chandra images of H II regions have unveiled many point sources dominated by hard X-ray emission (>2 keV), presumably due to the absorption of all soft emission. The X-ray luminosities of these objects are similar to those of low-mass protostars ($\log L_x$ [erg s^{-1}] ~ 31). Whether the emission is caused by magnetic processes as on young low-mass stars, or whether the strong winds play a role as on more evolved high-mass stars, is unclear.

In some cases, enhanced extended background emission was found that is spatially and spectrally different from the point source population, and was interpreted as diffuse emission [51]. Analysis of the associated low signal-to-noise X-ray spectra has not allowed to unambiguously determine the production mechanism for the diffuse component. Its coincidence with the presence of early O stars and the comparatively soft spectrum suggest that it originates from shocks in which the wind energy of the hot stars is thermalized.

Herbig-Haro (HH) objects are optically luminous nebulae in star forming regions, which form at the shock front where a protostellar jet hits the ambient medium. Since the first X-ray detection of the Herbig-Haro object HH-2 with Chandra in 2001 [37], a handful of X-ray sources associated with HH objects have been discovered. In some cases, the high spatial resolution of Chandra was essential for locating the X-ray emitter within the jet/protostar environment. The observed X-ray luminosities ($\log L_x$ [erg/s] $\sim 29 \ldots 30$) and temperatures ($\sim 1 - 10$ MK) are consistent with predictions for X-ray emission from shocked gas. Different shock geometries have been proposed, which cannot be distinguished on basis of the existing observations.

10.5 Stellar Wind Sources

The detection of X-ray emission from hot stars was one of the first science results obtained from observations with the Einstein Observatory. Systematic studies of hot main-sequence stars based on the extensive RASS data have shown that stars with spectral type earlier than B3 emit a fraction of $\sim 10^{-7}$ of their bolometric luminosity in the X-ray band (see Fig. 10.20). Stars of later spectral type show significantly larger L_x/L_{bol} ratios as well as a larger spread of X-ray luminosities, suggesting a qualitative difference in the underlying emission mechanism.

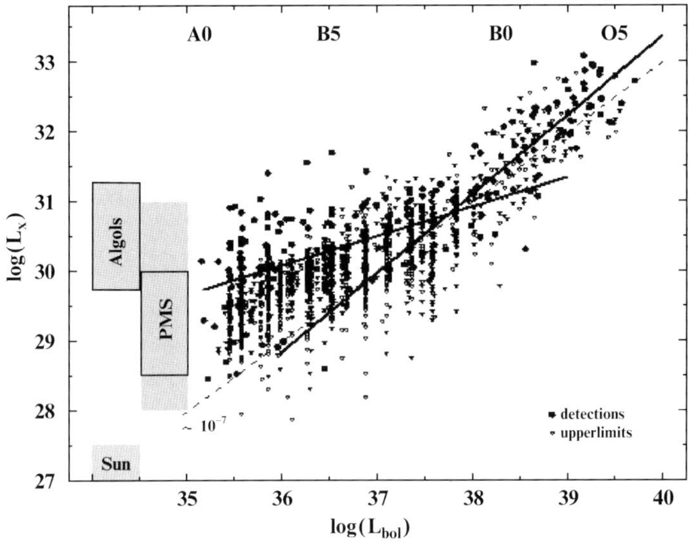

Fig. 10.20 X-ray luminosity vs. bolometric luminosity for a flux-limited sample of OB stars observed during the RASS. *Solid lines* represent regression fits for $L_{\rm bol} < 10^{38}\,{\rm erg\,s^{-1}}$ and $L_{\rm bol} > 10^{38}\,{\rm erg\,s^{-1}}$, and the *dashed line* shows the canonical relation for hot star X-ray emission of $\log(L_{\rm x}/L_{\rm bol}) = -7$. Typical ranges for some classes of late-type stars are indicated by the *bars* on the left (*Fig. 4 from [4]*)

Spectral analysis even of the low spectral resolution Einstein Observatory and ROSAT data show that OB stars produce relatively soft ($kT \sim 0.5\,{\rm keV}$) X-ray emission, which is significantly less variable than the X-ray flux observed from low-mass stars. The observed X-ray emission of early-type stars is widely attributed to a myriad of small shocks, expected to form due to instabilities in the radiatively-driven winds of these hot stars [28]. To arrive at X-ray producing temperatures, the shock velocities must reach a few hundreds kilometers per second, a value easily achieved in O star winds, where typical wind terminal velocities of up to 3000 km s^{-1} are found. Integrating over a multitude of small shocks yields an approximately constant X-ray luminosity, thus explaining the low level of observed variability.

The development of a radiatively accelerated wind depends both on a star's radiation field indicated by its spectral type (or $T_{\rm eff}$) and its gravity. For low temperatures (e.g., $T_{\rm eff} < 10000\,{\rm K}$), only weak-gravity stars can have winds, whereas on the main-sequence, for high temperatures ($T_{\rm eff} > 20,000\,{\rm K}$), winds can exist even for high-gravity stars. In practice, the transition between the weak-wind and fast-wind regimes near the main-sequence takes place for $T_{\rm eff} \simeq 14,000 - 18,000\,{\rm K}$, i.e., between spectral types B3 and B5, in rough agreement with the observed transition in the $L_{\rm x}/L_{\rm bol}$ diagram.

An important question that could not be addressed with the Einstein Observatory and ROSAT data is the production site of the X-ray emission. With terminal velocities of O star winds extending up to $3000\,\mathrm{km\,s^{-1}}$, one expects line broadening depending on the location of the X-ray emission site within the wind. Such line broadening can be resolved with the spectrometers onboard XMM-Newton and Chandra, but a discussion of these new results would go beyond the scope of this chapter.

10.6 Stars with Magnetic Winds

The picture of X-ray emission from stars with winds becomes more complex once magnetic fields are involved. Hybrid models invoking winds disturbed in large-scale magnetospheres have been developed to explain both radio and X-ray emission from early-type stars and have, in particular, been used to describe the observation of hard X-rays from some hot stars that cannot be explained by shocks forming in radiatively driven winds. Direct observational evidence for the existence of magnetic fields on early-type stars is scarce and mostly restricted to the class of chemically peculiar stars, often showing field strengths of a few kG [24]. In "normal," i.e., nonchemically peculiar stars, the strong rotational broadening of their photospheric lines has in most cases prevented the detection of fields with direct spectropolarimetry or Zeeman splitting, indicating that they have much weaker fields, if any. Magnetic fields are often held responsible for the observed cyclic variability of the wind properties usually observed in the UV as well as for the non-thermal radio emission seen in some O stars. Although traditionally magnetic fields on hot stars are believed to be fossil remnants of the star formation process, numerical modeling has shown that dynamos may actually operate in their convective cores [5]. The major difficulty the theory has to face is the transport of magnetic flux throughout the thick radiative layer to the stellar surface where such fields are accessible to observations.

Hybrid models for X-ray emission from early-type stars involving winds and magnetic fields come in two flavors as far as the mechanism for X-ray production is concerned. Both approaches assume an originally dipolar field channeling the stellar wind outward along its lines of force. In the first approach, X-ray emission is the result of magnetic reconnection in a current sheet that forms in the equatorial plane where gas pressure opens the field lines [26]. This scenario can explain the observed nonthermal radio emission of magnetic chemically peculiar stars, and yields expected X-ray luminosities of $\log L_{\rm X}\,[\mathrm{erg\,s^{-1}}] \sim 31 - 34$. These predictions include in particular the range of X-ray luminosities observed from OBA stars.

In the second approach, the magnetically channeled wind is forced down to the equatorial plane, where the winds from the two hemispheres collide and a quasistationary large-scale shock is formed [3]. In this so-called magnetically confined wind shock (MCWS) model, temperatures exceed 10^7 K, radiative cooling of the heated plasma goes along with comparatively hard X-ray emission, and X-ray luminosities even higher than in instability-driven wind shocks can be achieved. Depending on the relative amount of wind and magnetic field energy, the wind may open up the

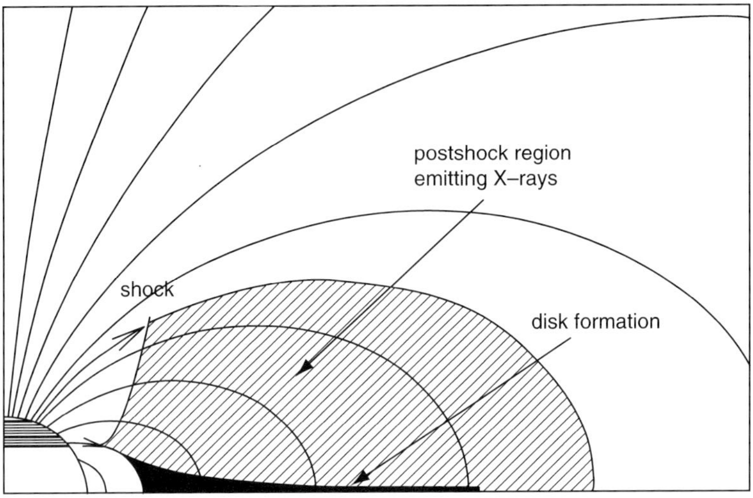

Fig. 10.21 MCWS model devised to explain the X-ray emission from magnetic early-type stars (*Fig. 7 of [3]*). In this representation, X-ray emission comes from a postshock region, which is confined by the magnetic field. Variations of the scenario predict that the field lines open up near the equator under the influence of the wind pressure, and additional X-rays are produced in the outflow region

field lines near the magnetic equator beyond a critical radius. X-ray photons may be emitted both from *closed* magnetic field structures (see Fig. 10.21 representing the original MCWS scenario) or from the open field equatorial outflow at large radii. The cooling material accumulates in a disk near the magnetic equator and may alter the observable X-ray spectrum by absorption.

The MCWS model was originally devised to explain the X-ray emission of the Ap star IQ Aur, but it has also been successfully applied to the young O star θ^1 Ori C [2]. θ^1 Ori C is the only O star with a directly measured magnetic field (1 000 G). It shows a high-amplitude periodic variation of its X-ray intensity, which is extraordinary for an early-type star. Figure 10.22 displays the lightcurve of θ^1 Ori C observed with the ROSAT HRI. Overlaid on the data is the variation predicted by the MCWS model for different viewing angles of the magnetosphere and cooling disk. A similar modulation of the X-ray emission following the period of its magnetic field variations has also been reported from the late-B star HD 133880 with a magnetic field of $\approx$2.5 kG. Although observed to be a rather rare phenomenon, these rotational modulations of the X-ray emission from early-type stars cannot be explained in terms of purely wind-driven X-ray production.

Magnetic fields would also naturally explain the X-ray flaring on early-type stars. Large X-ray flares were observed with ROSAT and XMM-Newton on the chemically peculiar star σ Ori E (spectral type B2 Vp), which has an inferred polar magnetic field in excess of 10 kG. However, whether these flares originate from the magnetic B star or from an as yet unseen late-type companion is subject of some debate

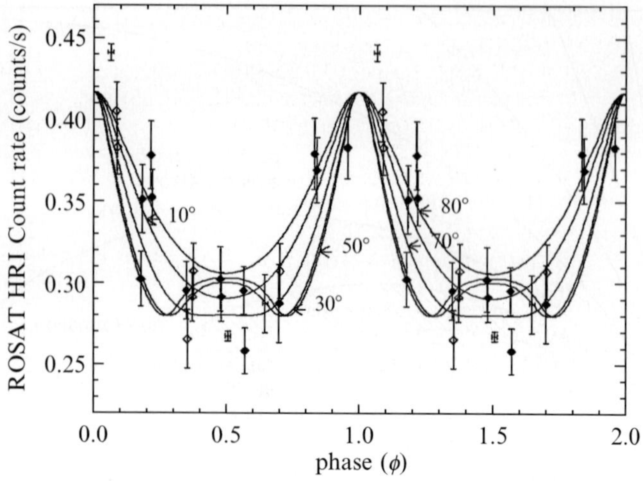

Fig. 10.22 Rotationally modulated X-ray emission from θ^1 Ori C observed by the ROSAT HRI and predictions by the MWCS model for different viewing angles (*Fig. 2 of [2]*)

at the moment. We finally note that X-ray emission has recently been detected from the famous Babcock's star, the noncompact star with the largest hitherto measured magnetic field. These new observational findings are very intriguing and may in fact point at a closer than previously assumed relationship between the X-ray emission from early- and late-type stars.

References

1. Ayres, T.R., Brown, A., Harper, G.M., 2003, Astrophys. J., 598, 610
2. Babel, J., Montmerle, T. 1997a, Astrophys. J. Lett., 485, L29
3. —. 1997b, Astron. Astrophys., 323, 121
4. Berghöfer, T.W., Schmitt, J.H.M.M., Danner, R., Cassinelli, J.P. 1997, Astron. Astrophys., 322, 167
5. Charbonneau, P., MacGregor, K.B. 2001, Astrophys. J., 559, 1094
6. Durney, B.R., DeYoung, D.S., Roxburgh, I.W. 1993, Sol. Phys., 145, 207
7. Favata, F., Micela, G. 2003, Space Sci. Rev., 108, 577
8. Favata, F., Schmitt, J.H.M.M. 1999, Astron. Astrophys., 350, 900
9. Feigelson, E.D., Montmerle, T. 1999, Annu. Rev. Astron. Astrophys., 37, 363
10. Flaccomio, E., Damiani, F., Micela, G., et al. 2003, Astrophys. J., 582, 398
11. Giardino, G., Favata, F., Micela, G., Reale, F. 2004, Astron. Astrophys., 413, 669
12. Gliese, W., Jahreiss, H. 1991, Heidelberg: Astronomisches Recheninstitut
13. Güdel, M. 2004, Astron. Astrophys. Rev., 12, 71
14. Hünsch, M., Schmitt, J.H.M.M., Sterzik, M.F., Voges, W. 1999, Astron. Astrophys. Suppl., 135, 319
15. Haisch, B., Strong, K.T., Rodono, M. 1991, Annu. Rev. Astron. Astrophys., 29, 275
16. Hamaguchi, K., Terada, H., Bamba, A., Koyama, K. 2000, Astrophys. J., 532, 1111

17. Hamaguchi, K., Yamauchi, S., Koyama, K. 2005, Astrophys. J., 618, 360
18. Hoffleit, D. 1982, New Haven: Yale University Press
19. Hünsch, M., Schmitt, J.H.M.M., Schröder, K., Zickgraf, F. 1998, Astron. Astrophys., 330, 225
20. Hünsch, M., Konstantinova-Antova, R., Schmitt, J.H.M.M., Schröder, K.-P., Kolev, D., de Medeiros, J.-R., Lèbre, A. & Udry, S., 2004, in Proceedings IAU Symposium 219, ed., Dupree, A. K. and Benz, A. O., p 223
21. Imanishi, K., Nakajima, H., Tsujimoto, M., Koyama, K., Tsuboi, Y. 2003, Publ. Astron. Soc. Jpn., 55, 653
22. Königl, A. 1991, Astrophys. J., 370, L39
23. Koyama, K., Hamaguchi, K., Ueno, S., Kobayashi, N., Feigelson, E.D. 1996, Publ. Astron. Soc. Jpn., 48, L87
24. Landstreet, J.D. 1992, Astron. Astrophys. Rev., 4, 35
25. Linsky, J. L. 1985, Sol. Phys., 100, 333
26. Linsky, J.L., Drake, S.A., Bastian, T.S. 1992, Astrophys. J., 393, 341
27. Linsky, J.L., Haisch, B.M. 1979, Astrophys. J. Lett., 229, L27
28. Lucy, L.B., White, R.L. 1980, Astrophys. J., 241, 300
29. Martín, E.L., Bouy, H. 2002, New Astron., 7, 595
30. Mohanty, S., Basri, G., Shu, F., Allard, F., Chabrier, G. 2002, Astrophys. J., 571, 469
31. Montmerle, T., Grosso, N., Tsuboi, Y., Koyama, K. 2000, Astrophys. J., 532, 1097
32. Neuhäuser, R. & Comerón, F. 1998, Science, 282, 83
33. Palla, F., Stahler, S.W. 1993, Astrophys. J., 418, 414
34. Pallavicini, R., Golub, L., Rosner, R., et al. 1981, Astrophys. J., 248, 279
35. Parker, E.N. 1955, Astrophys. J., 122, 293
36. Pizzolato, N., Maggio, A., Micela, G., Sciortino, S., Ventura, P. 2003, Astron. Astrophys., 397, 147
37. Pravdo, S.H., Feigelson, E.D., Garmire, G., et al. 2001, Nature, 413, 708
38. Preibisch, T., Zinnecker, H. 2002, Astron. J., 123, 1613
39. Rädler, K.-H., Wiedemann, E., Brandenburg, A., Meinel, R., Tuominen, I. 1990, Astron. Astrophys., 239, 413
40. Rutledge, R. E., Basri, G., Martin, E. L., Bildsten, L., 2000, Astrophys. J., 538, L141
41. Schmitt, J.H.M.M. 1997, Astron. Astrophys., 318, 215
42. Schmitt, J.H.M.M., Kürster, M. 1993, Science, 262, 215
43. Schmitt, J.H.M.M., Liefke, C. 2004, Astron. Astrophys., 417, 651
44. Schmitt, J.H.M.M., Ness, J.-U., Franco, G. 2003, Astron. Astrophys., 412, 849
45. Skumanich, A. 1972, Astrophys. J., 171, 565
46. Stassun, K.G., Ardila, D.R., Barsony, M., Basri, G., Mathieu, R.D. 2004, Astron. J., 127, 3537
47. Stelzer, B. 2004, Astrophys. J. Lett., 615, L153
48. Stelzer, B., Huélamo, N., Hubrig, S., Zinnecker, H., Micela, G. 2003, Astron. Astrophys., 407, 1067
49. Stelzer, B., Neuhäuser, R. 2001, Astron. Astrophys., 377, 538
50. Tout, C.A., Pringle, J.E. 1995, Mon. Not. Roy. Astron. Soc., 272, 528
51. Townsley, L.K., Feigelson, E.D., Montmerle, T., et al. 2003, Astrophys. J., 593, 874
52. Vaiana, G.S., Cassinelli, J.P., Fabbiano, G., et al. 1981, Astrophys. J., 245, 163
53. Vilhu, O. 1984, Astron. Astrophys., 133, 117
54. Zinnecker, H., Preibisch, T. 1994, Astron. Astrophys., 292, 152

11 White Dwarfs

K. Werner

11.1 Introduction

Thermal soft X-ray emission is detected from many hot hydrogen-rich white dwarfs (spectral type DA) with an effective temperature in excess of 20 000 K. Most of the objects with $T_{\rm eff} < 40000$ K have virtually pure hydrogen atmospheres while the majority of the hotter ones emit X-ray fluxes lower than predicted by hydrogen model atmospheres and, therefore, must contain heavier elements as absorbers. From ROSAT and EUVE observations it was concluded that trace metals sustained by radiative levitation increase the atmospheric X-ray opacity.

ROSAT also detected a smaller number of helium-rich white dwarfs (spectral type DO), as well as some of their immediate progenitors, the PG 1159 stars, whose photospheres are dominated by C, He, and O. Their effective temperatures exceed 100 000 K and 140 000 K, respectively. The hottest object (H1504+65, $T_{\rm eff} = 200\,000$ K) is unique and possibly a naked C–O stellar core. It exhibits an extremely rich absorption-line spectrum as revealed by EUVE and Chandra.

The origin of hard X-ray emission possibly detected in a number of hot white dwarfs remains unclear.

11.2 Discovery of X-Rays from White Dwarfs

White dwarfs (WD) are the final stage of evolution of low- and intermediate-mass stars up to about $8\,M_\odot$. It is believed that over 97% of the stars in the Galaxy will eventually end up as white dwarfs. On the basis of optical spectroscopy, WDs are grouped in two distinct classes, DA and non-DA white dwarfs. The optical spectra of WDs of type DA exhibit pure hydrogen Balmer line spectra whereas non-DA WDs show pure helium line spectra. The purity of WD atmospheres is the consequence of gravitational settling of heavy elements because of the high surface gravity. The vast majority of WDs evolves from asymptotic giant branch (AGB) stars passing through the phase of central stars of planetary nebulae. In particular, the progenitors of DA and non-DA white dwarfs are thought to be hydrogen-rich and hydrogen-poor central stars, respectively. The H-rich central stars evolve according to canonical stellar evolution theory, while the H-deficient central stars are probably the result of a late

J.E. Trümper and G. Hasinger (eds.), *The Universe in X-Rays.*

helium-shell flash suffered by a WD in its early cooling phase. As a consequence of this late flash, the WD is transformed back into a Red Giant ("born-again AGB star") and retraces the post-AGB evolution a second time. The H-deficiency is caused by the flash, which induces convective mixing of the whole stellar envelope. Hydrogen is ingested and burned while He, C, and O-rich material is dredged up to the surface. When the star enters the WD phase again, helium as the lightest element floats atop and the star turns into a non-DA white dwarf.

According to model atmosphere calculations, the youngest, hence hottest, WDs are emitters of thermal soft X-ray radiation. In particular, this holds for DA white dwarfs. Hydrogen is strongly ionised so that the soft X-ray opacity is strongly reduced. This, and the absence of any other absorbing species allows leaking of X-ray photons from the hot deep interior of WDs. However, as we will describe in the following, traces of metals can be retained in hot DA atmospheres by radiative levitation and strongly increase the X-ray opacity.

While the majority of known X-ray emitting white dwarfs are of spectral type DA, a small number of hot non-DA white dwarfs (spectral type DO) as well as their immediate progenitors, the H-deficient central stars of spectral type PG 1159, were detected (or even discovered) by ROSAT. We will discuss these objects in a separate section.

Thermal radiation from hot WDs has been detected in a wavelength range of roughly 30–700 Å, which covers the soft X-ray (say, $\lambda < 100$ Å) and EUV regions ($\lambda > 100$ Å) of the electromagnetic spectrum. Therefore, although we will concentrate in this review on observations performed with X-ray telescopes, we will also discuss results from EUV instruments, namely the Wide Field Camera (WFC) aboard ROSAT and the Extreme Ultraviolet Explorer (EUVE).

Sirius B and HZ 43 were the first white dwarfs from which X-rays were detected, using the Astronomical Netherlands Satellite (ANS) and the Third Small Astronomy Satellite (SAS-3) [21, 34]. HZ 43 and Feige 24 were detected with the EUV Telescope during the *Apollo-Soyuz* mission [29, 31]. HZ 43 also was the first WD of which a soft X-ray spectrum (albeit of poor energy resolution) was obtained in the 50–200 Å range, with a rocket-borne instrument [9]. These very first X-ray observations were useful to constrain the temperatures of the hot WDs, which were very poorly known at that time.

The first X-ray sky survey, which was sensitive to photon energies below 1.5 keV (>8 Å) and potentially able to detect hot WDs, was performed with the low energy detectors of the A-2 experiment aboard the HEAO-1 satellite [38]. Five DAs were detected (e.g., [26]) plus one strong X-ray source, which later turned out to be the unique non-DA H1504+65 (see later).

New insight into the physics of WD atmospheres was obtained with the first X-ray space observatories, Einstein and EXOSAT, launched in 1978 and 1983, respectively. Aside from three WD spectra recorded with the EXOSAT TGS (Transmission Grating Spectrometer), all observations were photometric. Soft X-rays were detected from four WDs with the Einstein IPC (Imaging Proportional Counter), and two with the HRI (high resolution imager), both instruments having a single pass band (3–100 Å). Generally, the observed X-ray fluxes were significantly lower than predicted [26, 41]. This was interpreted as the presence of homogeneously mixed helium in the DA atmospheres, increasing the X-ray opacity compared with pure hydrogen atmospheres. There was weak evidence for an increasing helium abundance with increasing T_{eff} and it was discussed that radiative levitation is responsible for this.

Twenty WDs were observed photometrically with EXOSAT CMA (Channel Multiplier Array) using several filters with high sensitivity in the range 44–400 Å. The previous Einstein results were essentially confirmed and it was claimed that He/H number ratios of up to 10^{-5} in the atmospheres can explain the measured X-ray fluxes [23, 28, 40]. On the other hand, it was shown that the data are fit as well with models assuming very thin pure hydrogen layers, with varying thickness from star to star, on a helium-envelope [27].

Of particular importance was the spectral observation of three WDs with the EXOSAT TGS. The spectra of HZ 43 and Sirius B could be fitted well by pure H model atmospheres. In contrast, the flux distribution of Feige 24 required the presence of metals as absorbers. Helium could be excluded, because the He II ground state edge (228 Å) was not detected. This was the first proof that metals can be present in hot WD atmospheres [48]; however, the spectral resolution was insufficient to identify unambiguously any particular species. In the end it was not clear if the metal-free atmospheres of HZ 43 and Sirius B or the metal-contaminated atmosphere of Feige 24 are representative for WDs at large. The presence of metals in Feige 24, which is a binary system, could be either due to radiative levitation or due to wind accretion from the WD companion.

11.3 ROSAT

Taking the point of view that objects like Feige 24 are the exception and that all DA white dwarfs have pure hydrogen atmospheres, it was estimated that about 5500 WDs should be detected in the ROSAT all sky survey [1]. One of the biggest surprises in the field of stellar X-ray astronomy provided by ROSAT was that this number was not even close. X-ray emission from hot white dwarfs is the exception rather than the rule [20]. Only 175 WDs were detected in the PSPC (Position Sensitive Proportional Counter) survey. Most of them (161) are of type DA, with an additional eight PG 1159 stars, three DOs (one is unconfirmed), and three DAOs (mixed H/He optical spectra).

Among the first WDs whose ROSAT all sky survey data were analysed was G191-B2B. Like many other hot DAs observed with EXOSAT, it showed an X-ray flux deficit, which was interpreted as either due to a stratified H/He atmosphere or due to the presence of trace metals in the photosphere. ROSAT photometry with the PSPC integrated count rate (25–100 Å) and the two survey filters of the WFC (60–140 Å and 112–200 Å bandwidth) revealed that trace metals must be a substantial source of opacity [4]. This was confirmed by an EUV spectrum obtained with a rocket-borne instrument, which shows no sign of the He II 228 Å ground-state edge [60].

Larger samples from the ROSAT all sky survey and the subsequent pointed phase of the mission were analysed in a similar manner by several groups. Barstow et al. [3] presented a sample of 30 previously catalogued DAs and concluded that most objects with $T_{\rm eff} > 50\,000$ K contain trace heavy elements in their atmospheres as a result of radiative levitation. Most DAs with $T_{\rm eff} < 40\,000$ K contain little or no material in their atmospheres other than H, indicating that radiative levitation is unimportant in the cooler WDs. Many DAs with $T_{\rm eff} = 50\,000$–$60\,000$ K were not detected by ROSAT and no DA with $T_{\rm eff} > 60\,000$ K. It was speculated that radiative levitation of metals becomes so strong with increasing $T_{\rm eff}$ that the soft X-ray and EUV flux is effectively blocked. Similar conclusions were drawn by other studies [24,32,61]. In the latter one it was shown that the PSPC pulse height distribution can be used to determine consistently the DA's $T_{\rm eff}$ and the interstellar H column density along the line of sight.

It is important that a number of objects with pure H atmospheres with $T_{\rm eff}$ significantly *above* 40 000 K were discovered in the sample presented by Marsh et al. [32]. Few such soft X-ray emitters were known before, the most famous one is HZ 43 [37]. Apparently some still unidentified competing physical process counteracts the equilibrium of radiative levitation and gravitational settling of metals in these particular white dwarfs.

The ROSAT survey has led to the discovery of a significant number of previously uncatalogued WDs [20], identified by the combined effort of several groups. [33] announced the identification of 69 previously unknown WDs detected in the WFC survey. Such follow-up observations revealed three DAs with $T_{\rm eff}$ in excess of 60 000 K, one of it (RE J1738+665) at 88 000 K, making it one of the hottest known DAs [5, 22]. High resolution UV spectroscopy with the *International Ultraviolet Explorer* confirmed that many of the hot DAs contain significant quantities of Fe and Ni in their photospheres [49,53].

Hard X-ray emission (>0.5 keV, i.e. <25 Å) from single white dwarfs is not expected. However, a hard spectral component peaking at 0.8 keV was detected from the central star of the Helix planetary nebula (NGC 7293) [30], which is a hot white dwarf of spectral type DAO with $T_{\rm eff}$ exceeding 100 000 K [36, 44]. A systematic search in the ROSAT X-ray point source catalog of [59] revealed that five other single WDs emit hard X-rays [14, 39]. Two of them are prominent objects and also very hot: KPD 0005 + 5106 (the hottest known DO WD, $T_{\rm eff}$ = 120 000 K, see below) and the prototype of the PG 1159 class PG 1159-035 ($T_{\rm eff}$ = 140 000 K [55]). The remaining three hard X-ray sources are DA white dwarfs [39]. It is intriguing that 50% of the hard X-ray emitting objects are among the hottest WDs. It is speculated that the emission is of thermal origin from very deep photospheric layers; however, this is not confirmed by current models. High-resolution deep X-ray observations are needed to verify the positional coincidence and to study the nature of the hard X-ray emission.

11.4 X-Ray Spectroscopy with EUVE, Chandra, and XMM-Newton

The next significant step forward to understand soft X-ray and EUV spectra of white dwarfs was taken with spectroscopic observations in the 80–760 Å range performed by EUVE, with a moderate resolution of $\lambda/\Delta\lambda \approx 200$. Many chemical elements in hot white dwarfs have no strong spectral lines in the UV or optical and can only be detected in the EUV/soft X-ray region by their line and bound-free opacities. The spectral resolution of EUVE was however still too low, so that an unambiguous identification of any spectral line or absorption edge in a DA could not be successful. The blanketing of millions of Fe and Ni lines, whose exact wavelength position is unknown, poses the greatest problem. We cannot discuss all results from EUVE spectroscopy in detail and restrict ourselves to a summary of the most important achievements. An excellent overview is given e.g., in [2].

Basically, all of the observed single DA white dwarfs show no evidence for the presence of photospheric helium. The nondetection of the He II resonance line series and the ground state edge in the 228–304 Å range allows to determine a rather tight upper limit for the helium abundance (He/H $\approx 10^{-6}$), which is two orders of magnitudes tighter than possible from optical or UV spectroscopy. For HZ 43 an upper limit of $3 \cdot 10^{-7}$ was found [7], which, however, must be relaxed to $3 \cdot 10^{-6}$ if non-LTE models are used, because these predict weaker He II lines than LTE models [35]. This upper limit fits to diffusion calculations as presented in [47], which predict that radiative levitation can sustain just that amount of He in the atmosphere of HZ 43.

GD 50 is an exceptional single DA white dwarf ($T_{\rm eff}$ = 40 300 K, surface gravity $\log g = 9$) showing a strong He II Lyman series, from which He/H $= 2.4 \cdot 10^{-4}$ was derived [50]. This value is about four orders of magnitudes larger than predicted by diffusion theory. GD 50 is an ultramassive WD, one of the most massive known (1.2 $M_\odot$ [8]) and believed to be the result of a stellar merger. It is probably a fast rotator and, hence, helium could be sustained in the atmosphere by large meridional currents induced by rotation.

Explaining the observed EUVE spectra of DAs that have metal absorbers is very difficult. It is virtually impossible to determine uniquely the mix of metals responsible for the observed flux distribution. In objects with relatively weak abundances of light metals (C, N, O, Si of the order 10^{-6}–10^{-7}), it is possible to derive at least upper limits for these species [7]. Such fits can be guided by abundance determinations from high-resolution UV spectroscopy, but often the results from EUV and UV analysis give discrepant results. The reason most probably is the use of homogeneously mixed atmosphere models, thus one neglects the vertical stratification of metals. The situation becomes even more difficult in the hottest DAs that have atmospheres with high heavy element abundances (e.g., Fe/H $= 2.5 \cdot 10^{-6}$, which is almost the solar value, in G191-B2B [45]). Their EUVE spectra are dominated by extensive blanketing of millions of lines from Fe and Ni.

The only rigorous way out of this complicated situation is the construction of a new type of model atmospheres, which considers self-consistently the vertical stratification of chemical abundances. This is done by solving the equations that describe the equilibrium between gravitational settling and radiative acceleration of ions through photon absorption in spectral lines and continua [12]. This implies that the only model parameters are $T_{\rm eff}$ and $\log g$, which can in principle be determined from optical or UV spectra by evaluating the hydrogen Balmer and Lyman lines, respectively. The metal abundances are no longer free parameters but are computed on physical grounds. The modeling needs to be performed under non-LTE conditions, because it is well known that deviations from LTE can have significant effects on hot WD spectra. Such models were first constructed by [18] and successfully applied to the EUVE spectrum of G191-B2B. Consequently, a systematic analysis of EUVE spectra of 26 DA white dwarfs was performed by [42, 43]. Generally, good agreement between observed and modeled spectra is achieved (Fig. 11.1); however, the models fail for a number of individual objects. The reason for this failure might be the neglect of physical processes in the models that affect the assumed equilibrium between settling and levitation. The most likely of these processes are very weak mass-loss (spectroscopically undetectable) from the WD and accretion of matter from the interstellar medium.

Let us shortly remark on white dwarfs in detached binaries (so-called Sirius-like systems) because they are not treated elsewhere in this book. These WDs do not significantly contribute to the optical flux if the spectral type of their companion is K or earlier. They can be discovered, however, in the UV and/or soft X-ray ranges because they are the brighter component. Only a handful of Sirius-like systems was known before; the WDs were detected serendipitously by the observation of eclipses or signatures in the UV spectra taken with IUE. Through the ROSAT WFC and EUVE surveys, over 20 new Sirius-like systems were discovered [11]. Hence,

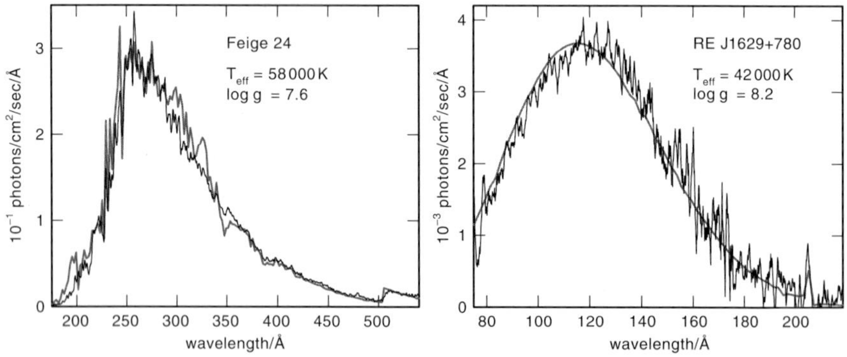

Fig. 11.1 Spectra of two DA white dwarfs taken with the *Extreme Ultraviolet Explorer* together with model fits (*thick lines*). *Left*: The soft X-ray flux of Feige 24 is effectively blocked below 200 Å by trace metals, which are kept in the hot atmosphere by radiative levitation. *Right*: The cooler WD RE J1629+780 is essentially metal free and, therefore, emits significant flux even below 100 Å. Data taken from [42]

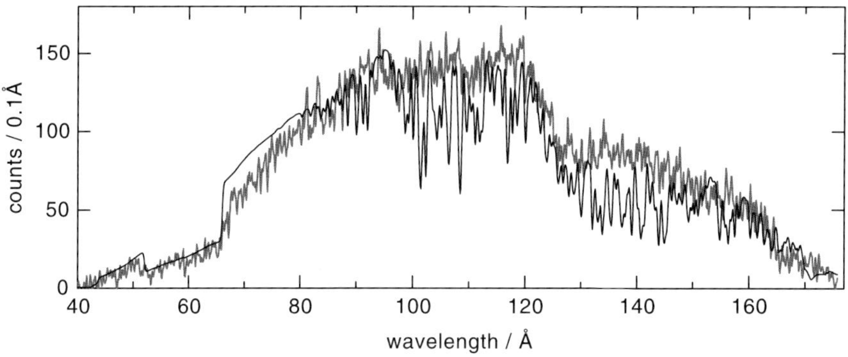

Fig. 11.2 Chandra spectrum of LB1919 (*thin line*). Overplotted is a model including iron and nickel lines (from [58])

these hidden white dwarfs represent an important, previously unaccounted for, contribution to the total space density of white dwarfs. Of particular significance is the detection of three bright B stars with hot WD companions [10]. These companions are important since they set an observational limit on the maximum mass for WD progenitors, and thus can potentially be used to investigate the high-mass end of the initial-final-mass relation and the WD mass-radius relation.

Observations of isolated WDs with Chandra and XMM-Newton are scarce. The Chandra LETG (Low Energy Transmission Grating) spectrum of HZ 43 (50–170 Å) together with EUV and UV spectra is perfectly matched by a pure hydrogen atmosphere, confirming previous results. The same holds for Sirius B. More challenging is the LETG spectrum of GD 246, a hot DA with significant trace metal abundances as derived from detailed studies with EUVE and (F)UV telescopes. Thanks to the superior spectral resolution of Chandra over EUVE, individual spectral lines (from Fe and Ni) could be identified in the soft X-ray spectrum of a DA star for the first time [46]. Figure 11.2 shows the Chandra LETG spectrum of the hot DA LB1919, which is dominated by a forest of Fe and Ni lines. XMM-Newton has observed two DAs, G191-B2B and GD 153, for calibration purposes [13] but no detailed results are available, yet. G191-B2B, as discussed earlier, contains trace metals and, therefore, is a less suitable calibration object than the pure-hydrogen WD GD 153.

11.5 Hydrogen-Deficient White Dwarfs

Hot helium-rich white dwarfs (spectral type DO) must be much hotter than DAs to emit soft X-rays blueward of the He II ground state edge (<228 Å). Only at $T_{\rm eff}$ as high as 100 000 K the He II opacity is sufficiently reduced because the ionisation balance shifts in favor of He III. Therefore, only two of the hottest (previously known) DOs were detected in the ROSAT PSPC survey [51]: PG 1034+001 (which was also detected by EXOSAT [40], $T_{\rm eff}$ = 100 000 K) and KPD 0005+5106, the hottest known DO white dwarf ($T_{\rm eff}$ = 120 000 K). The PSPC spectrum of the latter was interpreted as emission from a relatively cool corona about the WD [19]. A recent analysis of Chandra observations employing new model atmospheres, however, suggests a thermal photospheric origin of the observed 20–80 Å flux [16].

One out of two cooler DOs that lies in a region with very low interstellar H I density was discovered in the ROSAT WFC survey through the detection of EUV flux redward of the He II 228 Å edge: RE J0503-289 [6] ($T_{\rm eff}$ = 70 000 K). This object is remarkable because it is the only DO detected in both the WFC and EUVE sky surveys. Although many strong absorption features can be identified in the EUVE spectrum and attributed to He, C, and O, the spectral fit with models is unsatisfactory [56]. The main problem is the strong blanketing by nickel lines of which the majority has inaccurately known wavelength positions. Also, an unknown mix of other metals probably affects the EUV flux distribution. Fits to the EUV spectrum give results for $T_{\rm eff}$ and metal abundances that are inconsistent with optical and UV spectra. Ultimately, self-consistent diffusion models like in the case of DAs should be attempted; however, these models up to now completely fail to explain the metal abundances observed in the UV spectra of DO white dwarfs [17].

Only one other DO was detected in the EUV at $\lambda > 228$ Å, namely HD149499B, although it escaped detection in the WFC and EUVE sky surveys. This WD is important because it is the coolest known DO, and in the HRD it is located at the hot boundary of the DB gap in which no He-rich WD is known. Hence it marks the location where He-rich atmospheres are transformed into H-rich ones and in fact, this WD contains a significant amount of hydrogen in its atmosphere (H/He = 0.22). The reason for the occurrence of the DB gap is a longstanding and still unanswered question. Therefore, the precise parameters of this star are of considerable interest, but optical spectroscopy is problematic because of the bright companion. Flux was detected in an EUVE spectrum at 230–360 Å and $T_{\rm eff}$ could be determined with very high precision (49 500 K [25]). This result is in full agreement with optical and FUV spectroscopy so that the situation is satisfactory, in contrast to that of RE J0503-289 discussed earlier. Apparently, the surprisingly low metal content in HD149499B allows to derive consistent results with current models composed only of H and He.

PG 1159 stars, the DO progenitors, have high C and O abundances in addition to helium in their photospheres so that they are thermal X-ray emitters only at even higher temperatures ($T_{\rm eff} \geq 140\,000$ K). Eight objects were detected in the ROSAT PSPC survey, two are remarkable new discoveries, which turned out to be among the hottest objects of their class [51]. One of them, RX J2117.1+3412, is a bright central star of a planetary nebula and still very luminous (about 10^4 $L_\odot$) and hot ($T_{\rm eff}$ = 170 000 K). It is a nonradial g-mode pulsator (a GW Vir variable) and became a prime target for asteroseismology analyses [15]. PG 1520 + 525, also a central star, is the only PG 1159 star (except for H1504+65, see below) for which a reasonable

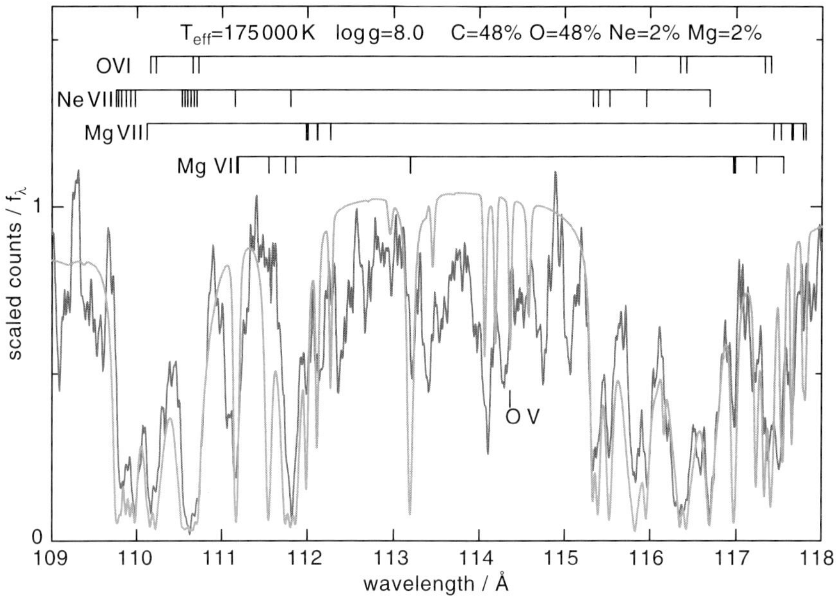

Fig. 11.3 Detail from the Chandra spectrum of H1504+65 (*thin line*). Overplotted is a model with parameters as given in the figure. Data from [57]

EUVE spectrum could be obtained. Flux is detected in a very narrow region, 100–130 Å, and it was used to estimate $T_{\rm eff}$ = 150 000 K. The star is a nonpulsator so that this result in combination with the temperature of the pulsating prototype PG 1159-035 (= GW Vir, 140 000 K) constrains the location of the blue edge of the GW Vir instability strip in the HRD [52]. This result was confirmed later by a Chandra LETG observation [58].

The most bizarre X-ray emitting WD is H1504+65. It was discovered as one of the brightest soft X-ray sources in the sky in the HEAO-1 survey [38]. Optical follow-up spectroscopy revealed a hot WD closely related to the PG 1159 class. It is the hottest WD known ($T_{\rm eff}$ = 200 000 K) and has a unique surface photospheric composition. It is H- *and* He-free and mainly composed of C and O by equal parts. The EUVE spectrum shows flux in the 75–150 Å range and is dominated by strong and broad O VI and Ne VII absorption lines [54]. Chandra has taken a spectacular LETG spectrum, arguably the richest X-ray absorption line spectrum recorded so far (Fig. 11.3). Many lines from several ionisation stages of Ne and Mg can be seen and high abundances were derived: Ne = Mg = 2% by mass. It is speculated that H1504+65 is a naked C–O white dwarf or even a WD with a O–Ne–Mg core resulting from carbon burning [57].

References

1. Barstow, M.A. 1989, in G. Wegner (ed.), White Dwarfs, Springer, Berlin, Heidelberg, New York, 156
2. Barstow, M.A., Holberg, J.B. 2003, Extreme Ultraviolet Astronomy, Cambridge University Press, Cambridge
3. Barstow, M.A., Fleming, T.A., Diamond, C.J., Finley, D.S., Sansom, A.E., Rosen, S.R., Koester, D., Marsh, M. C., Holberg, J.B., Kidder, K. 1993, Mon. Not. Roy. Astron. Soc., 264, 16
4. Barstow, M. A., Fleming, T. A., Finley, D. S., Koester, D., Diamond, C. J. 1993, Mon. Not. Roy. Astron. Soc., 260, 631
5. Barstow, M.A., Holberg, J.B., Marsh, M.C., Tweedy, R.W., Burleigh, M.R., Fleming, T.A., Koester, D., Penny, A.J., Sansom, A.E. 1994, Mon. Not. Roy. Astron. Soc., 271, 175
6. Barstow, M.A., Holberg, J.B., Werner, K., Buckley, D.A.H., Stobie, R.S. 1994, Mon. Not. Roy. Astron. Soc., 267, 653
7. Barstow, M.A., Dobbie, P.D., Holberg, J.B., Hubeny, I., Lanz, T. 1997, Mon. Not. Roy. Astron. Soc., 286, 58
8. Bergeron, P., Kidder, K.M., Holberg, J.B., Liebert, J., Wesemael, F., Saffer, R.A. 1991, Astrophys. J., 372, 267
9. Bleeker, J.A.M., Davelaar, J., Deerenberg, A.J.M., Huizenga, H., Brinkman, A.C., Heise, J., Tanaka, Y., Hayakawa, S., Yamashita, K. 1978, Astron. Astrophys., 69, 145
10. Burleigh, M.R., Barstow, M.A. 2000, Astron. Astrophys., 359, 977
11. Burleigh, M., Barstow, M., Good, S., Bond, H.E., Holberg, J. 2003, in D. De Martino, R. Silvotti, J.-E. Solheim, R. Kalytis (eds.), White Dwarfs, Kluwer, Dordrecht, 341
12. Chayer, P., Fontaine, G., Wesemael, F. 1995, Astrophys. J. Suppl. Ser., 99, 189
13. Chen, B., Schartel, N., Kirsch, M.G.F., Smith, M.J.S., Altieri, B., Pollock, A.M.T. 2004, Memorie della Societa' Astronomical Italiana (MmSAI), 75, 561
14. Chu, Y.-H., Gruendl, R.A., Williams, R.M., Gull, T.R., Werner, K. 2004, Astron. J., 128, 2357
15. Córsico, A.H., Althaus, L.G., Miller Bertolami, M.M., Werner, K. 2007, Astron. Astrophys., 461, 1095
16. Drake, J.J., Werner, K. 2005, Astrophys. J., 625, 973
17. Dreizler, S. 1999, Astron. Astrophys., 352, 632
18. Dreizler, S., Wolff, B. 1999, Astron. Astrophys., 348, 189
19. Fleming, T.A., Werner, K., Barstow, M.A. 1993, Astrophys. J., 416, L79
20. Fleming, T.A., Snowden, S.L., Pfeffermann, E., Briel, U., Greiner, J. 1996, Astron. Astrophys., 316, 147
21. Hearn, D.R., Richardson, J.A., Bradt, H.V.D., Clark, G.W., Lewin, W.H.G., Mayer, W.F., McClintock, J.E., Primini, F.A., Rappaport, S.A. 1976, Astrophys. J., 203, L21
22. Holberg, J.B., Barstow, M.A., Buckley, D., Chen, A., Dreizler, S., Marsh, M.C., O'Donoghue, D., Sion, E.M., Tweedy, R.W., Vauclair, G., Werner, K. 1993, Astron. Astrophys., 416, 806
23. Jordan, S., Koester, D., Wulf-Mathies, C., Brunner, H. 1987, Astron. Astrophys., 185, 253
24. Jordan, S., Wolff, B., Koester, D., Napiwotzki, R. 1994, Astron. Astrophys., 290, 834
25. Jordan, S., Napiwotzki, R., Koester, D., Rauch, T. 1999, Astron. Astrophys. 318, 461
26. Kahn, S.M., Wesemael, F., Liebert, J., Raymond, J.C., Steiner, J.E., Shipman, H.L. 1984, Astrophys. J., 278, 255
27. Koester, D. 1989, Astrophys. J., 342, 999
28. Koester, D., Beuermann, K., Thomas, H.-C., Graser, U., Giommi, P., Tagliaferri, G. 1990, Astron. Astrophys., 239, 260
29. Lampton, M., Margon, B., Paresce, F., Stern, R., Bowyer, S. 1976, Astrophys. J., 203, L71
30. Leahy, D.A., Zhang, C.Y., Kwok, S. 1994, Astrophys. J., 422, 205
31. Margon, B., Lampton, B., Bowyer, S., Stern, R., Paresce, F. 1976, Astrophys. J., 210, L79
32. Marsh, M.C., Barstow, M.A., Buckley, D.A., Burleigh, M.R., Holberg, J.B., Koester, D., O'Donoghue, D., Penny, A.J., Sansom, A.E. 1997, Mon. Not. Roy. Astron. Soc., 287, 705

33. Mason, K.O., Hassall, B.J.M., Bromage, G.E., et al. 1995, Mon. Not. Roy. Astron. Soc., 274, 1194
34. Mewe, R., Heise, J., Gronenschild, E.H.B.M., Brinkman, A.C., Schrijver, J., den Boggende, A.J.F. 1975, Astrophys. J., 202, L67
35. Napiwotzki, R. 1995, in D. Koester, K. Werner (eds.) White Dwarfs, Springer, Berlin, Heidelberg, New York, 132
36. Napiwotzki, R. 1999, Astron. Astrophys., 350, 101
37. Napiwotzki, R., Barstow, M.A., Fleming, T., Holweger, H., Jordan, S., Werner, K. 1993, Astron. Astrophys., 278, 478
38. Nugent, J.J., Jensen, K.A., Nousek, J.A., Garmire, G.P., Mason, K.O., Walter, F.M., Bowyer, C.S., Stern, R.A., Riegler, G.R. 1983, Astrophys. J. Suppl., 51, 1
39. O'Dwyer, I.J., Chu, Y.-H., Gruendl, R.A., Guerrero, M.A., Webbink, R.F. 2003, Astrophys. J., 125, 2239
40. Paerels, F.B.S., Heise, J. 1989, Astrophys. J., 339, 1000
41. Petre, R., Shipman, H.L., Canizares, C.R. 1986, Astrophys. J., 304, 356
42. Schuh, S.L. 2005, Dissertation, University of Tübingen, Germany
43. Schuh, S.L., Dreizler, S., Wolff, B. 2002, Astron. Astrophys., 382, 164
44. Traulsen, I., Hoffmann, A.I.D., Rauch, T., Werner, K., Dreizler, S., Kruk, J.W. 2005, in D. Koester, S. Moehler (eds.), 14th European Workshop on White Dwarfs, ASP Conference Series, 334, 325
45. Vennes, S., Lanz, T. 2001, Astrophys. J., 553, 399
46. Vennes, S., Dupuis, J. 2002, in E.M. Schlegel, S.D. Vrtilek (eds.), The High Energy Universe at Sharp Focus: Chandra Science, ASP Conference Series, 262, 57
47. Vennes, S., Pelletier, C., Fontaine, G., Wesemael, F. 1988, Astrophys. J., 331, 876
48. Vennes, S., Chayer, P., Fontaine, G., Wesemael, F. 1989, Astrophys. J., 336, L25
49. Vennes, S., Chayer, P., Thorstensen, J.R., Bowyer, S., Shipman, H.L. 1992, Astrophys. J., 392, L27
50. Vennes, S., Bowyer, S., Dupuis, J. 1996, Astrophys. J., 461, L103
51. Werner, K. 1996, in H.U. Zimmermann, J.E. Trümper, H. Yorke (eds.), Röntgenstrahlung from the Universe, MPE Report, 263, 205
52. Werner, K., Dreizler, S., Heber, U., Rauch, T. 1996, in S. Bowyer, R.F. Malina (eds.), Astrophysics in the Extreme Ultraviolet, Kluwer, Dordrecht, 229
53. Werner, K., Dreizler, S. 1994, Astron. Astrophys., 286, L31
54. Werner, K., Wolff, B. 1999, Astron. Astrophys., 347, L9
55. Werner, K., Heber, U., Hunger, K. 1991, Astron. Astrophys., 244, 437
56. Werner, K., Deetjen, J. L., Rauch, T., Wolff, B. 2001, in J.L. Provencal, H.L. Shipman, J. MacDonald, S. Goodchild (eds.), White Dwarfs, ASP Conference Series, 226, 55
57. Werner, K., Rauch, T., Barstow, M.A., Kruk, J.W. 2004, Astron. Astrophys., 421, 1169
58. Werner, K., Drake, J.J., Rauch, T., Schuh, S., Gautschy, A. 2007, in R. Napiwotzki, M.R. Burleigh (eds.), 15th European Workshop on White Dwarfs, ASP Conference Series, in press (astro-ph/0610013)
59. White, N., Giommi, P., Angelini, L. 2000, The WGA Catalogue of ROSAT Point Sources, GSFC, Greenbelt
60. Wilkinson, E., Green, J.C., Cash, W. 1992, Astrophys. J., 397, L51
61. Wolff, B., Jordan, S., Bade, N., Reimers, D. 1995, Astron. Astrophys., 294, 183

12 X-Ray Emission of Cataclysmic Variables and Related Objects

K. Beuermann

12.1 Historical Introduction

Cataclysmic variables (CVs) are mass-transfer binaries, in which the accretor is a white dwarf. The secondary can be a normal star with a hydrogen-rich envelope or a second white dwarf. Energy released in the gravitational field of the accreting white dwarf powers the phenomena observed in cataclysmic variables, while subsequent nuclear processing of the accreted matter leads to various types of nova events and supersoft X-ray sources (SSS). Much of the physics of CVs and the related objects is basically similar to that of X-ray binaries, in which the accretor is a neutron star or a black hole. When Sco X-1 was discovered in 1961, CVs were already known [11, 73, 93], and after its optical identification with a faint blue star, Shklovsky suggested that in Sco X-1 "we observe a binary system similar to WZ Sagittae in which one of the components is a neutron star" [79]. The nature of the bright galactic X-ray sources was finally settled by the exciting results obtained with Uhuru (1970–1973) as reminisced by Giacconi [19]. At least three CVs, 4U1228–29 (EX Hya) [94], 4U1809+50 (AM Her), and 4U 1849–31 (V1223 Sgr), were found among the fainter sources in the 4U-catalog [14]. The optical identification of the faint X-ray sources from Uhuru and other early missions was a tedious process, however, because of the comparatively low angular resolution and sensitivity of the early X-ray missions and the visual faintness of the CVs. A breakthrough was the identification of 4U 1809+50 with the long-known 12–15 mag variable star AM Her and Tapia's [84] discovery that AM Her was circularly polarized. The object was immediately interpreted as containing a strongly magnetic accreting white dwarf and became the prototype of the subclass of *polars*, a term coined by Krzeminski [41].

Nova events are regular cataclysms experienced by CVs during their long life. Close-binary supersoft X-ray sources (CBSS) were predicted in 1978 as "nonejecting novae" [78, see also [64]], discovered in 1981 with Einstein [49], but undisputedly accepted as such only after detailed studies with ROSAT [21, 87, 88]. Soft X-ray emission from a nova outburst was first seen in GQ Mus (N Mus 1983) [40]. Because of absorption in the Galactic plane, few SSS were found in the Galaxy, but some were subsequently identified among known CVs [21]. More recently, population studies of SSS have been conducted in external galaxies with ROSAT, Chandra, and XMM-Newton [24, 87].

J.E. Trümper and G. Hasinger (eds.), *The Universe in X-Rays.*

12.2 The Zoo of CVs

CVsz can be classified according to the nature of the mass accretor and the mass donor. The accretor is, by definition, a white dwarf, the donor can be a normal nuclear burning star, a substellar object (brown dwarf), or a white dwarf. In many systems, the nuclear burning stars are only moderately evolved and I refer to them loosely as main sequence stars. The physics of the accreting white dwarf is similar in symbiotic binaries, in which the secondary is an evolved star on the first giant branch or on the asymptotic giant branch (AGB). The phenomenology of these systems, however, is dominated by the emissions of the secondary or the wind and X-rays are often severely quenched. I refer to symbiotic binaries only in passing.

Table 12.1 lists the number of known members of a CV subclass, the fraction discovered by their X-ray emission, the name of the subclass, and the observed range of the X-ray vs. optical flux ratio F_X/F_V. The present census contains about 400 nonmagnetic and 130 magnetic CVs with known orbital period. *Nonmagnetic CVs* include all systems in which the white dwarf is not specifically recognized as magnetic. They accrete via an accretion disk and the observed soft and hard X-rays originate from the boundary layer (BL) between disk and white dwarf [43, 65, 93, and references therein]. *Polars (AM Her stars)* and *intermediate polars (IPs/DQ Her stars)* contain magnetic white dwarfs. In polars, the rotation period of the white dwarf is locked to the orbital motion of the binary and the formation of an accretion disk is prevented by the field. In intermediate polars, the white dwarf rotates freely and an accretion disk may or may not exist. The terms *intermediate polar* and *DQ Her star* are used synonymously by some authors, while others distinguish them by the white dwarf spin period. Only a handful of *double degenerates/AM CVn stars* is known and none of them harbors a magnetic white dwarf, although such systems may exist.

The majority of nonmagnetic CVs have been discovered in optical surveys, of which the Sloan Digital Sky Survey (SDSS) is a major source for the more recent new discoveries. Quite differently, 75% of all polars known at the end of the ROSAT era were discovered by their X-ray and XUV emission, most of them in the

Table 12.1 Subdivision of CVs according to the nature of donor and accretor

Accretor	Synchro-nized	Total number	X-ray discov. (%)	Donor Star: MS-Star/Brown Dwarf Subclass	F_X/F_V	White Dwarf Subclass	F_X/F_V
Non-magn.	No	400	~10	non-mag CV	0.01–1	AM CVn	0.1–5
Magnetic	No	35	65	IP/DQ Her	0.1–10	?	
Magnetic	Yes	80	75	polar/AM Her	1–100	?	

F_X/F_V is the ratio of the X-ray flux in the ROSAT PSPC band vs. the visual flux in a 1 000 Å wide band [3]
"?" denotes that no example of this subclass is known

ROSAT-All-Sky-Survey (RASS). Although all CV subtypes have been detected as X-ray sources, they are characterized by vastly different X-ray vs. optical flux ratios F_X/F_V. In disk accretors, half of the gravitational energy is released in the accretion disk and the remainder appears either as emission from the BL or is added to the rotational energy of the white dwarf. The escape of X-ray photons from the BL may be severely impeded by absorption within the BL, in the accretion disk or in a wind emanating from the disk, all to the effect to lower F_X/F_V. In IPs/DQ Her stars, the inner accretion disk is disrupted by the magnetic field, but absorption may still occur in the magnetically guided accretion flow. In polars, on the other hand, a disk is missing, the soft X-ray flux escapes more or less uninhibited by internal absorption, and F_X/F_V assumes large values.

12.3 Accretion Geometries

Matter overflowing the Roche lobe of the secondary tends to form an accretion disk around the white dwarf. Depending on the magnetic field strength of the white dwarf, the disk extends to its surface ($r = R$) or is disrupted by the field at $r_i > R$. Accretion can effectively take place only when $r_i < r_{co}$, with $r_{co} = (GM/\Omega_*^2)^{1/3}$ the corotation radius and Ω_* the angular velocity of the white dwarf. A disk can form only when $r_i < r_{circ}$, where r_{circ} is the circularization radius at which the Keplerian specific angular momentum equals that of matter passing through the inner Lagrangian point L_1. For $r_i \gtrsim r_{circ}$, a ring-like structure forms or the stream couples directly to the magnetosphere [62].

Disruption of the disk occurs when the tangential magnetic stresses overcome the viscous stresses. The resulting inner disk radius may be expressed in terms of the magnetospheric radius r_μ for spherical accretion by $r_i = \eta r_\mu = \eta\,(2GM)^{-1/7}\dot{M}^{-2/7} B_0^{4/7} R^{12/7}$, where M is the white dwarf mass, $\dot{M}$ the accretion rate, and B_0 the surface field strength of the white dwarf in the orbital plane. From discussions of the disk-field interaction [10, 52], one finds $\eta \simeq 0.5 - 1.5$ for $r_i/r_{co} = 1.0 - 0.3$ somewhat larger than the originally suggested and frequently quoted value $\eta \simeq 0.52$ [20]. Approximate field strengths separating nonmagnetic CVs and IPs on the one hand or IPs and polars on the other hand are obtained by equating r_i to R or to the binary separation a, respectively. Hence, the disk remains unaffected by the field for

$$B_0 \lesssim 6\dot{M}_9^{1/2}\ \text{kG}, \qquad (12.1)$$

while the CV may synchronize as a polars for

$$B_0 \gtrsim 13\dot{M}_9^{1/2} P_4^{7/6} (1+q)^{7/12}\ \text{MG}, \qquad (12.2)$$

where $\dot{M}_9$ is the accretion rate in units of $10^{-9}\,M_\odot\,\text{yr}^{-1}$, P_4 is the orbital period in units of 4 h, $q = M_{sec}/M$ is the mass ratio, and we have used $M = 0.6M_\odot$ with

$R = 9 \times 10^8$ cm. The latter limit appears reasonable since the smallest field strength in a short-period polar is around 7 MG [98]. The former limit suggests that many CVs considered as nonmagnetic may, in fact, harbor white dwarfs with magnetic field strengths too small to be directly detected and to effectively channel the flow far out of the orbital plane. The estimate suggests also that "nonmagnetic" CVs may switch from an IP-like accretion geometry in quiescence to equatorial accretion in the high-$\dot{M}$ outbursts.

12.3.1 Nonmagnetic CVs

If the field cannot effectively channel the flow near the white dwarf, an equatorial BL forms in which the disk material is braked from the Keplerian angular velocity Ω_K to the stellar value Ω_*. If the BL is optically thin as expected for dwarf novae in quiescence, the accreted matter is heated roughly to (or to a fraction of) the virial temperature $kT_{vir} = \mu m_u GM(1-\omega^2)/3R \simeq 49\,(1-\omega^2)(M/M_\odot)^{1.9}$ keV, where μ is the mean molecular weight (0.617 for solar composition), m_u is the unit mass, $\omega = \Omega_*/\Omega_K(R)$, and the mass radius relation for white dwarfs with $T_{eff} \simeq 20000$ K has been approximated by $R \simeq 5.8 \times 10^8\,(M/M_\odot)^{-0.9}$ cm. Shocks, even multiple shocks, may occur in the flow. While the structure of the BL is not yet fully understood, it does represent some kind of cooling flow with temperatures ranging from T_{vir} (or a fraction thereof) down to the equatorial surface temperature of the white dwarf. With increasing $\dot{M}$, the BL becomes optically thick and its luminosity and blackbody temperature are given by

$$L_{BL} = GM\dot{M}(1-\omega^2)/2R \qquad (12.3)$$

$$kT_{bb} = (L_{BL}/4\pi R^2 f)^{1/4} \qquad (12.4)$$

$$\simeq 11.4\,(1-\omega^2)^{1/4}(\dot{M}_8/f)^{1/4}(M/M_\odot)^{0.92}\ \text{eV}$$

where the emitting area is a narrow equatorial belt of width $2H$ and fractional area $f = 2\pi R \times 2H/4\pi R^2 = H/R$, which is of the order of 0.01. Internal absorption is quite likely to be important as is absorption and scattering in the wind that emanates from a luminous CV disk. If the inclination is low, spectral flux will be scattered *out* of the line of sight and if the inclination is high and the BL is hidden from view, flux will be scattered *into* the line of sight [54]. The sketch in the left-hand diagram of Fig. 12.1 gives a rough idea of the geometry [65].

12.3.2 Magnetic CVs

In magnetic CVs, the field is sufficiently strong to channel the accretion flow to the polar region(s) of the white dwarf. Radial accretion along polar field lines is an approximation to this case and is depicted in the center graph of Fig. 12.1. The emission is dominated by thermal emission (mostly hard-ray bremsstrahlung) in the limit

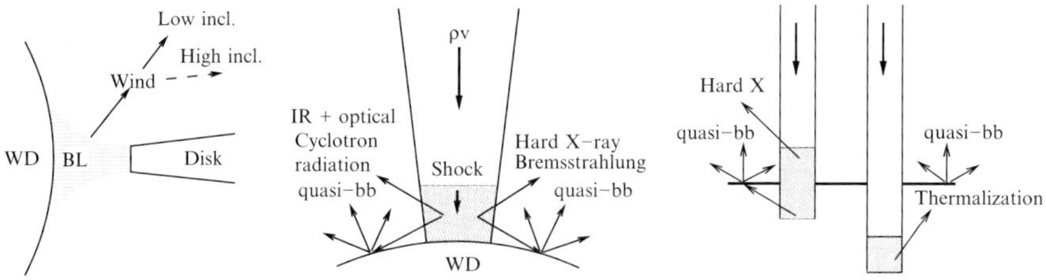

Fig. 12.1 The accretion geometry and the observed radiation components. *Left:* Boundary layer in nonmagnetic CVs. *Center:* Cooling flow behind free-standing shock for radial accretion in a magnetic CV with low mass-flow density (i.e., specific accretion rate). *Right:* Partially and completely buried shocks for intermediate and high mass flow densities, respectively

of low field strength B and high mass flow density $\dot{m}$.[1] In the shock, the plasma is heated to the shock temperature $kT_s = (3\mu m_u/16k)v_{ff}^2 \simeq 55\,(M/M_\odot)^{1.9}$ keV, where $v_{ff} = (2GM/R)^{1/2}$ the free-fall velocity [13]. The flow is largely optically thin, but becomes optically thick in the strongest emission lines. Besides thermal X-ray emission, IR/optical cyclotron radiation is an important cooling agent, which dominates for high B and low $\dot{m}$. In the cyclotron-dominated regime, the peak plasma temperature drops to only a few kiloelectronvolt and the shock height collapses to small values [2,13,100]. In the non-hydrodynamical regime at $\dot{m} \lesssim 10^{-4}B_7^{2.6}$ g cm^{-2}s^{-1} [13], with B_7 the field strength in units of 10 MG, a shock does not form and the outermost layers of the atmosphere of the white dwarf are directly heated by the kinetic energy of the ions, a situation called bombardment solution [44,99]. Again, the plasma temperature is lower than for a bremsstrahlung-dominated shock.

The stand-off distance of the bremsstrahlung-dominated shock is $h_{sh} = 6.0 \times 10^{-19}\,(2GM/R)^{3/2}/\dot{m}$ in cgs-units, which holds as long as the one-dimensional approximation with $h_{sh} \ll R$ is valid [13]. For $M = 0.6M_\odot$, $h_{sh} \simeq 4.5 \times 10^7/\dot{m}$ cm. At the same time, the bottom of the cooling flow is depressed by the ram pressure to a depth $d_{ram} \simeq 10^6\,\dot{m}^{0.4}$ cm below the photosphere of the white dwarf [2,15]. As a consequence, the entire cooling flow disappears below the level of the photosphere for $h_{sh} < d_{ram}$ or $\dot{m} \gtrsim 15$ g cm^{-2}s^{-1} [2]. Such buried shocks are schematically depicted in the right-hand graph in Fig. 12.1. Below, we show examples of spectra calculated for different $\dot{m}$ values (Fig. 12.7). In the buried high-$\dot{m}$ shocks, the bremsstrahlung flux from the postshock plasma is severely reduced by photoabsorption in the surrounding atmosphere and the corresponding fraction of the accretion luminosity is thermalized and reappears as soft quasi-blackbody emission from the heated atmosphere [2, 15, 44]. Thermalization as well as heating of the atmosphere by irradiation from tall, low-$\dot{m}$ shocks [36, 45] combine to produce the large excesses of soft over hard X-ray emission observed from many, but by no means from all polars [2,69] (see also Sect. 12.6).

[1] The quantity $\dot{m}$ (cgs unit g cm^{-2}s^{-1}) is also called specific accretion rate.

The emission properties of polars and IPs are basically similar, while the radiative transfer properties differ. Accretion from a disk or ring in IPs causes the accreting matter to form an azimuthally extended curtain, which photoabsorbs soft X-rays over much of the hemisphere as seen from the white dwarf. Broad-band X-ray spectral analyses of IPs suggest that the accretion stream is highly structured and may best be modeled by a range of optical depths present in the line of sight to the source [61]. The narrow accretion stream in polars, on the other hand, blocks soft X-rays only over a very restricted solid angle and allows a largely unrestricted view of the radiation that escapes from the immediate vicinity of the accretion spot.

12.4 X-Ray and EUV Emission from Nonmagnetic CVs

Observations performed with the Einstein and ROSAT satellites led to the consensus that the X-ray emission from nonmagnetic CVs falls increasingly short of the predictions of simple BL theories as the accretion rate and the luminosity of the accretion disk increase [65, 89, 92]. The observed HeII $\lambda\lambda 4686, 1640$ Å line fluxes suggest the presence of a higher intrinsic soft X-ray flux than actually seen and led to the conclusion that part of the EUV flux is hidden from view. Such interpretation is supported by the fact that the F_X/F_V ratio decreases with increasing inclination. In quiescence, a thermal X-ray component with a temperature of a few kiloelectronvolts dominates. The analysis of this source in the eclipsing CVs HT Cas and OY Car demonstrated that the observed emission arises from very close to the white dwarf. In the XMM-Newton EPIC light curve of OY Car in Fig. 12.2, the length of the X-ray totality is identical to that of the optical light of the white dwarf, but its ingress and egress times are significantly shorter than that of the white dwarf. This

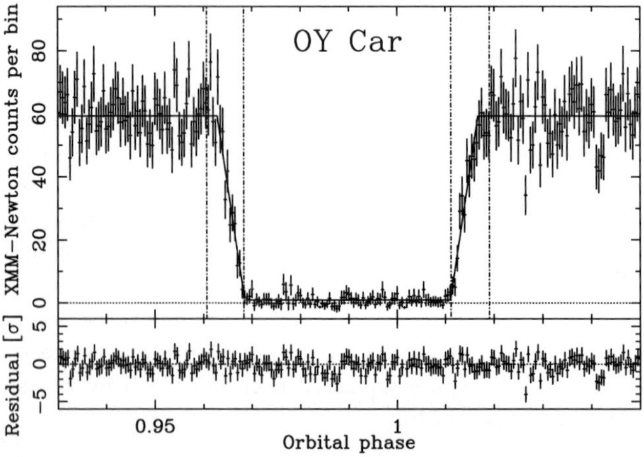

Fig. 12.2 Folded XMM-Newton eclipse light curve of OY Car (June 2000). The *dash-dotted vertical lines* indicate the optical contact points marking ingress and egress of the white dwarf [96, private communication]

difference suggests that the X-ray emission arises from a region on the white dwarf, which is displaced toward its upper rotational pole [96]. Having noted (see (12.1)) that a moderate magnetic field of the white dwarf can lift the flow out of the plane, the offset of the X-ray source appears plausible. The special geometry may also prevent excessive internal absorption of the X-ray emission by the accretion disk at the inclination of OY Car of $i \simeq 83°$.

Wavelength dependent X-ray light curves through dwarf nova outbursts represent an independent powerful approach to study the BL emission. Results include the X-ray, EUV, and optical coverage of the low-inclination long-period system SS Cyg [97] and the worldwide campaign to observe the unexpected 2001 superoutburst of the eclipsing short-period system WZ Sge [42, 55]. Earlier outburst observations of OY Car [56], U Gem [50], and VW Hyi [95] complement the picture. In outburst, the rise of the optical light precedes that of the EUV by the time the heat wave needs to transit the disk. The optically thin X-ray source is quenched and replaced by the optically thick EUV component. The fainter X-ray emission during outburst cannot be uniquely assigned to a single source, but contains components from regions with different excitation conditions absorbed by different column densities. This may include emission from an accretion disk corona, possibly heated by magnetic activity [97]. At the end of the outburst, the EUV subsides, the collisionally excited thermal BL X-rays reappear, and the system returns to the emission of the quiescent state. HST observations of the cooling white dwarf in the 2001 outburst of WZ Sge indicate that the entire white dwarf is heated, but initially its temperature distribution is not uniform, with the equatorial belt likely being hotter [51].

The intense EUV component in outburst finds a comparatively simple explanation as the optically thick BL emission. It is directly observable in low-inclination systems, while at high inclination only radiation scattered in the intense disk wind is seen [56]. With appropriate object-dependent wind parameters, the seemingly quite different EUV outburst spectra of the high-inclination system WZ Sge and of the low-inclination system SS Cyg can be successfully modeled (Fig. 12.3) [54, 55]. The mass loss rate by the wind in SS Cyg approaches the maximum rate $\dot{M}_{\rm max} = L/cv_{\rm wind,\infty}$.

The luminosity of the boundary layer is reduced below the standard value if the white dwarf is rotating (3). While in U Gem, the luminosities of BL and disk agree [50], Mauche finds for SS Cyg a disconcertingly low value of $L_{\rm BL}/L_{\rm disk} = 0.05^{+0.18}_{-0.03}$ [55]. This result suggests that the white dwarf rotates with a very short period of $\sim$9 s, which agrees with the period of its quasi-coherent oscillations. The origin of the soft X-ray oscillations seen in quiescence and in outburst is not finally settled (for a review that concentrates on the X-ray emission see [43]).

12.5 X-Rays from Intermediate Polars

The defining feature of IPs is a coherent pulsation at a period $P_{\rm s}$ shorter than the orbital period $P_{\rm orb}$ [62, this reference contains also a list of confirmed IPs]. This short period usually dominates at X-ray energies and is interpreted as the spin period

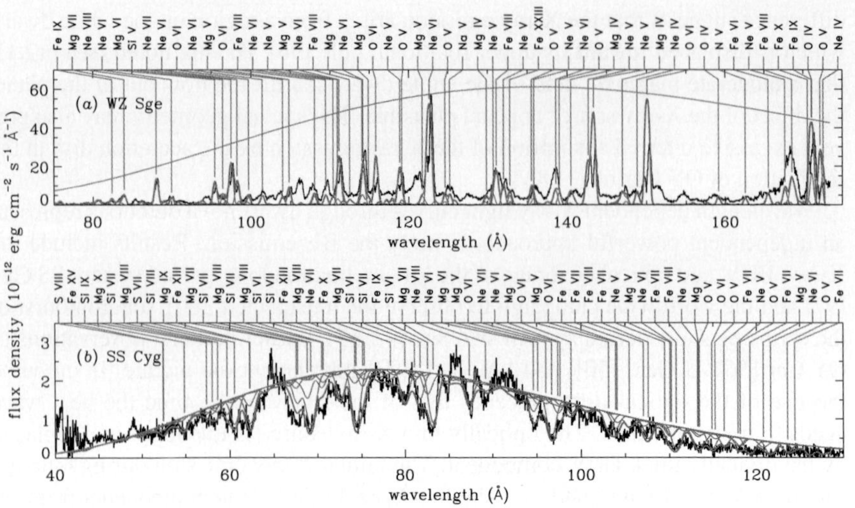

Fig. 12.3 Chandra LETG spectra of WZ Sge in superoutburst (*top panel*) and SS Cyg in outburst (*bottom panel*). The data are shown in *black* and the model spectrum in *grey* with the strongest lines indicated (from [54])

of the white dwarf. This interpretation is convincingly confirmed when reprocessed optical/ultraviolet radiation at the beat period $P_b = (P_s^{-1} - P_{orb}^{-1})^{-1}$ is also detected (see [66, 93] for further possible periodicities). The magnetic nature of the accreting white dwarf is taken for granted, although in most cases a direct observational proof is missing, with the exception of the few systems found to be polarized at red or near IR wavelengths. An evolutionary link to polars has long been claimed and is likely since synchronization is bound to take place in some systems as their orbital separation decreases with time [62, and references therein].

A comprehensive study based on EXOSAT results [61] summarized the knowledge we had of the spin modulated X-ray emission of IPs before the ROSAT era and demonstrated the importance of partial absorption, X-ray reflection from the white dwarf surface, self occultation, and a structured emission flow. Subsequent pulse profile studies at different photon energies have greatly added to our understanding of the complex accretion and emission geometries. With ROSAT, the link to polars was strengthened by the discovery of about a dozen soft X-ray emitting IPs [25]. The X-ray spectra of some of these closely resemble those of polars, consisting of an optically thin thermal hard X-ray and an optically thick quasi-blackbody soft X-ray component. The superior sensitivity of Chandra and XMM-Newton permits detailed spin-phase resolved spectroscopy of these objects. Figure 12.4 shows the LETG wavelength-integrated light curve of the IP PQ Gem and in Fig. 12.6, later, we compare the mean LETG spectra of PQ Gem and the prototype polar AM Her [8]. Both spectra display a thermal hard X-ray component and a superimposed soft component. For PQ Gem, the latter can be approximated by a 40 eV blackbody absorbed by a column density of 1.3×10^{20} cm^{-2}, for AM Her, more than a single bb-component

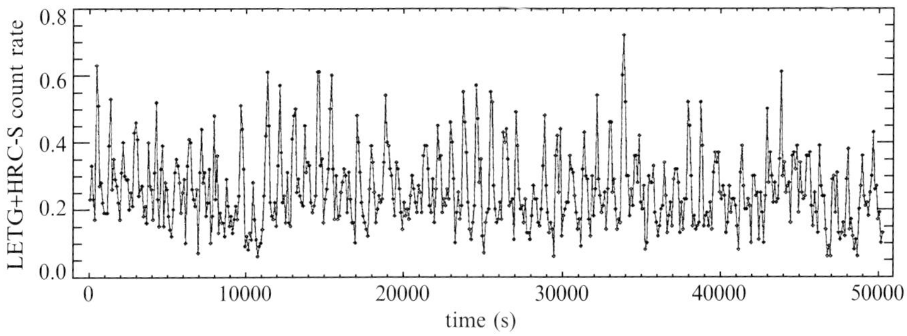

Fig. 12.4 Chandra LETG soft X-ray light curve of the intermediate polar PQ Gem. The 50000 s exposure covers 2.7 orbital periods and shows the individual 833 s pulsations (from [8, and private communication by V. Burwitz])

and a complex absorber [63] is needed. The OVII 21.6 – 22.1Å complex is nicely resolved in the original LETG data [8].

Near the white dwarf, accretion in IPs is quasiradial and the central graph in Fig. 12.1 applies. Information on the structure of the region emitting the thermal component may be gathered by detailed emission line studies. The Chandra HETG spectra of cooling flows exhibit a continuum and strong lines of H- and He-like ions of the abundant elements, the Fe K-shell lines, and the entire Fe L-shell complex. The line spectra of the little absorbed IPs EX Hya and V603 Aql can be nicely modeled by cooling flows with near-solar abundances. The same is true for the non-magnetic CVs U Gem and SS Cyg in quiescence, indicating the similarity of the hard X-ray emission regions in the two types of objects [58]. The line spectra of the more strongly internally absorbed IPs, V1223 Sgr, AO Psc, and GK Per, on the other hand, are basically different in that they are dominated by lines produced by photoionization [58]. The spectral resolution of the grating spectrometers on board of Chandra and XMM-Newton allows for the first time detailed plasma diagnostic studies of the emission regions in the brighter X-ray emitting CVs. Mauche et al. [57] have studied the density, temperature, and photoionization dependency of various line ratios and found the FeXXII $I(11.92\,\text{Å})/I(11.77\,\text{Å})$ line ratio to be particularly useful for CV studies with a critical density of about $5 \times 10^{13}\,\text{cm}^{-3}$ and a low sensitivity to temperature and photoionization. Figure 12.5 shows a small section of the Chandra MEG spectrum of EX Hya, which contains the two FeXXII lines (along with an FeXXIII line at 11.74 Å). The $I(11.92\,\text{Å})/I(11.77\,\text{Å})$ line ratio implies a density in the emission region of $n_\mathrm{e} = 1.0 \pm^{2.0}_{0.5} \times 10^{14}\,\text{cm}^{-3}$ at a temperature of $T_\mathrm{e} \simeq 1.2 \times 10^7$ K [57]. If one combines this result with the shock temperature for the $0.5 M_\odot$ white dwarf [5, and references therein], one finds a preshock mass flow density of $\sim 3 \times 10^{-3}\,\text{g}\,\text{cm}^{-2}\text{s}^{-1}$, a key plasma parameter characterizing the accretion flow that produces the FeXXII lines in EX Hya. The example demonstrates the potential of the plasma-diagnostic tools now available.

Observations with the Chandra HETG can easily resolve the H-like, He-like, and the fluorescence lines of the Fe Kα emission line complex and are on the verge of

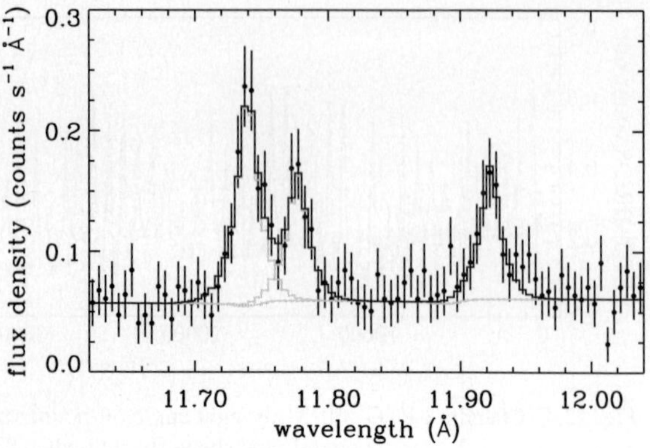

Fig. 12.5 Detail of the Chandra HETG spectrum of EX Hya in the vicinity of the FeXXII 3 → 2 lines, binned to 0.005 Å (from [57])

resolving the He-like resonance, intercombination, and forbidden lines, as well as the dielectronic satellite lines in the brighter IPs including EX Hya [32]. The absence of Doppler shifts in these lines suggests that they are emitted from near the bottom of the cooling flows. Hence, the flows are not buried in the photosphere of the white dwarf, consistent with the small value of $\dot{m}$ found above for EX Hya. Observations of EX Hya with the Chandra HETG have also, for the first time, allowed to measure the orbital motion of the white dwarf in a CV using X-ray emission lines [31].

12.6 X-Rays from Polars

The majority of the presently known polars was discovered in the ROSAT All-Sky-Survey by their soft X-ray and EUV emission (see Table 12.1). The dominance of soft X-ray emission is a property of many, but by no means of all polars [2, 69, 70]. The Chandra LETG spectrum of AM Her in Fig. 12.6 [8, and private communication by V. Burwitz] shows the presence of a strong soft X-ray component in its high state. The soft X-rays disappear in the low state, probably because the temperature drops and the emission retreats from the accessible window [2, see also Fig. 12.7, below]. In the high state, the hard thermal component consists of bremsstrahlung with emission lines of the abundant elements superimposed (e.g., OVII at 22 Å). Most of the bolometric flux, however, is in an optically thick component that can be approximated by a blackbody component with a dominant temperature near $kT_{\rm bb} \simeq$ 29 eV and no obvious emission or absorption lines. The blackbody approximation is reasonable as shown also by the EUVE spectra of several polars [53]. Although most polars were discovered as soft ROSAT sources, the sample of hard *RASS* sources contains additional polars [77], and data from XMM-Newton suggested that a larger fraction of polars than considered earlier is *not* dominated by soft X-rays [69, 70]. In addition, very low accretion rate polars were discovered in the Hamburg quasar and the Sloan surveys by their cyclotron emission. They were not seen as soft X-ray sources in the *RASS* and are convincingly interpreted as magnetic pre-CVs, which

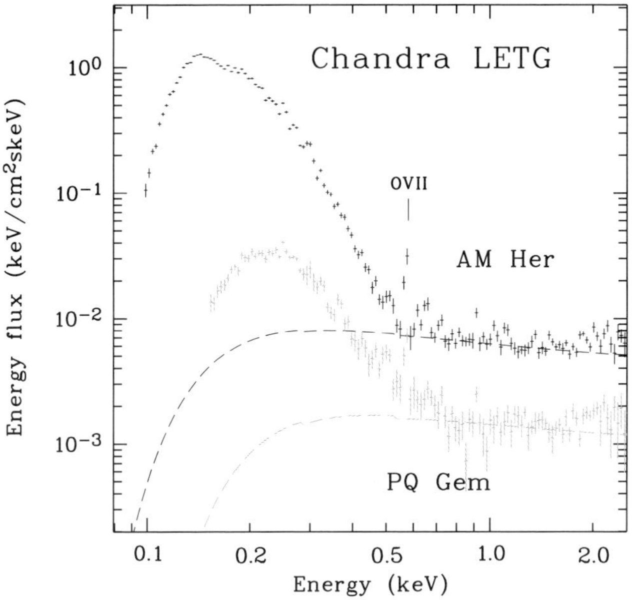

Fig. 12.6 Spectral energy distributions of AM Her and PG Gem based on Chandra LETG spectra binned into 100 bins per decade in energy. The model for the hard X-ray bremsstrahlung component is included (*dashes lines*) [8, and private communication V. Burwitz]

have synchronized already but transfer matter at a low rate from the wind of the secondary via the magnetic field to the white dwarf and have not yet started Roche lobe overflow [75].

The early model of a white dwarf atmosphere heated by bremsstrahlung from the postshock plasma suggested a ratio of blackbody vs. bremsstrahlung luminosities of about unity [36, 45], substantially less than observed in AM Her. Much of the emission from the heated atmosphere was, furthermore, found to appear in the FUV rather than in soft X-rays [17, 39]. Remedy comes, as noted in Sect. 12.3, from the concept of buried shocks [2, 15, 44]. Shock burying (Fig. 12.1) occurs only for high mass flow densities $\dot{m}$ and the emerging overall spectral energy distribution depends, therefore, strongly on the $\dot{m}$-spectrum incident on the white dwarf. Figure 12.7 shows representative spectra calculated for the parameters of AM Her, $B = 14\,\mathrm{MG}$, an angle of $\Theta = 60^\circ$ against the field direction, and $\dot{m}$ from 10^{-3} to $100\,\mathrm{g\,cm^{-2}s^{-1}}$, each for an emitting area of $10^{15}\,\mathrm{cm}^2$ at 100 pc distance [2]. They are based on radiation-hydrodynamic calculations [13] and approximately account for absorption due to burying and for reprocessing. The reprocessed radiation from irradiation by low-$\dot{m}$ tall shocks appears in the UV rather than in the soft X-ray regime. The blackbody assumption used here is a crude approximation and a more correct treatment of irradiated atmospheres is presented in [39]. We find that the observed overall spectral flux distribution of AM Her can be synthesized from model spectra as shown in Fig. 12.7, with the lowest $\dot{m}$ being responsible for the

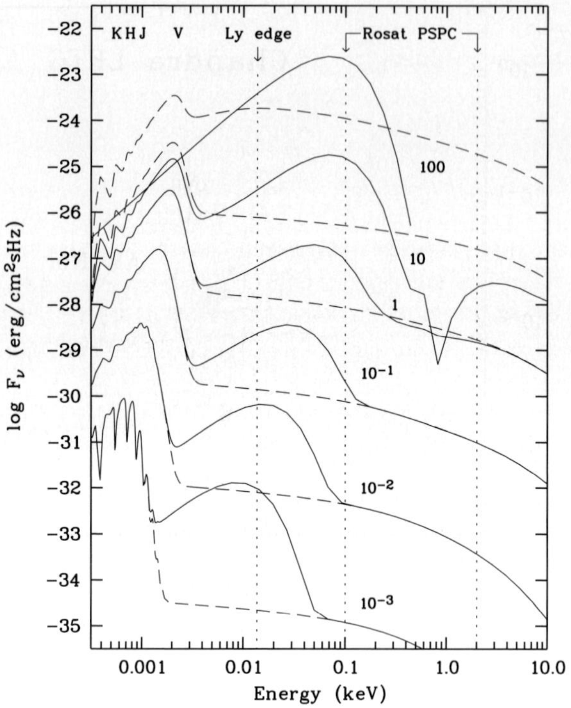

Fig. 12.7 Representative spectra for $B = 14$ MG, an angle $\Theta = 60°$ against the field direction, and mass flow densities $\dot{m} = 10^{-3}, 10^{-2}, 10^{-1}, 1, 10,$ and $100\,\mathrm{g\,cm^{-2}s^{-1}}$ (from bottom to top), each calculated for an emitting area of 10^{15} cm^2 at 100 pc distance. *Dashed curves* are the spectra emitted by the postshock plasma, *solid curves* the emergent spectra that include absorption and the reprocessed component from the white dwarf photosphere (from [2])

observed cyclotron emission, $\dot{m}$ values around $1\,\mathrm{g\,cm^{-2}s^{-1}}$ producing the observed bremsstrahlung, and the highest $\dot{m}$ being responsible for the soft X-rays via thermalization, though with substantial overlap. In conclusion, a large soft excess indicates high $\dot{m}$ values and buried shocks, a weaker soft X-ray flux generally lower $\dot{m}$. The $\dot{m}$-spectrum of AM Her in its high state rises toward high $\dot{m}$ values, while it is rather flat in the low state [2]. These differences arise from either a gating mechanism in the magnetosphere or are imprinted on the stream already in the L_1 point. Near the white dwarf, a high field strength acts as a bottle neck and increases $\dot{m}$, which explains the generally softer X-ray emission of high-field polars. On the other end of the $\dot{m}$-scale, the magnetic pre-CVs seem to be characterized not only by a low total accretion rate $\dot{M}$, but also by a low mass flow density $\dot{m}$ and, hence, a comparatively large area of the accretion spot as defined by the field lines that connect to the secondary star.

Further information on the structure of the X-ray emission region in polars can be gathered by timing studies (light curves) and by X-ray spectroscopic studies. Light curves in the kiloelectronvolt regime are consistent with the notion that the observed bremsstrahlung originates from tall shocks, i.e., those subcolumns which are not buried. The soft X-ray light curves at photon energies $E < 0.5\,\mathrm{keV}$ provide information on the structure of the hot photosphere. In the simplest notion, the hot spots in the white dwarf photosphere are modeled as blackbody-emitting surface elements. That this concept is too simple was shown already by the EXOSAT soft X-ray light curve of AM Her in its "reversed" mode, in which it emitted soft X-rays from a second, normally inactive pole. The observed light curve of the self-eclipsing source was found to be box-like with superimposed fluctuations, suggesting that the source was about as tall as wide and raised above the photosphere by some 10^7 cm [29]. Light curves of bright polars obtained with EUVE and especially with the ROSAT PSPC provided better statistics and suggested mounds rising to a height of a few percent of the white dwarf radius [76, 81]. Such mounds can be explained by local heating of the photosphere surrounding individual matter-loaded flux tubes [47] and by splashes by which the photosphere reacts to the termination of a high-$\dot{m}$ filament or "blob"-like a compressed spring.

X-ray spectroscopy has reached a level that allows detailed diagnostics of the optically thin hard X-ray emitting post-shock plasma. What was said above for IPs applies for polars, too. It is already possible to perform radial velocity studies of the emitting ions in the postshock region [8] and one may envisage to sample the velocity structure of the flow as a function of temperature using the variation in excitation as the flow settles.

12.7 Accretion Rates

The accretion luminosity of a nonrotating white dwarf

$$L_{\mathrm{acc}} \simeq \frac{GM\dot{M}}{R} \qquad (12.5)$$

is radiated quasi-instantaneously from the disk or from near its surface. From (12.5) one can derive the accretion rate $\dot{M}$ if the mass and radius of the white dwarf and the bolometric accretion luminosity are known. Deriving L_{acc} requires knowledge of the distance, of the overall spectral energy distribution, and (at least an estimate) of the wavelength-dependent angular distribution of the emitted radiation. The obvious observational difficulties explain why accretion rates are notoriously uncertain. Attempts, based on accurate *HST FGS* parallaxes and X-ray/FUV/optical fluxes are described in [5, 6].

An entirely different approach considers compressional heating of the accreting white dwarf and derives $\dot{M}$ from the effective temperature of the white dwarf, which has the invaluable property of being independent of the distance. For bright X-ray sources such as polars, one obtains a check on $\dot{M}$ derived from the X-ray fluxes and the notoriously uncertain distances. Accretion of ΔM compresses the nondegenerate envelope of the white dwarf and drives a corresponding mass at its bottom into degeneracy, increasing the core mass by ΔM. While the mass of the accreted envelope, $M_{\rm acc} = 10^{-6} - 10^{-3}\,M_\odot$, is negligible compared with the core mass $M_{\rm c}$, the geometrical thickness $\Delta R = R - R_{\rm c}$ of the envelope, is not and the white dwarf radius R exceeds the core radius $R_{\rm c}$ noticeably,[2] $R \simeq 1.06 R_{\rm c}$ $(1.17 R_{\rm c})$ for an $0.6 M_\odot$ white dwarf with a hydrogen envelope of $10^{-4}\,M_\odot$ and $T_{\rm eff} = 10^4$ (3×10^4) K. Accretion at a rate $\dot{M}$ compresses the envelope and produces an equilibrium luminosity

$$L_{\rm env} = 4\pi R^2 \sigma T_{\rm eff}^4 = \left(1 - \frac{\nabla_{\rm env}}{\nabla_{\rm ad}}\right) GM\dot{M} \left(\frac{1}{R_{\rm c}} - \frac{1}{R}\right), \qquad (12.6)$$

which is rooted in the envelope and appears in addition to the external accretion luminosity of (12.5). The term $GM\dot{M}(1/R_{\rm c} - 1/R)$ describes the energy release by the addition of the accreted mass to the core. The fraction $(1 - \nabla_{\rm env}/\nabla_{\rm ad}) \simeq 0.48$ is available to be radiated, while $\nabla_{\rm env}/\nabla_{\rm ad}$ serves to heat the accreted matter to the core temperature as it sinks. Here, $\nabla = \partial \log T/\partial \log P$ is the logarithmic gradient of the temperature variation with pressure in the envelope and the indices "env" and "ad" refer to the actual gradient in the envelope and the adiabatic gradient, respectively.[3] Equation (12.6) has been used by [18] to interprete the low temperature of the white dwarf in the long-period polar RX J1313-32 in terms of a low accretion rate. A similar form was also discussed in [80] to explain the temperatures of accreting white dwarfs in CVs. Since the cooling time of the envelope is of the order of a million years, compressional heating measures a long-term mean accretion rate.

A comprehensive study of compressional heating, including an internally consistent treatment of the path to ignition of hydrogen burning, has been presented by Townsley & Bildsten [85, 86]. Their numerical results on the dependence of the effective temperature of the white dwarf on $\dot{M}, M_{\rm acc}$ and (negligibly) on M [85, their Fig. 1] can be approximated as

$$T_{\rm eff} \simeq 14400\,\dot{M}_{10}^{0.27} f^{0.05}\ K\,, \qquad (12.7)$$

where $\dot{M}_{10}$ is the accretion rate in units of $10^{-10}\,M_\odot\,{\rm yr}^{-1}$ and f the mass of the envelope accumulated since the last nova outburst in units of the ignition mass $M_{\rm ign}$ leading to the next nova event. Figure 12.8 displays a collection of HST-derived effective temperatures of white dwarfs in CVs [1] selected to avoid as far as possible the effects of recent heating in dwarf nova outbursts. The right-hand ordinate provides the equivalent long-term mean accretion rates from (12.7) using $f = 0.5$ corresponding to a time midways between two nova outbursts. The fact that the white dwarf experiences short-term heating episodes in nova and, if applicable, in dwarf

---

[2] The core has approximately the Chandrasekhar radius for $\mu_{\rm e} = 2$. $\Delta R = R - R_{\rm c} \simeq kT_{\rm c}/(\nabla_{\rm env}\mu m_{\rm u} G M_{\rm c}) \simeq 5 H_{\rm P}$, with $\nabla_{\rm env} \simeq 0.21$ and $H_{\rm P}$ the barometric pressure scale height for the temperature $T_{\rm c}$ at the bottom of the envelope.

[3] Note that $L_{\rm env}$ vanishes if rapid accretion leads to adiabatic heating and a convective envelope. For a monatomic gas, the adiabatic gradient is $\nabla_{\rm ad} = 2/5$, the radiative gradient is $\nabla_{\rm ad} \simeq 0.21$ and depends on the opacity law $\kappa(T,P)$.

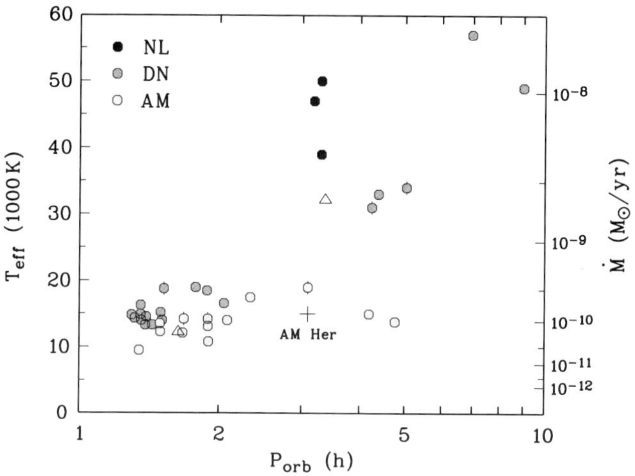

Fig. 12.8 Observed effective temperatures of white dwarfs in CVs vs. orbital period (mostly from [1]). Different subtypes of CVs are indicated by the shading of the *symbols*. Statistical error bars are typically smaller than the symbol size. The right-hand scale gives a long-term average $\langle \dot{M} \rangle$ from (12.7). Also shown is the 20-yr average $\dot{M}$ of AM Her [30] as a cross and the luminosity-derived $\dot{M}$ of EX Hya and V1223 Sgr [5, 6] as *open triangles*

nova outbursts, in addition to the long-term compressional heating suggests that one has to interprete the results with caution. Nevertheless, the result looks convincing in the sense that the derived accretion rates of short-period CVs – in particular, of the polars – agree with those expected for gravitational radiation and the accretion rates for the long-period dwarf novae increase with orbital period as expected from magnetic braking. The low long-term $\dot{M}$ of the long-period polars may be due to a lower efficiency of the angular momentum loss by the wind the secondary in the presence of the strong field of the primary. We recall that the so derived $\dot{M}$ values represent averages over the Kelvin-Helmholtz cooling time scale of the envelope. For comparison, we have added the 20-yr average accretion rate for AM Her, $\dot{M} \simeq 1.2 \times 10^{-10} M_{\odot}\,\mathrm{yr}^{-1}$ [30], which represents a mean between high and low states (cross). The T_{eff}-derived accretion rate (lightly filled circle) is higher by about a factor of two and suggests that AM Her has spent the last million years on average in a state rather corresponding to the present high than the low state. Also added are the present accretion rates of the intermediate polars EX Hya [5] and V1223 Sgr [6] derived from accurate parallaxes and the bolometric – that is largely X-ray – fluxes (triangles). They fall in the same range as the temperature-derived rates and support the basic validity of the compressional heating model. The high $\dot{M}$-values of the three novalike variables near an orbital period of 3 h demonstrate that there is no simple $\dot{M}$–P relationship. These systems as well as long-period dwarf novae reach up to beyond $10^{-8} M_{\odot}\,\mathrm{y}^{-1}$ into the region where surface hydrogen burning can occur.

12.8 Novae and Close-Binary Supersoft Sources (CBSS)

In this section, I address accreting white dwarfs, which experience intermittent or continuous near-surface hydrogen burning. Burning occurs intermittently in novae and more or less continuously in close-binary supersoft X-ray sources (CBSS). The physics and X-ray emission of novae is more extensively addressed by Krautter (this volume). Continuous burning of accreted matter is of paramount interest as a means to drive a white dwarf above the Chandrasekhar mass limit into a SN Ia event [26, 82].

The observationally defined "supersoft" X-ray sources (SSS) represent a mixed bag. They include single hot stars as the PG1159 stars, and also white dwarfs, which have experienced a late helium shell flash and become EUV sources in a "borne-again" scenario [33, 64]. In binary stars, on the other hand, surface hydrogen burning on the white dwarf is not a singular phenomenon, but rather is rekindled over again or kept going by continued accretion. This subclass includes classical novae, recurrent novae, symbiotic novae, and CBSS.

12.8.1 The Relation between CVs, Novae, and CBSS

Close binaries with accreting white dwarf appear as CVs (or related objects) as long as hydrogen burning is not ignited in the accreted matter. They appear as novae when the gas is electron-degenerate at ignition leading to a thermonuclear runaway and they become CBSS when the gas is still nondegenerate at ignition and stable burning ensues. The outcome depends on the thermal history of the white dwarf, i.e., on the present core temperature and the time scale of previous accretion. Compression raises the temperature and pressure at the base of the envelope until the ignition takes place at $T_{\rm ign} \sim 1.5 \times 10^7$ K and $P_{\rm ign} = 10^{18} - 10^{20}$ dyne cm^{-2} [101]. For the long-term mean accretion rates typical of CVs driven by gravitational radiation and magnetic braking (Figs. 12.8 and 12.9), ignition cannot be avoided. Only if the rate drops continually below a few times $10^{-13}\,M_\odot\,{\rm yr}^{-1}$, does cooling win over compressional heating and a cold degenerate configuration results [101].

Degeneracy is measured by the parameter ψ with $e^{\psi} \propto P/T^{5/2}$ and the condition $\psi = 0$ or $T_{\rm crit} = 3 \times 10^7\,(P/10^{19}\,{\rm dyne\,cm^{-2}})^{2/5}$ K delineates the beginning of strong degeneracy. Stable burning can commence if the base temperature at ignition significantly exceeds $T_{\rm crit}$, a strong nova explosion ensues if it falls significantly below $T_{\rm crit}$. The transition is smooth in the sense that moderate degeneracy causes weak shell flashes with little or no mass ejection.

The mass of the accreted envelope above a level with hydrostatic pressure $P_{\rm ign,19}$ in units of 10^{19} dyne cm^{-2} is $M_{\rm acc} \simeq 3 \times 10^{-5}\,P_{\rm ign,19}(M/M_\odot)^{-5}\,M_\odot$, where the mass–radius relation of white dwarfs has been approximated as $R \propto M^{-1}$. The envelope mass at ignition decreases steeply with increasing M, and so does the recurrence time scale $\tau_{\rm rec} = M_{\rm acc}/\dot{M}$ of the shell flashes. Example, for a $1.25\,M_\odot$ white

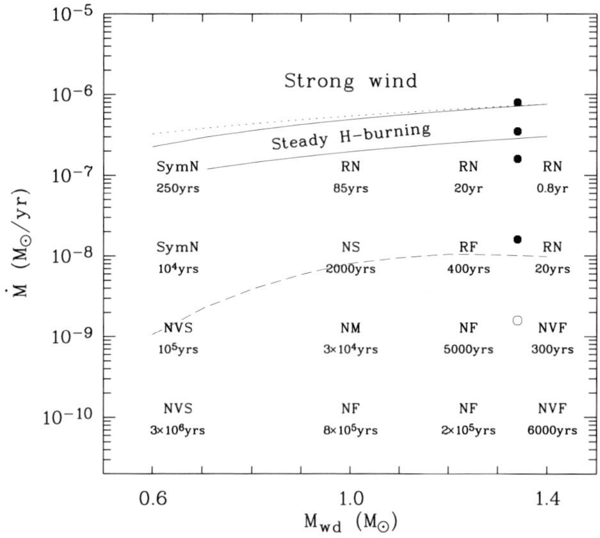

Fig. 12.9 Characteristics of hydrogen burning on accreting white dwarfs as a function of the white dwarf mass and the accretion rate. *Curves* denote the Eddington luminosity for a hydrogen mass fraction $X = 0.70$ (*dotted*), the band of steady hydrogen burning [34] (*solid*), and the nova dud line [48], below which strong nova explosions occur (*dashed*). Also indicated are the types of outburst [68] and models of continuously burning very hot white dwarfs [82] (*filled circles*)

dwarf accreting at $\dot{M} = 10^{-7} M_\odot$, $P_{\rm ign,19} \simeq 0.2$ [101] and the recurrence time becomes as short as $\tau_{\rm rec} \simeq 20$ yrs.

Figure 12.9 summarizes the results on the outcome of nuclear burning on accreting white dwarfs. The continuous-burning zone is limited toward large $\dot{M}$ by the core-mass luminosity relation $L_{\rm crit} \simeq 4.6 \times 10^4 (M/M_\odot - 0.26) L_\odot$ [34], which can be converted to a maximum accretion rate $\dot{M}_{\rm crit} = L_{\rm crit}/\varepsilon X = 6.6 \times 10^{-7} (M/M_\odot - 0.26) M_\odot {\rm yr}^{-1}$, by division by the energy yield for hydrogen burning $\varepsilon \simeq 6 \times 10^{18}$ erg s^{-1} and the assumed hydrogen mass fraction of the accreted matter taken to be $X \simeq 0.70$ (upper solid curve). This limit is close to the Eddington luminosity and accretion rate[4] (dotted curve). Above $\dot{M}_{\rm crit}$, a giant envelope was thought to form [16], quenching the short-wavelength emission from the burning zone. The existence of a quasi-static atmosphere seems to be impossible, however, given the large opacity for matter with near-solar metalicity and instead a radiation-driven wind develops [26]. This argument is based on the OPAL opacities [35], which were revised significantly upward from the older Los Alamos values. Toward smaller $\dot{M}$, the width of the continuous burning zone is about a factor of 2.5 in accretion rate [16], below which weak flashes of intermittent burning occur. Such objects do not eject

[4] The dotted curve in Fig. 12.9 represents the electron-scattering Eddington luminosity divided by εX. Without burning, the Eddington accretion rate rises to the conventional value $\dot{M}_{\rm edd} = L_{\rm edd} R/GM$.

mass and can appear either as CBSS (QR And = RX J0019.8+2156 [4, 22] may be such a case) or as symbiotic or recurrent nonejecting novae and temporary supersoft X-ray sources. Shell flashes eject less mass than accreted, possibly down to the nova dud line [48] (dashed curve). Included in Fig. 12.9 is the outcome of nova model calculations giving the type of nova and the rounded recurrence times [68, 101]. The abbreviations refer to symbiotic and recurrent novae (SymN and RN), which mostly eject no matter, and to classical novae of the different speed classes, very slow novae (NVS), slow (NS), medium slow (NM), fast (NF), and very fast novae (NVF). The dividing line between SymN/RN and classical nova events approximately agrees with Livio's earlier estimate of the nova dud line [48]. The Prialnik and Kovetz calculations [68, 101] were performed for white dwarfs with core temperatures up to 5×10^7 K. If the accreting white dwarf is still hotter, hydrogen ignition will occur further away from the degeneracy line and stable burning can proceed at lower accretion rates. Starrfield [82] has published models for hot $1.35 M_\odot$ white dwarfs, which burn hydrogen steadily as far down as $1.6 \times 10^{-8} M_\odot\,\mathrm{yr}^{-1}$ (solid dots). At an $\dot{M}$ still lower by a decade (open circle), the star experienced a strong helium shell flash, which leads to mass loss or disruption of the star before it reaches the carbon deflagration mass of $1.38 M_\odot$. In summary, there seems to be a wider range of conditions for quasi-steady hydrogen burning than indicated by the narrow band in Fig. 12.9 labeled "steady burning" and the problem of fine-tuning the accretion rate is probably less severe than thought previously. The results are promising with respect to the possibility to drive white dwarfs with $M > 0.9 M_\odot$ toward the Chandrasekhar mass into a deflagration SN Ia event [26, 82].[5]

CVs accreting at roughly $\dot{M} \lesssim 10^{-9} M_\odot\,\mathrm{yr}^{-1}$ expel more mass in a classical nova outbursts than they accrete in the quiescent intervals [68, 101]. As a result, the white dwarf cannot secularly gain mass by accretion. Furthermore, it retains little or no hydrogen after the outburst and the phase of supersoft X-ray emission of the postnova may be short or even missing [74].

12.8.2 Close-Binary Supersoft X-Ray Sources (CBSS)

One of the important imprints left behind by ROSAT is the observational definition of a class of supersoft X-ray sources with blackbody temperatures $kT < 100$ eV [21, 24, 87]. A prime property of most SSS seems to be a short duty cycle of X-ray bright states as evidenced by the fact that observations of M31 with Chandra and XMM-Newton failed to find most SSS discovered previously with ROSAT [24]. On the other hand, some CBSS, as CAL83 and CAL87 in the Large Magellanic Cloud were discovered already with Einstein [49] and have been observed as soft X-ray

[5] Carbon ignition in a CO white dwarf or oxygen ignition in an ONeMg white dwarf leads to a deflagration supernova event if the deflagration wave is sufficiently fast, whereas a collapse and neutron star formation occur when the deflagration wave is too slow. In addition, electron capture on ^{24}Mg and ^{20}Ne in an ONeMg white dwarf leads to a collapse if it starts before oxygen ignition [60].

sources for more than two decades. Hence, the long-term time variability differs for individual sources and a full discussion requires optical identification and dedicated observations.

CAL 83 and CAL 87 are known to reside in binaries with periods of order day. Members of this subclass are also RX J0513.9-6951 in the LMC, the Galactic binaries QR And (RX J0019.8+2156) and MR Vel (RX J0925.7-4758), and probably V Sge commonly classified as a long-period CV and the recurrent nova U Sco with outburst intervals of 8–43 yrs. Some other CVs have properties similar to V Sge and, hence, more potential or temporary CBSS may hide among the known CVs [83]. An attractive explanation of the properties of all these stars is thermal time scale mass transfer from a secondary more massive than the white dwarf [7, 37, 88], although one has to caution that for most objects the masses of the secondaries have not yet been measured. Nevertheless, they seem to fit into the picture described in the last section, are more or less variable, and have accretion rates around $10^{-7}\,M_{\odot}\,\mathrm{yr}^{-1}$, except for U Sco, which may have $\dot{M}$ as low as $10^{-8}\,M_{\odot}\,\mathrm{yr}^{-1}$, given the $\sim$20-yr outburst period and a very high ejection velocity (compare Fig. 12.9). For CAL 83 and U Sco, there is some spectroscopic evidence that the donor is helium-rich, which points to an origin of the white dwarf from a path other than the usual AGB evolution [7, 27].

A particularly detailed long-term time-variability study has been performed on the LMC CBSS RX J0513.9-6951. Figure 12.10 shows the result of the ROSAT HRI and optical monitoring throughout the X-ray bright interval along with the optical data folded over the cycle length [72]. The monitoring revealed the existence of quasi-regular X-ray and optical high and low states of opposite sense, suggesting that the underlying clockwork is a relaxation process. About 120 days of an optical high and X-ray off state are followed by a sharply defined $\sim$40-day optical low and X-ray on state. The oscillating behavior of RX J0513.9-6951 has been interpreted as a limit cycle brought about by variations of the accretion rate around $\dot{M}_{\mathrm{crit}}$. The Reinsch et al. model [72] is based on the viscous time scales for the irradiation-induced drainage of the outer accretion disk and the refilling of the inner disk hole created by the expanded white dwarf envelope. In this model, the mass transfer rate from the secondary may stay constant, with the accretion disk acting as a buffer. The more recent Hachisu & Kato model [28] starts from the similarity between CBSS and novae and takes account of the strong wind observed in several CBSS in the form of blue and red-shifted Balmer satellite lines. It associates the cyclic variations of the accretion rate with the interaction between the wind and the atmosphere of the secondary star, which modulates the transfer rate. Both models are typical relaxation scenarios. Other CBSS, like CAL 83 show a similar, but less regular optical/X-ray variability [23].

Another subclass of CBSS is that of symbiotic binaries, which have orbital periods of the order year and a giant donor of about solar mass or less. Examples are SMC3 (=RX J0048.4-7332) [38], the Einstein source 1E 1339.8+2837 in the globular cluster M3, and the stars AG Dra and RR Tel in the Galactic disk [21, 37]. These

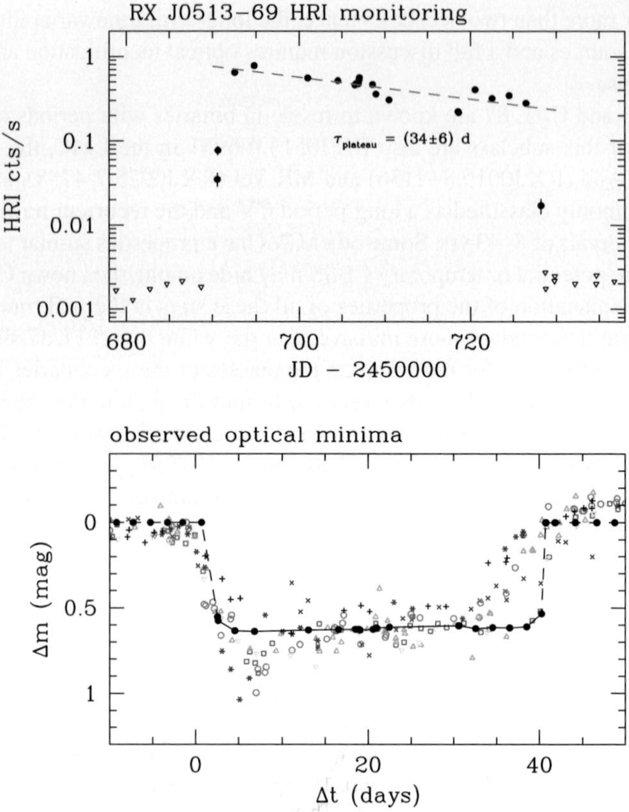

Fig. 12.10 The X-ray and optical light curves of RX J0513-69 throughout an outburst cycle (from [72])

sources also fit into the picture presented in the last section and are quite likely candidates for the SSS discovered in elliptical galaxies and in the bulges of spirals [12].

Other SSS do not fit into these groups because they have well established periods of only a few hours, typical of CVs, as the SMC source 1E 0035.4-7230 with $P = 4.1$ h [21, 37] and possibly the LMC source RX J0439.8-6809, which has a proposed orbital period of 3.4 h [21, see, however, [90]]. The accretion rate required to power these stars exceeds those typical of CVs and may be induced by the intense irradiation of the secondary star [91]. Still other sources display a transient behavior and do not fit into any of these patterns.

Progress in our understanding of the dynamical processes in CBSS and a more physical classification of the individual observed SSS requires more specific observations and substantial theoretical effort. In particular, time-dependent high-resolution X-ray spectroscopy is a promising avenue. Such observations have become possible with the transmission and reflection gratings on board of Chandra and XMM-Newton, respectively. The interpretation of such observations requires

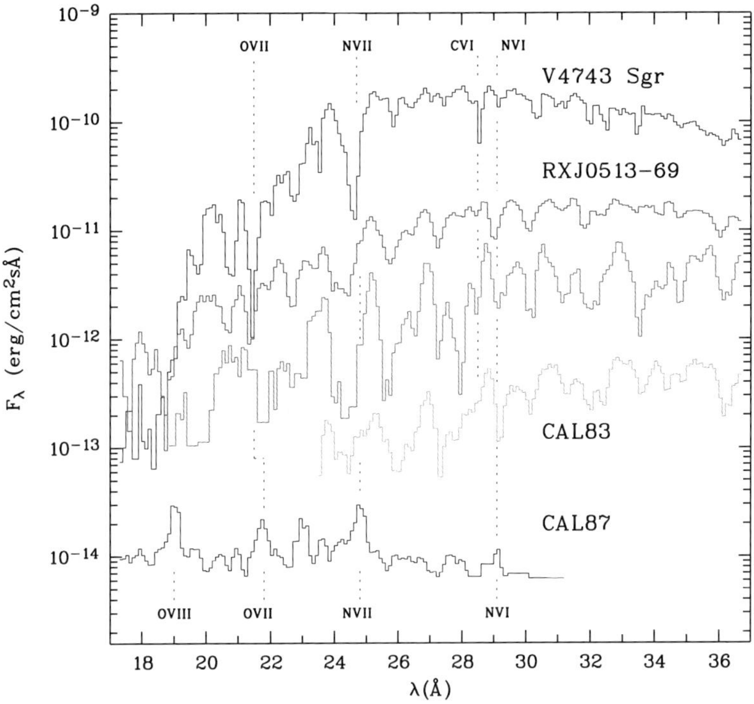

Fig. 12.11 Observed Chandra LETGS/HRC-S spectra of V4743 Sgr (N Sgr 2002 No. 3 (top), RX J0513-69 (second and third from top, multiplied by factors of 10 and 8, respectively), CAL 83 (second from bottom), and CAL 87 (bottom). The spectra have been binned into 0.1 Å bins and gently smoothed with a box car over 3 bins. They are not corrected for interstellar absorption. Noisy parts have been omitted in the lower two spectra

the extension of the art of NLTE-modeling of expanding atmospheres to the soft X-ray regime [46, 67, 71]. Figure 12.11 compares the observed spectra of the CBSS RX J0513.9-6951 [9], CAL 83 [46] and CAL 87 [6] with that of the X-ray bright phase of V4743 Sgr (N Sgr 2002) [59]. V4743 Sgr shows an absorption line spectrum with a few emission features superimposed [67, 71]. The upper of the two spectra of RX J0513.9-6951 is the mean of two exposures near X-ray maximum, the lower was taken during the rise from the X-ray off phase. The spectral structure of RX J0513.9-6951 and CAL83 suggests that strong blends and the dynamic properties of the atmosphere combine to produce what is observed. CAL 87, on the other hand, shows an emission line spectrum. Combined with the X-ray faintness of the eclipsing system, this result suggests that we do no see the accreting white dwarf directly, but only scattered photons from the wind or from the accretion disk rim. This interpretation is supported by the general similarity of the absorption and emission line spectra of the CBSS observed at different inclinations with the corresponding

[6] Chandra LETG/HRS-C observation ID 1896, August 13, 2001, PI J. Greiner.

spectra of dwarf novae in outburst shown in Fig. 12.3. At present, the interpretation of these spectra is just beginning. In the near future, high-resolution X-ray spectroscopy combined with advanced atmospheric modeling will substantially improve our understanding of the dynamical processes in CVs, novae, and CBSS – and quite certainly raise new questions as well.

References

1. Araujo-Betancor, S., Gänsicke, B.T., Long, K. et al. Astrophys. J., 622, 589
2. Beuermann, K. 2004, ASP Conf. Ser., 315, 187
3. Beuermann, K., Thomas, H.-C. 1993, Adv. Space Res. 13(12), 115
4. Beuermann, K., Reinsch, K., Barwig, H. et al. 1995, Astron. Astrophys., 294, L1
5. Beuermann, K., Harrison, T.E., McArthur, B.E., Benedict, G.F., Gänsicke, B.T. 2003, Astron. Astrophys., 412, 821
6. Beuermann, K., Harrison, T.E., McArthur, B.E., Benedict, G.F., Gänsicke, B.T. 2004, Astron. Astrophys., 419, 291
7. Bitzaraki, O.M., Rovithis-Livaniou, H., Tout, C.A., van den Heuvel, E.P.J. 2004, Astron. Astrophys., 416, 263
8. Burwitz, V., Reinsch, K., Haberl, F. et al. 2002, ASP Conf. Ser., 261, 137
9. Burwitz, V., Reinsch, K., Greiner, J. et al. 2007, ASP 40, 1294
10. Campbell, C.G. 1997, Magnetohydrodynamics in Binary Stars, Kluwer, Dordecht
11. Crawford, J.A., Kraft, R.P. 1956, Astrophys. J., 123, 44
12. DiStefano, R., Kong, A.K.H. 2003, Astrophys. J., 592, 884
13. Fischer, A., Beuermann, K. 2001, Astron. Astrophys., 373, 211
14. Forman, W., Jones, C., Cominsky, L. et al. 1978, Astrophys. J. Suppl. 38, 357
15. Frank, J., King, A.R., Lasota, J.-P. 1988, Astron. Astrophys., 193, 113
16. Fujimoto, M.Y. 1982, Astrophys. J., 257, 767
17. Gänsicke, B.T., Beuermann, K., de Martino, D. 1995, Astron. Astrophys., 303, 127
18. Gänsicke, B.T., Beuermann, K., de Martino, D., Thomas, H.-C. 2000, Astron. Astrophys., 354, 605
19. Giacconi R. 2002, Nobel Laureate Lecture
20. Ghosh, P., Lamb, F.K. 1979, Astrophys. J., 234, 296
21. Greiner, J. 1996, Supersoft X-Ray Sources, 472, 299
22. Greiner, J., Wenzel, W. 1995, Astron. Astrophys., 294, L5
23. Greiner, J., Di Stefano, R. 2002, Astron. Astrophys., 387, 944
24. Greiner, J., Di Stefano, R., Kong, A., Primini, F. 2004, Astrophys. J., 610, 261
25. Haberl, F., Motch, C. 1995, Astron. Astrophys., 297, L37
26. Hachisu, I., Kato, M., Nomoto, K. 1996, Astrophys. J., 470, L97
27. Hachisu, I., Kato, M., Nomoto, K., Umeda, H. 1999, Astrophys. J., 519, 314
28. Hachisu, I., Kato, M. 2003, Astrophys. J., 590, 445
29. Heise, J., Brinkman, A.C., Gronenschild, E. et al. 1985, Astron. Astrophys., 148, L14
30. Hessman, F.V., Gänsicke, B.T., Mattei, J.A. 2000, Astron. Astrophys., 361, 952
31. Hoogerwerf, R, Brickhouse, N.S., Mauche, C. 2004, Astrophys. J., 610, 411
32. Hellier, C., Mukai, K. 2004, Mon. Not. Roy. Astron. Soc., 352, 1037
33. Iben, I. Jr. 1982, Astrophys. J., 259, 244
34. Iben, I. Jr., Tutukov, A.V. 1996, Astrophys. J. Suppl., 105, 145
35. Iglesias, C.A., Rodgers, F.J. 1993, Astrophys. J., 412, 752
36. King, A.R., Lasota, J.P. 1979, Mon. Not. Roy. Astron. Soc., 188, 653
37. Kahabka, P., van den Heuvel, E.P.J. 1997, Annu. Rev. Astron. Astrophys., 35, 69
38. Kahabka, P. 2004, Astron. Astrophys., 416, 57

39. König, M., Beuermann, K., Gänsicke, B.T. 2006, Astron. Astrophys., 449, 1129
40. Krautter, J., this volume
41. Krzeminski, W., Serkowski, K. 1977, Astrophys. J., 216, L45
42. Kuulkers, E., Knigge, C., Steeghs, D. et al. 2002, ASP Conf. Ser., 261, 443
43. Kuulkers, E., Norton, A., Schwope, A., Warner, B. 2004, in W.H.G. Lewin and M. van der Klis (eds.), Compact Stellar X-Ray Sources, Cambridge University Press, Cambridge
44. Kuijpers, J., Pringle, J.E. 1982, Astron. Astrophys., 114, L4
45. Lamb, D.Q., Masters, A.R. 1979, Astrophys. J., 234, L117
46. Lanz, T., Telis, G.A., Audard, M. et al. 2005, Astrophys. J., 619, 517
47. Litchfield, S.J., King, A.R. 1990, Mon. Not. Roy. Astron. Soc., 247, 200
48. Livio, M. 1992, Astrophys. J., 393, 516
49. Long, K.S., Helfand, D.J., Grabelsky, D.A. 1981, Astrophys. J. 248, 925
50. Long, K.S., Mauche, C.W., Raymond, J.C. et al. 1996, Astrophys. J. 469, 841
51. Long, K.S., Sion, E.M., Gänsicke, B.T., Szkody, P. 2004, Astrophys. J. 602, 948
52. Matt, S., Pudritz, R.E. 2005, Mon. Not. Roy. Astron. Soc., 356, 167
53. Mauche, C.W. 1999, ASP Conf. Ser., 157, 157
54. Mauche, C.W. 2004, RevMexAA, 20, 174
55. Mauche, C.W. 2004, Astrophys. J., 610, 422
56. Mauche, C.W., Raymond, J.C. 2000, Astrophys. J., 541, 924
57. Mauche, C.W., Liedahl, D.A., Fournier, K.B. 2003, Astrophys. J., 588, L101
58. Mukai, K., Kinkhabwala, A., Peterson, J.R., Kahn, S.M., Paerels, F. 2003, Astrophys. J., 586, L77
59. Ness, J.U., Starrfield, S., Burwitz, V. et al. 2003, Astrophys. J., 594, L127
60. Nomoto, K., Kondo, Y. 1991, Astrophys. J., 367, L19
61. Norton, A.J., Watson, M.G. 1989, Mon. Not. Roy. Astron. Soc., 237, 853
62. Norton, A.J., Wynn, G.A., Somerscales, R.V. 2004, Astrophys. J., 614, 349
63. Paerels, F., Hur, M.Y., Mauche, C.W., Heise, J. 1996, Astrophys. J., 464, 884
64. Paczy'nski, B., Żytkow, A.N. 1978, Astrophys. J., 222, 604
65. Patterson, J., Raymond, J.C. 1985, Astrophys. J., 292, 535 and 550
66. Patterson, J. 1994, Publ. Astron. Soc. Pac., 106, 209
67. Petz, A., Hauschildt, P.H., Ness, J.-U., Starrfield, S. 2005, Astron. Astrophys., 431, 321
68. Prialnik, D.M.M., Kovetz, A. 1995, Astrophys. J., 445, 789
69. Ramsay, G., Cropper, M. 2004, Mon. Not. Roy. Astron. Soc., 347, 497
70. Ramsay, G., Cropper, M., Wu, K. et al. 2004, Mon. Not. Roy. Astron. Soc., 350, 1373
71. Rauch, T., Werner, K., Orio, M. 2005, AIP Conf. Proc., 774, 361
72. Reinsch, K., van Teeseling, A., King, A.R., Beuermann, K. 2000, Astron. Astrophys., 354, L37
73. Robinson, E.L. 1976, Annu. Rev. Astron. Astrophys., 14, 119
74. Sala, G., Hernanz, M. 2005, Astron. Astrophys., 439, 1061
75. Schmidt, G.D., Szkody, P., Vanlandingham, K.M. et al. 2005, Astrophys. J., 630, 1037
76. Schwope, A.D., Schwarz, R., Sirk, M., Howell, S.B. 2001, Astron. Astrophys., 375, 419
77. Schwope, A.D., Brunner, H., Buckley, D. et al. 2002, Astron. Astrophys., 396, 895
78. Shara, M.M., Prialnik, D., Shaviv, G. 1978, Astron. Astrophys., 61, 363
79. Shklovsky, I.S. 1967, Astrophys. J., 148, L1
80. Sion, E.M. 1999, Publ. Astron. Soc. Pac., 111, 532
81. Sirk, M.M., Howell, S.B. 1998, Astrophys. J., 506, 824
82. Starrfield, S., Timmes, F.X., Hix, W.R., Sion, E. et al. 2004, Astrophys. J., 612, L53
83. Steiner, J.E., Diaz, M.P. 1998, Publ. Astron. Soc. Pac., 110, 276
84. Tapia, S. 1977, Astrophys. J., 212, L125
85. Townsley, D.M., Bildsten, L. 2003, Astrophys. J., 596, L227
86. Townsley, D.M., Bildsten, L. 2004, Astrophys. J., 600, 390
87. Trümper, J.E., Hasinger, G., Aschenbach, B. et al. 1991, Nature, 349, 579
88. van den Heuvel, E.P.J., Bhattacharja, D., Nomoto, K., Rappaport, S.A. 1992, Astron. Astrophys., 262, 97

89. van Teeseling, A., Beuermann, K., Verbunt, F. 1996, Astron. Astrophys., 315, 467
90. van Teeseling, A., Reinsch, K., Hessman, F.V., Beuermann, K. 1997, Astron. Astrophys., 323, L41
91. van Teeseling, A., King, A.R. 1999, Astron. Astrophys., 338, 957
92. Vrtilek, S.D., Silber, A., Raymond, J.C., Patterson, J. 1994, Astrophys. J., 425, 787
93. Warner, B. 1995, Cataclysmic Variable Stars, Cambridge University Press, Cambridge
94. Watson, M.G., Sherrington, M.R., Jameson, R.F. 1978. Mon. Not. Roy. Astron. Soc., 184, 79P
95. Wheatley, P.J., Verbunt, F., Belloni, T. et al. 1996, Astron. Astrophys., 307, 137
96. Wheatley, P.J., West, R.G., 2003, Mon. Not. Roy. Astron. Soc., 345, 1009
97. Wheatley, P.J., Mauche, C.W., Mattei, J.A. 2003, Mon. Not. Roy. Astron. Soc., 345, 49
98. Wickramasinghe, D., Ferrarion, L. 2000, Publ. Astron. Soc. Pac., 112, 873
99. Woelk, U., Beuermann, K. 1992, Astron. Astrophys., 256, 498
100. Woelk, U., Beuermann, K. 1996, Astron. Astrophys., 306, 232
101. Yaron, O., Prialnik, D., Shara, M.M., Kovetz, A. 2005, Astrophys. J., 623, 398

13 Classical Novae

J. Krautter

13.1 Introduction

Classical nova explosions are the third most violent explosions that can occur in a galaxy, exceeded only by a supernova explosion and a γ-ray burst. With total liberated energies of more than 10^{45} erg s^{-1} novae are less energetic than supernovae but nova outbursts are far more frequent in a galaxy than supernovae. According to the now commonly accepted standard model, nova explosions happen in cataclysmic binary systems that are close binaries with one member, a white dwarf, and the other member, a main sequence or slightly evolved star that fills its Roche lobe. Mass transfer from the secondary to the white dwarf through the inner Lagrangian point leads to the formation of an accretion disk. Ultimately, because of viscous processes, the material ends up on the surface of the white dwarf. The accreted layer of hydrogen-rich material on top of the white dwarf will grow in thickness and temperature and density will increase. If a certain critical pressure is reached at the bottom of the accreted envelope (which must be, at least partially, degenerate), explosive thermonuclear burning of hydrogen via the CNO cycle will start. The evolution of this "thermonuclear runaway" (*TNR*) depends upon the mass and luminosity of the white dwarf, the mass accretion rate, and the chemical composition of the accreted layer (cf. e.g., [33]). The temperature at the hydrogen burning zone will grow to values exceeding 10^8 K. Luminosities are of the order of the Eddington luminosity L_{Edd}, peak luminosities in the early phases after the onset of the TNR can even be in excess of L_{Edd} by factors up to ten. Convection turns on and transports the energy generated to the surface of the accreted envelope. Radioactive decay of β^+-active nuclei eventually provides the energy to eject the nova shell. Ejection velocities are between several hundred and several thousand kilometer per second. According to the TNR model the entire character of the outburst, light curve, ejection velocities, and timescale of the outburst evolution depend on the amount of CNO nuclei present in the accreted envelope (Starrfield & Sparks [34]). One of the predictions of the TNR model was that novae should have strong overabundances of CNO nuclei, which was later verified to a full extent. The CO nuclei are supposed to be mixed from the white dwarf into the accreted envelope. Abundance studies of novae revealed the existence of a new class of white dwarfs; about 25% of all novae show strong overabundances of O, Ne, and Mg that come from a white dwarf with an ONeMg core. Novae are empirically classified by the decay time t_3, the

J.E. Trümper and G. Hasinger (eds.), *The Universe in X-Rays.*

time it takes for the visual brightness to decline from maximum brightness by three magnitudes, which is a measure of the violence of the outburst. Fast novae have $t_3 < 100^d$, slow novae $t_3 \geq 100^d$. According to the number of outbursts observed, one distinguishes between classical and recurrent novae. For recurrent novae two or more outburst were recorded, while for classical novae only one outburst was observed. In the following, we shall concentrate on classical novae that are much more frequent than recurrent novae. However, it should be mentioned that on longer timescales all novae are believed to be recurrent.

X-ray observations have turned out to be a very powerful tool to study the outburst of novae, since the X-ray regime is best suited to study the hot phases in a nova outburst. X-ray observations have provided many fundamental and in part unexpected results. However, so far the picture that emerged from X-ray observations of novae is far less systematic than the one from other spectral regimes, since unlike in the optical, infrared, or ultraviolet regime, only few objects were observed in X-rays.

The progress in our knowledge of X-rays from novae is closely related to the advances in sensitivity and spectral resolution of subsequent gererations of X-ray satellites. There are three milestones in the observational X-ray history of novae: the first nova in outburst was detected with EXOSAT, ROSAT and its much superior sensitivity established many basic X-ray properties, and the next big step forward came with XMM and Chandra particularly through their grating spectrometers with a resolution of up to a thousand ($R/\Delta R$). Reviews of X-ray observations of novae in outburst were given by Krautter [13, 14], Ögelman & Orio [22], and Orio [29].

13.2 Sources of X-Rays

There are several mechanisms for X-ray emission from a nova in outburst. The most obvious one is thermal radiation from the hot white dwarf. First, one expects thermal radiation during the very early phases of the outburst. After the energy created by the nuclear reactions has reached the surface of the white dwarf, the effective surface temperature increases rapidly. With a luminosity of the order of L_{Edd} and a white dwarf radius one expects temperatures of several hundred thousand K, depending on the white dwarf mass (and, thereby, radius). The nova emits strong soft X-rays with the spectral energy distribution (SED) of a hot stellar atmosphere. After the ejection of the nova shell the temperature of the expanding pseudophotosphere decreases rapidly with increasing radius and the X-ray flux drops, since the expanding envelope becomes opaque to X-rays. Since the duration of the soft X-ray emission during this so called "fireball phase" is very short, of the order of a few hours only, no X-rays have ever been observed during this phase. Also in future it will be very difficult to observe a nova in X-rays during the fireball phase.

A second phase of X-ray emission from the hot white dwarf is found in later phases of the outburst. As the TNR model predicts, only part of the ejected shell reaches velocities higher than escape velocity. The remaining material quickly returns to quasistatic equilibrium and forms an envelope around the cataclysmic binary system with the initial dimensions of a giant star. The matter in this quasistatic envelope supplies the fuel for the ongoing hydrogen burning (via the CNO cycle), which takes place at constant luminosity on top of the white dwarf. As evolution proceeds during this "constant bolometric luminosity phase," the residual hydrogen-rich envelope matter will be consumed by hydrogen burning, radiation driven mass loss, and dynamical friction. The radius of the stellar photosphere shrinks in time at a constant luminosity of $\sim L_{\mathrm{Edd}}$ so that the temperature increases up to several hundred thousand Kelvin, depending on the white dwarf mass. The peak of the nova's observed SED shifts from visual to ultraviolet and finally to the X-ray energy band. The nova becomes again a strong X-ray emitter with a soft SED of a hot stellar atmosphere. The duration of the phase of constant bolometric luminosity should be an inverse function of the white dwarf mass [12, 35].

The second general mechanism is X-ray emission from the circumstellar material surrounding the nova system where the expanding nova shell and/or a nova wind interact with preexisting material or with each other. Here one expects strong shocks giving rise to a thermal bremsstrahlung SED with temperatures up to several kiloelectronvolt. Balman, Krautter and Ögelman [2] have summarized the different ways by which the X-rays can be produced by shocks.

Two further possible mechanisms for X-ray radiation should be shortly mentioned: After reestablishment of accretion one expects X-ray emission typical for cataclysmic variables in quiescence with a thermal bremsstrahlung spectrum. Some evidence for such an X-ray emission was found by Hernanz and Sala [8] for V2487 Oph (1998). A mechanism of a totally different nature was suggested by Livio et al. [15], namely that downgradation of gamma rays produced by radioactive decay of Na^{22} could give rise to X-rays. However, no evidence for this mechanism has been found so far.

13.3 EXOSAT: A Rather Noisy Beginning

The story of X-ray observations of classical novae in outburst began on April 20, 1984, when GQ Mus (1983) was discovered 460 days after outburst at a 4.5 σ level with the low energy telescope (0.04–2 keV) and the CMA onboard EXOSAT by Ögelman, Beuermann, and Krautter [23]. Because of the missing spectral resolution of the CMA, no spectral information could be obtained. The count rate was compatible with either a shocked shell of circumstellar gas emitting 10^7 K thermal bremsstrahlung radiation at 10^{35} erg s^{-1} or with a hot white dwarf remnant emitting at $3.5 \cdot 10^{37}$ erg s^{-1}. Similar results were obtained from three more observations of GQ Mus and from several observations of QU Vul (1984 #1) and PW Vul (1984 #2) [24]. The data of QU Vul and PW Vul that were both observed during early phases indicated that the count rate increased during the early phases of the outburst. The results were consistent with hydrogen burning of a hot white dwarf at constant bolometric luminosity as predicted by the TNR model.

13.4 ROSAT: Basic Properties

A breakthrough came when ROSAT started its observation in 1990. With the position sensitive proportional counter (PSPC), which had an energy range 0.1–2.4 keV, X-ray data of much higher sensitivity could be obtained than from any comparable X-ray telescope/detector before. Even if the energy resolution of the PSPC was poor, some spectral information could be obtained and many basic X-ray properties of novae could be detected. Several observational programs mark the process in our understanding of X-ray properties of novae.

The most fundamental results were obtained from the observations of V1974 Cyg (1992), a moderately fast ($t_3 \sim 35^d$) ONeMg nova, which was discovered in outburst on February 20. With a maximum brightness $V_{\rm max} \sim 4.4$ mag it was the brightest nova since V1500 Cyg (1975). Observations with the PSPC onboard of ROSAT started on April 20, 63 days after outburst, as soon as the nova had entered the ROSAT observing window (Krautter et al. [11]). Subsequently, V1974 Cyg was observed by Krautter et al. [12] with ROSAT in the 0.1–2.4 keV energy range over a period of nearly two years on a total of 18 occasions.

Three phases can be distinguished in the lightcurve presented in Fig. 13.1: From days 63 to 147 during an initial rise phase the count rates are low, however, the count rate is strongly increasing from 0.03 to 0.37 counts s^{-1}. Three months later,

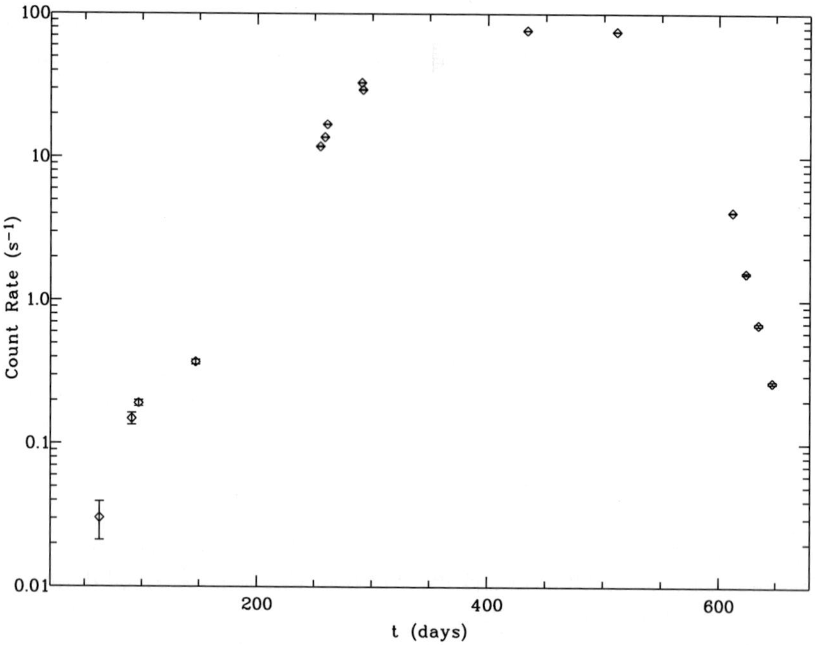

Fig. 13.1 X-ray lightcurve log counts s^{-1} vs. days after outburst of V1974 Cyg [12]

on day 255, the beginning of the second phase, the "plateau phase," a big surprise came when the count rate had enormously increased to 11.3 counts s^{-1}. The count rate increased further to a peak of about 76 counts s^{-1} on day 434 and remained essentially the same on the next observation on day 511. The last phase, the decline phase, starts between days 511 and 612 and is characterized by a strong decline of the count rate down to 0.2 counts s^{-1} on day 653. Because of the large gaps of about 3 months between the observing windows, there is an uncertainty in the beginning and the end of the individual phases.

The spectral properties are reflected in the three phases of V1974 Cyg. SED characteristics for each of the three phases is presented in Fig. 13.2 [2]. During the rise phase, the SEDs are hard with essentially no photons below 0.7 keV with the hardest SED found on day 63. Subsequently, the SED got softer, the first indications

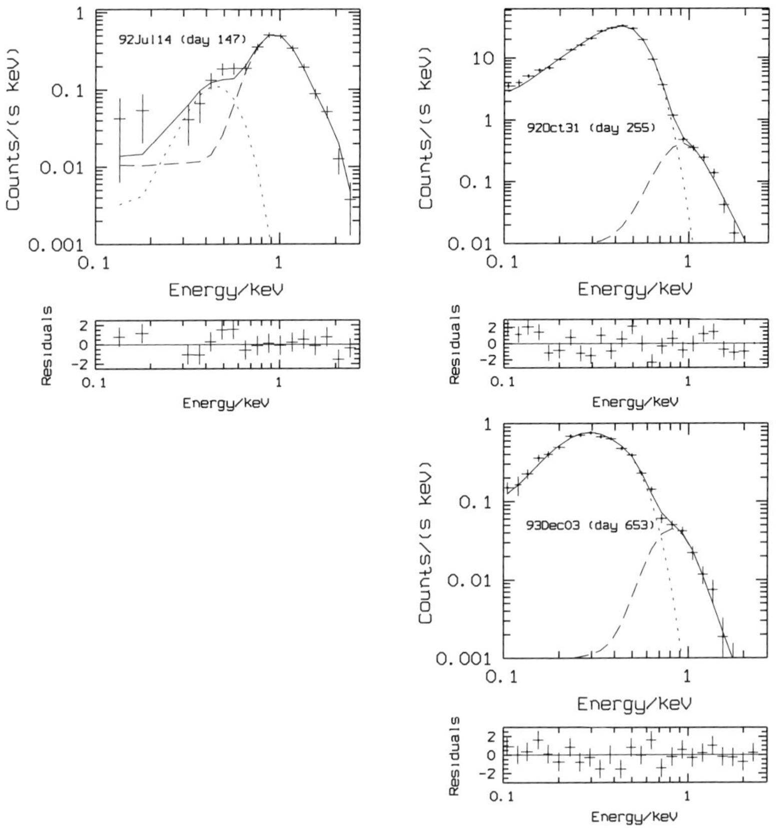

Fig. 13.2 Spectral energy distributions of V1974 Cyg in the ROSAT PSPC band on three different epochs. The actual PSPC data are indicated with *crosses*. The data are fitted with O-Ne-enhanced WD atmosphere models (soft component) and Raymond-Smith plasma emission models (hard components). The lower figures show the residuals between the data and the models in standard deviations [2]

for a soft component showed up on day 97. A dramatic change occurred between days 147 and 255, since on the latter date a strong soft component had appeared. The SED exhibited now, like for all other spectra obtained during the plateau and the decline phase, the characteristics of a super-soft source (SSS) (e.g., [9]). With a maximum count rate of 76.5 counts s^{-1} on day 511 V1974 Cyg was by far the strongest SSS ever observed by ROSAT. The hard component that had shown up in the early spectra was present during the whole plateau phase. During the decline phase, the soft component decreased strongly and the hard component became more prominent again in relation to the soft component.

The origin and temporal behavior of the soft component can be best interpreted in terms of the TNR model of the nova outburst. As Krautter et al. [12] had shown, fits with blackbody energy distributions did not give any reasonable results. This result can be generalized to all other SSS [9]. Balman et al. [2] analyzed the ROSAT data with LTE white dwarf model atmospheres with O–Ne overabundances by MacDonald and Vennes [17]. They could show that, as predicted by the TNR model, during the plateau phase $T_{\rm eff}$ increased at constant luminosity from $\sim 4\times 10^5$ K up to to a peak temperature of $\sim 6\times 10^5$ K, while the white dwarf radius decreased by about a factor of 2. During the decline phase, both L and $T_{\rm eff}$ decreased indicating that hydrogen burning on top of the white dwarf had turned off. Using a cooling timescale of about 6 months Krautter et al. [12] estimated a mass of $\sim 10^{-5}$ $M_\odot$ for the hydrogen-exhausted, remnant envelope of the white dwarf.

Balman, Krautter and Ögelman also carried out a detailed analysis of the hard component applying Raymond-Smith thermal plasma emission models. Although the plasma temperature decreased from an initial $\sim$5–10 keV to a more or less constant level around $\sim$1 keV, the flux increased to a maximum around day 150 with a subsequent decline. The peak unabsorbed flux corresponded to a luminosity of (0.8–2.0)$\times 10^{34}$ erg s^{-1} at a distance of 2–3 kpc. The temporal evolution and the plasma temperature suggest a shock origin of the hard X-ray flux. Balman, Krautter, and Ögelman favor a model in which a fast wind from the nova collides with either a preexisting gaseous shell or material ejected at the initial phase of the nova.

The reason for the absence of any soft X-ray emission during the early phases is the initial high hydrogen density $N_{\rm H}$ which decreased by about a factor of 10, until it reached a constant level around day 255. Lloyd et al. [16] found a similar result for the fast nova V838 Her (1991), which was observed 5 days after visual maximum. Like V1974 Cyg only a hard component with no photons below 0.7 keV was found at a count rate of 0.16 counts s^{-1}.

The second milestone was GQ Mus, the first nova of which X-rays were discovered in outburst in 1984. GQ Mus was also detected by ROSAT, 6 years after its last EXOSAT observation and 8.5 years after maximum (day 3118) at a count rate of 0.143±0.006 counts s^{-1} in the ROSAT All-sky Survey (RASS). At subsequent observations carried out in the pointed mode the count rate dropped from 0.127 counts s^{-1} on day 3322 [25] via 0.007 counts s^{-1} on day 3871 to <0.003 (3σ) counts s^{-1} [31]. Balman and Krautter [1] showed that the hydrogen burning in GQ Mus had most likely ceased already on day 3322. With a duration between 8.5

and 9.1 years GQ Mus exhibited an extraordinarily long phase of hydrogen burning, the longest known so far. The much longer turn-off timescale of GQ Mus, when compared with V1974 Cyg, indicates that the mass of the white dwarf in GQ Mus is lower than that one in V1974 Cyg. GQ Mus had also some unsusual properties in the optical spectral range [10]. For many years its emission line spectrum exhibited a very high ionization with the coronal lines [FeXI] $\lambda 7892$ as strong and [FeX] $\lambda 6374$ about 2.5 as strong as Hα, respectively, which has not been observed – at least to the knowledge of the author – so far in any other astronomical object. As Krautter and Williams showed, the high ionization lines were due to photoionization by a hot (several 10^5 K) underlying source. In a spectrum taken on March 10, 1993, after the turn-off of the hydrogen burning, the coronal lines had disappeared and the ionization was much lower. This shows that the ionization stage of the optical spectrum could be used generally as a qualitative indicator for the turn-off of the hydrogen burning.

To get more information on the duration of the phase of constant luninosity, Orio et al. [26] did a systematic search for supersoft-sources in the ROSAT archive. They analyzed 350 pointed and serendipitous observations of 108 different classical and recurrent novae in outburst and in quiescence. They found only three novae with a super-soft spectrum, the already mentioned V1974 Cyg and GQ Mus and, in addition, N LMC 1995. For a time of up to 10 years after explosion, 30 galactic and 9 LMC novae were in the ROSAT sample. In the post-ROSAT phase a few more novae with soft X-ray emission were detected by recent X-ray satellites (see later), but it is clear that the vast majority of all postnovae were not observed as soft X-ray sources. One has to ask whether this missing soft X-ray emission is real, i.e., novae switch off hydrogen burning after a relatively short time, or whether it is due to a selection effect. A plausible selection effect could be that the interstellar hydrogen column density is so high that the soft X-ray radiation gets absorbed. However, as Nickel [21] showed, 21 novae, which were observed in the ROSAT All-sky Survey within 10 years after the outburst and for which extinction data were available, have an average extinction $E_{B-V} \sim 0.5$ mag. With this value, only for a few novae the hydrogen column densities should be so high that all the soft X-ray radiation gets absorbed. Even more striking is that with one exception no LMC nova was found where the average extinction E_{B-V} is below 0.2 mag. Circumstellar extinction should not play any role, since most novae were observed months or even years after maximum when the expanding envelope should have become transparent for soft X-ray radiation. On the basis of these data, a selection effect can be safely excluded. One can conclude that the majority of all novae switch off after a relatively short time. Of course, a precise number cannot be given, but it seems safe to conclude that for most novae hydrogen burning turns off after less than 2 years.

The turn-off time scale as function of the white dwarf mass was calculated by Starrfield et al. [35] who assumed that the accreted envelope is ejected by a radiation driven wind. They find a strong inverse dependence of the turn-off time scale from the white dwarf mass. For instance, a nova with a 1.00 $M_\odot$ white dwarf should turn-off after about 100 years whereas a 1.25 $M_\odot$ white dwarf has a much shorter turn-off time of about two years. The short turn-off timescales found from the X-ray observations is a clear evidence that most novae have white dwarfs with masses well above $\sim$1.0–1.1 $M_\odot$. Such a result had been predicted by the TNR model, since according to this model the critical mass that is needed to ignite the thermonuclear reactions and to start the runaway decreases strongly with increasing white dwarf mass [33]. Statistically, novae with high mass white dwarfs should be found much more frequently than novae with white dwarfs with lower masses.

13.5 Chandra and XMM: High Resolution and New Surprises

While only few observations were obtained with ASCA, RXTE, and BeppoSAX, another big step forward came with Chandra and XMM, which have higher sensitivity than ROSAT, grating spectrometers, which allow for X-ray observations with a resolution of up to $(R/\Delta R) \sim 1\,000$, and an energy range that extends, at the hard end, far beyond ROSAT's 2.4 keV.

Already the first nova observed with Chandra, V382 Vel (1999), clearly demonstrated the impact of the new high resolution facilities. V382 Vel, a fast ($t_3 \sim 10^d$) ONeMg nova was with a maximum brightness $V_{\rm max}$=2.6 mag one of the brightest novae of the last century and was extensively observed in many different spectral regimes. X-ray observations with RXTE, ASCA, and BeppoSAX started immediately after discovery [18, 27]. During the early phases a hard component was found, which Mukai and Ishida attributed to emission from shocks within the nebular ejecta. On day 185, Orio et al. [28] found with BeppoSAX observations a strong super-soft component, which was still present when V382 Vel was observed on day 223 for the first time with Chandra and ACIS [4].

The first high resolution observation (R $\sim$ 600) of any nova in outburst were carried out by the same team on day 268 with the low energy transmission grating (LETG) onboard of Chandra [20]. Figure 13.3 shows the LETG spectrum. Surprisingly, the strong soft component had disappeared and was replaced by an emission line spectrum. Below 50 Å a marginal continuum emission is found consistent with a black-body spectrum with a temperature of T = 2.7×10^5 K. The most prominent emission lines are N VI, O VII, O VIII, Ne IX, Ne X, Mg X, Mg XI, and Si XII. No Fe lines were found. Ness et al. conclude that the abundances of all elements showing lines are significantly enhanced in the ejected material and that Fe shows normal abundance (to hydrogen), i.e., no underabundance. The emission lines are broadened and exhibit a complex line profile. From the Gaussian line width, Ness et al. estimate an expansion velocity of $\sim$1 200 km s^{-1} which is somewhat lower than the one found from UV spectra [32]. Since the emission lines found in the spectrum are formed under a range of temperature conditions, Ness et al. could get some rough information on the temperature distribution responsible for the formation of the lines.

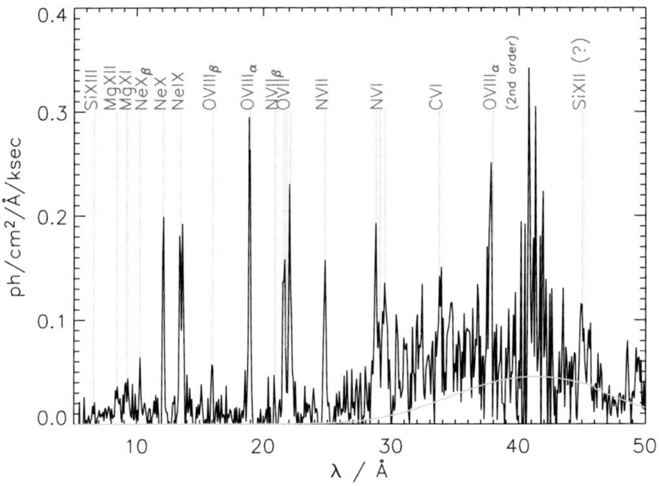

Fig. 13.3 Chandra LETG spectrum of V382 Vel (background subtracted). A continuum model is overplotted representing a diluted thermal black-body spectrum [20]

The absence of the soft-component on day 268 indicates that hydrogen burning must have turned off between days 223 and 268, probably shortly after day 223, since even the longest possible cooling time of about 6 weeks is extremely short. The total duration of the hydrogen burning was with only 7.5–8 months very short. This indicates that the white dwarf in V382 Vel has a high mass that is consistent with its ONeMg nature and the short decay time t_3.

The next chapter in the story of X-ray observations was written by V1494 Aql (1999), a fast CO nova with a t_3 of 13 days. The first two observations with Chandra's low resolution detector ACIS on days 134 and 187 yielded a hard spectrum with emission lines, but no soft component. After a strong soft component had appeared on day 247, two high-resolution observations with LETG+HRC were carried out on days 300 and 303 with exposure times of 8 and 17 ks, respectively. The high-resolution spectrum was the first one even obtained from a nova in its supersoft phase and features never seen before showed up. The spectra are dominated by a strong soft continuum component on which features are superimposed, which look on a first glance like emission lines. However, so far none of these features have been identified with any known emission lines. Therefore, it cannot be excluded that the spectrum is in reality an absorption spectrum where the emission features are only those part with less local absorption. UV spectra with such a character were observed during the early fireball phase [7]. However, during the early fireball phases one observes the initial expansion of the opaque shell that was certainly not the case 10 months after the outburst started. The application of suitable NLTE models to the observed SED would help to clarify this situation; however, this has not been done so far. The soft component had disappeared on day 726. Since no other X-ray observation had taken place in between, the duration of the hydrogen burning phase between 10 months and <2 years is rather badly defined.

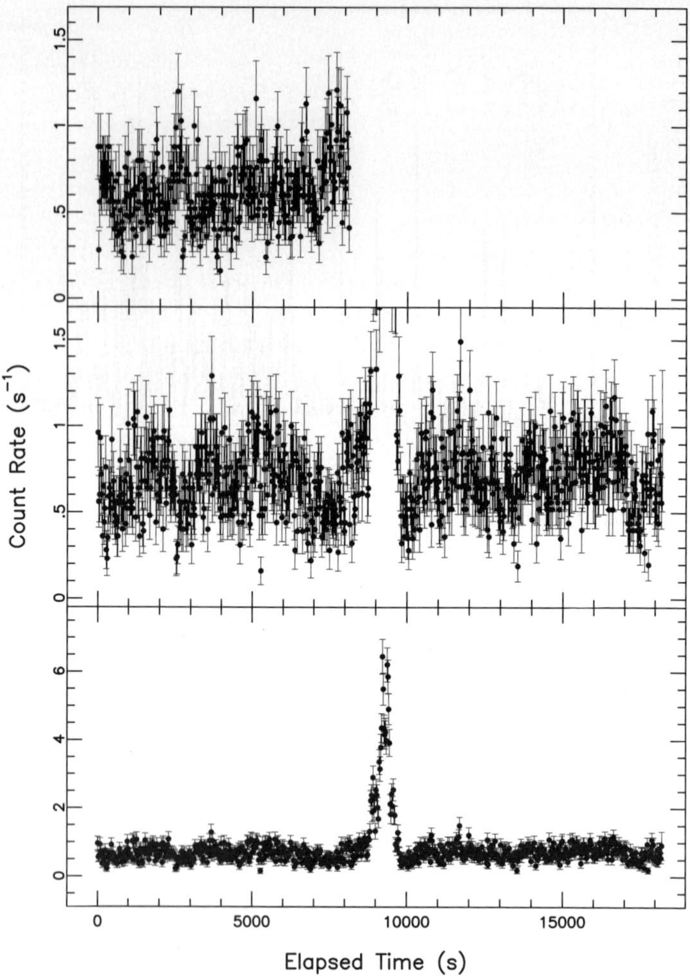

Fig. 13.4 Chandra LETG+HRC-S X-ray light curve of V1494 Aql. *Top*: light curve obtained on day 300. *Middle*: Light curve obtained on day 303. *Bottom*: Same as middle panel but data scaled by the full range of the count rate to highlight the burst [5]

An analysis of the X-ray lightcurve of V1494 Aql by Drake et al. [5] yielded several unusual features. Figure 13.4 shows the two X-ray lightcurves, which were extracted by Drake et. al from the bright zeroth order of the LETG+HRC-S spectrum.

Three diffent types of variability are found in these lightcurves: (i) A stochastic irregular short term variability is present on timescales of a few minutes. This kind of variability seems to be present in all X-ray lightcurves of novae obtained during the super-soft phase. For instance, from BeppoSAX observations of V382 Vel Orio et al. [28] found irregular flickering and a decrease of the flux in the 0.1–0.7 keV range by a factor of two within less than 1.5 h. The flux remained faint for some 15 min; no significant spectral changes were found. So far no convincing explanation for this variability could be found; Orio et al. exclude absorption due to an ejected clump of matter since in this case the hardness ratio of the spectrum should have changed. (ii) Very obvious in the lower panel is a short time X-ray burst, which lasted about 1 000 s. At its peak, the count rate of the burst was about a factor of six higher than the mean level before and after the burst. During the burst, the X-ray count rate showed a complex rise and fall with several maxima and minima, two main flares with possibly a precursor and a trailer. During the burst, the spectrum is slightly harder than during the rest of the observations. So far the nature of this outburst remains a puzzle. (iii) A timing analysis of the combined 25 ks observations revealed the existence of periodic variations with a period $P \sim 2\,499$ s. This period was found in independent analyses of both the zeroth-order and the dispersed spectrum. To check whether the periodicity found might be an instrumental artifact, i.e., related to spacecraft motion or dithering, Drake et al. performed identical periodogram analyses on HZ 43 and Sirius B. Since in neither of these analyses was any periodicity found, Drake et al. concluded that the observed periods are real and not instrumental. In addition to the 2 499 s period, several other periods were found. This suggests that the periodic variations are not due to rotation of the white dwarf. Drake et al. interpret this result as the discovery of nonradial g^+-mode pulsations in the hot, rekindled white dwarf that is driven by the κ/γ effects in the partial ionization zones of C and O near the surface of the white dwarf. The hot, luminous white dwarf in a nova evolving from explosion to quiescence has a structure that resembles that of the central stars of planetary nebulae. The power spectrum and the X-ray lightcurve of V1494 Aql are very similar to the hot central star of several planetary nebulae [3].

V1494 Aql did not remain a unique case. Figure 13.5 shows the lightcurve of a 25 ks observation with Chandra/LETG+HRC of V4743 Sgr (2002), another fast nova with $t_3 < 15$ days [19]. The observation was carried out on day 180 after the nova had entered the super-soft state.

Immediately obvious are large-amplitude variations from ~30 to 60 counts s^{-1} with a period of 1 325 s followed by a decline in the total count rate after ~13 ks of observations. The count rate dropped from ~40 counts s^{-1} to practically zero within ~6 ks and stayed low for the remaining 6 ks of the observations. Besides the period of 1 325 s a timing analysis revealed two harmonic overtones at 668 and 448 s. Orio [29] reports that a lightcurve obtained from XMM-Newton observations 2 months after the Chandra observations yielded a rich power spectrum with two main periods of 1 308 and 1 374 s. As the middle and lower panels of Fig. 13.5 show, the spectral hardness ratio changed from maxima to minima in correlation with the oscillations and became significantly softer during the decay. Neither the interpretation of the periodic oscillations nor of the decline of the soft flux is straightforward. The strong amplitude of the oscillations is difficult to reconcile with pulsations of the white dwarf. Maybe a combination of rotation and pulsation of the white dwarf was the cause of the complicated lightcurve. For the decline of the count rate an eclipse would be the most obvious interpretation. However, the duration of the decline

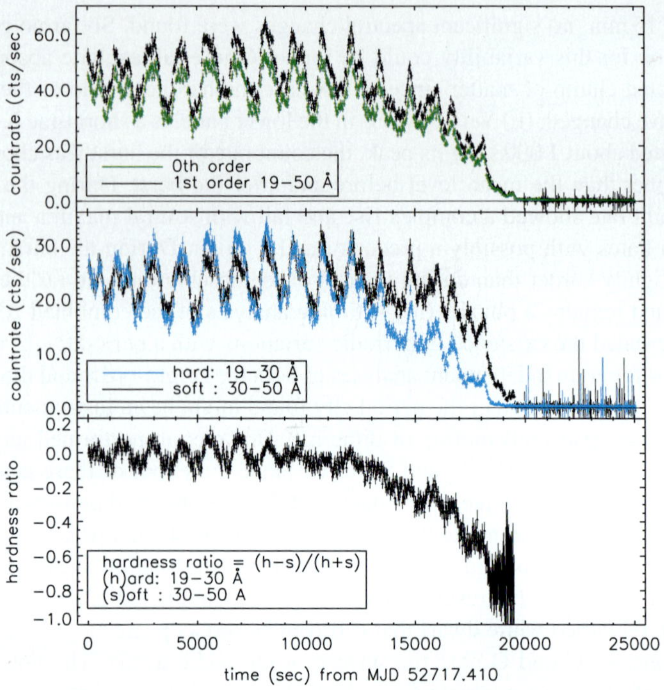

Fig. 13.5 Light curve of a 25 ks exposure of V4743 Sgr extracted in the designed wavelength intervals. *Top*: Complete wavelength ranges in zeroth and first order. *Middle*: Light curve broken into "hard" and "soft" components of the spectrum. *Bottom*: Time evolution of the hardness ratio [19]

(>6 ks) is much too long for a typical cataclysmic variable binary period of several hours. If it was an eclipse, V4743 Sgr should have a longer period of ~2 days like the old nova GK Per (1918), which is an intermediate polar with hard X-ray variations. An alternate solution could be a third component in the system. The measurement of the orbital binary period will help to clarify this question. The spectrum obtained after the decline of the count rate showed emission lines of CVI, NVI, and NVII, which are broadened by ~800–1 200 km s^{-1}. Three months later, at the XMM-Newton observation and at subsequent Chandra observations, the count rate had indreased again and was at about the same level as at the first obsererervation on day 180.

The spectrum of V4743 Sgr (Fig. 13.6) exhibited a strong super-soft continuum with absorption features, which could be identified by Ness et al. as H- and He-like lines of CV, CVI, NVI, NVII, and OVII, and are blueshifted by ~2 400 km s^{-1}. This velocity is twice the expansion velocity of 1 200 km s^{-1} found for the material ejected early in the outburst. Petz et al. [30] carried out model calculations with PHOENIX, a 1D spherical, expanding, line blanketed, full NLTE model atmosphere [6]. A best fit yielded an effective temperature of $T_{\rm eff} = 5.8\times10^5$ K, an expansion velocity of 2 500 km s^{-1} and a bolometric luminosity of 2×10^{38} erg s^{-1}.

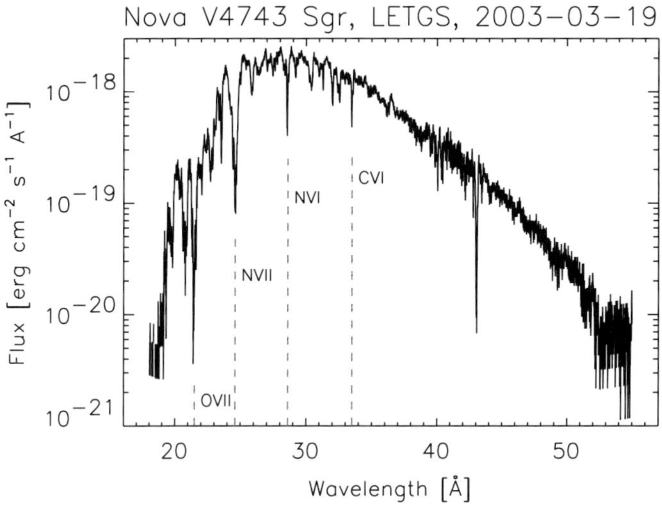

Fig. 13.6 Chandra LETGS spectrum of V4743 Sgr. The strongest absorption lines are indicated. Probably also weak emission lines are present [19]

13.6 Concluding Remarks

During the 21 years of the exciting history of X-ray observations of novae, results have changed a from very noisy first detection just above the significance limit to high resolution X-ray spectra. The X-ray regime is crucial for the observational coverage of novae, since many phenomena are only observable in the X-ray regime. X-ray observations allow us to study the hot white dwarf and shocks in the expanding envelope, to derive temperatures and luminosities and to carry out plasma diagnostics of the line emitting regions.

However, the picture we have is far from complete and systematic. Only few novae have so far been observed in X-rays and for only one nova, V1974 Cyg, does a fairly complete coverage of the whole outburst cycle exist. All other observations are more or less "snapshots." One of the main characteristics of the X-ray results obtained so far is that each nova was different from the predecessors and nearly each new nova observed in X-rays brought new surprises and, in part, totally unexpected phenomena like X-ray bursts and periodic variations. Many of these phenomena are still unexplained, in particular the variations of the soft X-ray flux defied any generally accepted explanation to this date. A problem of more general kind is that during the early phases the soft flux is not observable because of the high opacity in the expanding envelope.

For the future more, and above all, systematic X-ray observations over the whole outburst are needed. Only X-rays allow us to look directly at those areas in a nova where the essential physical processes are going on, i.e., where the energy is produced. All other wavelength regions can yield only indirect evidence from these regions.

References

1. Balman, S., Krautter, J. 2001, Mon. Not. Roy. Astron. Soc., 326, 1441
2. Balman, S., Krautter, J., Ögelman, H. 1998, Astrophys. J., 499, 395
3. Bond, H.E., Kawaler, S.D, Ciardullo, R., Stover, R., Kuroda, T., Ishida, T., Ono, T., Tamura, S., Malasan, H., Yamasaki, A., Hashimoto, O., Kambe, E., Takeuti, M., Kato, T., Kato, M., Chen, J.-S., Leibowitz, E.M., Roth, M.M., Soffner, T., Mitsch, W. 1996, Astron. J., 112, 2699
4. Burwitz, V., et al. 2002, in M. Hernanz, J. José (eds.), Classical Nova Explosions: International Conference on Classical Nova Explosions, 377
5. Drake, J., Wagner, R.M., Starrfield, S., Butt, Y., Krautter, J., Bond, H.E., Della Valle, M., Gehrz, R.D., Woodward, C.E., Evans, A., Orio, M., Hauschildt, P., Hernanz, M., Mukai, K., Truran, J.W. 2003, Astrophys. J., 548, 448
6. Hauschildt, P. 1992, Journal of Quantitative Spectroscopy and Radiative Transfer, 47, 433
7. Hauschildt, P.H., Wehrse, R., Starrfield, S., Shaviv, G. 1992, Astrophys. J., 393, 307
8. Hernanz, M., Sala, G. 2002 Science, 298, 393
9. Kahabka, P., van den Heuvel, E.P.J. 1997, Annu. Rev. Astron. Astrophys., 35, 69
10. Krautter, J., Williams, R.E. 1989, Astrophys. J., 341, 968
11. Krautter, J., Ögelman, H., Starrfield, S., Trümper, J., Wichmann, R. 1993, Ann. Isr. Phys. Soc., 10, 28
12. Krautter, J., Ögelman, H., Starrfield, S., Wichmann, R., Pfeffermann, E. 1996, Astrophys. J., 456, 788
13. Krautter, J. 2002, in M. Hernanz, J. José (eds.), Classical Nova Explosions: International Conference on Classical Nova Explosions, 345
14. Krautter, J. 2008, Chap. 10 in M. Bode, A. Evans (eds.), Classical Novae, 2nd edn., Cambridge University Press, Cambridge, in press
15. Livio, M., Mastichiadis, A., Ögelman, H., Truran, J.W. 1992, Astrophys. J., 394, 217
16. Lloyd, H.M., O'Brien, T.J., Bode, M.F., Predehl, P., Schmitt, J.H.M.M., Trümper, J., Watson, M.G., Pounds, K.A. 1992, Nature, 356, 222
17. MacDonald, J., Vennes, S. 1991, Astrophys. J., 373, L51
18. Mukai, K., Ishida, M. 2001, Astrophys. J., 551, 1024
19. Ness, J.-U., Starrfield, S., Burwitz, V., Wichmann, R., Hauschildt, P., Drake, J., Wagner, R.M., Bond, H.E., Krautter, J., Orio, M., Hernanz, M., Gehrz, R.D., Woodward, C.E., Butt, Y., Mukai, K., Balman, S., Truran, J.W. 2003, Astrophys. J., 594, L127
20. Ness, J.-U., Starrfield, S., Jordan, C., Krautter, J., Schmitt, J.H.M.M., Mon. Not. Roy. Astron. Soc., 364, 1015
21. Nickel, U. 1995, Examensarbeit Ruprecht-Karls-Universität Heidelberg
22. Ögelman, H., Orio, M., 1995, in A. Bianchini, M. della Valle, M. Orio (eds.), Cataclysmic Variables, Astrophys. Space Sci. Lib., 205, 11
23. Ögelman, H., Beuermann, K., Krautter, J. 1984, Astrophys. J., 287, L31
24. Ögelman, H., Krautter, J., Beuermann, K. 1987, Astron. Astrophys., 177, 110
25. Ögelman, H.B., Orio, M., Krautter, J., Starrfield, S. 1993, Nature, 361, 331
26. Orio, M., Covington, J., Ögelman, H. 2001a, Astron. Astrophys., 373, 542
27. Orio, M., Parmar, A.N., Benjamin, R., Amati, L., Frontera, F., Greiner, J., Ögelman, H., Mineo, T., Starrfield, S., Trussoni, E. 2001b, Mon. Not. Roy. Astron. Soc., 326, L13
28. Orio, M., Parmar, A.N., Greiner, J., Ögelman, H., Starrfield, S., Trussoni, E. 2002, Mon. Not. Roy. Astron. Soc., 333, L11
29. Orio, M. 2004, RevMexAA (Serie de Conferencias), 20, 182
30. Petz, A., Hauschildt, P.H., Ness, J.-U., Starrfield, S., Astron. Astrophys., 431, 321
31. Shanley, L., Ögelman, H., Gallagher, J., Orio, M., Krautter, J. 1995, Astrophys. J., 438, L95
32. Shore, S.N., Schwarz, G., Bond, H.E., Downes, R.A., Starrfield, S., Evans, A., Gehrz, R.D., Hauschilt, P.H., Krautter, J., Woodward, C.E. 2003, Astron. J., 125, 1507
33. Starrfield, S. 1989, in M. Bode, A. Evans (eds.), Classical Novae, New York, Wiley, 39
34. Starrfield, S., Sparks, W.M. 1987, Astrophys. Space Sci., 131, 397
35. Starrfield, S., Truran, J.W., Sparks, W.M., Krautter, J. 1991, in R. Malina, S. Bowyer (eds.), Extreme Ultraviolet Astronomy, Pergamon, New York, 168

14 Pulsars and Isolated Neutron Stars

W. Becker, F. Haberl, and J. Trümper

14.1 Introduction: Historical Overview

The idea of *neutron stars* can be traced back to the early 1930s, when Subrahmanyan Chandrasekhar discovered that there is no way for a collapsed stellar core with a mass more than 1.4 times the solar mass, $M_\odot$, to hold itself up against gravity once its nuclear fuel is exhausted. This implies that a star left with $M > 1.4\ M_\odot$ (the *Chandrasekhar limit*) would keep collapsing and eventually disappear from view.

After the discovery of the neutron by James Chadwick in 1932, scientists speculated on the possible existence of a *star composed entirely of neutrons*, which would have a radius of the order of $R \sim (\hbar/m_n c)(\hbar c/G m_n^2)^{1/2} \sim 3 \times 10^5$ cm. In view of the peculiar stellar parameters, Lev Landau called these objects "unheimliche Sterne" (weird stars), expecting that they would never be observed because of their small size and expected low optical luminosity.

Walter Baade and Fritz Zwicky were the first who proposed the idea that neutron stars could be formed in *supernovae*. First models for the structure of neutron stars were worked out in 1939 by Oppenheimer and Volkoff (Oppenheimer-Volkoff limit). Unfortunately, their pioneering work did not predict anything astronomers could actually observe, and the idea of neutron stars was not taken serious by the astronomical community. Neutron stars, therefore, had remained in the realm of imagination for nearly a quarter of a century, until the discovery of pulsars and of accreting binary neutron stars with Uhuru [16, 50, 94].

The discovery of the first radio pulsar was very soon followed by the discovery of the two most famous pulsars, the fast 33 ms pulsar in the Crab Nebula and the 89 ms pulsar in the Vela supernova remnant. The fact that these pulsars are located within supernova remnants provided striking confirmation that neutron stars are born in core collapse supernovae from massive main sequence stars. These exciting radio discoveries triggered subsequent pulsar searches at nearly all wavelengths. Since those early days of pulsar astronomy more than 1 750 radio pulsars have been discovered by now (see, e.g., the ATNF pulsar database [66]).

Many radio pulsars had been observed by the mid-seventies, and two of them, the Crab and Vela pulsars, had been detected at high photon energies. Although the interpretation of both isolated and accreting pulsars as neutron stars with enormous magnetic fields, $\sim 10^{12}$ G, had been generally accepted, the first direct measurement

J.E. Trümper and G. Hasinger (eds.), *The Universe in X-Rays.*

of a neutron star magnetic field came from an X-ray observation: an electron cyclotron line feature discovered in the accreting binary pulsar Her X-1 indicated a polar magnetic field of $\sim 4 \times 10^{12}$ G [99] (cf. chapter 15).

Particularly, important results on isolated neutron stars, among many other X-ray sources, were obtained with HEAO-2, widely known as the Einstein X-ray observatory. Einstein investigated the soft X-ray radiation from the previously known Crab and Vela pulsars and resolved the compact nebula around the Crab pulsar [2]. It discovered pulsed X-ray emission from two more very young pulsars, PSR B0540−69 in the Large Magellanic Cloud and PSR B1509−58, having periods of 50 and 150 ms, respectively. Interestingly, these pulsars were the first ones to be discovered in the X-ray band and only subsequently at radio frequencies. Einstein also detected X-rays from three middle-aged radio pulsars, PSR B0656+14, B1055−52, B1951+32, and the X-ray counterparts of two nearby old radio pulsars, PSR B0950+08 and B1929+10. In addition, many supernova remnants were mapped – 47 in our Galaxy and 10 in the Magellanic Clouds and several neutron star candidates were detected as faint, soft point sources close to the center of the supernova remnants RCW 103, PKS 1209−51/52, Puppis-A and Kes 73.

Some additional information on isolated neutron stars was obtained by EXOSAT (European X-ray Observatory Satellite). In particular, it measured the soft X-ray spectra of the middle-aged pulsar PSR B1055−52 and of a few neutron star candidates in supernova remnants (e.g., PKS 1209−51/52).

The situation improved drastically in the 1990s because of the results from ROSAT, ASCA, EUVE, BeppoSAX, and RXTE, as well as Chandra and XMM-Newton launched close to the millennium.

The complement to ROSAT, covering the harder X-ray band 1–10 keV, was ASCA launched in 1993. The EUVE (Extreme Ultraviolet Explorer) was launched in 1992 and was sensitive in the range 70–760 Å. It was able to observe several neutron stars at very soft X-rays, 0.07–0.2 keV.

The contributions to neutron star research, provided by the instruments aboard BeppoSAX, sensitive in the range of 0.1–200 keV, and RXTE (Rossi X-ray Timing Explorer), both launched in the mid-90s, were particularly useful for studying X-ray binaries, including accretion-powered pulsars.

At present Chandra, with its outstanding subarcsecond imaging capability, and XMM-Newton, with its unprecedently high spectral sensitivity and collecting power, provide excellent new data.

In the following, we will summarize the current knowledge of X-ray emission properties of neutron stars based on these missions, browsing through the various categories from young Crab-like pulsars to very old radio pulsars, including recycled millisecond pulsars as well as neutron stars showing pure thermal emission. Before doing so, however, we will briefly review the various emission processes discussed to be the source for their observed X-ray emission.

14.2 Physics and Astrophysics of Isolated Neutron Stars

Neutron stars represent unique astrophysical laboratories, which allow us to explore the properties of matter under the most extreme conditions observable in nature.[1] Studying neutron stars is, therefore, an interdisciplinary field, where astronomers and astrophysicists work together with a broad community of physicists. Particle, nuclear and solid-state physicists are strongly interested in the internal structure of neutron stars, which is determined by the behavior of matter at densities above the nuclear density $\rho_{\rm nuc} = 2.8 \times 10^{14}$g cm^{-3}. Plasma physicists are modeling the pulsar emission mechanisms using electrodynamics and general relativity. It is beyond the scope of this article to describe in detail the current status of the theory of neutron star structure or the magnetospheric emission models. We rather refer the reader to the literature [17, 20, 36, 68, 106] and provide only the basic theoretical background relevant to Sect. 14.3, which summarizes the observed high-energy emission properties of rotation-powered pulsars and radio-quiet neutron stars.

14.2.1 Rotation-Powered Pulsars: The Magnetic Braking Model

Following the ideas of Pacini [81, 82] and Gold [37, 38], the more than 1750 radio pulsars detected so far can be interpreted as rapidly spinning, strongly magnetized neutron stars radiating at the expense of their rotational energy. This very useful concept allows one to obtain a wealth of information on basic neutron star/pulsar parameters just from measuring the pulsar's period and period derivative. Using the Crab pulsar as an example will make this more clear. A neutron star with a canonical radius of $R = 10$ km and a mass of $M = 1.4\ \mathrm{M}_\odot$ has a moment of inertia $I \approx (2/5)MR^2 \approx 10^{45}$ g cm^2. The Crab pulsar spins with a period of $P = 33.403$ ms. The rotational energy of such a star is $E_{\rm rot} = 2\pi^2\, I\, P^{-2} \approx 2 \times 10^{49}$ erg. This is comparable with the energy released in thermonuclear burning by a usual star over its entire live. Very soon after the discovery of the first radio pulsars, it was noticed that their spin periods increase with time. For the Crab pulsar, the period derivative is $\dot{P} = 4.2 \times 10^{-13}$ s s^{-1}, implying a decrease in the star's rotation energy of $\mathrm{d}E_{\rm rot}/\mathrm{d}t \equiv \dot{E}_{\rm rot} = -4\pi^2 I\dot{P}P^{-3} \approx 4.5 \times 10^{38}$ erg s^{-1}. Ostriker & Gunn [78] suggested that the pulsar slow-down is due to the braking torque exerted on the neutron star by its magneto-dipole radiation, which yields $\dot{E}_{\rm brake} = -(32\pi^4/3c^3)\, B_\perp^2\, R^6\, P^{-4}$ for the energy loss of a rotating magnetic dipole, where $B_\perp$ is the component of the equatorial magnetic field perpendicular to the rotation axis. Equating $\dot{E}_{\rm brake}$ with $\dot{E}_{\rm rot}$, we find $B_\perp = 3.2 \times 10^{19}(P\dot{P})^{1/2}$ Gauss. For the Crab pulsar, this yields $B_\perp = 3.8 \times 10^{12}$ G. From $\dot{E}_{\rm rot} = \dot{E}_{\rm brake}$ one further finds that $\dot{P} \propto P^{-1}$, for a given $B_\perp$. This relation can be generalized as $\dot{P} = kP^{2-n}$, where k is a constant, and n is the so-called braking index ($n = 3$ for the magneto-dipole braking). Assuming that

[1] Although black holes are even more compact than neutron stars, they can only be observed through the interaction with their surroundings.

the initial rotation period P_0 at the time t_0 of the neutron star formation was much smaller than today, at $t = t_0 + \tau$, we obtain $\tau = P/[(n-1)\dot{P}]$, or $\tau = P/(2\dot{P})$ for $n = 3$. This quantity is called the characteristic spin-down age. It is a measure for the time span required to lose the rotational energy $E_{\rm rot}(t_0) - E_{\rm rot}(t)$ via magneto-dipole radiation. For the Crab pulsar one finds $\tau = 1\,258$ yrs. As the neutron star in the Crab supernova remnant is the only pulsar for which its historical age is known (the Crab supernova was observed by Chinese astronomers in 1054 AD), we see that the spin-down age exceeds the true age by about 25%. Although the spin-down age is just an estimate for the true age of the pulsar, it is the only one available for pulsars other than the Crab, and it is commonly used in evolutionary studies (e.g., neutron star cooling).

A plot of observed periods vs. period derivatives is shown in Fig. 14.1, using the pulsars from the ATNF online pulsar database [66]. Such a P-$\dot{P}$ diagram is extremely useful for classification purposes. The colored symbols represent those pulsars that were detected at X-ray energies until fall 2006. The objects in the upper right corner represent the soft-gamma-ray repeaters (SGRs) and anomalous X-ray pulsars (AXPs) , which have been suggested to be magnetars (neutron stars with ultra strong magnetic fields).

Although the magnetic braking model is generally accepted, the *observed* spin-modulated emission, which gave pulsars their name, is found to account only for a small fraction of $\dot{E}$. The efficiencies, $\eta = L/\dot{E}$, observed in the radio and optical bands are typically in the range $\sim 10^{-7} - 10^{-5}$, whereas they are about $10^{-4} - 10^{-3}$ and $\sim 10^{-2} - 10^{-1}$ at X-ray and gamma-ray energies, respectively. It has, therefore, been a long-standing question of how rotation-powered pulsars lose the bulk of their rotational energy.

The fact that the energy loss of rotation-powered pulsars cannot be fully accounted for by the magneto-dipole radiation is known from the investigation of the pulsar braking index, $n = 2 - P\ddot{P}\dot{P}^{-2}$. Pure dipole radiation would imply a braking index $n = 3$, whereas the values observed so far are $n = 2.515 \pm 0.005$ for the Crab, $n = 2.8 \pm 0.2$ for PSR B1509−58, $n = 2.28 \pm 0.02$ for PSR B0540−69, 2.91 ± 0.05 for PSR J1911−6127, 2.65 ± 0.01 for PSR J1846-0258, and $n = 1.4 \pm 0.2$ for the Vela pulsar. The deviation from $n = 3$ is usually taken as evidence that a significant fraction of the pulsar's rotational energy is carried away by a pulsar wind, i.e., a mixture of charged particles and electromagnetic fields, which, if the conditions are appropriate, forms a *pulsar-wind nebula* observable at optical, radio, and X-ray energies. Such pulsar-wind nebulae (often called plerions or synchrotron nebulae) are known so far *only* for a few young and powerful (high $\dot{E}$) pulsars and for some center-filled supernova remnants, in which a young neutron star is expected, but only emission from its plerion is detected.

Thus, the popular model of magnetic braking provides plausible estimates for the neutron star magnetic field $B_\perp$, its rotational energy loss $\dot{E}$, and the characteristic age τ, but it does not provide detailed information about the physical processes that operate in the pulsar magnetosphere and are responsible for the broad-band spectrum, from the radio to the X-ray and gamma-ray bands. Forty years after the discovery of pulsars, the physical details of their emission mechanisms are still barely known.

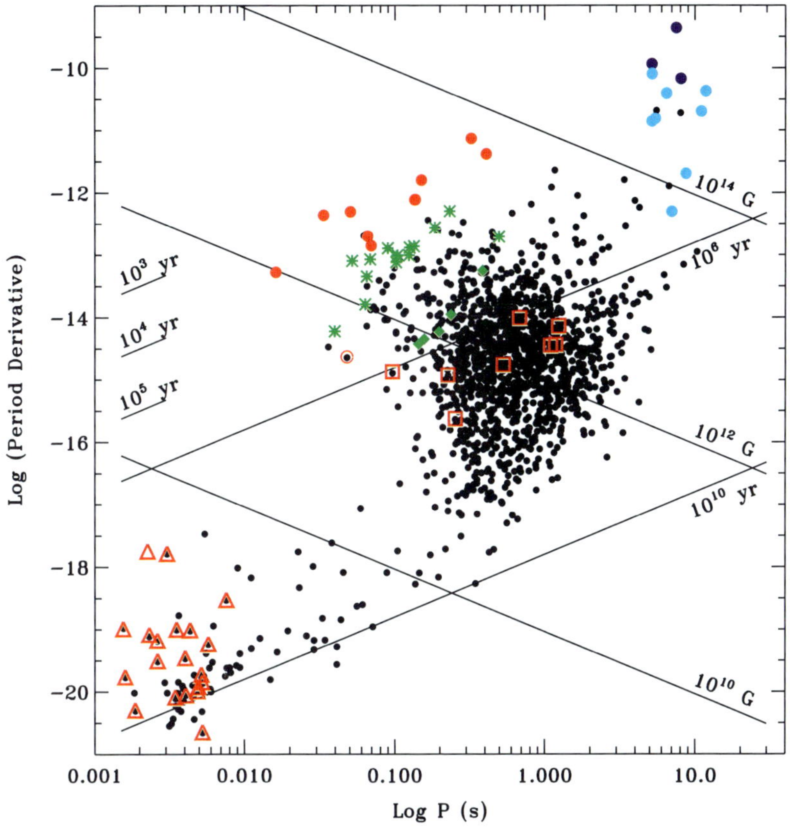

Fig. 14.1 The $P - \dot{P}$ diagram: distribution of rotation-powered pulsars (*small black dots*) over their spin parameters. The *straight lines* correspond to constant ages $\tau = P/(2\dot{P})$ and magnetic field strengths $B_{\perp} = 3.2 \times 10^{19}(P\dot{P})^{1/2}$ as deduced in the frame of the magnetic braking model. Separate from the majority of ordinary-field pulsars are the millisecond pulsars in the lower left corner and the high magnetic field pulsars – soft gamma-ray repeaters (*dark blue*) and anomalous X-ray pulsars (*light blue*) in the upper right. Although magnetars and anomalous X-ray pulsars are not rotation-powered, they are included in this plot to visualize their estimated superstrong magnetic fields. X-ray detected pulsars are indicated by colored symbols. *Red filled circles* indicate the Crab-like pulsars. *Green stars* indicate Vela-like pulsars, *green diamonds* the X-ray detected cooling neutron stars, and *red squares* million years old pulsars

As a consequence, there exist a number of magnetospheric emission models, but no generally accepted theory.

14.2.2 High-Energy Emission Models

Although rotation-powered pulsars are most widely known for their radio emission, the mechanism of the radio emission is poorly understood. However, it is certainly different from those responsible for the high-energy (infrared through gamma-ray) radiation observed from them with space observatories. It is well known that the radio emission of pulsars is a coherent process, and the coherent curvature radiation has been proposed as the most promising mechanism (see [68] and references therein). On the other hand, the optical, X-ray, and gamma-ray emission observed in pulsars must be incoherent. Therefore, the fluxes in these energy bands are directly proportional to the densities of the radiating high-energy electrons in the acceleration regions, no matter which radiation process (synchrotron radiation, curvature radiation, or inverse Compton scattering) is at work at a given energy. High-energy observations thus provide the key for the understanding of the pulsar emission mechanisms. So far, the high-energy radiation detected from rotation-driven pulsars has been attributed to various thermal and nonthermal emission processes including the following:

- Nonthermal emission from charged relativistic particles accelerated in the pulsar magnetosphere (cf. Fig. 14.2). As the energy distribution of these particles follows a power-law, the emission is also characterized by power-law-like spectra in broad energy bands. The emitted radiation can be observed from optical to the gamma-ray band.
- Extended emission from pulsar-driven synchrotron nebulae. Depending on the local conditions (density of the ambient interstellar medium), these nebulae can be observed from radio through hard X-ray energies.
- Photospheric emission from the hot surface of a cooling neutron star. In this case, a modified black-body spectrum and smooth, low-amplitude intensity variations

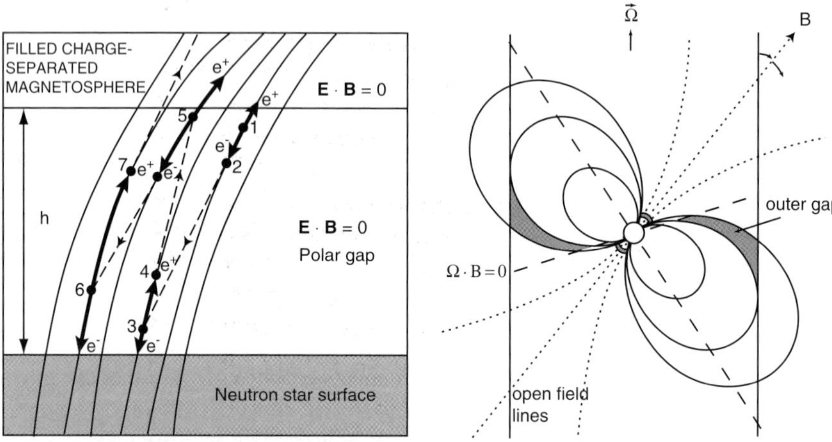

Fig. 14.2 Geometry of the acceleration zones as they are defined in the polar cap model (*left*), according to Ruderman and Sutherland [90], and outer gap model (*right*), according to Cheng, Ho and Ruderman [28, 29]. The polar cap model predicts "pencil" beams emitted by particles accelerated along the curved magnetic field lines. According to the outer gap model, the pulsar radiation is emitted in "fan" beams. Being broader, the latter can easier explain two (and more) pulse components as observed in some X-ray and gamma-ray pulsars

with the rotational period are expected, observable from the optical through the soft X-ray range.

- Thermal soft X-ray emission from the neutron star's polar caps which are heated by the bombardment of relativistic particles streaming back to the surface from the pulsar magnetosphere.

In almost all pulsars, the observed X-ray emission is due to a mixture of different thermal and nonthermal processes. Often, however, the available data do not allow to fully discriminate between the different emission scenarios. This was true for ROSAT, ASCA, and BeppoSAX observations of pulsars and is – at a certain level – still true in Chandra and XMM-Newton data.

In the following subsections, we will briefly present the basics on the magnetospheric emission models as well as material relevant to thermal emission from the neutron star surface.

14.2.2.1 Magnetospheric Emission Models

So far, there is no consensus as to where the pulsar high-energy radiation comes from (see for example [68]). There exist two main types of models – the *polar cap models*, which place the emission zone in the immediate vicinity of the neutron star's polar caps, and the *outer gap models*, in which this zone is assumed to be close to the pulsar's light cylinder[2] to prevent materializing of the photons by the one-photon pair creation in the strong magnetic field, according to $\gamma + B \rightarrow e^+ + e^-$ (see Fig. 14.2). The gamma-ray emission in the polar cap models forms a hollow cone centered on the magnetic pole, producing either double-peaked or single-peaked pulse profiles, depending on the observer's line of sight. The outer gap model was originally proposed to explain the bright gamma-ray emission from the Crab and Vela pulsars ([28, 29]) as the efficiency to get high-energy photons out of the high B-field regions close to the surface is rather small. Placing the gamma-ray emission zone at the light cylinder, where the magnetic field strength is reduced to $B_L = B\,(R/R_L)^3$, provides higher gamma-ray emissivities, which are in somewhat better agreement with the observations. In both types of models, the high-energy radiation is emitted by relativistic particles accelerated in the very strong electric field, $\mathscr{E} \sim (R/cP)B$, generated by the magnetic field corotating with the neutron star. These particles are generated in cascade (avalanche) processes in charge-free gaps, located either above the magnetic poles or at the light cylinder. The main photon emission mechanisms are synchrotron/curvature radiation and inverse Compton scattering of soft thermal X-ray photons emitted from the hot neutron star surface.

In recent years the polar-cap and outer-gap models have been further developed, incorporating new results on gamma-ray emission from pulsars obtained with the Compton Gamma-Ray Observatory. At the present stage, the observational data can be interpreted with any of the two models, albeit under quite different assumptions

[2] The light cylinder is a virtual cylinder whose radius, $R_L = cP/(2\pi)$, is defined by the condition that the azimuthal velocity of the corotating magnetic field lines is equal to the speed of light.

on pulsar parameters. The critical observations to distinguish between the two models include measuring the relative phases between the peaks of the pulse profiles at different energies. Probably, the AGILE and GLAST gamma-ray observatories, which are supposed to be launched in November 2007, will provide valuable information to further constrain both models.

14.2.2.2 Thermal Evolution of Neutron Stars

Neutron stars are formed at very high temperatures, $\sim 10^{11}$ K, in the imploding cores of supernova explosions. Much of the initial thermal energy is radiated away from the interior of the star by various processes of neutrino emission (mainly, Urca processes and neutrino bremsstrahlung), leaving a one-day-old neutron star with an internal temperature of about $10^9 - 10^{10}$K. After $\sim$100 yr (typical time of thermal relaxation), the star's interior (densities $\rho > 10^{10}$ g cm^{-3}) becomes nearly isothermal, and the energy balance of the cooling neutron star is determined by the following equation:

$$C(T_\mathrm{i})\,\frac{\mathrm{d}T_\mathrm{i}}{\mathrm{d}t} = -L_\nu(T_\mathrm{i}) - L_\gamma(T_\mathrm{s}) + \sum_k H_\mathrm{k}\,,$$

where T_i and T_s are the internal and surface temperatures, $C(T_\mathrm{i})$ is the heat capacity of the neutron star. Neutron star cooling thus means a decrease of thermal energy, which is mainly stored in the stellar core, because of energy loss by neutrinos from the interior ($L_\nu = \int Q_\nu\,\mathrm{d}V$, Q_ν is the neutrino emissivity) plus energy loss by thermal photons from the surface ($L_\gamma = 4\pi R^2 \sigma T_s^4$). The relationship between T_s and T_i is determined by the thermal insulation of the outer envelope ($\rho < 10^{10}$ g cm^{-3}), where the temperature gradient is formed. The results of model calculations, assuming that the outer envelope is composed of iron, can be fitted with a simple relation

$$T_\mathrm{s} = 3.1\,(g/10^{14}\ \mathrm{cm\ s^{-2}})^{1/4}\ (T_\mathrm{i}/10^9\ \mathrm{K})^{0.549} \times 10^6\,\mathrm{K}$$

where g is the gravitational acceleration at the neutron star surface [43]. The cooling rate might be reduced by heating mechanisms H_k, like frictional heating of superfluid neutrons in the inner neutron star crust or some exothermal nuclear reactions.

Neutrino emission from the neutron star interior is the dominant cooling process for at least the first 10^5 yrs. After $\sim 10^6$ years, photon emission from the neutron star surface takes over as the main cooling mechanism. The thermal evolution of a neutron star after the age of $\sim$10–100 yr, when the neutron star has cooled down to $T_\mathrm{s} = 1.5 - 3 \times 10^6$ K, can follow two different scenarios, depending on the still poorly known properties of super-dense matter (see Fig. 14.7). According to the so-called *standard cooling scenario*, the temperature decreases gradually, down to $\sim 0.3 - 1 \times 10^6$ K, by the end of the neutrino cooling era and then falls down exponentially, becoming lower than $\sim 0.1 \times 10^6$ K in $\sim 10^7$ yr. In this scenario, the main neutrino generation processes are the modified Urca reactions, $n + N \rightarrow p + N + e + \bar{\nu}_e$ and $p + N + e \rightarrow n + N + \nu_e$, where N is a nucleon (neutron

Table 14.1 Nuclear reactions and their neutrino emissivity as a function of neutron star temperature [84]

Neutrino emissivity used in neutron star cooling models		
Process	Nuclear reaction	Emissivity (erg s^{-1} cm^{-3})
Direct URCA-Process	$n \rightarrow p + e^- + \bar{\nu}_e$ $p + e^- \rightarrow n + \nu_e$	$\sim 10^{27} \times T_9^6$
π-condensate	$n + \pi^- \rightarrow n + e^- + \bar{\nu}_e$ $n + e^- \rightarrow n + \pi^- + \nu_e$	$\sim 10^{26} \times T_9^6$
Quark-URCA-Process	$d \rightarrow u + e^- + \bar{\nu}_e$ $u + e^- \rightarrow d + \nu_e$	$\sim 10^{26}\ \alpha_c\ T_9^6$
Kaon condensate	$n + K^- \rightarrow n + e^- + \bar{\nu}_e$ $n + e^- \rightarrow n + K^- + \nu_e$	$\sim 10^{24} \times T_9^6$
Modified URCA-Process	$n + n \rightarrow n + p + e^- + \bar{\nu}_e$ $n + p + e^- \rightarrow n + n + \nu_e$	$\sim 10^{20} \times T_9^8$
Direct coupled Elektron–neutrino-process	$\gamma + e^- \rightarrow e^- + \nu_e + \bar{\nu}_e$ $\gamma_{plasmon} \rightarrow \nu_e + \bar{\nu}_e$ $e^+ + e^- \rightarrow \nu_e + \bar{\nu}_e$	$\sim 10^{20} \times T_9^8$
Neutron–neutron and Neutron–proton-bremsstrahlung	$n + n \rightarrow n + n + \nu + \bar{\nu}$ $n + p \rightarrow n + p + \nu + \bar{\nu}$	$\sim 10^{19} \times T_9^8$
Elektron–ion–neutrino-Bremsstrahlung	$e^- + (Z,A) \rightarrow$ $e^- + (Z,A) + \nu_e + \bar{\nu}_e$	$\propto T_9^6$

T_9 is the temperature in units of 10^9 K. Each particle (n, p, e^-), which takes part in a reaction contributes to the temperature dependence with a T and each neutrino with a T^3. The reactions denoted as direct-Urca, π-condensate, Quark-URCA-process, and Kaon condensate are taken into account in the so-called accelerated cooling models. They have an order of magnitude higher neutrino emissivity in comparison with the other nuclear reactions. The higher the neutrino emissivity is the more efficient is the neutron star cooling

or proton) needed to conserve momentum of reacting particles (cf. Table 14.1). In the *accelerated cooling scenarios*, associated with higher central densities (up to 10^{15} g cm^{-3}) and/or exotic interior composition (e.g., pion condensation, quark-gluon plasma), a sharp drop of temperature, down to $0.3 - 0.5 \times 10^6$ K, occurs at an age of $\sim 10 - 100$ yr, followed by a more gradual decrease, down to the same $\sim 0.1 \times 10^6$ K at $\sim 10^7$ yr. The faster cooling is caused by the direct Urca reactions, $n \rightarrow p + e + \bar{\nu}_e$ and $p + e \rightarrow n + \nu_e$, allowed at very high densities.

The neutron star models used in these calculations are based on a *moderate* equation of state that opens the direct Urca process for $M > 1.35 M_\odot$, the stars with lower M undergo the standard cooling. Recent studies have shown that both the standard and accelerated cooling can be substantially affected by nucleon superfluidity in the stellar interiors (see [83, 100, 112] for comprehensive reviews). In particular, many cooling curves exist intermediate between those of the standard and accelerated scenarios, depending on the properties of nucleon superfluidity, which are also poorly known.

Thus, the thermal evolution of neutron stars is very sensitive to the composition and structure of their interiors, in particular, to the equation of state at super-nuclear densities. Therefore, measuring surface temperatures of neutron stars is an important tool to study super-dense matter. Since typical temperatures of such neutron stars correspond to the extreme UV – soft X-ray range, the thermal radiation from cooling neutron stars can be observed with X-ray detectors sufficiently sensitive at $E \lesssim 1$ keV.

14.3 High-Energy Emission Properties of Neutron Stars

As a result of observations with the satellite observatories Einstein, ROSAT, ASCA, BeppoSAX, Chandra, and XMM-Newton, 80 rotation-powered pulsars were detected at X-ray energies until the end of 2006. Thus, in 7 yrs of operation XMM-Newton and Chandra have more than doubled the number of detected X-ray pulsars compared to what was known at the end of the ROSAT mission in December 1999 [11]. Table 14.2 reflects the progress made in detecting pulsars of various categories at X-ray energies in recent years. This progress clearly goes along with the increase of sensitivity and angular resolution of the available X-ray telescopes. XMM-Newton with its super collecting power allows to obtain timing and spectral information even from faint and millions of years old pulsars, while Chandra stands for subarcsecond angular resolution, which made it possible to detect and study neutron stars located in source confused regions such as globular clusters and supernova remnants.

Table 14.2 Progress in detecting rotation-powered pulsars with X-ray observatories (status as of December 2006)

Pulsar age (years)	Pulsar category	Einstein	ROSAT ASCA	XMM-Newton Chandra
$\leq 10^4$	Crab-like	3	5	9
$10^4 - 10^5$	Vela-like	1	9	15
$10^5 - 10^6$	Cooling NS		5	6
$10^6 - 10^8$	Old and nearby	1	3	8
	binary		1	3
$\geq 10^8$	ms-Pulsar		11	39
	Σ detected:	5	33	80

While Einstein had only the sensitivity to see pulsed X-ray emission from the youngest and brightest pulsars and to detect a few others at the limit of its sensitivity, ROSAT/ASCA and XMM-Newton/Chandra allowed for the first time to study the emission mechanisms of rotation-powered pulsars based on a broader sample and of various categories

Fortunately, with the increase in sensitivity of todays observatories a growing number of neutron stars are detected in more than just one waveband (e.g., at radio, optical, EUV, X- and gamma-rays), making it possible for the first time to carry out multiwavelength studies of the pulsar emission. This is a big advantage as the physical processes that cause the emission in different wavelength bands are obviously related to each other. Multiwavelength studies thus provide a much broader view into the physical processes operating in the neutron star magnetosphere than interpretating emission properties observed in a single wave band only.

14.3.1 Young Neutron Stars in Supernova Remnants

X-ray observations allow us to find both supernova remnants (SNRs) and the compact objects that may reside within them. In fact, neutron stars and neutron star candidates have been found in a small fraction of the 265 known galactic SNRs[3] [40]. About 35 of these compact stellar remnants in SNRs are radio pulsars, others are radio-silent (or, at least, radio-quiet) neutron stars, which were found as faint point-like X-ray sources near to the geometrical center of their supernova remnant.

Being in orbit for more than half of their nominal lifetime almost all young radio pulsars have been observed and detected by either XMM-Newton and/or Chandra (cf. Fig. 14.1). The young rotation-powered pulsars can be divided in two groups, Crab-like and Vela-like pulsars, according to somewhat different observational manifestations apparently associated with the evolution of pulsar properties with age. The radio-silent neutron stars include anomalous X-ray pulsars (AXPs), soft gamma-ray repeaters (SGRs), and "quiescent" neutron star candidates in SNRs. There is growing evidence that AXPs and SGRs are indeed magnetars (see [109] for a review). Magnetars are neutron stars with an ultra strong magnetic field ($B \geq 10^{14}$G), which is supposed to be the source of the detected high energy radiation. A common property of these objects is that their periods are in a narrow range of $5 - 12$ s, substantially exceeding typical periods of radio pulsars. Although no gamma-ray emission has been detected from AXPs, SGRs occasionaly emit soft gamma-ray bursts of enormous energy (up to 10^{42}–10^{44} erg), a property, which gave this sources their name.

14.3.1.1 Crab- and Vela-Like Pulsars

On July 4, 1054 A.D., Chinese astronomers noted a *guest star* in the constellation Taurus. As we know today, this event marked the arrival of light from the deadth of a massive main sequence star that underwent a core collapse when its internal thermal energy produced by the nuclear fusion processes was not sufficient anymore to counteract the gravitational force against the star's collapse. The cloud of gas that

[3] http://www.mrao.cam.ac.uk/surveys/snrs/

we observe today at the position of this *guest star* is the Crab supernova remnant. In the optical band the nebula has an extent of 4×6 arcmin, corresponding to $\sim 7 \times 10$ light years for a distance of 2 kpc.

What we observe from the Crab nebula in X-rays is not the thermal emission from the ejecta-driven blast wave of the supernova, though, but the emission from charged particles that emit synchrotron radiation as they move along magnetic field lines. In X-rays the nebula has the form of a torus with jets, wisps and a counter-jet, having an overall extent of 2×2 arcmin in the sky. For the 2 kpc distance, the radius of the torus is 0.38 pc, that of the inner ring is 0.14 pc [107, 108]. A composite image showing the Crab nebula and pulsar in the radio, optical and X-ray band is shown in Fig. 14.3.

In studying this system it became clear very early that the observed nonthermal emission required a continuous input of energetic charged particles to keep the nebula emitting. It was the question of the Crab nebula's central engine that caused

Fig. 14.3 A composite image of the Crab nebula and its central pulsar as produced by emission detected in the radio (*green*), optical (*red*), and X-ray (*blue*) wavebands

Pacini (1967) a few month before the discovery of radio pulsars to propose that a fast spinning and strongly magnetized neutron star could be the required source that supplies the energy into the nebula.

Indeed, the 33 ms pulsar in the Crab supernova remnant, PSR B0531+21, was the first rotation-powered pulsar from which high energy radiation was detected. Being the strongest rotation-powered pulsar with the highest spin-down energy it was considered – until recent years – to be the proto type for all young neutron stars of age $10^3 - 10^4$ years. Because of this and its favorable brightness, it was studied in all frequency bands and by almost every observatory suitable to do so. The pulsar's characteristic double peaked pulse profile and its energy spectrum have been measured in detail throughout almost the entire electromagnetic spectrum. A compilation of pulse profiles as observed from the radio to the X-ray bands is shown in Fig. 14.4.

Despite this strong interest and a wealth of data that have been taken from the pulsar since its discovery, it only recently became clear in deep Chandra observations that the X-ray emission from the Crab pulsar is actually 100% pulsed [95] and that the radio, optical, and X-ray pulses are not fully phase aligned as suggested by

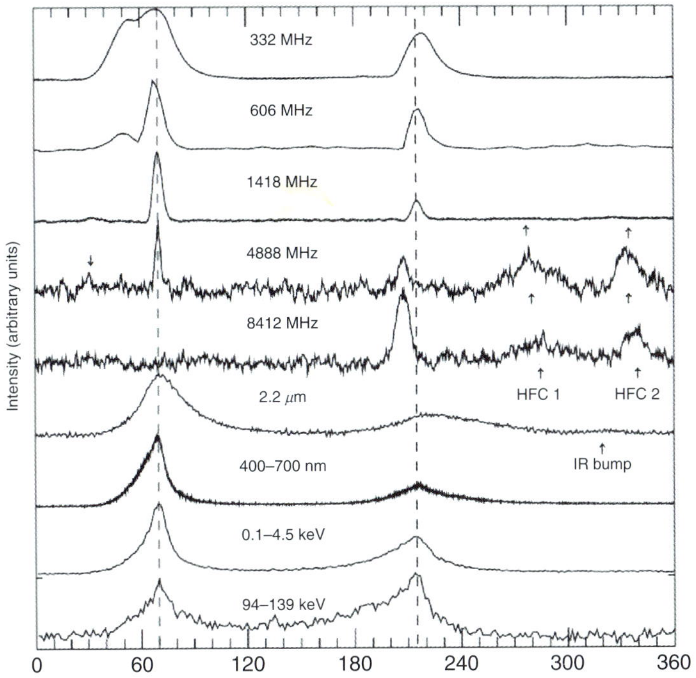

Fig. 14.4 The Crab pulsar's characteristic pulse profiles as observed at various frequency bands (from [71]). The phases of low (LFC) and high frequenc (HFC) radio pulse components are indicated. The *dashed lines* indicate the phase of the main pulse component and the interpulse. The profiles have been arbitrarily aligned by the peak of the main pulse

the high-energy emission models. Indeed, the X-ray pulses lead the optical pulses by the small amount of $\sim 60\,\mu s$ and the optical the radio pulses by $260\,\mu s$ [6]. Mapping these pulse arrival time differences to photon travel-time differences means that the Crab's X-ray pulses may come from locations in the pulsar magnetosphere closer to the neutron star's surface than the radio emission, which then might be emitted at higher altitudes.

As far as the pulsar's emission mechanisms are concerned, it is very well established that magnetospheric emission from charged particles, accelerated in the neutron star magnetosphere along the curved magnetic field lines, dominates the radiation not only from the Crab pulsar but from almost all young rotation-powered pulsars with ages $\lesssim 5000$ yrs (cf. 14.2.2). Accordingly, the radiation of Crab-like pulsars is characterized by a power-law spectrum, $dN/dE \propto E^{-\alpha}$, as the energy distribution of the particles that emit this radiation follows a power-law in a broad energy range. For the Crab pulsar, the slope of its flux spectrum slowly increases with photon energy – the photon index varies from $\alpha = 1.1$ at $E \sim 1$ keV to $\alpha = 2.1$ at $E \sim 10^{10}$ eV.

Only recently a pulsar was detected, which seems to conflict with this empirical evidence of nonthermal dominated emission in young rotation-powered pulsars. PSR J1119-6127, which is located in the SNR G292.2-0.5, has an age of $\sim$1 600 yrs and a deduced magnetic field strength of $B \sim 4.1 \times 10^{13}$G. The latter is close to the quantum critical field of $B_{QED} = m_e^2 c^3/e\hbar = 4.4 \times 10^{13}$G and close to the magnetar range. There is strong evidence that its spectrum is dominated by thermal radiation corresponding to a temperature of $\sim 2.4 \times 10^6$ K and an emitting radius of $\sim$3.4 km while its pulsed fraction in the $0.5 - 2.0$ keV band is as high as $\sim 74\% \pm 14\%$. Although it still requires better data to finally identify the X-rays of this pulsar as being of thermal origin, it is an interesting question of whether the presence of the strong magnetic field causes a completely different emission scenario than observed in other young and Crab-like pulsars.

Pulsars with a spin-down age of $\sim 10^4 - 10^5$ yrs are often referred to as Vela-like pulsars, because of their apparent similar emission properties. Among the 14 pulsars of this group that have been detected in X-rays, four of them (the Vela pulsar PSR B0833−45, PSR B1706−44, B1046−58, and B1951+32) are gamma-ray pulsars, and only the Vela pulsar has been detected in the optical band. In some respects, these objects appear to be different from the Crab-like pulsars. In particular, their optical radiation is very faint compared with that of the very young pulsars, and the overall shape of their high-energy spectra looks different. For instance, the closest ($d \approx 300$ pc) and, hence, best-investigated Vela pulsar has an optical luminosity four orders of magnitude lower than the Crab pulsar [75], whereas its rotation energy loss is only a factor of 65 lower. Its pulse profile at various wavelength is very complex and difficult to associate with the various possible emission mechanisms [64]. The pulsed fraction in the soft X-ray range, $\approx$7%, is much lower than that observed from Crab-like pulsars.

In contrast to the young Crab-like pulsars, the soft X-ray spectrum of the Vela pulsar has a substantial thermal contribution with an apparent temperature

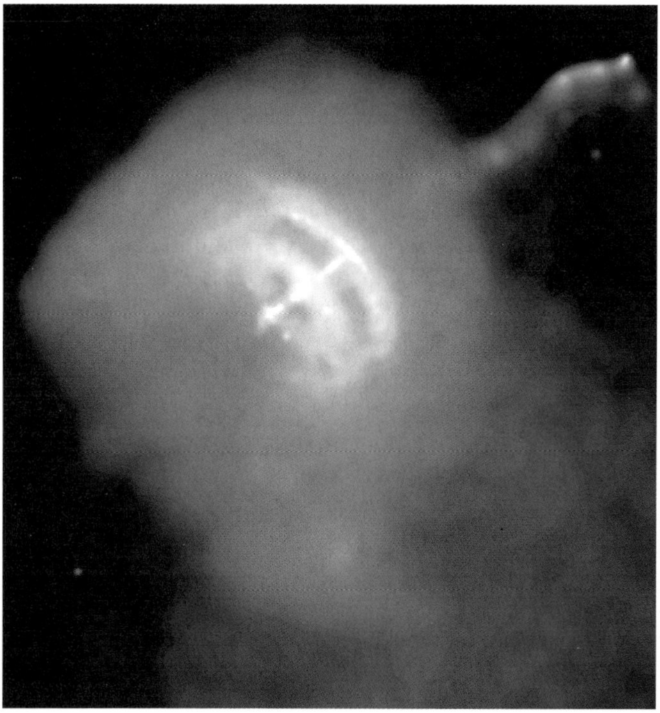

Fig. 14.5 The Vela pulsar and its plerion as observed by the Chandra ACIS-I detector. A torus and jets are seen similar as in the Crab plerionic nebula. The symmetry axis is almost aligned to the pulsar's proper motion direction

of $\approx 10^6$ K [79]. On the other hand, the spatial structure of the Vela plerion strongly resembles the inner Crab nebula – it also has a torus-like structure, an inner ring and jets (see Fig. 14.5). The symmetry axis of the nebula, which can be interpreted as the projection of the pulsar's rotation axis onto the sky plane, is roughly coaligned with the direction of proper motion. This is similar as observed in the Crab pulsar although the missalignement there is $26^\circ \pm 3^\circ$. The idea of a torus configuration formed by a shock-confined pulsar wind was first introduced by Aschenbach and Brinkmann [2] as a model to explain the shape of the inner Crab nebula. The discovery of a similar torus-like structure in the Vela synchrotron nebula indicated that this model may be applicable to many young pulsars. According to this model, the torus-like structure and its geometrical orientation with respect to the direction of the pulsar's proper motion arise because the interaction of the postshock plasma with the ambient medium compresses the plasma and amplifies the magnetic field ahead of the moving pulsar. This, in turn, leads to enhanced synchrotron emission with the observed torus-like shape.

Thus, young rotation-powered pulsars are in general surrounded by pulsar-powered nebulae (plerions) and/or supernova ejecta. Presumably, their magnetospheric emission extends from at least the infrared to gamma-ray energies, with typical photon indices varying between $\alpha \approx 1-2$ (about $1.4-1.7$ in the soft X-ray range). As the plerionic emission is synchrotron radiation its spectrum is a power law. In the Crab plerion, Willingale et al. [108] found that the shape of the spectrum changes as a function of distance from the pulsar. He fitted the power law slope of the torus ($\alpha = 1.8 \pm 0.006$), the jet ($\alpha = 2.1 \pm 0.013$), and the outer nebula regions ($\alpha = 2.34 \pm 0.006$). Similar results were obtained by Chandra, measuring the hardness ratio distribution throughout the nebula [107]. For the pulsar, a photon spectral index of $\alpha = 1.63 \pm 0.09$ is observed. The spectral difference between the jet and the torus is found to be likely due to an intrinsically steeper electron spectrum of the jet. The outer regions of the nebula show the steepest spectrum, which is likely to be due to enhanced synchrotron losses of the electrons during their ride from the pulsar to the outskirts.

14.3.1.2 Central Compact Objects in Supernova Remnants

For many years, it has been generally believed that all young neutron stars have similar emission properties as those observed in Crab- and Vela-like pulsars, i.e., emitting strongly pulsed radiation caused by nonthermal emission processes in the neutron star's magnetosphere. Several recent observations of compact X-ray sources in supernova remnants, however, suggest that this picture is incomplete and indeed no longer justified: it has been shown that there are other manifestations of young neutron stars, e.g., as anomalous X-ray pulsars, soft gamma-ray repeaters or simply as faint point-like X-ray source in a supernova remnant. Most of these sources were identified by their high X-ray to optical flux ratios, others simply by their locations near to the expansion centers of supernova remnants, strongly suggesting that they are indeed the compact stellar remnants formed in the supernova events.

The group of SNRs, which are known to host a radio-quiet but X-ray bright central compact object (CCO), is listed in Table 14.3.

Whether this group of CCOs forms a homogenous class of sources such as the rotation-powered pulsars is currently an open question and is actually difficult to answer in view of the small number of known objects. All sources in common is that (1) they are located in supernova remnants of age $\leq 10^4$ yrs, (2) their X-ray luminosities are all in the braked $10^{32} - 10^{34}$ erg s^{-1}, (3) down to an extent of ≤ 1 arcs none of them has been seen to maintain a plerionic X-ray nebula such as the Crab, Vela, or other young pulsars, and (4) no radio or optical counterpart could be detected from any CCO by now.

None of these properties is distinctive enough to justify the interpretation that all these sources form an own class of objects (e.g., Geminga is radio-silent as well). Interestingly, though, is that all CCOs share very similar spectral properties and those are markedly different from what is observed in young rotation-powered pulsars. The X-ray spectra of virtually all CCOs are very well modeled by either one or two component blackbody model with $T_{bb} = (2-7) \times 10^6$ k and a size of the projected emitting area in the range $R_{bb} \sim (0.3-5)$ km. Alternatively, spectral models consisting of a blackbody and a power law provide valid fits as well. The inferred slope

Table 14.3 List of X-ray detected radio-quiet and optically dim central compact objects in supernova remnants (status: January 2007)

CCO	Hosting SNR	Age (kyr)	d (kpc)	P	log L_x (erg s^{-1})	Ref.
CXOU J232327.9+584843	Cas-A	~0.3	~3	...	~32.94	[33, 93]
RX J0852.0−4622	Vela-Jr	~2	~2.2	...	~32.83	[3, 7]
RX J1713.7−3946	G347.3-0.5	~10	~6	...	~32.63	[63, 87]
1E 1613−5055	RCW 103	~2	~3.3	6.67 h	variable	[32, 101]
RX J0822−4300	Pupis-A	~2	~2.2	0.22* s	~33.69	[53, 54, 86]
1E 1207.4-5209	PKS 1209-51	~10	~2.1	0.424 s	~33.39	[48, 110]
CXOU J185238.6+004020	Kes 79	~9	~7	0.105 s	~33.07	[39, 92]

The X-ray luminosity is computed for the energy band 0.5−10 keV and the specified distances
*The periodicity of RX J0822−4300 awaits confirmation

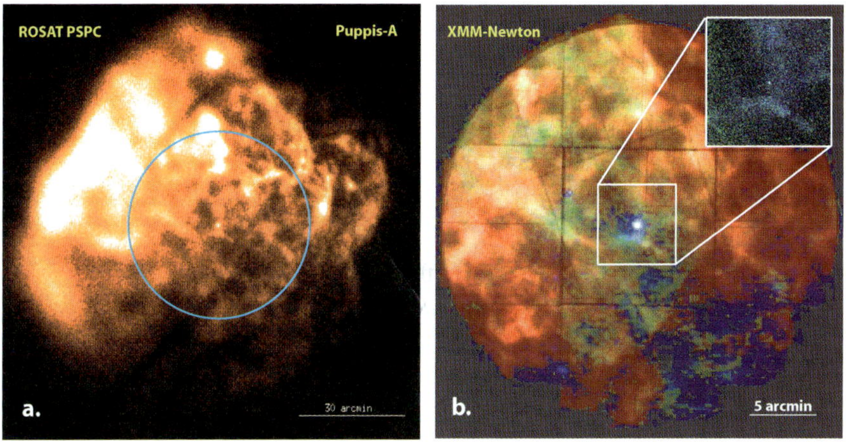

Fig. 14.6 (**a**) Composite ROSAT HRI image of the Puppis-A supernova remnant. The *blue ring* indicates the 30 arcmin central region, which has been observed by XMM-Newton. (**b**) XMM-Newton MOS1/2 false color image of the central region of Puppis-A (*red*: 0.3 – 0.75 keV, *green*: 0.75 – 2 keV, and *blue*: 2 – 10 keV). The central source is the CCO RX J0822−4300. The inset shows the squared region as observed by the Chandra HRC-I. (Image from [53])

of the power law component, though, is ~4 – 5, which is steeper than the photon-index $\Gamma = 1 - 3$ observed for rotation-powered pulsars [13]. However, as the true nature of the CCOs and their emission mechanisms are unknown it might not be justified to use the steepness of the power law to reject these models as unphysical. Worth to mention in this context is that the spectra observed from AXPs also require a blackbody plus power law model of similar properties.

Despite the common spectral emission properties, there are distinct differences in the temporal emission properties of some CCOs. There is strong evidence that 1E 1613−5055 in RCW 103 and 1E 1207.4-5209 in PKS 1209-51/52 are actually binaries composed of a compact object and a low-mass star in an eccentric orbit [10, 32, 110]. For 1E 1613−5055 in RCW 103, a strong periodic modulation at 6.67 ± 0.03 h has been found in long XMM-Newton observations along with changes in the X-ray flux by factors 10–100 while 1E 1207.4-5209 in PKS 1209-51/52 shows erratic behavior in both its pulsed fraction and period derivative [114]. X-ray pulsations have also been observed from CXOU J185238.6+004020 in Kes 79 [39] and RX J0822−4300 in Puppis-A [53] and both results do not support a scenario of steady spin-down. For RX J0822−4300, e.g., the period time derivative calculated from the separation of the epochs of the two available XMM-Newton data sets is $\dot{P} = (2.112 \pm 0.002) \times 10^{-10}$s s^{-1}, which is among the largest spin-down rates in the neutron star population. The largest known $\dot{P}$ was inferred for SGR 1806-20, $\dot{P} = (8-47) \times 10^{-11}$s s^{-1}, [60, 111]. If the identifications of P and $\dot{P}$ for RX J0822−4300 are correct, it implies a nonsteady spin-down behavior. It should be noted that these results lend evidence to the interpretation that CCOs, AXPs, and SGRs are all magnetars. There are several AXPs and few SGRs that have been associated with supernova remnants [34, 67]. On the basis of *Spitzer* and ground-based K_s-band images, Krause et al. (2005) suggested that a mid-twentieth century flare from CXOU J232327.9+584843 in Cas-A could have been the source of the apparent 24 μm light echo filaments observed ~20 arcmin north and south of the Cas-A remnant. However, with the available data the relation of CCOs with magnetars cannot be concluded without certainty.

Recently, it became possible for the first time to measure the proper motion of a CCO and to confirm that its back projected birth place is in agreement with the remnants explosion center, thus providing the first confirmation that CCOs are indeed the compact remnants formed in their hosting supernova. Using two Chandra data sets, which span an epoch of 1952 days, Hui and Becker [54] found the position of RX J0822−4300 in Puppis-A (c.f. Fig. 14.6) different by 0.57 ± 0.18 arcs, implying a proper motion of $\mu = 107 \pm 34$ mas/yr. For a distance of 2.2 kpc, this proper motion is equivalent to a recoil velocity of 1120 ± 360 km s^{-1}. Since both the magnitude and direction of the proper motion are in agreement with the birth place of RX J0822−4300 being near to the optical expansion center of the supernova remnant.

It is finally worth to mention that, by now, the relatively small number of discovered members of this class might be due to observational selection effects only. From the observers point of view, it is much easier to detect and identify active pulsars than these quiet compact sources observable only in the soft X-ray band. Also, once a supernova remnant disappears after about 10^5 yrs, it is almost impossible to find and identify its left over CCO. It is, therefore, very plausible that, in fact, CCOs may be more common than young Crab- and Vela-like radio pulsars.

14.3.2 Cooling Neutron Stars

The thermal radiation from neutron star surfaces was first detected in an unambiguous way with ROSAT from the Vela pulsar, PSR B0656+14, Geminga and PSR

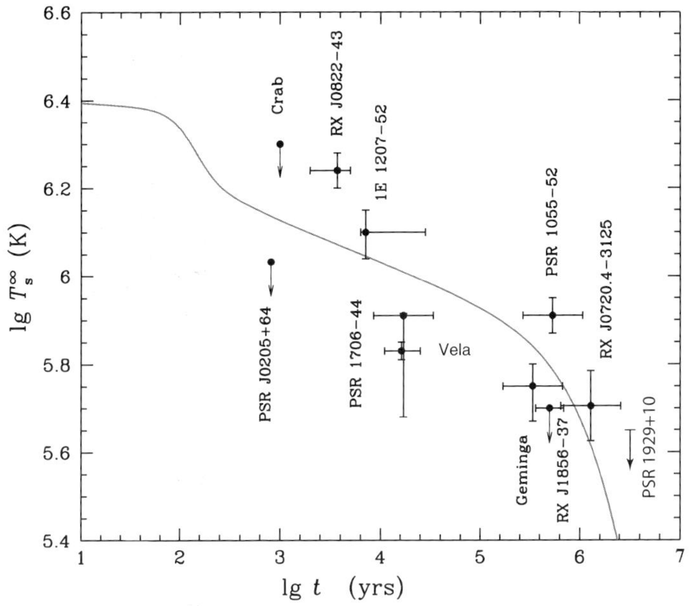

Fig. 14.7 Observations of surface temperatures and upper bounds for several isolated neutron stars. The *solid line* is the basic theoretical cooling curve of a nonsuperfluid neutron star with $M = 1.3 M_\odot$ [112]. The upper limit on the temperature of the old pulsar PSR 1929+10 was added [8]

B1055-52 [78, 79], and a few other sources. ROSAT also discovered seven neutron stars showing pure thermal emission in X-rays (see next paragraph). More sources and many more details on spectra and time variability were obtained more recently with Chandra and XMM-Newton [9]. Figure 14.7 shows a comparison of observed neutron star temperatures as a function of age compared with the results of standard cooling theory as summarized by Yakovlev and Pethick [112]. This comprehensive article also contains more detailed comparisons of observations and predictions from a variety of cooling models.

Addressing cooling neutron stars it might be worth to closer consider the group of isolated radio-quiet neutron stars. ROSAT observations revealed the existence of a group of seven of these bright and soft X-ray sources [44, 73, 97]. The X-ray spectra of the "magnificent seven," as they are sometimes called, are purely thermal and blackbody-like. Derived blackbody temperatures are in the range of $(0.5-1.2)\times 10^6$ K. They do not show hard spectral components or radio emission (with maybe the exception of RBS 1223) making them the best candidates for "genuine" isolated neutron stars with the least disturbed view to their stellar surface. With the known distance to RX J1856.5−3754 an X-ray luminosity of $\sim 10^{31}$ erg s^{-1} is inferred. Because of this intrinsically faint emission, the ROSAT-discovered isolated neutron stars are also called X-ray dim isolated neutron stars.

Table 14.4 X-ray and optical properties of nearby radio-quiet isolated neutron stars

Object	T (10^6 K)	Flux[3] (erg cm^{-2} s^{-1})	Period (s)	Pf[4] (%)	Optical (mag)	PM (mas yr^{-1})
RX J0420.0−5022	0.51	3.3×10^{-13}	3.45	13	B = 26.6	
RX J0720.4−3125	0.99-1.10	9.7×10^{-12}	8.39	8-15	B = 26.6	97
RX J0806.4−4123	1.11	2.4×10^{-12}	11.37	6	B >24	
RXS J1308.8+2127[1]	1.00	3.2×10^{-12}	10.31	18	m_{50ccd} = 28.6	
RX J1605.3+3249	1.11	6.1×10^{-12}	–	–	B = 27.2	145
RX J1856.5−3754	0.73	1.3×10^{-11}	7.06	1.5	V = 25.7	332
RXS J2143.0+0654[2]	1.17	2.8×10^{-12}	9.44	4	R >23	

(1) = RBS 1223; (2) = RBS 1774; (3) X-ray flux in the 0.1−2.4 keV band, (4) pulsed fraction

For six objects a periodic modulation was discovered in the X-ray flux, which indicates the rotation of the neutron star. Spin periods in the range of 3.4−11.4 s are longer than found for the bulk of radio pulsars, although also a few radio pulsars are known with similar pulse periods. The pulsed fraction in the X-ray flux varies from a few percent up to 18% in the case of RBS 1223. Table 14.4 summarizes X-ray and optical properties of the seven known radio-quiet isolated neutron stars. New X-ray observations with Chandra and XMM-Newton and deep optical imaging revealed new interesting and unexpected results that are described in the following.

Low interstellar absorption columns of these objects and high proper motions for the three brightest ones indicate that they represent a local population at distances up to a few hundred parsec, which is supported by the parallax of 120 pc observed for the brightest member of the class, RX J1856.5−3754. The high proper motions make accretion from the interstellar medium inefficient and favor the picture of neutron star cooling. Tracing back the apparent sky trajectories inferred from the proper motion measurements of RX J1856.5−3754, RX J1605.3+3249 and possibly also RX J0720.4−3125 points into the direction of the nearby OB association Sco OB2 [72]. The time it took for the neutron stars to move from this potential birth place to their current position is consistent with ages of around 10^6 years (RX J1856.5−3754: 0.5×10^6). Such an age is expected from their current temperature under the assumption of standard neutron star cooling scenarios. The Sco OB2 complex is the closest OB association and part of the Gould Belt. Population synthesis studies [88] for isolated neutron stars support the idea that the "magnificent seven" are of local origin dominated by the production of the Sco OB2 association.

The X-ray spectra obtained with the ROSAT PSPC were compatible with blackbody radiation (e.g., RX J0720.4−3125 [45]), only modified by moderate photoelectric absorption by the interstellar medium. From high-resolution observations with the grating spectrometers on board of Chandra and XMM-Newton, it was expected to see spectral features originating in the atmospheres of the neutron stars. However, the first Chandra observations of RX J1856.5−3754, collecting more than 5 days of LETGS high-resolution data, showed a featureless spectrum best represented by a blackbody model. Moreover, after the first optical identifications, it became immediately clear that the radio-quiet isolated neutron stars are optically brighter by typically a factor of 5–10 than expected from the extrapolation of their X-ray spectrum using the blackbody model. Hydrogen or Helium atmosphere models on the other hand produce a spectrum with Planckian shape in the X-ray band and reproduce the observed spectrum but predict a far too high optical flux. To reduce this discrepancy in the optical brightness heavy elements would be required which, on the other hand, produce features not observed in the X-ray spectra [85].

The problems to explain the overall energy distribution from optical to X-rays wavelengths with neutron star atmosphere or blackbody models led to the suggestion of a nonuniform temperature distribution on the stellar surface. Hot spots at the magnetic poles would account for the X-ray spectrum, while the cooler and larger part of the emitting surface could explain the optical emission. They also can explain the observed low amplitude smooth X-ray pulsations. Such hot spots could arise from anisotropic heat transport from the neutron star interior through the crust in a strong magnetic field or by polar cap heating by bombardment of particles from the magnetosphere.

These isolated neutron stars must have very strong magnetic fields, based on a number of arguments: From period measurements distributed over more than 10 years by ROSAT, XMM-Newton, and Chandra, the first accurate value for the spin period change was obtained for RX J0720.4−3125 [57]. Under the assumption of the magnetic dipole model this allows an estimate for the magnetic field strength of 2.4×10^{13} G. This model can also be used to estimate the magnetic field of RX J1856.5−3754, using the observed period and the age of the star (see earlier). This results in $B = 3\times10^{13}$ G. For such strong magnetic fields cyclotron absorption lines from protons or heavier nuclei are expected in the 0.1-1.2 keV band. Indeed strong evidence for a broad absorption line was first found for RBS 1223 [46] and later detected in other sources (cf. Table 14.5 and for an example see Fig. 14.8). For the two pulsars RBS 1223 and RX J0720.4−3125 the depth of the absorption line was found to vary with pulse phase [46, 47].

Table 14.5 Magnetic field estimates

Object	dP/dt (10^{-13} ss^{-1})	E_{cyc} (eV)	B_{db} (10^{13} G)	B_{cyc} (10^{13} G)
RX J0420.0−5022	<92	~330?	<18	6.6?
RX J0720.4−3125	0.698(2)	260	2.4	5.2
RX J0806.4−4123	<18	400–450?	<14	8.0−9.1
RBS 1223	1.120(3)	100–300	3.4	2−6
RX J1605.3+3249		450–480		9.1−9.7
RX J1856.5−3754	<19		4.2$^{(*)}$	
RBS 1774		~700?		~14

$^{(*)}$ based on $P = 7.06$ s and an age of $\sim5\times10^5$ years

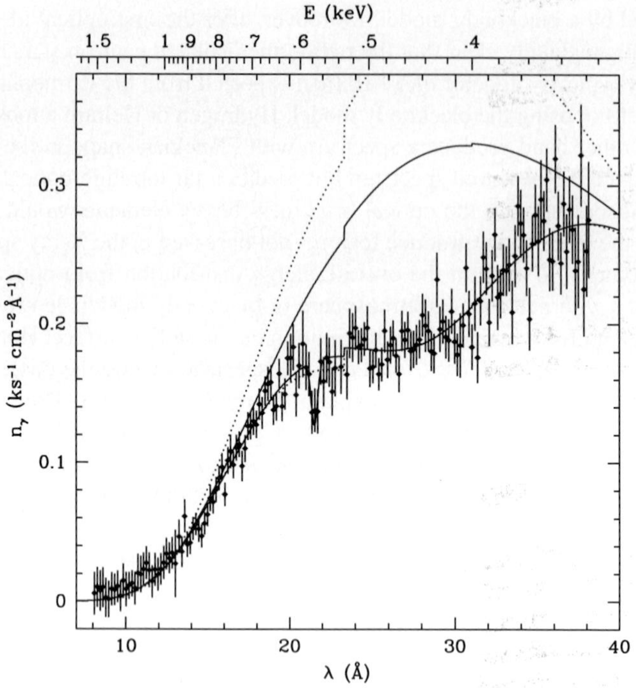

Fig. 14.8 RGS spectrum of RX J1605.3+3249 (reproduced from [58]). The upper histograms show blackbody continua

For a line centered at the energy E caused by proton cyclotron absorption, the magnetic field strength is given by $B = 1.6\times10^{11}\ E(eV)/(1\text{-}2GM/c^2R)^{1/2}$ G. For a standard neutron star with $M = 1.4M_\odot$ and $R = 10$ km, the corresponding magnetic field strengths are listed in Table 14.5. If the line were produced by heavier nuclei, the magnetic field would be larger by a factor A/Z∼2. Also given are the magnetic field strengths B_{db} of RX J0720.4−3125 and RBS 1223 inferred from the spin down measurements. These values are probably more uncertain as they depend on how close the model assumption of a vacuum magnetic dipole is to the real situation. Upper limits for the period derivative from two other objects only allow restrictions on B_{db} to smaller than $\sim2\times10^{14}$ G.

The brightest object among the magnificent seven, RX J1856.5−3754, is unique in many respects. Apart from the featureless spectrum and the small amplitude (∼1.5%) of periodic modulation in the X-ray flux (with P=7.06s [96]), it possesses an emission nebula with cometary-like morphology. The nebula was discovered on H_α images [59] with an apex approximately 1″ ahead of the neutron star and aligned with the direction of the proper motion of the star. Under the assumption that the age of the star is $\sim5\times10^5$ yrs (consistent with a likely birthplace in the Sco OB2 association) and using a period of 7.06 s the magnetic dipole braking model allows an estimate for the magnetic field strength of $\sim4.1\times10^{13}$ G. We conclude that there

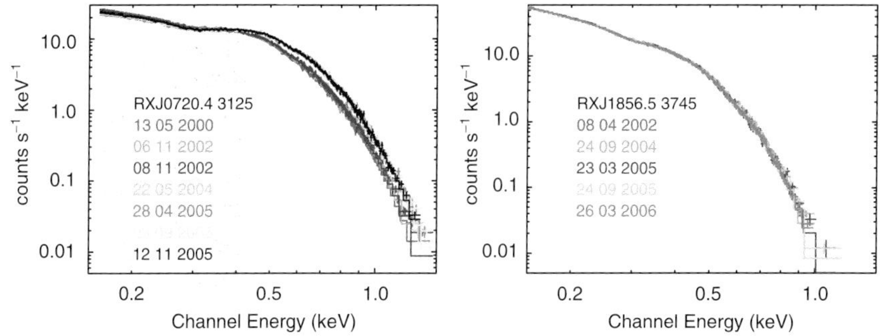

Fig. 14.9 EPIC-pn spectra of RX J0720.4−3125 and RX J1856.5−3754 modeled with blackbody continuum and in the case of RX J0720.4−3125 an additional broad absorption line with Gaussian shape

is strong evidence that the magnetic fields of the magnificent seven are of the order of a few times 10^{13} G.

The discovery of spectral changes from RX J0720.4−3125 on time scales of years was a big surprise. For no other cooling neutron star detected in X-rays such a behavior is known. The object was frequently observed as calibration target for XMM-Newton to monitor the low-energy response of the EPIC instruments. Read-out modes and filters were alternated for the EPIC cameras and a spectral hardening was first noticed in the RGS spectra [24]. Figure 14.9 shows the spectral evolution of RX J0720.4−3125 as measured by EPIC-pn from observations with the same instrumental setup. For comparison, five spectra obtained from RX J1856.5−3754 are fully consistent with each other, demonstrating the stability of both source and instrument.

Simultaneously with the spectral change observed from RX J0720.4−3125 the pulsed fraction as measured by EPIC increased from about 8% to 15%. Already from the first XMM-Newton observation in May 2000 it was found that the hardness-ratio[4] is modulated with the spin period of the neutron star. However, the hardness-ratio pulse profile is shifted with respect to the intensity pulse profile, showing that the spectrum is hardest just before intensity minimum and softest before intensity maximum. In October 2003, this phase offset has reversed, the hardness ratio pulse profile then lagging the full intensity pulse profile.

The analysis of the spectra from different epochs has shown that the blackbody temperature increased from 1.0×10^6 K in May 2000 to 1.1×10^6 K in May 2004. Simultaneously the equivalent width (a measure of the depth of the absorption line with respect to the continuum) of the absorption line increased by a factor of nearly 10 from $\sim$6 to 50 eV. It was suggested that the behavior of RX J0720.4−3125 is caused by precession of the neutron star and that we saw more directly onto the hot spot at the magnetic pole in 2004. However, in this case it is puzzling that the

[4] The hardness ratio is calculated as ratio of count rates in two different energy bands and can be used as measure for spectral changes.

derived size of the emitting area decreased from 2002 to 2004. Further monitoring of this interesting object with Chandra and XMM-Newton is in progress and should reveal if the spectral changes are periodic.

14.3.2.1 The Radius of RX J1856.5−3754 and the Equation of State at Supernuclear Densities

Neutron stars are giant atomic nuclei of stellar dimensions, bound by gravity. Because of the effects of gravity, the density of matter in the cores of neutron stars must be larger than that of atomic nuclei, but the composition and equation of state (EOS) of matter at supernuclear densities is still poorly understood, largely due to the fact that these conditions cannot be tested directly by colliding beam experiments at high energies. The equations under discussion show a dispersion of a factor of six or more of the pressure at a given density (e.g., [19, 62]). Therefore, observations of neutron stars constraining the nuclear physics parameters are very important. Since a given EOS leads to a specific mass–radius relationship, the relevant prime observables are the mass and the radius of a neutron star. Unluckily, there is not a single neutron star for which both the mass and the radius have been determined yet. While masses can be directly determined for neutron stars in binary systems, and have reached a high accuracy in the case of binary neutron stars, estimates of neutron star radii have been based on more indirect and model dependent methods:

1. Luminosities of X-ray bursts, assuming that they are close to the Eddington limit (e.g. [102])
2. Frequency of quasi-periodic oscillations, assuming that they reflect the Kepler frequency of accreted matter close to the neutron star (e.g. [69])
3. Limit of rotational stability of neutron stars by observing coherent high frequency oscillations in X-ray bursts (e.g. [27])
4. Measuring the light curve of coherent X-ray burst oscillations (e.g. [4])
5. "Radiation radii" of neutron stars using their thermal emission (e.g. [98])

Another method implies the measurement of the gravitational redshift of atomic lines in neutron stars with low magnetic fields ($<10^{10}$ G) where magnetic shifts are not important. Several papers have claimed the detection of lines in X-ray bursts (e.g., [31, 74]). Unfortunately, the real existence of these features has remained in doubt because they were not confirmed by later observations. But this method is very promising, because in principle it allows to determine three key parameters simultaneously: mass, radius, and distance of the neutron star [80].

In the following, we will concentrate on method No. 5 and its application to RX J1856.5−3754. To derive a radiation radius from observed thermal spectra, one needs to know the radiative properties of the atmosphere and the distance to the source. In the case of RX J1856.5−3754 an astrometric distance has been derived from HST observations, yielding 117 $\pm$ 12 pc [104] or more recently 140 $\pm$ 40 pc [51]. The X-ray spectrum of the neutron star measured with high resolution and large photon statistics with the Chandra LETG can be very well fitted by a blackbody with kT=0.63 $\pm$ 0.2 keV, devoid of any spectral features [23]; the optical/UV-spectrum measured by HST and VLT shows an ν^2 dependence typical for a Rayleigh-Jeans spectrum. However, its intensity is a factor $\sim$5 higher than the extrapolation of the X-ray blackbody, indicating that the latter comes from a small hot spot (with a radius of 4.4 $\pm$ 0.1 km $\times$ d/117pc) [23], while the optical/UV is emitted by the bulk surface. A double component fit to the overall spectrum yields a blackbody radius R_{bb}=16.5 km, while a fit with a continuous temperature distribution across the stellar surface gives $R_{\rm bb}$=16.8 km, both for a distance of 117 pc. It has been argued that these figures represent lower limits to the real radiation radius using the argument that a blackbody is the most efficient radiator for a given temperature distribution [22, 98]. Thus, we conclude that the radiation radius is R $>$ 16.5 km. A model involving a thin (1 g cm^{-2}), magnetic ($3-4\times10^{12}$ G), partially ionized hydrogen atmosphere on top of a condensed matter surface gives a radiation radius of 17 km assuming a distance of 140 pc [51]. However, this model is not fully self-consistent and assumes a magnetic field that is one order of magnitude below the one estimated from observations (cf. Table 14.5).

From the radiation radius R (seen by an observer at infinity) the true stellar radius $R_{\rm o}$ can be calculated using the relation $R = R_{\rm o}(1-R_{\rm s}/R_{\rm o})^{-1/2}$, where $R_{\rm s} = 2GM/c^2$ is the Schwarzschild radius of the neutron star. The resulting lower limit to $R_{\rm o}$ as a function of mass is shown in Fig. 14.10, using the lower limit for R_o of 16.5 km. This implies a rather stiff EOS and is inconsistent with RX J1856.5$-$3754 being a quark star.

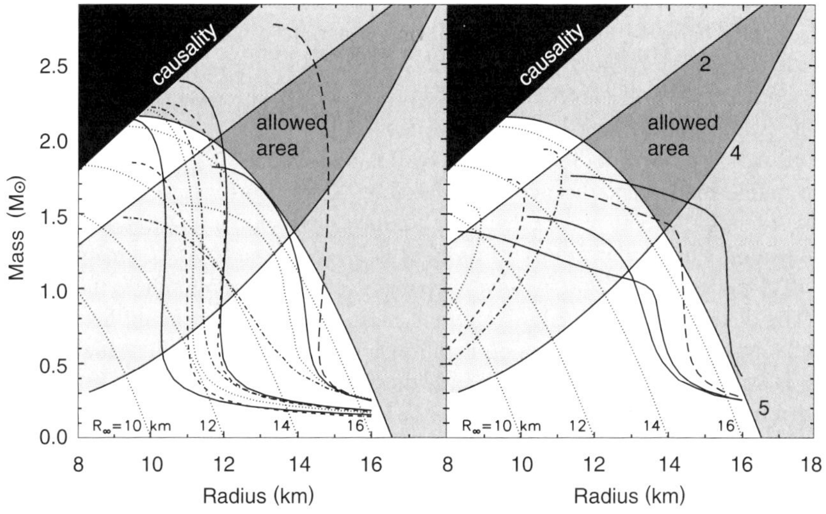

Fig. 14.10 Constraints provided by neutron star observations for the mass–radius relation: The numbers attached to the different curves refer to the methods 1–5 mentioned in the text. The mass–radius relations for various equations of state for nucleon/hyperon stars (*left*) and for kaon condensates/strange quark matter (*right*) are taken from [62]

14.3.3 Millisecond Pulsars

In the P-$\dot{P}$ parameter space, millisecond pulsars (ms-pulsars) are distinguished from the majority of ordinary-field pulsars by their short spin periods of $\lesssim$20 ms and small period derivatives of $\approx 10^{-18} - 10^{-21}$. In the frame of the magnetic braking model this corresponds to very old spin-down ages of typically $10^9 - 10^{10}$ yrs and low magnetic field strengths of $\sim 10^8 - 10^{10}$ G (cf. Fig. 14.1). More than ~75% of the known disk ms-pulsars are in binaries, usually with a low-mass white dwarf companion, compared to ≅ 1% binaries among the ordinary pulsars. This gives support to the idea that these neutron stars have been spun-up by angular momentum transfer during a past mass accretion phase [1, 5, 18]. Further evidence for this came from the discovery of seven accreting ms-pulsars, which seem to confirm this scenario (see [105] for a review). Presumably, these pulsars were originally among ordinary pulsars that would have turned off because of the loss of their rotational energy if they were not in close binaries (cf. Fig. 14.11). Millisecond pulsars are, therefore, often called "recycled" pulsars to better distinguish them from fast spinning pulsars seen in young supernova remnants.

By the end of 2006, about 10% of the 1 765 known radio pulsars fall into the category of ms-pulsars, i.e., are recycled [66]. The majority of them (almost 130) are located in 24 globular clusters [26], which apparently provide a favorable environment for the recycling scenario. Of these globular cluster ms-pulsars 54 (41%) are solitary, the others are in binaries. Interestingly, the ratio of solitary to binary ms-pulsars is almost identical to the 40% observed in the population of galactic disk ms-pulsars. The formation of solitary recycled pulsars is not well-understood, but it is widely believed that either the pulsar's companion was evaporated (a process that is believed to be at work in the PSR 1957+20 ms-pulsar/binary system) or the system was tidally disrupted after the formation of the ms-pulsar.

Recycled pulsars had been studied exclusively in the radio domain until the 1990s, when ROSAT, ASCA, EUVE, RXTE, and BeppoSAX were launched. The first millisecond pulsar discovered as pulsating X-ray source was PSR J0437−4715 [14], a nearby 5.75 ms pulsar that is in a binary orbit with a low-mass white dwarf companion. Further detections followed, which, by the end of 2006 sum up to ~40% of all X-ray detected rotation-powered pulsars (cf. [21] and references therein).

The data quality available from them, though, is far from being homogenous. While from several ms-pulsars high quality spectral, temporal and spatial information is available, many others, especially those on globular clusters, are just detected with a handful of events. Nevertheless, the improvements in sensitivity by Chandra and XMM-Newton provided a step forward in classifying the ms-pulsars' X-ray emission properties, indicating that there is a dichotomy between thermal and nonthermal dominated emitters, similar to what is observed from nonrecycled pulsars.

X-ray emission observed from ms-pulsars, which have a spin-down energy of $\dot{E} \geq 10^{35}$ erg s^{-1}, i.e., PSR J0218+4232, PSR B1821−24, and PSR B1937+21, is caused by nonthermal radiation processes [9, 61, 76]. This is confirmed from their power law spectra (photon-index α in the range 1.5–2) and pulse profiles that show

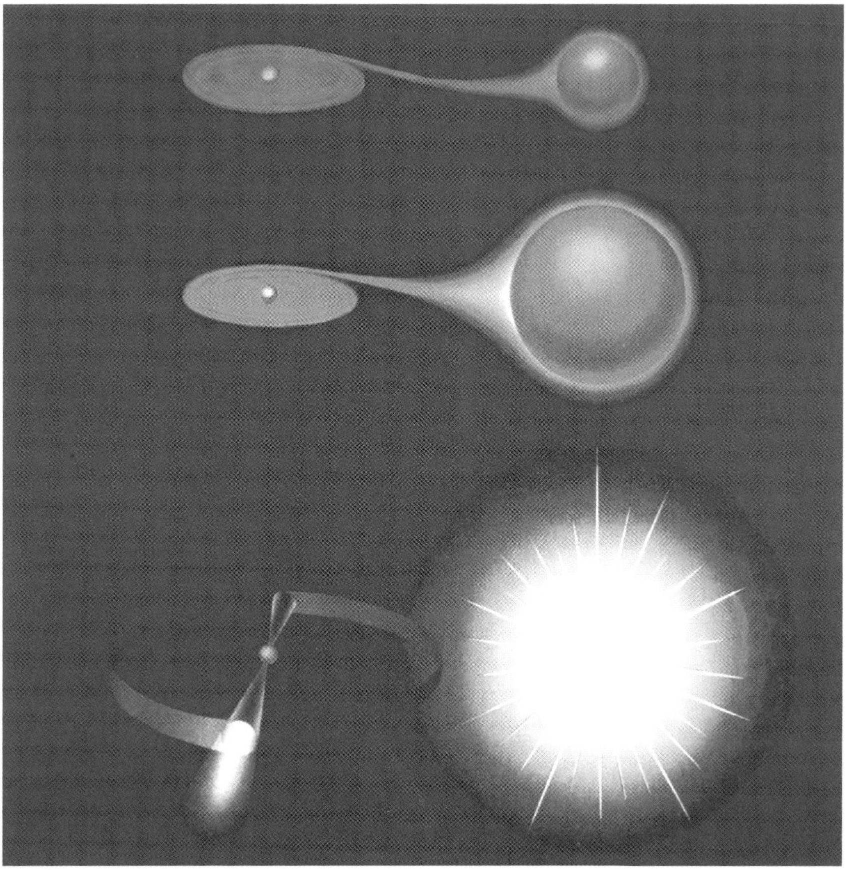

Fig. 14.11 Weakly-magnetized neutron stars that accrete matter from low-mass companion stars form the ∼150 currently known low-mass X-ray binaries. These systems are believed to be the progenitors of "recycled" pulsars. Along with the accretion of matter angular momentum transfer from the companion star takes place which spins-up the neutron star to millisecond periods. As the companion star evolves a solitary ms-pulsar or a ms-pulsar binary system is left

narrow peaks and have pulsed fractions of up to ∼90–100% (cf. Fig. 14.12). In common to these pulsars is that all show relatively hard X-ray emission, which made it possible to study some of them already with ASCA, BeppoSAX, and RXTE. For example, emission from PSR B1821−24 is detected by RXTE up to ∼20 keV [70] and PSR J0218+4232 is a candidate for a gamma-ray pulsar [61].

For the remaining ms-pulsars ($P \geq 4$ ms, $\dot{E} \sim 10^{33-34}$ erg s^{-1}), the X-ray emission is found to be much softer. Their X-ray spectra can be described by compound models consisting of a blackbody plus a power law component. The latter is required to describe the emission beyond 2–3 keV. For PSR J0437−4715 that is the brightest ms-pulsar detected in X-rays and thus is the one for which the best photon statistics is available, a three component spectral model is required consisting of a

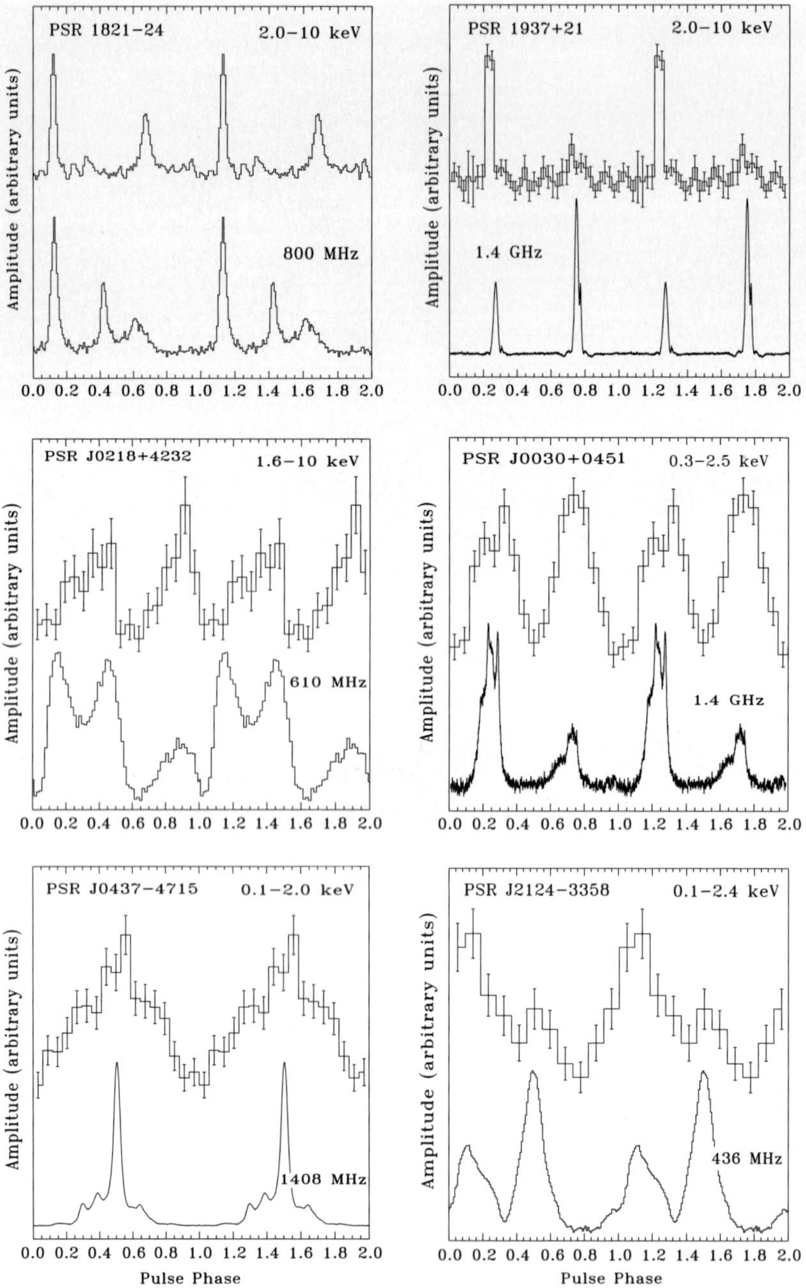

Fig. 14.12 X-ray and radio pulse profiles for the six brightest ms-pulsars. Two full pulse cycles are shown for clarity. The relative phase between the radio and X-ray pulses is only known for PSR 1821−24, B1937+21, 0218+4232, and PSR J0437-4715 with sufficient accuracy. The phase alignment in all other cases is arbitrary

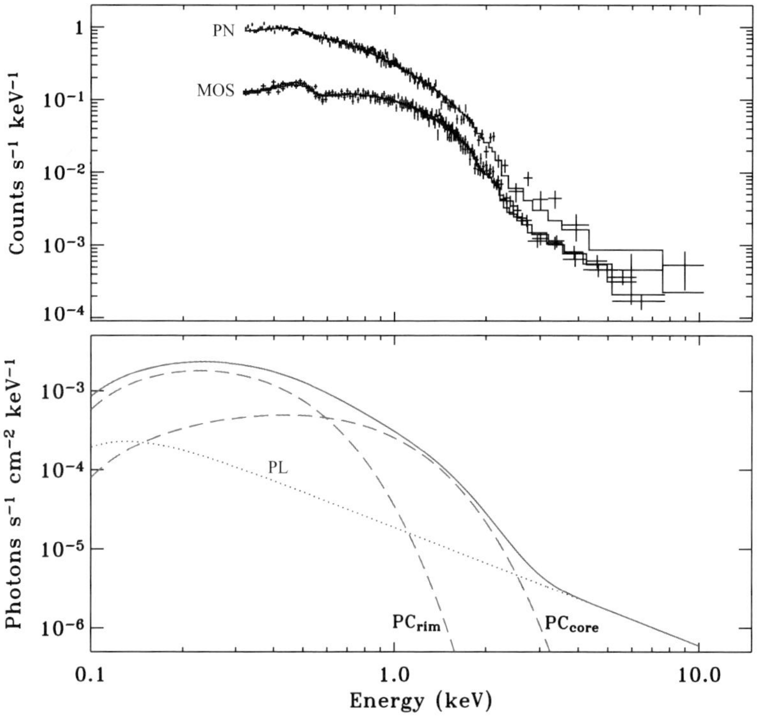

Fig. 14.13 X-ray spectrum of PSR J0437-4715. The *solid curves* show the best fitting model that is the sum of a power law (PL) and a two temperature blackbody model labeled as *core* and *rim*

two temperature blackbody plus a power law model. The X-ray spectrum of PSR J0437-4715 as detected with XMM-Newton is shown in Fig. 14.13.

The relatively small blackbody radii found by these spectral fits suggest that the thermal emission is coming from one or two heated polar-caps whereas the power law component describes the nonthermal radiation emitted from accelerated particles in the corotating magnetosphere. The prototypical ms-pulsar of this group, which is still the one for which the best data are available, is the nearest and brightest millisecond pulsar PSR J0437−4715. It was already evident in the ROSAT and ASCA data that its X-ray emission consists of at least two different spectral components [12–14]. Chandra and XMM-Newton data have further constrained its emission properties [21, 115]. The two thermal components are interpreted as emission from a hot polar cap, having a nonuniform temperature distribution with a hot core ($T_{\rm core} = 1.4 \times 10^6$ K, $R_{\rm core} = 0.4$ km) and a cooler rim ($T_{\rm rim} = 0.5 \times 10^6$ K, $R_{\rm rim} \sim 2.6$ km). The power law component yields a photon index of $\alpha \sim 2.0$. The size of the polar cap is found to be roughly in agreement with the theoretical predictions. Defined as the area of open field lines in which the bombardment by relativistic particles is expected, it is $R_{\rm pc} = R(R\Omega/c)^{1/2}$. Assuming $R = 10$ km for the neutron star radius and taking $\Omega = 1.09 \times 10^3$ for the pulsars angular frequency yields $R_{pc} = 1.9$ km for a polar cap radius of PSR J0437−4715.

Interaction between relativistic pulsar winds (which carry away the rotational energy of pulsars) and the surrounding interstellar medium is expected to create detectable diffuse emission. If the physical conditions are appropriate this emission takes the form of a pulsar bow-shock nebula (cf. Fig. 14.14).

By now, such diffuse emission is seen in H_α from the black widow pulsar PSR B1957+20 [52, 91], from PSR J0437-4715 [15], and from PSR J2124−3358 [35]. Diffuse X-ray emission associated with these bow-shock nebulae could only be detected from PSR B1957+20 [15, 91] and from the solitary ms-pulsar PSR J2124−3358 [55]. For the latter the emission extends from the pulsar to the north-west by $\sim$0.5 arcmin (cf. Fig. 14.14). Adopting the pulsar distance of $\sim$250 pc, the tail has a length of $\sim 1.1 \times 10^{17}$ cm. The spectrum of this nebular emission can be modeled with a power-law of photon index 2.2 ± 0.3, in line with the emission originating from accelerated particles in the post shock flow. Comparable deep observations to those of PSR J2124−3358 and PSR B1957+20 have been performed by XMM-Newton in previous years on almost all X-ray bright ms-pulsars. For PSR J0437−4715, PSR J0030+0451, and PSR J1024−0719, which all have spin parameters similar to that of PSR J2124−3358, no diffuse emission was detected down to a $3-\sigma$ limiting flux of $\sim 4-7 \times 10^{-15}$ erg s^{-1} cm^{-2} [55], suggesting that the formation of bow-shocks depends not on the pulsars spin-parameters but might be a function of the ISM density and pulsar proper motion only.

The majority of the detected ms-pulsars resides in globular clusters. The first millisecond pulsar discovered in a globular cluster was PSR B1821-24 in M28 [65]. Its inferred pulsar parameters make it the youngest ($P/2\dot{P} = 3.0 \times 10^7$ yrs) and most powerful ($\dot{E} = 2.24 \times 10^{36} I_{45}$ erg s^{-1}) pulsar among all known MSPs. Since the

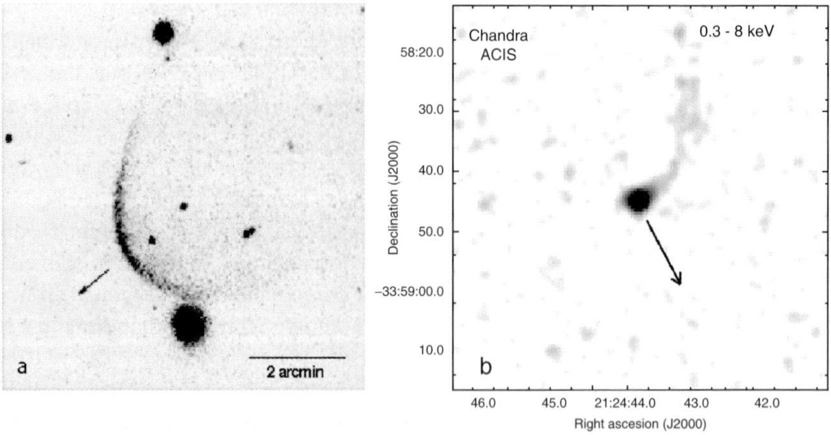

Fig. 14.14 (**a**) The bow-shock around PSR J0437-4715 as visible in H_α. (**b**) Chandra image of PSR J2124−3358 and its diffuse, arc-like X-ray emission associated with the pulsar's bow-shock. The pulsars proper motion directions are indicated

Einstein era, it has been clear that globular clusters contain various populations of X-ray sources of very different luminosities [49]. The stronger sources ($L_x \approx 10^{36} - 10^{38}$ erg s^{-1}) were seen to exhibit X-ray bursts, which led to their identification as low-mass X-ray binaries (LMXBs). The nature of the weaker sources, with $L_x \leq 3 \times 10^{34}$ erg s^{-1}, however, was more open to discussion [30, 56]. Although many weak X-ray sources were detected in globulars by ROSAT [56, 103], their identification has been difficult because of low photon statistics and strong source confusion in the crowded globular cluster fields, except for a few cases. Of particular interest are the results obtained from Chandra observations of PSR B1821-24 in M28 [9] and on 47 Tuc = NGC 104. From the latter, Grindlay [42] reported the detection of 108 sources within a region corresponding to about five times the 47 Tuc core radius. Nineteen of the soft/faint sources were found to be coincident with radio-detected millisecond pulsars (MSPs), and Grindlay [41, 42] concluded that more than 50% of all the unidentified sources in 47 Tuc are MSPs. This conclusion is in line with theoretical estimates on the formation scenarios of short-period (binary) pulsars in globular clusters [89]. The application of the Chandra X-Ray Observatory sub-arcsecond angular resolution along with the temporal resolution provided by its HRC-S detector allowed to detect X-ray pulsations at a $\sim 4\,\sigma$ level from four of these sources in a recent 830 ks deep observation [25].

14.3.4 Summary

X-ray astronomy has made great progress in the past several years thanks to telescopes with larger effective areas and greatly improved spatial, temporal, and spectral resolutions. The next generation of proposed instruments like eROSITA, Simbol-X, XEUS, and Constellation-X are supposed to bring again a major improvement in sensitivity, which will make these instruments even more suitable for pulsar and neutron star astronomy.

References

1. Alpar M.A., Cheng A.F., Ruderman M.A., Shaham J., 1982, Nature, 300, 728
2. Aschenbach, B., Brinkmann, W., 1975, Astron. Astrophys., 41, 147
3. Aschenbach, B., 1998, Nature, 396, 141
4. Bhattarcharyya, S., Strohmayer, T.E., Miller, M.C., Markwardt, C.B., 2005, Astrophys. J., 619, 483
5. Bhattacharya, D., van den Heuvel, E.P.J., 1991, Phys. Rep., 203, 1
6. Becker, W., Tennant, A., Kanbach, G., et al., 2007, submitted to Astron. Astrophys.,
7. Becker, W., Hui, C.Y., Aschenbach, B., Iyudin, A., 2007b, Astron. Astrophys., astro-ph/0607081
8. Becker, W., Kramer, M., Jessner, A., et al., 2006, Astrophys. J., 645, 1421
9. Becker, W., Swartz, D.A., Pavlov, G.G., et al., 2003, Astrophys. J., 594, 798

10. Becker, W., Aschenbach, B., 2002, in W. Becker, H. Lesch, J. Trümper (eds.), Neutron Stars, Pulsars and Supernova Remnants, MPE-Report 278, 64, (astro-ph/0208466)
11. Becker, W., Pavlov, G.G., 2001, in J. Bleeker, J. Geiss, M. Huber (eds.), The Century of Space Science, Kluwer, Dordrecht, 721, astro-ph/0208356
12. Becker, W., Trümper, J., 1999, Astron. Astrophys., 341, 803
13. Becker, W., Trümper, J., 1997, Astron. Astrophys., 326, 682
14. Becker, W., & Trümper, J., 1993, Nature, 365, 528
15. Bell, J.F., Bailes, M., Manchester, R. N., Weisberg, J. M., Lyne, A.G., 1995, Astrophys. J., 440, L81
16. Bell, J. 1977, Ann. NY Acad. Sci., 302, 685
17. Beskin, V.S., Gurevich, A.V., Istomin, Ya.N., 1993, Physics of the Pulsar Magnetosphere, Cambridge University Press, ISBN 0-521-41746-5
18. Bisnovatyi-Kogan G.S., Komberg B.V., 1974, Sov. Astron., 18, 217
19. Blaschke, D., Grigorian, H., Voskresensky, D.N., 2001, Astron. Astrophys., 368, 561
20. Blaschke D., Glendenning, N.K., Sedrakian, A., 2001, Springer, ISBN 3540423400
21. Bogdanov, S., Grindlay, J.E., Heinke, C.O., et al., 2006, Astrophys. J., 646, 1104
22. Braje, T.M., Romani, R.W., 2002, Astrophys. J., 580, 1043
23. Burwitz, V., Haberl, F., Neuhäuser, R., et al., 2003, Astron. Astrophys., 399, 1109
24. de Vries, C. P., Vink, J., Méndez, M., & Verbunt, F. 2004, A&A, 415, L31
25. Cameron, P.B., Rutledge, R.E., Camilo, F., et al., 2007, APJ 660, 587
26. Camilo, F., Rasio, A.F., 2005, in F.A., Rasio, I.H., Stairs (eds.), Binary Radio Pulsars, ASP. Conf. Ser., 328, 147
27. Chakrabarty, D., et al. 2003, Nature, 424, 42
28. Cheng, K.S., Ho, C., Ruderman, M.A., 1986a, Astrophys. J., 300, 500
29. Cheng, K.S., Ho, C. Ruderman, M.A., 1986b, Astrophys. J., 300, 522
30. Cool, A.M., Grindlay, J.E., Krockenberger, M., & Bailyn, C.D., 1993, Astrophys. J., 410, L103
31. Cottam, J., Paerels, F., Mendez, M., 2002, Nature 420, 51
32. De Luca, A., Caraveo, P.A., Mereghetti, S., et al., 2006, Science, 313, 814
33. Fesen, R.A., Pavlov, G.G., Sanwal, D., 2006, Astrophys. J., 636, 848
34. Gaensler, B.M., 2004, Adv. Space. Res., 33, 645
35. Gaensler, B.M., Jones, D.H., Stappers, B.W., 2002, Astrophys. J., 580, L137
36. Glendenning, N.K., 2001, Compact Stars, Springer, ISBN 0387989773
37. Gold, T., 1968, Nature, 218, 731
38. Gold, T., 1969, Nature, 221, 25
39. Gotthelf, E.V., Halpern, J.P., Seward, F.D., 2005, Astrophys. J., 627, 390
40. Green D.A., 2006, 'A Catalogue of Galactic Supernova Remnants (2006 April version)', Astrophysics Group, Cavendish Laboratory, Cambridge, UK")
41. Grindlay, J.E., Camilo, F., Heinke, C.O., et al., 2002, Astrophys. J., 581, 470
42. Grindlay, J.E., Heinke, C.O., Edmonds, P.D., et al., 2001, Astrophys. J., 563, 53
43. Gudmundsson, E.H., Pethick, C.J., Epstein, R.I. 1983, Astrophys. J., 272, 286
44. Haberl, F. 2004, Advances in Space Research, 33, 638
45. Haberl, F., Motch, C., Buckley, D. A. H., Zickgraf, F. J., & Pietsch, W. 1997, A&A, 326, 662
46. Haberl, F., Schwope, A. D., Hambaryan, V., Hasinger, G., & Motch, C. 2003, A&A, 403, L19
47. Haberl, F., Zavlin, V. E., Trümper, J., & Burwitz, V. 2004, A&A, 419, 1077
48. Helfand, D.J., Becker, R.H., 1984, Nature, 307, 215
49. Hertz, P., Grindlay, J.E, 1983, Astrophys. J., 275, 105
50. Hewish, A., Bell, S.J., Pilkington, J.D.H., et al., 1968, Nature, 217, 709
51. Ho, W.C.G., Kaplan, D.L., Chang, P., et al., 2007, to appear in Mon. Not. R. Astron. Soc.
52. Huang. H.H., Becker,W., 2007, Astron. Astrophys., 463, L5, astro-ph/0701611
53. Hui, C.Y., Becker, W., 2006a, Astron. Astrophys., 454, 543
54. Hui, C.Y., Becker, W., 2006b, Astron. Astrophys., 457, L33
55. Hui,C.Y., Becker,W., A & A., 448, 13, 2006c

56. Johnston, H.M., & Verbunt, F., 1996, Astron. Astrophys., 312, 80
57. Kaplan, D. L. & van Kerkwijk, M. H. 2005, ApJ, 628, L45
58. van Kerkwijk, M. H., Kaplan, D. L., Durant, M., Kulkarni, S. R., & Paerels, F. 2004, ApJ, 608, 432
59. van Kerkwijk, M. H. & Kulkarni, S. R. 2001, A&A, 380, 221
60. Kouveliotou, C., Dieters, S., Strohmayer, T., et al., 1998, Nature, 393, 235
61. Kuiper, L., Hermsen, W., Verbunt, F., et al., 2002, Astrophys. J., 577, 917
62. Lattimer, J.M., Prakash, M., 2001, Astrophys. J. 550, 426
63. Lazendic, J.S., Slane, P.O., Gaensler, B.M., et al., 2003, Astrophys. J., 593, L27
64. Lommen, A., Donovan, J., Gwinn, C., et al., 2006, Astrophys. J. submitted, astro-ph/0611450
65. Lyne, A.G., Brinklow, A., Middleditch, J., et al., 1987, Nature, 328, 399
66. Manchester, R.N., Hobbs, G.B., Teoh, A., Hobbs, M., 2005, AJ, 129, 1993
67. Mereghetti, S., Chiarlone, L., Israel, G.L., Stella, L., 2002, in W. Becker, H. Lesch, J. Trümper (eds.), Neutron Stars, Pulsars and Supernova Remnants, MPE-Report 278, 29, astro-ph/0205122
68. Michel, F.C., 1991, Theory of Neutron Star Magnetospheres, University of Chicago Press, Chicago, ISBN 0-226-52331-4
69. Miller, M.C., 2003, in X-Ray Timing 2003: Rossi and Beyond, astro-ph/0312449
70. Mineo, T., Cusumano, G., Massaro, E., Becker, W., Nicastro, L., 2004, Astron. Astrophys., 423, 1045
71. Moffett, D.A., Hankins, T.H., 1996, Astrophys. J. 468, 779
72. Motch, C., Sekiguchi, K., Haberl, F., et al. 2005, A&A, 429, 257
73. Motch, C. 2001, in X-ray Astronomy, Stellar Endpoints, AGN, and the Diffuse X-ray Background, AIP Conference Proceedings, 244–253
74. Nakamura, N., Inoue, H., and Tanaka, Y., 1988, Publ. Astron. Soc. Jpn., 40, 209
75. Nasuti, F.P., Mignani, R., Caraveo, P.A., Bignami, G.F., 1997, Astron. Astrophys., 323, 839
76. Nicastro, L., Cusumano, G., Loehmer, O., et al., 2004, A & A, 413, 1065
77. Ostriker, J.P., Gunn, J.E., 1969, Astrophys. J., 157, 1395
78. Ögelman, H., 2005, in Alpar, Kiziloglu, van Paradijs (eds.), The Lives of the Neutron Stars, NATO ASI Series, Kluwer, Dordrecht, 101
79. Ögelman, H., Finley, J.P., Zimmerman, H.U., 1993, Nature, 361, 136
80. Özel, F., 2006, Nature, 441, 1115
81. Pacini, F., 1967, Nature, 216, 567
82. Pacini, F., 1968, Nature, 219, 145
83. Page, D., 2007, in W. Becker (ed.), Springer Lecture Notes on Pulsars and Neutron Stars: 40 Years After the Discovery
84. Page, D., Reddy, S., 2006, Annu. Rev. Nucl. Particle Syst., 56, 1, 327
85. Pavlov, G. G., Zavlin, V. E., Trümper, J., Neuhäuser, R. 1996, ApJ, 472, L33
86. Petre, R., Canizares, C.R., Kriss, G.A., Winkler, P.F., Jr., 1982, Astrophys. J., 258, 22
87. Pfeffermann, E., Aschenbach, B., 1996, in H.H. Zimmermann, J. Trümper, H. Yorke (eds.), Roentgenstrahlung from the Universe, MPE Rep. 263, 267
88. Popov, S. B., Colpi, M., Prokhorov, M. E., Treves, A., Turolla, R. 2003, A&A, 406, 111
89. Rasio, F.A., Pfahl, E.D., Rappaport, S., 2000, Astrophys. J., 532, 47
90. Ruderman, M., Sutherland, P.G., 1975, Astrophys. J., 196, 51
91. Stappers, B.W., Gaensler, B.M., Kaspi, V.M., et al., 2003, Science, 299, 1372
92. Seward, F.D., Slane, P.O., Smith, R.K., Sun, M., 2003, Astrophys. J., 584, 414
93. Tananbaum, H. 1999, IAU Circ. #7246
94. Tananbaum, H., Gursky, H., Kellogg, E.M., et al., 1972, Astrophys. J., 174, L143
95. Tennant, A.F., Becker, W., Juda, et al., 2001, Astrophys. J., 554, L173
96. Tiengo, A. & Mereghetti, S. 2007, ApJ, 657, L101
97. Treves, A., Turolla, R., Zane, S., & Colpi, M. 2000, Pub. Astron. Soc. Pacific, 112, 297
98. Trümper, J. E., Burwitz, V., Haberl, F., & Zavlin, V. E. 2004, Nuclear Physics B Proceedings Supplements, 132, 560
99. Trümper, J.E., Pietsch, W., Reppin, C., et al., 1978, Astrophys. J., 219, L105

100. Tsuruta S., 1998, Physics Reports, 292, 1
101. Tuohy, I.R., Garmire, G.P., 1980, Astrophys. J., 239, L107
102. van Paradijs, J., 1982, Astron. Astrophys., 107, 51
103. Verbunt, F. 2001, Astron. Astrophys., 368, 137
104. Walter, F. M., Lattimer, J. 2002, Astrophys. J. 576, L 145
105. Wijnand, R., 2005, in *Nova Science Publishers: Pulsars New Research*, astro-ph/0501264
106. Weber, F. 1999, Pulsars as Astrophysical Laboratories for Nuclear and Particle Physics, Institute of Physics, ISBN 0-7503-0332-8
107. Weisskopf, M.C., Hester, J.J., Tennant, A.F., et al. 2000, Astrophys. J., 536, L81
108. Willingale, R., Aschenbach, B., Griffiths, R.G., et al., 2001, Astron. Astrophys., 365, L212
109. Woods, P.M., Thompson, C., 2006, in *Compact stellar X-ray sources*, eds W.Lewin & M. van der Klis, p. 547-586, Cambridge University Press, ISBN 978-0-521-82659-4, astro-ph/0406133
110. Woods, P.M., Zavlin, V.E., Pavlov, G.G., 2006, in D. Page, R. Turolla, S. Zane (eds.), Isolated Neutron Stars: from the Interior to the Surface, astro-ph/0608483
111. Woods, P.M., Kouveliotou, C., Gögüs, E., et al. 2002, Astrophys. J., 576, 381
112. Yakovlev, D.G., Pethick, C.J., 2004, Ann. Rev. Astron. Astrophys. 42, 169
113. Yakovlev, D.G., Levenfish, K.P., Shibanov, Yu.A. 1999, Physics-Uspekhi, 169, 825
114. Zavlin, V.E., Pavlov, G.G., Sanwal, D. 2004, Astrophys. J., 606, 444
115. Zavlin, V.E., in D. Page, R. Turolla, S. Zane (eds.), Astrophysics and Space Science, Volume 308, Issue 1–4, 297

15 Accreting Neutron Stars

R. Staubert

15.1 Introduction

In this chapter neutron stars (NS) are discussed, which are powered through accretion from a companion star in a binary system. The companion is either a "normal star" or a White Dwarf. X-ray binaries where the mass receiving compact object is a White Dwarf or Black Hole, as well as isolated NS, which are powered by a different kind of energy supply (spin energy, magnetic field energy, thermal energy content or accretion from the interstellar medium), are discussed in other chapters of this book.

The transfer of material from the companion star, finally ending up on the surface of the NS where most of the gravitational energy is released, is by stellar wind or by Roche lobe overflow (see Fig. 15.1), or by the NS crossing the circumstellar disk of an OB or Be star.

The first extra-solar X-ray source, Sco X-1, discovered in 1962 by a rocket experiment [17] was later found to be a NS binary. Only the first X-ray satellite, Uhuru, with its first all sky X-ray survey in 1970, allowed to identify these objects as a class: of the 339 sources listed in the 4th Uhuru catalog [16] one third are associated with accreting NS in a binary system. Today we know a few hundred sources of this kind in our galaxy and the nearest neighbour galaxies [29, 30]. These sources are characterized by a high X-ray luminosity (10^{34} to 10^{38} erg s^{-1}), hard spectra, and a high degree of variability of different nature. They are important laboratories for the study of material under extreme physical conditions, such as high densities, high temperatures, high photon densities, or strong gravitational and magnetic fields, as well as for stellar evolution. The research on NS deals with fundamental physics, e.g., the nature of ultra-dense matter and its equation of state or general relativity in the immediate vicinity of the compact object where particles in close orbits move in curved space time.

Important books dealing with accreting neutron stars are: *X-ray Timing 2003* [1], *X-Ray Binaries* [2], *Accretion driven stellar X-ray sources* [5], *X-ray astronomy* [6], *Compact stellar X-ray sources* [8]. The textbook *Exploring the X-ray Universe* [3] contains three chapters on X-ray binaries. More specialized reviews and original references are cited later. An introduction to *Accretion Power in Astrophysics* is given by [4]. A catalogue of low and high mass X-ray binaries compiled by van Paradijs can be found in *X-Ray Binaries* [2] and in updated form by [29, 30].

J.E. Trümper and G. Hasinger (eds.), *The Universe in X-Rays*.

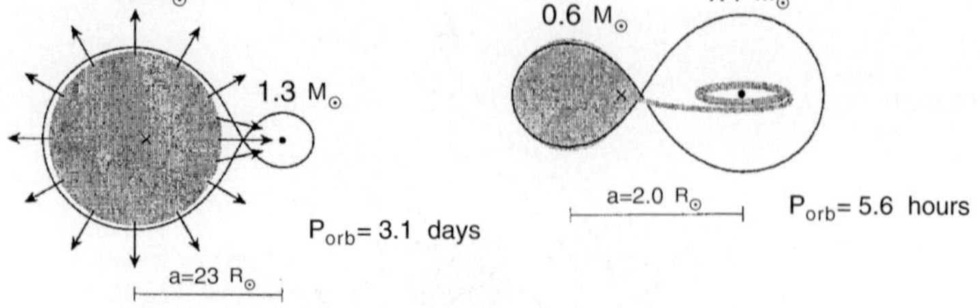

Fig. 15.1 Wind fed HMXB and Roche lobe overflow fed LMXB

15.2 Overview

Accreting NS appear in a large variety of types, persistent and transient, exhibiting various kinds of observational features, both in their spectral and in their temporal behavior. The two primary physical factors determining the properties of individual systems are (1) the *mass of the companion* and (2) the *magnetic field of the NS*. Further factors, usually intimately connected to (1) and (2) are (3) the *geometry of the binary orbit*, (4) the *mechanism, geometry, and magnitude of the accretion flow*, and (5) the *spin of the NS*. The mass of the neutron star itself is found in a rather narrow range around 1.4 solar masses. Characteristic observables are the overall luminosity, the shape of the spectrum and the time variability properties.

In this multi-parameter space accreting NS of different appearances occupy preferred regions. The first fundamental property is the mass of the companion. It is, therefore, useful and common to distinguish between *High Mass X-ray Binaries* (HMXB) and *Low Mass X-ray Binaries* (LMXB) (both of which include also systems with black hole candidates, see chapter 16). In HMXB the optical companion is a Be star or OB supergiant. They usually have long (days to tens of days) orbital periods and mass transfer is by stellar wind, while in the case of LMXB the companion is of spectral type A or later, the orbital periods are shorter (hours to days) and the mass transfer is by Roche Lobe overflow. High mass systems containing a Be star tend to have orbits with substantial eccentricity and the mass transfer is connected to the circumstellar structure (i.e., an equatorial disk) in these strongly rotating stars.

The second most important factor is the strength of the magnetic field (B) of the NS. In the case of a strong (10^{12} Gauss) field, the inner boundary of the accretion disk cannot reach close to the surface of the NS but will be truncated at the magnetospheric radius, defined by equal pressure of the gas and the B field. The accreted material is funneled along the magnetic field lines down to the polar regions of the NS surface where an X-ray emitting hot spot forms (Fig. 15.2). If the magnetic axis is inclined to the rotational axis, a *pulsar* may be observed. If instead the magnetic field is weak (10^9 Gauss or less), the accretion can happen over larger parts of the NS surface and the material can accumulate for some time until a nuclear burning flash

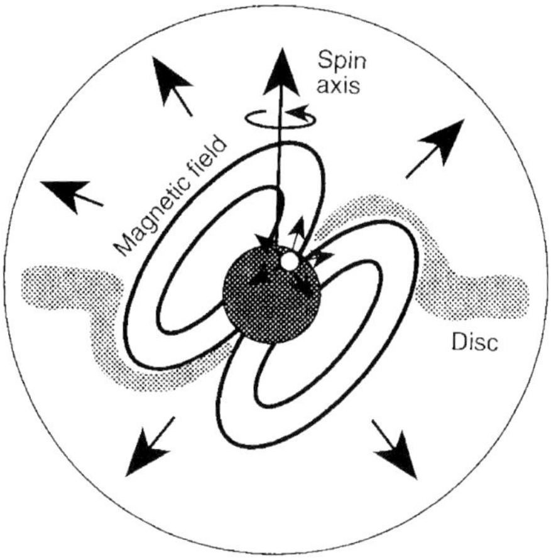

Fig. 15.2 Magnetic NS and its magnetosphere

may ignite, giving rise to X-ray bursts. Other types of X-ray bursts are generated when clumps of matter fall onto the NS surface within a short time.

15.2.1 The Zoo

There is a whole "zoo" of X-ray sources with respect to their observational appearances, which is of course governed by the physical conditions within the systems. The diagnostic tools for understanding the physics is provided through observations of the spectral and time variability properties of the sources over the widest possible ranges in wavelength and time scales, respectively. In the X-ray range, it has turned out that the study of the timing behavior, including spectral variability, is the tool with the highest diagnostic power. In reaching millisecond time resolution, the natural dynamical time scales in the immediate vicinity of NS are tested.

In the following, a brief introduction is given to some common types of X-ray sources containing accreting NS, describing their spectral and variability properties.

Transients are sources that are not persistent, but rather show up in episodes (days to weeks) of X-ray emission, sometimes recurrent in a regular or irregular way.

Eclipsing X-ray binaries allow a straight forward determination of their binary nature and an easy measurement of the binary period. They are usually found among the HMXB.

X-ray binary pulsars are characterized by a regular (coherent) modulation of the X-ray flux because of the spin of the NS. In these cases the binary nature is confirmed by the observation of Doppler shift of the pulse frequency, thereby allowing to measure the shape and physical parameters of the binary orbit. In this way also the mass function of the system is determined, allowing an estimate of the mass of the NS.

Bursters are sources that show repeated short duration (minutes) *bursts* of X-ray flux. These sources are all LMXB with weakly magnetized NS. One has learned to distinguish between two types of bursts: Type I, associated with *thermonuclear flashes* on the surface of the NS and Type II, first found in addition to Type I bursts in the so called *Rapid Burster*, which are interpreted as events of intermittent accretion because of disk instabilities.

QPO sources are sources that show so called *quasi periodic oscillations* (QPOs). In contrast to the coherent modulation of pulsars, this kind of variability reveals itself by broad peaks in the power spectral density function (PSD) obtained in a Fourier analysis of the timing data. Initially detected QPOs in LMXB had frequencies in the range of a few Hertz to a few tens of Hertz. They are usually associated with irregular variability ("noise" components in the PSD) at low and high frequencies. Similar variability properties are also observed in black hole candidates. The QPO sources are all LMXB and tend to exhibit characteristic spectral variations, describable by a smooth movement through a "color/color diagram" (Sect. 15.6). According to the shape of the patterns of the movements, such sources are called *Z–sources* or *Atoll–sources*. A more recent discovery, made possible by the Rossi X-ray Timing Explorer (RXTE), are the *kHz QPOs*, which provide information about the very close vicinity of a spinning, accreting NS.

15.2.2 Orbits and Super-Orbital Periods

The primary indicator to associate a newly detected X-ray source with an accreting NS is evidence for binary motion, which can come from one or more of the following observations: *eclipses*, *smooth periodic modulation*, *regular absorption dips*, *regular or semiregular X-ray outbursts*, *periodic pulsar arrival time modulation*, or *radial velocity variations*. The X-ray emitting compact object could still be a black hole (BH) or a white dwarf (WD). The distinction between NS and BH is not always easy (see chapter 16). If regular pulsations or Type I X-ray bursts are observed, the compact object is a NS. If a very soft spectrum (possibly with a power law type tail) and fast irregular variability is seen, a BH might be suspected. The most reliable signature for a BH, however, is a mass of the compact object in excess of $\sim 3\,M_\odot$. Similarly, a WD is identified by masses $\leq 1\,M_\odot$ and by their optical and UV appearance (see chapter 12). To distinguish between high and low mass NS binaries, the spectral type of the optically identified companion is used. In the case that orbital parameters are determined through pulse arrival times, the mass function is known. If neither of the above information is available, one would tentatively classify an object as a LMXB if a soft X-ray spectrum, Type I X-ray bursts, or an orbit of less than 12 h are observed, and as a HMXB if a hard (power law type) X-ray spectrum, X-ray pulsations, or an orbit greater than 12 h are observed.

In several systems modulation of the X-ray flux with periods (or quasiperiods) of the order of a few weeks to several months, much longer than the orbital period, is observed [43]. These *super-orbital* periods are generally attributed to the accretion disk which, through precessional motion, (semi)regularly blocks the line of sight to the X-ray emitting region.

15.2.3 Accretion Physics

The radiation observed from accreting NS is predominantly in the X-ray range (0.1 keV to beyond 100 keV) and mostly originates from the surface of the NS with contributions from the accretion disk. The radiation mechanism is mostly thermal Bremsstrahlung and black body emission with temperatures in excess of 10^7 K. Comptonisation by hot thermal electrons is thought to be responsible for spectra characterized by a hard power law and a high energy cut-off. Strongly magnetized NS tend to show one or more Cyclotron resonant scattering lines in their spectra. In the case of LMXB also UV and optical radiation from the accretion disk are observed. In interpreting the observed spectra, secondary effects like absorption and scattering in cold and warm (partially ionized) material, as well as fluorescence and recombination need to be taken into account.

Accreting NS provide a laboratory for accretion physics. The process of accretion, i.e., conversion of gravitational/potential energy into radiation energy, is the most effective energy conversion process (of order 10%, as compared to 0.7% for nuclear fusion). Protons falling down the deep potential well of a NS and being abruptly stopped at the surface of the NS (~10 km radius) reach velocities ~0.4 of the speed of light with the corresponding kinetic energy of ~200 MeV. The accretion luminosity is given by $L_{\rm acc} = \eta \left(\frac{GMm}{R}\right) \sim 10^{37} m_{17} \left(\frac{M}{M_\odot}\right) \left(\frac{R}{10\,km}\right)^{-1} erg\,s^{-1}$, where M is the mass of the NS, R its radius, η is the energy conversion efficiency, m is the mass accretion rate and m_{17} the same in units of $10^{17}\,{\rm g\,s^{-1}}$, and $M_\odot$ the mass of the sun. This means that a mass accretion rate of $10^{17}\,{\rm g\,s^{-1}}$ (equivalent to $10^{-9}\,M_\odot\,yr^{-1}$) yields an X-ray luminosity of $10^{37}\,{\rm erg\,s^{-1}}$. The thermalized kinetic energy causes high temperatures (10^6 to beyond 10^8 K), leading to thermonuclear burning in the case of sufficient density and to direct radiation in the X-ray range.

The material accreted from the companion carries angular momentum that needs to be dissipated before the material can fall down onto the NS surface. This happens in an accretion disk where particles circle the NS in quasi Keplerian orbits. Through viscosity, angular momentum is redistributed such that some material can spiral down toward the NS. In the case of low B field, the accretion disk can reach close to the NS surface, for high B field, however, the magnetic field will start to dominate the movement of the plasma at the magnetospheric radius and from there on guide the material along the field lines down to the polar regions of the NS surface (see Fig. 15.2). The interaction between the material and the magnetic field can be substantial, leading to transfer of angular momentum between the accreted material and the NS, causing, depending on the exact conditions, spin-up or spin-down of the NS.

15.3 High Mass X-ray Binaries: HMXB

The optical companions in HMXB are either Be stars ($M \geq 5M_{\odot}$) or OB supergiants ($M \geq 15M_{\odot}$). The NS is in a close orbit around the huge star, and the mass transfer is by accretion from the circumstellar disk of the Be star or by stellar wind from the supergiant, which may lose up to $10^{-5} M_{\odot} yr^{-1}$. The X-ray radiation emerging from the surface of the NS has to escape through a variable environment of circumstellar material in which it suffers absorption and scattering, observable by variable low energy cut-offs in the X-ray spectra. The ratio of X-ray to optical luminosity (L_x/L_{opt}) is of order unity and X-ray heating of the companion is not important. Many ($\sim$50%) of the NS in HMXB are highly magnetized and, therefore, appear as pulsars (Sect. 15.5), about half of which show cyclotron lines in their spectra. The continuum spectra are rather hard ($kT \geq 15\,keV$). The binary orbits range from 4.8 h to 187 days. Only a few of all HMXB ($\sim$4%) have a black hole as the compact object. The catalogue of HMXB [30] lists 130 objects.

15.4 Low Mass X-ray Binaries: LMXB

The optical companions of LMXB are generally of spectral type A to M with less than two solar masses. The mass transfer through Roche lobe overflow feeds an optically thick accretion disk, the X-ray emission of which often dominates the optical and UV appearance of the binary. The L_x/L_{opt} is typically 100–1000, except in those systems where an *accretion disk corona* (ADC) hides the central source leading to a much reduced L_x/L_{opt} of $\sim$20. Heating of the optical star by the compact X-ray source is generally significant, leading to an optical modulation from which the binary nature can be inferred. LMXB are less likely to show eclipses, but often show so called *dips*, which occur when the line of sight to the X-ray source is intercepted by the accretion stream or by material splashing off the accretion disk when the gas stream hits its outer edge. The periods of the binary orbits in LMXB (known only of one third of all systems) range from 0.19 h to 17 days. The catalogue of LMXB [29] shows 150 entries. About 20% of those ($\sim$30) are BH candidates. NS in LMXB are weakly magnetized (B $<10^{10}\,G$) and, therefore, they are usually not pulsars but rather bursters and QPO sources. There is, however, the class of *ms* pulsars that clearly contain weakly magnetized NS [7] (Sect. 15.6.4). A number of objects, historically named *Z*– or *Atoll*–sources, show interesting spectral/timing behavior. Roughly half of the known LMXB are transient X-ray sources (Sect. 15.6.1).

The spectra of LMXB are generally the sum of two components (a very soft and a hard component), with some showing an additional very hard tail.

One might expect a population of *intermediate mass X-ray binaries* (IMXB) with optical companions in the mass range (1–10 $M_\odot$). However, these systems are rare because of selection effects: the masses are not high enough to produce the required high mass loss rates in stellar winds and when they evolve through Roche lobe overflow the mass transfer phase is very short. Two known NS systems may be placed into this group: Her X-1 and Cyg X-2.

15.5 Strongly Magnetized Neutron Stars

Strongly magnetized accreting NS are uniquely recognized by the observation of coherent X-ray pulsations with $P_{\mathrm{pulse}} > 0.1\,s$. The accreting material, when reaching the magnetospheric radius, is coupled to the magnetic field lines and channeled along those lines down to the surface of the NS at the magnetic poles. The released gravitational energy creates a hot spot with $\sim 10^8$ K (Fig. 15.2). If the axis of rotation is inclined to the axis of the magnetic field (in the simplest form assumed to be a dipole field) any distant observer will see an X-ray pulsar, e.g., the X-ray flux is modulated at the spin frequency of the NS. In the X-ray spectra of half of these pulsars, cyclotron lines are observed, allowing the direct measurement of the field strength. These objects are referred to as *accreting binary pulsars* or *accretion powered X-ray pulsars*. Two comprehensive reviews are [33] and [11].

15.5.1 Classical X-Ray Pulsars

The first two X-ray pulsars, Cen X-3 (4.8 s) and Her X-1 (1.24 s), were discovered by Uhuru shortly after its launch in 1970. The identification as binary stars with accreting NS was in both cases by the observation of regular eclipses and by the regular modulation of the pulse arrival times (or Doppler shift of the pulse frequency) with the same binary period. The sinusoidal shape of the Doppler curve allowed to conclude that the orbits were highly circular, with upper limits on eccentricity. Characteristic features of X-ray pulsars were already recognized in these two objects, such as highly structured pulse profiles (see Fig. 15.5), high variability, including the 35-day modulation in Her X-1, long term changes of the pulse periods, as well as the general shape of the X-ray spectrum (a hard power law with an exponential cut-off). Today, we know more than 160 accreting binary X-ray pulsars (most of them in HMXB): $\sim$90 in our own galaxy, $\sim$50 in the SMC, $\sim$10 in the LMC and $\sim$10 in M31 and other galaxies of the Local Group. In addition there are eight so called *Anomalous X-ray Pulsars* (AXP) [25] with pulse periods in the narrow range of 5–12 s and large period derivatives. They show no companions and are thought to be powered not by accretion but by the energy released during the decay of an extremely strong magnetic field.

15.5.1.1 Spin Periods

The spin periods of accretion powered pulsars are found in three groups: the "classical" pulsars, showing a broad distribution between $\sim$1 and $\sim$1 000 s, a small group of three members with spin periods of 33, 61, and 69 ms and the (as yet 6) *accreting ms pulsars* with spin periods between 1.7 and 5.4 ms. The latter objects are LMXB with weakly magnetized NS and will therefore be discussed later. The first two groups contain highly magnetized NS and are mostly HMXB, except for the following four LMXB: Her X-1, 4U 1626-67, GX 1+4, GRO J1744-28, and 2A 1822-371.

For HMXB, the Corbet diagram [13] shows an interesting relationship between the spin period of the NS and the orbital period. Although for Be binaries this relationship is linear, there is no obvious dependence for OB supergiant binaries and for LMXB.

15.5.1.2 Spin-Up/Spin-Down

Shortly after the discovery of binary X-ray pulsars, it was recognized that their pulse periods were not constant. Figure 15.3 shows the long-term development of four prominent X-ray pulsars, demonstrating that both *spin-up* and *spin-down* of the NS occurs and this can vary on a wide range of time scales (down to a few days). The physical reason for this behavior is the interaction of the material of the accretion disk with the magnetic field of the NS in a *boundary layer* at the magnetospheric radius r_m (found by equating the ram pressure of a spherically symmetric inflow to the magnetic pressure of the B-field). The sign of the angular momentum transfer (to or from) the NS depends on the exact conditions in the region of interaction, in particular on the ratio of the corotation radius $r_{\rm co}$, where the magnetic field frozen to the NS has the same angular velocity as the accreting plasma in its Kepler rotation, to the magnetospheric radius $r_{\rm m}$. Obviously, when $r_{\rm m}$ equals $r_{\rm co}$ an equilibrium is reached and no angular momentum is transfered. This equilibrium period can be expressed as $P_{\rm eq} \sim (2.7\,s)\,\mu_{30}^{6/7}\,M_{\rm X}^{-2/7}\,R_6^{-3/7}\,L_{37}^{-3/7}$, μ_{30} is the magnetic moment of the NS in units of $10^{30}\,Gauss\,cm^3$, $M_{\rm X}$ is the mass of the NS in $M_\odot$, R_6 is the radius of the NS in 10^6 cm and L_{37} is the X-ray luminosity in units of 10^{37} erg s^{-1}.

If $r_{\rm m} < r_{\rm co}$ spin-up occurs, in that case the unperturbed Kepler rotation at the magnetospheric radius is faster than that of the B-field. If the magnetosphere expands (e.g., as a result of decreasing pressure due to a decreasing accretion rate) $r_{\rm m}$ can exceed $r_{\rm co}$ and spin-down results. This *accretion barrier*, also called the *propeller effect*, can completely stop the accretion such that the X-ray source shuts off. The complex details of these interactions are formulated in the accretion torque theory by Ghosh and Lamb [18, 19]. The observed correlation between

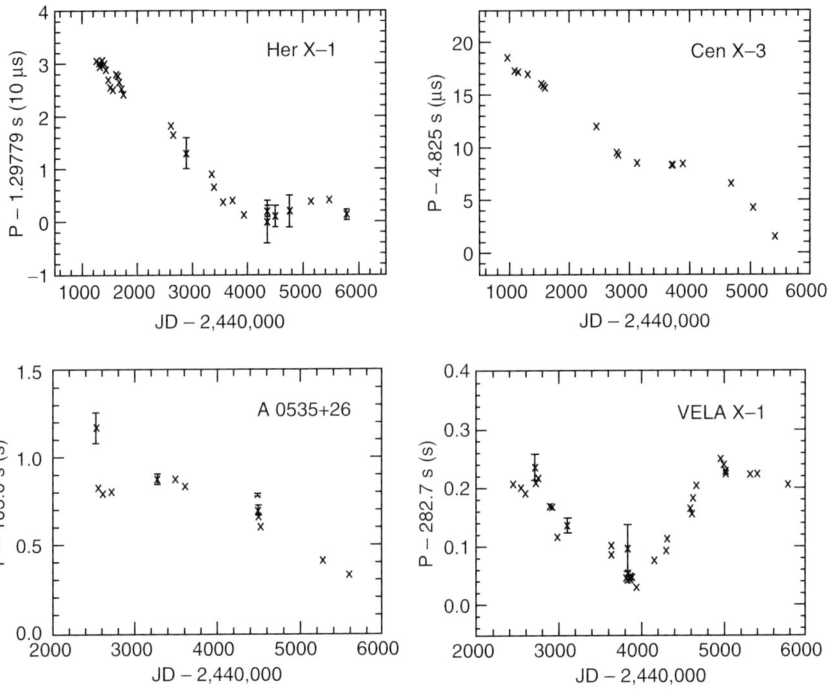

Fig. 15.3 Pulse period history of four X-ray binary pulsars (after [2])

the magnitude of spin-up $\mathrm{d}P/\mathrm{d}t$ and L_X is explained by this theory (Fig. 15.4). Since the observed X-ray luminosity L_X reflects the mass accretion rate, one can understand the fluctuations in pulse period as variations in the accretion rate. Most X-ray pulsars appear to operate near to their equilibrium period. Measuring P_{eq} then allows a direct estimate of the magnetic moment of the NS (e.g., using standard values for mass and radius of the NS).

15.5.1.3 Pulse Profiles

The pulse profiles of X-ray pulsars can look very different, from simple sinusoidal type shapes to double and multiple peaked profiles. Figure 15.5 gives a few examples. In a particular source, the profiles are energy dependent, generally showing "simpler" profiles at higher photon energies. They are also not always constant in time but can show dramatic changes, both systematically repeated or seemingly irregular. In analysing and modeling these profiles, a great deal can be learned about the geometry of the hot spots on the surface of the NS, their emission characteristics and the emission mechanism itself, as well as about the geometry of the observation, e.g., the line of sight with respect to the axis of rotation and the axis of the magnetic

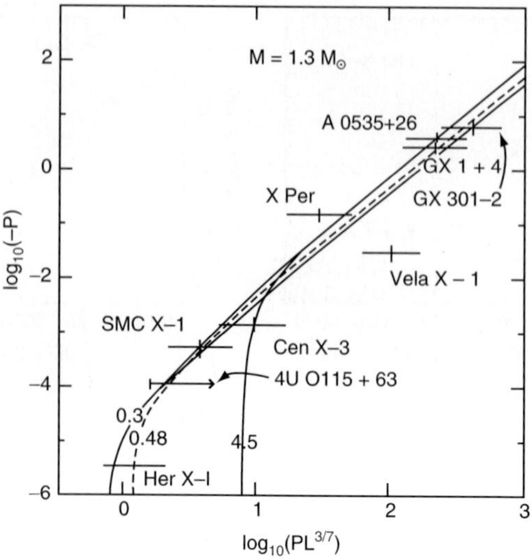

Fig. 15.4 Correlation between pulse period derivative $-\mathrm{d}P/\mathrm{d}t$ and $P \,\mathrm{x}\, L^{3/7}$ [19, 34]

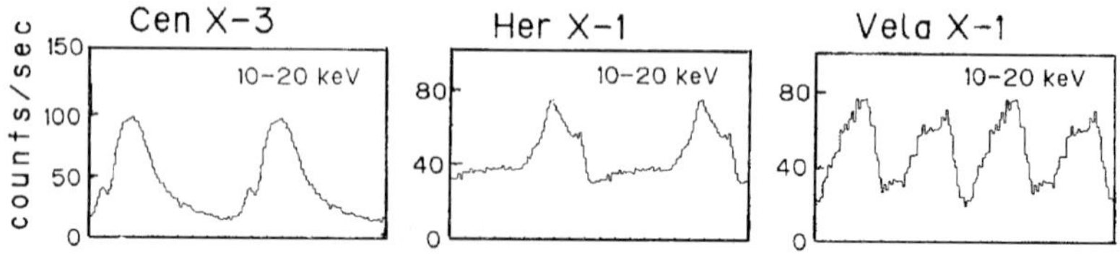

Fig. 15.5 Pulse profiles of selected X-ray pulsars (after [33])

field, and in fact about the structure of the magnetic field itself (dipole and multipole components). In doing the model calculations, one has to take into account the gravitational bending of the photon's path in the strong gravitational field of the NS.

15.5.1.4 Continuum Spectra and Cyclotron Lines

The X-ray continuum spectra of X-ray pulsars are rather characteristic, they are successfully described by a hard power law spectrum that is exponentially cut off. The modeling parameters are the power law photon index (typically between 0 and 1), the energy at which the cut-off sets in ($\sim$20 keV) and the folding energy of the exponential cut-off ($\sim$several kilo electrovolt). Various types of mathematical fitting functions exist. Physically, the spectra are thought to arise through the emission (thermal bremsstrahlung) of the hot regions at the base of the accretion column at the

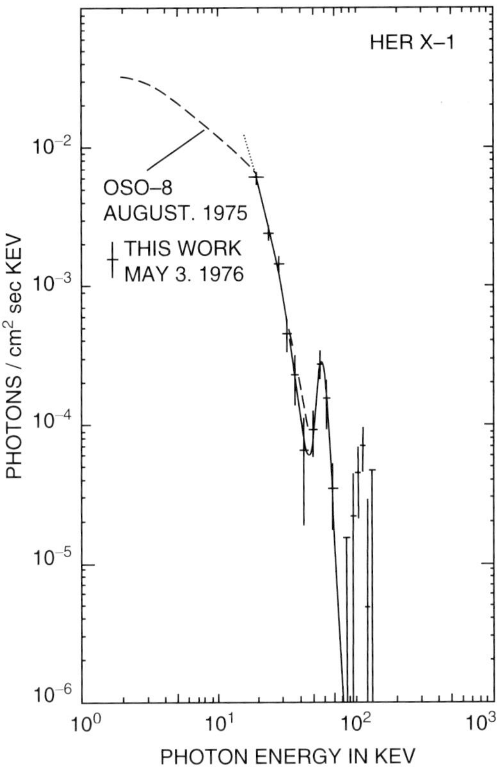

Fig. 15.6 The high energy X-ray spectrum of Her X-1 as observed by the MPE/AIT balloon detector, leading to the discovery of the first cyclotron line (Fig. 15.6 of [37])

magnetic poles of the NS (the *accretion mound*) and modifications by comptonization through fast electrons (thermal or nonthermal). Interestingly, the heuristical fit functions have so far been more successful than attempts to do physical modeling.

An important modification of such a spectrum was discovered through a balloon observation of Her X-1 where a line feature was detected and interpreted as due to cyclotron radiation [37] (see Fig. 15.6). The strong magnetic field in the polar region of a NS forces discrete energy levels (Landau levels) of electrons with respect to their motion perpendicular to the direction of the field. The cyclotron energy, E_C, defined as the difference between the (to first order equidistant) energy levels is given by $E_C = \hbar eB/m_e c \sim 11.6 B_{12}$ keV, where B_{12} is the field strength in units of 10^{12} Gauss. The early data did not allow to distinguish between emission and absorption. This was achieved through higher quality observations [26, 35, 42] and model calculations [41], which showed that *cyclotron lines* are formed through resonant scattering of photons trying to escape from the accretion mound, which has led to the term *CRSF – Cyclotron Resonant Scattering Feature*. The observed line energy is at $E = E_C/(1+z)$ with z being the gravitational redshift (~0.3). The CRSF at 37 keV in Her X-1, therefore, corresponds to a field strength of 4×10^{12} G. The indirect estimate using accretion torque theory (see earlier) leads for Her X-1 to a magnetic moment of 10^{30} Gauss cm^{-3}, equivalent to 2×10^{12} G (for a dipolar field). This is, keeping the model dependency in mind, consistent with the cyclotron line measurement. In any case, cyclotron lines provide the means for a direct measurement of the magnetic field strength at the emission site.

Today, at least 15 cyclotron line objects are known (for recent reviews see [23, 36]). In most objects just one line is seen, in three objects two lines are observed, in one object three lines are seen, the record holder with five lines is 4U 0115+63.

It has been found that the line spectra are changing with pulse phase. This type of analysis, *pulse phase spectroscopy*, has revealed that the cyclotron energy, E_C, in most objects varies substantially (by up to 20%) throughout the pulse. During the rotation of the NS, the observer apparently sees emission regions with different magnetic field strength. Such analysis has still to be performed in a systematic way on all cyclotron line objects using archival data of RXTE and BeppoSAX. New high resolution observations by INTEGRAL in coordination with theoretical calculations offer the prospect for a deeper understanding.

15.5.1.5 Aperiodic Variability in Pulsars

It had taken some time to recognize that many of the classical accretion powered pulsars also show rapid X-ray variability, which can in part be described in terms of quasi-periodic oscillations (QPO) mostly seen in LMXB [38] (as discussed in Sect. 15.6). The physics is probably similar in both cases (see later), most likely involving special regions of the accretion flow, e.g., the magnetospheric boundary at the inner edge of the accretion disk. In at least 13 pulsars (of all types) QPOs have now been seen, with frequencies mainly in the 2–200 mHz range, in the case of Cen X-3 possibly reaching the kilohertz regime [14].

15.6 Weakly Magnetized Neutron Stars

Neutron stars with weak magnetic fields ($B \sim 10^{8-10}$ G) are all found in LMXB. They are most likely old systems with lifetimes that have allowed the originally strong field of the NS to decay (for a discussion on magnetic field decay see [9]). The low mass and the corresponding long development time scale of the companion are in agreement with this picture. Most weakly magnetized NS do not show regular pulsations but rather other types of variability behavior such as *X-ray bursts*, *quasi periodic oscillations* (QPOs) and strong low and high frequency *noise*. However, the *Bursting Pulsar*, the new class of *accreting ms X-ray pulsars* and those bursters showing *ms burst oscillations* or *kHz twin QPOs* are examples of weakly magnetized NS in which the spin of the NS is displayed.

15.6.1 Z- and Atoll-Sources

Even though X-ray bursters are the largest subgroup of LMXB, we start with a smaller group of objects, some of which are also bursters, with highly interesting spectral and timing properties. They are divided into two types:

- High luminosity ($>10^{37}\,erg\,s^{-1}$) *Z–sources* (6 objects), showing *quasi periodic oscillations* (QPOs)
- Low luminosity ($<10^{37}\,erg\,s^{-1}$) *Atoll–sources* (18 objects), mostly the ones showing *X-ray bursts*. Several of them are now known to also show low frequency QPOs

Both types show strong low and high frequency noise components. The Z/Atoll-classification, based on the *correlated* spectral and timing behavior of a sample of bright LMXB observed with EXOSAT, was introduced by [22]. For reviews on aperiodic variability in X-ray sources (BH and LMXB) see [38, 39]. Figure 15.7 shows

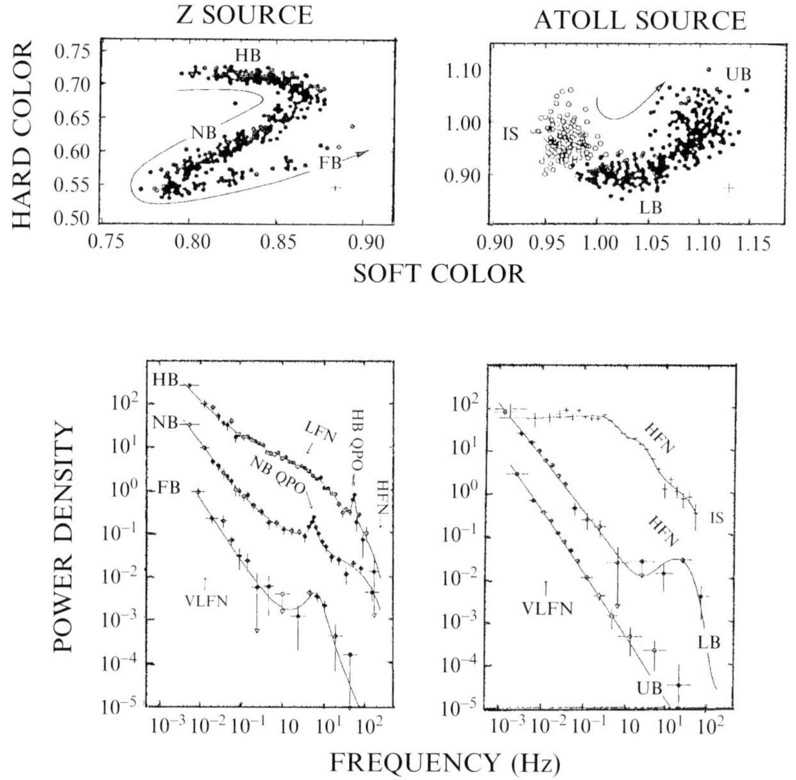

Fig. 15.7 Color/color diagrams (*top*) and Power Spectral Densities (*bottom*) for Z-sources (*left*) and for Atoll-sources (*right*). Soft color is (3–5 keV)/(1–3 keV) and hard color is (6.5–18 keV)/(5–6.5 keV) (after [2])

color/color diagrams[1] (CD) and power spectral distributions for Z–sources (left) and Atoll-sources (right). The instantaneous state of an individual Z- or Atoll-source is characterized by its current position in the CD. The physical reason is most likely the mass accretion rate $\dot{M}$, changes of which simultaneously affect the X-ray spectrum and the X-ray variability. The arrows in Fig. 15.7 indicate the direction in which the accretion rate is thought to increase. So, a source never jumps from one position to another in such a diagram, but always moves along the track according to luminosity. It seems that Z-sources generally have higher magnetic field ($10^{9-10}\,G$), higher mass accretion rate and and longer orbital periods ($>10\,h$) than Atoll-sources. For the very large amount of details with respect to spectral and timing behavior within the various branches in the CD (which cannot be discussed here) see e.g., [38,41].

15.6.1.1 The Physics of QPOs

Regarding the physics behind QPOs, many models were proposed ([38] and references therein, [40]). A recent comprehensive review on *Rapid X-ray varibility* is given in [39]. For moderate accretion rates the so-called *beat frequency model* has been widely accepted. In this model the QPO frequency is identified with the beat frequency ν_B between the frequency of Kepler rotation ν_K of the material at the inner edge of the accretion disk and the spin frequency of the NS ν_S: $\nu_B = \nu_K - \nu_S$. The idea is that "clumps" of material can enter the magnetosphere preferentially at certain points (e.g., near the magnetic poles). If ν_S and ν_K are different the favorable situation occurs with the beat frequency between these two. Even though the relationship between the position in the CD and $\dot{M}$ is not really known, certain constraints on the magnetic field and the spin period of the NS can be obtained by applying standard accretion torque theory. Spin periods between 3 and 20 ms were predicted in this way [20]. It took another 6 years until the first *accreting ms pulsar* with coherent pulsations of 2.5 ms was finally discovered by RXTE in 1998 (see below).

15.6.2 kHz QPOs

In February 1996, only 2 months after its launch the Rossi X-Ray Timing Explorer (RXTE) discovered kilohertz quasi-periodic oscillations (*kHz QPOs*) in the well known LMXB Sco X-1 [39]. They appear as two simultaneous peaks (*twin peaks*) in the power spectrum (Fig. 15.8). Today, some 20 LMXBs (both Z and Atoll type) with QPO peaks between 200 and 1 300 Hz are known. The frequency of both peaks usually increases with increasing X-ray flux, while the separation stays nearly constant. For reviews on observations and the basic physical interpretation see [39] and [40]. Variations in this frequency range had long been expected because this is close to the natural dynamic time scale around compact objects like NS or stellar black holes. The orbital frequency around a NS is given by

[1] "color" is defined as the ratio of count rates in adjacent energy intervals

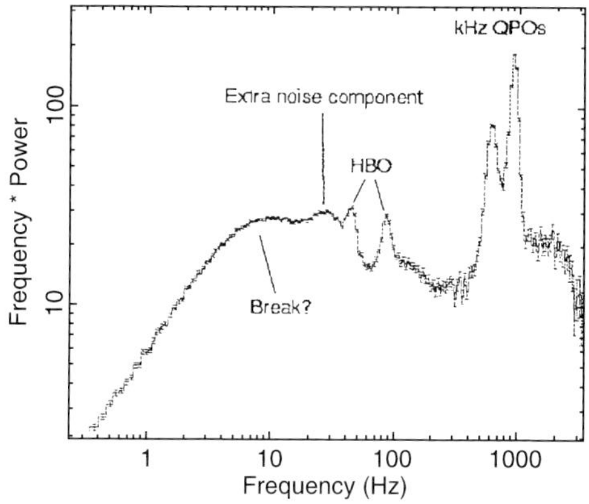

Fig. 15.8 Power Density Spectrum of Sco X-1 (after [1])

$\nu_{\rm orb} = 1\,200\,{\rm Hz}\,(r_{\rm orb}/15\,{\rm km})^{-3/2} m_{1.4}^{1/2}$, with $m_{1.4}$ being the mass of the NS in units of $1.4\,M_\odot$. The maximum frequency is then reached at the innermost stable circular orbit (ISCO), which in a Schwarzschild geometry is given by 3 times the Schwarzschild radius ($R_{ISCO} = 3\,R_s = 6\,GM/c^2 = 12.5\,{\rm km}\,m_{1.4}$): $\nu_{ISCO} = 1580\,{\rm Hz}\,/m_{1.4}$. The spin-orbit beat–frequency model, as already discussed earlier for low frequency QPOs, can provide a basic understanding of the observed features: calling the upper and lower frequency (of the two twin peaks) ν_2 and ν_1, respectively, and identifying ν_2 with ν_{orb} and ν_1 with the beat frequency between $\nu_{\rm orb}$ and $\nu_{\rm spin}$, then $\nu_1 = \nu_{beat} = \nu_2 - \nu_{spin}$. This would identify the spin frequency with the difference between the two observed kilohertz peaks, in agreement with the expectation that $\nu_{\rm spin}$ is constant. Only a single sideband ($\nu_{\rm orb} - \nu_{\rm spin}$) is observed, as expected from a rotational interaction, with spin and orbit in the same sense. In the *sonic point beat-frequency model*, the preferred orbital radius is essentially identified with the inner edge of the accretion disk. The above interpretation has received very strong support by the observations of *burst oscillations* (Sect. 15.6.3) and by the discovery of bursting *ms X-ray pulsars*, which are believed to show the NS spin frequency directly (see later; it appears, that for some sources the difference between the twin kHz QPO frequencies is $\nu spin$ for others it is $\nu spin/2$).

15.6.3 Bursters

In weakly magnetized NS, the accretion disk can come close to the surface of the NS and the material can trickle down to this surface in a quasi radial-symmetric way. The liberated gravitational energy gives rise to some steady X-ray emission. If the temperature and the density is high enough, the accreted material, mostly hydrogen, can peacefully fuse into helium. When sufficient helium is produced and critical values of density and temperature are reached, helium burning can start. This

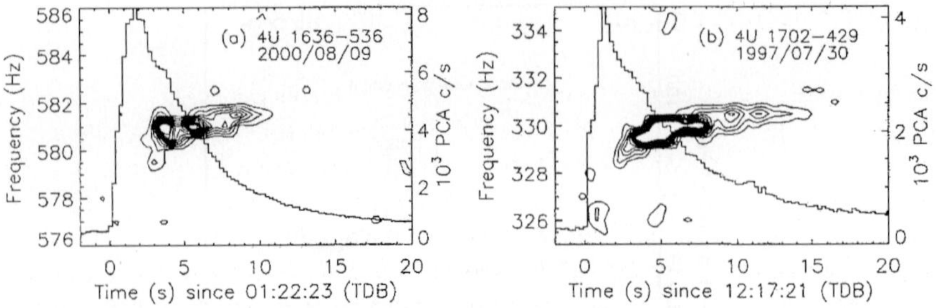

Fig. 15.9 Thermonuclear bursts and frequency development of burst oscillations [32]

process, however, is unstable and resembles an explosion event, producing what we observe as an *X-ray burst*: a fast rise (<1–10 s) in X-ray flux with a subsequent exponential decay (with a duration typically between 10 s and a few minutes), see Fig. 15.9. The time between bursts is typically from 1 h to a few hours. A review about X-ray bursts is given in [28].

The first such X-ray bursts were observed in 1975 through observations with the Dutch X-ray satellite ANS from a neutron star inside globular clusters [21], which was soon followed by the discovery of a dozen X-ray bursters with SAS-3 [27]. Today, about 65 X-ray burst sources are known, situated mainly in globular clusters and in the galactic bulge region. They are members of an old population. The above picture, called the *thermonuclear flash model*, was soon developed on the basis of solid physical evidence. The measured spectra were consistent with black-body emission, and it was found that the characteristic temperature decreases when the burst fades, also evidenced by a faster decay at higher photon energies. Furthermore, the absolute fluxes and temperatures measured toward the end of many bursts from objects in the galactic bulge (providing an estimate of the distance to the sources) were consistent with the radiation coming from objects of 10–15 km diameter [28], just what is expected from a NS. Since a solid surface is needed for the fusion to take place, the X-ray burst sources must be NS, not black holes. If the luminosity of the steady flux is compared with the integrated burst flux a factor of ~100 is found, which is close to the expected ratio of gravitational energy to nuclear energy of the accreted material. Finally, an interesting correlation [28] was found between the time interval since the previous burst and the strength of the burst: the longer the waiting time is, the more nuclear fuel can accumulate for the next burst. There is evidence, particularly in strong bursts, that there is a photospheric radius expansion during the early phases of the burst where the X-ray luminosity reaches the Eddington limit. Any additional energy available is put into kinetic and potential energy of the expanding atmosphere. Bursts from thermonuclear flashes with the characteristics described earlier are called *Type I bursts*. From seven LMXB *superbursts* have been observed with decay times between 1 and 6 h and ~100 times more energetic than regular bursts. They are thought to be due to unstable burning of carbon rather than helium [24].

15.6.3.1 The Rapid Burster

The necessity to distinguish between two types of bursts came with the discovery of an object called the *Rapid Burster (RB)* [28, 38]. As the name suggests, the rate of bursts is very high with intervals between bursts as short as $\sim$7 s (and no longer than $\sim$1 h). These bursts do not show the characteristic softening during their decay, as found in Type I bursts. They are, therefore, called *Type II bursts*. In addition to Type II bursts, the *RB* generally also shows Type I bursts with the usual characteristics. But there can be periods with only one type of bursts or with no bursts at all. In Type II bursts the strength of the burst correlates with the time interval to the *next* burst, not the previous one as in Type I bursts. If the RB is burst active, the integrated X-ray flux of Type II X-ray bursts from the *RB* is about 120 times greater that that of the Type I bursts [28], suggesting gravitational energy behind Type II bursts. For some time, the *RB* was the only object showing Type II bursts, now we know a second object, GRO 1744–28 [15], also called the *Bursting Pulsar* (see later).

Many models have been proposed to explain Type II bursts (for references and a discussion see [28]). They all involve some sort of gating mechanism, which – due to an instability – allows to (partially) empty a reservoir of material which can then suddenly fall down to the surface of the NS and cause the X-ray burst. The observed linear correlation of the energy of the burst and the time to the next burst for the RB is very suggestive of a *relaxation oscillator*: when in one burst the reservoir was emptied to a high degree, it will take a longer time to replenish it again, while a partial emptying will result in a short waiting time to the next burst. It is, however, not entirely clear whether the storage medium is the accretion disk itself or the magnetosphere and which type of instability triggers the gate to open. The *RB* and GRO 1744–28 must distinguish themselves from all other bursters (showing ordinary Type I bursts) by special physical parameters or conditions. The strength of the magnetic field may be one parameter, but then GRO 1744–28 is a pulsar, the RB is not.

15.6.3.2 The Bursting Pulsar

For about 20 y, the Rapid Burster (RB) was the only object showing Type II bursts, until BATSE on December 2, 1995 detected such bursts from a transient X-ray source close to the Galactic Center (GRO J1744–28) [15]. On top of this, the new source established itself as a new type of X-ray source: one day before the onset of the burst activity coherent X-ray pulsations with a period of 0.467 s were discovered [46]. GRO J1744–28 was dubbed the *Bursting Pulsar (BP)*. Until its discovery, it was believed that coherent pulsations and X-ray burst activity were mutually exclusive in X-ray sources, since the former were associated with highly magnetized NS, the latter with weakly magnetized NS. Most likely, the *BP* has a magnetic field of intermediate strength ($\sim 10^{11}$ G) and undergoes strong magnetically guided accretion (albeit gated, as in the *RB*), which suppresses thermonuclear (Type I) burst activity. Coherent pulsations at the spin period are seen in the persistent flux as well as in bursts [46]. Today, we know two sisters of the *BP* – objects showing coherent pulsations and (Type I) burst oscillations at the same frequency. However, they belong to the recently identified class of *accreting ms X-ray pulsars* (Sect. 15.6.4).

15.6.3.3 X-Ray Burst Oscillations

If a thermonuclear burst ignites on the surface of a rotating NS, one might expect to see signs of the rotation because of nonisotropic emission due to patchy burning or the influence of the magnetic field. The first oscillation with a drifting frequency around 363 Hz was observed by RXTE in early 1996 in a Type I burst from 4U 1728-34 [40]. Today, 13 objects (out of the 65 bursters) have been found to exhibit oscillations in Type I bursts [32]. The frequencies range from 270 to 620 Hz. Figure 15.9 shows examples of burst profiles with the development of the frequency of the oscillations throughout the bursts. There is an upward drift (by a few %) toward an asymptotic frequency. Taking a simple model for the frequency drift into account, the oscillations can be accepted as coherent. Also, in a given source the end frequency is stable (within measurement uncertainties) from burst to burst, even if they are years apart. The modulation is generally sinusoidal (with very little harmonic content). These findings have led to the interpretation that the observed end frequency is equal to the rotational frequency of the NS. This conclusion is strongly supported by burst oscillations of two bursting millisecond pulsars (Sect. 15.6.4) in which exactly the same frequency is measured in the persistent flux and in the thermonuclear bursts. It is, however, not clear why nonisotropies in the emission of the bursting ms pulsars remain for $\sim$15 s, while it is believed that the nuclear explosion engulves the entire NS surface within 1 s from the ignition, and why the frequency shows this development in time [32].

It is interesting to note, that among the 13 known ms burst oscillators eight objects also show twin kilohertz QPOs. However, only for four of those is the frequency difference between the two kHz QPO peaks nearly equal to the frequency observed during the bursts, in the other four cases it is close to half the frequency during the burst. Interestingly, the former group has $\nu_{\mathrm{burst}} < 400\,\mathrm{Hz}$, while for the latter $\nu_{\mathrm{burst}} > 400\,\mathrm{Hz}$ [31].

15.6.4 Accreting ms Pulsars

While millisecond oscillations in bursts were discovered shortly after the launch of RXTE, it took more than 2 y until RXTE in April 1998 finally also found coherent ms pulsations at a frequency of 401 Hz in the persistent flux of the transient and bursting LMXB SAX J1808.4-3658 [44] (a reanalysis of earlier BeppoSAX data revealed a marginal detection of a 401 Hz oscillation in a burst). During the third outburst of this source in October 2002 oscillations with 401 Hz were clearly seen in all four bursts observed [12], establishing that the burst oscillations trace the spin of the NS. In addition, twin kHz QPOs were discovered with $\Delta\nu$ equal to 200 Hz (half of the burst frequency) [45]. Today, a group of six objects with millisecond coherent periodicity in their persistent flux are known. They are all transients. Two of them, SAX J1808.4-3658 and XTE J1814-338, show also millisecond burst oscillations.

These *accreting ms X-ray pulsars* provide the long sought link to the rotation powered ms radio pulsars [7]. The respective magnetic fields of both classes are in the range of $(1-10)\ 10^8$ Gauss. For such fast spinning objects, a stronger magnetic field would prevent accretion. The conventional wisdom is that the ms radio pulsars have developed from NS in LMXB, which had been spun up over a long time by accretion – we now see the progenitors. For rotation powered X-ray pulsars see [10].

15.7 Summary

Accreting NS in their wide range of appearances are an important class of X-ray sources. They have been discussed mainly from an observational point of view. Several issues could not be covered within this chapter. For instance, NS in binaries are an important source of information on stellar evolution, of the evolution of Globular Clusters and the Galaxies in which they are found as a whole. There are for instance clear differences in the types of accreting NS binaries between our galaxy and the SMC and LMC, the low metallicity of which is thought to be responsible for the existence of exceptionally bright X-ray binaries. Among the known accreting NS systems, there are several exotic systems, like Aql X1 (showing both NS and BH behavior) or Cyg X-3 (which may not contain a NS but a BH), or systems with extremely rich and complex sets of information (and corresponding lack of physical understanding), like the Z– and Atoll–sources, the Bursting Pulsar, Her X-1 or others. Finally, there are many features – spectral as well as timing – which are found in a quite similar way in NS and in black holes, implying that the existence or not of a solid surface and a stellar magnetic field may not be so important. Likewise, the contributions of the research on accreting neutron stars to the determination of the most fundamental physical parameters of neutron stars, the mass, and radius, leading to constraints for the equation of state (EOS), as well as observations with relevance to general relativity effects in the immediate vicinity of the compact objects, are not discussed here. These aspects are discussed in the chapter 14 from the point of view of single neutron stars.

References

1. X-Ray Timing 2003, Eds. Kaaret, P., Lamb, F.K., Swank, J.H. 2004, American Inst. of Physics, AIP Conf. Proc. 714, ISBN 0-7354-0183-4
2. X-Ray Binaries, Eds. Lewin, W., van Paradijs, J., van den Heuvel, E. 1995, Cambridge Astrophysics Series 26, ISBN 0 521 41684

3. Exploring the X-ray Universe, Charles, P.A., Seward, F.D. 1995, Cambridge University Press ISBN 0 521 43712 1
4. Accretion Power in Astrophysics, Frank, J., King, A., Raine, D. 1992, Cambridge Astrophysics Series 21, ISBN 0 521 40863 6
5. Accretion driven stellar X-ray sources, Eds. Lewin, W., van den Heuvel, E. 1983, Cambridge Univ. Press, ISBN 0 521 24521 4
6. X-ray astronomy, Eds. Giacconi, R., Gursky, H. 1974, D. Reidel Publ. Co., 1974, Astrophysics and Space Science Library (ASSL) 43, ISBN 90 277 0387 6
7. X-Ray Binaries and Recycled Pulsars, Eds. van den Heuvel, E.P.J., Rappaport, S. 1992, Kluwer, Dordrecht, NATO ASI Ser. C 377
8. Compact stellar X-ray sources, Eds. Lewin, W., van der Klis, M. 2006, Cambridge Astrophysics Series 39, ISBN-13 978-0-521-82659-4
9. Battacharya, D., Srinivasan, G., In: [2], p.495
10. Becker, W., Trümper, J. 1999, A&A 341, 803
11. Bildsten, L., et al. 1997, ApJS 113, 367
12. Chakrabarty, D. 2003, Nature 424, 42
13. Corbet, R.H.D. 1984, A&A 141, 91
14. Finger, M.H. In: [1], p.342
15. Fishman, G.J., et al. 1995, IAU Circ 6272,
16. Forman, W., et al. 1978, ApJS 38, 357
17. Giacconi, R., et al. 1962, Phys. Rev. Lett. 9, 439
18. Gosh, P., Lamb, F.K. 1979, ApJ 232, 259
19. Gosh, P., Lamb, F.K. 1979, ApJ 234, 296
20. Gosh, P., Lamb, F.K. In: [7], p.487
21. Grindlay, J., Gursky, H., Schnopper, H. 1976, ApJ 205, L127
22. Hasinger, G., van der Klis, M. 1989, A&A 225, 79
23. Heindl, W.A., et al. In: [1], p.323
24. in't Zand, J.J.M., et al. In: [1], p.253
25. Kaspi, V.M., Gavriil, F.P. In: [1], p.281
26. Kunz, M. 1995, PhD thesis Univ. Tübingen
27. Lewin, W., Joss, P. 1981, Space Sci. Rev., 28, 3
28. Lewin, W. In: [2], p. 175
29. Liu, Q.Z., et al. 2000, A&A Suppl. 147, 25
30. Liu, Q.Z. et al. 2001, A&A 368, 1021
31. Miller, M.C. In: [1], p.365
32. Muno, M.P. In: [1], p.239
33. Nagase, F. 1989, PASP 41, 1
34. Rappaport, S., Joss, P.C. 1977, Nature 266, 683
35. Soong, Y., et al. 1990, ApJ 348, 641
36. Staubert, R. 2003, Chin. J. Astron. Astrophys. Vol. 3, Suppl., 270
37. Trümper, J., et al. 1978, ApJ 219, L105
38. van der Klis, M. In: [2], p.252
39. M. van der Klis, In: [8], p.39
40. van der Klis, M. 2000, ARAA 38, 717
41. Nagel, W. 1981, ApJ 251, 288
42. Voges, W., et al. 1982, ApJ 263, 803
43. White, N.E., Nagase, F., Parmar, A.N. 1995, In: [2], p. 1
44. Wijnands, R., van der Klis, M. 1998, Nature 394, 344
45. Wijnands, R., et al. 2003, Nature 424, 44
46. Woods, P.M., et al. 2000, ApJ 540, 1062

16 Black-Hole Binaries

Y. Tanaka

16.1 Introduction

Binary systems that are composed of a black hole and a normal star are called *black-hole binaries*. In this chapter, the observational evidence for black holes and the specific properties of the black-hole binaries are discussed.

A black hole is a singular object in the framework of general relativity. In an early theoretical work based on general relativity, Oppenheimer and Snyder [55] showed that a sufficiently massive star, when all thermonuclear energy is exhausted, would collapse indefinitely and disappear inside a sphere of a limiting radius from which even photons could not escape (the "event horizon"). This radius is called the Schwarzschild radius $R_s = 2R_g$, where $R_g = MG/c^2$ is the gravitational radius, and M, G, and c are the mass, the gravitational constant, and the light velocity, respectively. Such an object was later called a black hole. However, black holes had remained only in theoretical interests for a long time.

The real existence of stellar mass black holes came to light with the birth of X-ray astronomy (1962) [21]. It began in early seventies with surprising discoveries that the bright X-ray source Cyg X-1 is a binary system and that its X-ray emitting compact object is much more massive than a neutron star (Sect. 16.2). Since then, many stellar-mass black holes have been found among binary X-ray sources: X-ray binaries (hereafter abbreviated to XBs). It is to be emphasized that X-ray observations have played a unique role in the study of black holes, since X-ray observation is practically the only means to discover black-hole binaries. The presently known stellar-mass black holes were all discovered from XBs. The discovery of black holes was certainly one of the highlights of astronomy in the twentieth century.

The observational identification of black holes is not straightforward. The direct proof for a black hole would be to demonstrate the presence of an event horizon in a compact (gravitationally collapsed) object. This requires (1) to confirm complete absence of a material surface, and (2) to find the general relativistic effects that are genuinely unique near the event horizon, such as particle motion near the speed of light, a large gravitational redshift, bending of the light path. Observed facts in support of these have been accumulating, but not yet established as indisputable evidence.

So far, the most reliable evidence for a black hole has come from the mass determination by showing that the mass of the compact object exceeds $3M_\odot$. To

J.E. Trümper and G. Hasinger (eds.), *The Universe in X-Rays.*

the best of our current knowledge, any compact object more massive than $3M_{\odot}$, the firm mass upper limit of a stable neutron star in general relativity, is believed to have no other fate than to collapse into a black hole.

Once the optical counterpart of an XB is identified, optical observation allows mass estimation of the compact object (Sect. 16.3). At present, 20 compact objects in XBs are known to have a mass greater than $3M_{\odot}$ (see Table 16.1). On this basis, they are considered to be "secure" black holes. (Note that black holes can in principle have any mass, hence black holes of $<3M_{\odot}$, if they exist, are missed with this criterion.) Even so, the genuine general relativistic tests are not as yet perfect. In this sense, one can say that the presence of stellar mass black holes is virtually solid, though not strictly proven.

A wealth of observational results on XBs has become available in the last few decades. Accordingly, studies of the X-ray properties of XBs have much advanced, and fair understanding of the nature of their X-ray emission has been obtained. In particular, from the study of the secure ($>3M_{\odot}$) black-hole binaries, it is found that most of them share common X-ray properties that are distinctly different from XBs containing a neutron star ("neutron-star XBs"), as discussed in Sect. 16.4. On the basis of these properties, one can select black hole "candidates" from the X-ray observations alone. In addition, recent multiwavelength observations from radio through gamma-rays also provide important clues to black holes.

Notably, most of the black-hole binaries are not persistently bright in X-rays, but they are transient sources undergoing an X-ray outburst only for a short while (Sect. 16.4.2). It is, therefore, obvious that many more black-hole binaries exist in our Galaxy, though most of them are X-ray quiet.

The above topics and other relevant topics on stellar-mass black holes are discussed in the following sections, based on the results available as of late 2005. The references are admittedly limited. For more details, we refer to previous reviews, e.g., Tanaka and Lewin [76], Tanaka and Shibazaki [78], and a more recent extensive review by McClintock and Remillard [39], and the references therein.

16.2 X-Ray Binaries

Soon after the opening of X-ray astronomy in 1962, the concept of an accreting compact object emerged to account for the large X-ray luminosity. It is well established that bright X-ray sources with X-ray luminosity $L_{\mathrm{X}} \gtrsim 10^{36}$ erg s^{-1}in the range 1–10 keV, excluding supernova remnants, are binary systems containing a gravitationally collapsed object, either a neutron star or a black hole. As explained in Sect. 16.4.1, they are powered by mass accretion from a companion star, and the key to the high X-ray luminosity is an extremely deep gravitational potential well of the collapsed object.

X-ray binaries (XBs) are divided into two distinct groups according to the type of the companion star. Those with high-mass O or B stars are called high-mass XBs

(HMXBs), and the others with low-mass ($\lesssim M_{\odot}$) stars, mostly K, M stars, are called low-mass XBs (LMXBs).

Historically, the first discovery of the binary nature was made for Cyg X-1 in 1972. When its optical counterpart was identified with the O-type supergiant HDE226868, Webster & Murdin [82] and Bolton [7] discovered a sinusoidal Doppler motion of the O-type star, which is a clear evidence for a binary system. The invisible companion must be an X-ray emitting compact object. Thus, Cyg X-1 is a HMXB.

Their results contained another big surprise. The mass of the compact object estimated from the mass function (see Sect. 16.3) most likely exceeded $3M_{\odot}$. They independently considered that the compact object was a black hole! This discovery of great impact excited various critical discussions. However, none of the alternative possibilities survived. Cyg X-1 still remains as one of the secure black-hole binaries.

16.3 Black Holes Identified from Mass Functions

The mass function $f(M)$ is given by the following equation:

$$f(M) = M_{\mathrm{X}}^3 \sin^3 i/(M_{\mathrm{X}} + M_c)^2 = PK^3/2\pi G,$$

where M_{X} and M_c are the masses of the X-ray emitting compact object and of the companion star, respectively, and i is the inclination of the binary orbit. P is the orbital period, and K denotes the amplitude of the Doppler curve that gives the line-of-sight component of the radial velocity of the companion. Both P and K are optically measurable quantities. For HMXBs, the optical light from the companion dominates. For LMXBs, the companions are very faint, hence optical search is practically impossible. When a transient X-ray outburst occurs, it accompanies a nova-like optical brightening, which allows identification of the optical counterpart (see Sect. 16.4.2). After the source returns to X-ray quiescence, the optical light from the companion can be observed.

Once $f(M)$ is obtained, the actual mass of the compact object can be estimated if i and M_c are known. The method of obtaining these quantities from spectroscopic and photometric measurements has been well developed (e.g. [58]), although sometimes subject to systematic uncertainties.

Cyg X-1 had been the sole black hole candidate for the following 10 yrs until two bright X-ray sources in the Large Magellanic Cloud, LMC X-3, and LMC X-1, were added. Both of them were optically identified with early-type stars. The mass functions obtained indicated that the masses of the compact objects in these two sources were also larger than $3M_{\odot}$. These three are all high-mass ($M_c \gg M_{\odot}$) systems.

In 1986, the research of black-hole XBs entered into a new era. McClintock and Remillard [38] discovered that the compact object of a LMXB transient A0620–00 was definitely more massive than 3 $M_\odot$. A0620–00 underwent a transient X-ray outburst in 1975. It became the brightest X-ray source in the sky, three times as bright as Sco X-1, and faded back to quiescence (see the lightcurve in Fig. 16.1). The secondary was found to be a K-dwarf ($M_c < M_\odot$, hence LMXB). Note that, as evident from the equation, $f(M)$ gives an absolute lower limit of the mass of the compact object. Since the measured mass function itself was close to 3 $M_\odot$ (see Table 16.1), there was very little doubt that $M_X > 3M_\odot$. This was an epoch-making discovery not only because it presented a convincing case for a black hole, but also because it was the first in low-mass binary systems and furthermore it was a transient source.

In the following 10 yrs, X-ray outbursts of several LMXBs were successively detected with the X-ray satellites such as Ginga, Granat, and CGRO. Surprisingly, most of them turned out to contain black holes based on the mass functions (see Table 16.1). In particular, since RXTE became operational in 1996, the detection rate of transients has increased significantly. The RXTE all-sky monitor (ASM) surveys up to ~80% of the sky every day, and it is sensitive enough to detect bright X-ray transients practically from the entire Galaxy.

Currently (as of 2005), there are 20 secure black-hole binaries that satisfy $M_X > 3M_\odot$. They are listed in Table 16.1. Remarkably, there are only three persistent black-hole binaries (Cyg X-1, LMC X-1, and LMC X-3), and they are all HMXBs. The others are all LMXBs, and also transient sources detected during X-ray outbursts (see Table 16.1). As a matter of fact, a large fraction of LMXB transients have turned out to be black-hole binaries (see Sect. 16.4.2). Two black-hole binaries, 4U 1543–47 and SAX J1819.3–2525, have relatively massive ($\sim 3M_\odot$) secondaries [59], yet distinctly less massive than O/B-stars.

One finds from Table 16.1 that excluding the three HMXBs, 14 LMXBs out of 17 show $f(M) \gtrsim 3\,M_\odot$. For them, the $f(M)$ value alone is sufficient to conclude $M_X > 3\,M_\odot$, though additional quantities are required for deriving the actual value of M_X. The evidence for $M_X > 3\,M_\odot$ is also very strong for the remaining three LMXBs. In clear contrast, for those binaries of which the compact objects are known to be neutron stars, the M_X values estimated from $f(M)$ are almost all close to 1.4 $M_\odot$ (the Chandrasekhar limit, e.g., [9]).

16.4 X-Ray Properties

So far, the identification of black holes in XBs relied on the criterion of $M_X > 3\,M_\odot$. It is still considered to be the most reliable black hole discriminator. On the other hand, continued efforts have been made to find unique signatures of black-hole binaries in the observed properties, in particular the X-ray properties. At present, there are enough black-hole XBs established from the mass function (Table 16.1) to investigate their properties and to compare with those of neutron-star XBs, as discussed in the following sections.

Table 16.1 Black-hole binaries confirmed with the mass function

Source name		Year[a]	Type[b]	$f(M)$ ($M_\odot$)	M_X ($M_\odot$)	X-ray [c] spectrum
Cyg X-1			H, P	0.244±0.005	6.9–13.2	S + PL
LMC X-3			H, P	2.3 ±0.3	5.9–9.2	S + PL
LMC X-1			H, P	0.14±0.05	4.0–10.0	S + PL
J0422 + 32	V518 Per	'92	L, T	1.19±0.02	3.2–13.2	PL
0620–003	V616 Mon	'17, '75	L, T	2.72±0.06	3.3–12.9	S + PL
1009–45	MM Vel	'93	L, T	3.17±0.12	6.3–8.0	S + PL
J1118 + 480	KV Uma	2000	L, T	6.1 ±0.3	6.5–7.2	PL
1124–684	GU Mus	'91	L, T	3.01±0.15	6.5–8.2	S + PL
1354–645	BW Cir	'71,'87,'97	L, T	5.75±0.3	>7.8	S + PL
1543–475	IL Lup	'71,'83,'92	L, T	0.25±0.01	7.4–11.4	S + PL
J1550–564	V381 Nor	'98	L, T	6.86±0.71	8.4–10.8	S + PL
J1650–500		2001	L, T	2.73±0.56	4–7.3	PL
J1655–40	V1033 Sco	'94	L, T	2.73±0.09	6.0–6.6	S + PL
1659–487	GX 339–4	(P~460d)	L, T	5.8±0.5	>5.8	S + PL
1705–250	V2107 Oph	'77	L, T	4.86±0.13	5.6–8.3	S + PL
J1819.3–2525	V4641 Sgr	'99	L, T	3.13±0.13	6.8–7.4	S + PL
J1859 + 226	V406 Vul	'99	L, T	7.4±1.1	7.6–12	S + PL
1915 + 105	V1487 Aql	'92—	L, T	9.5±3.0	10–18	S + PL
2000 + 251	QZ Vul	'88	L, T	5.01±0.12	7.1–7.8	S + PL
2023 + 338	V404 Cyg	'38,'56,'89	L, T	6.08±0.06	10.1–13.4	PL

[a] The year of outburst, including earlier records as optical novae.
[b] H: high-mass binary. L: low-mass binary: P: persistent. T: transient.
[c] X-ray spectral type near the maximum luminosity. S: soft thermal. PL: power-law.
References: see [39], except for 1354–645 [8] and J1650-500 [60].

16.4.1 Mass Accretion

X-ray binaries are powered by mass accretion. The current concept of the mass accretion process, which determines the X-ray properties of XBs, is summarized later. In HMXBs, the stellar wind of a massive early-type companion feeds mass to the compact object. The mass accretion mechanism in LMXBs is different. LMXBs are contact binary systems in which the companion fills its own Roche equipotential lobe. As the binary separation tends to become smaller (hence the Roche lobe tends to shrink), matter from the companion overflows through the inner Lagrangian point L_1 into the gravitational potential well of the compact object. This is called Roche-lobe overflow.

Matter from the companion star has angular momentum and cannot fall directly onto the compact object. It forms a disk-like structure circulating around the compact object. This is called an accretion disk. The gas rotates in the Keplerian orbit with velocity $(GM_x/r)^{1/2}$ at a distance r from the compact object. Since the inner layer moves faster than the outer layer, the gas is in differential rotation. In such a disk, viscosity acts to transport angular momentum outward, allowing gas to fall gradually inward. At the same time, viscous energy is dissipated for heating the gas. Thus, the gravitational potential energy is converted to rotational energy and thermal energy, as the gas falls into the deep potential well. Half of the gravitational energy released goes to rotational energy and the other half to thermal energy. The accretion process and the accretion disk structure have been studied extensively over the past decades. Shakura and Sunyaev [69] gave the most comprehensive description, which is often referred to as the standard accretion disk model.

In the case of a nonspinning black hole (Schwarzschild hole), the accretion disk extends to $3R_s$, which is the radius of the innermost stable circular orbit (R_{in}). Inside the innermost disk, matter falls freely into the black hole. If the black hole is spinning (Kerr hole) in the same direction as the disk rotation, the disk can extend further inward. For an extreme Kerr hole, the innermost disk radius equals R_g. When the mass accretion rate is $\dot{M}$, the amount of gravitational energy transformed to thermal energy is $GM\dot{M}/2R_{in}$, which is the source of radiation. Eventually $\sim$6% of the rest mass energy of the accreting matter, $\dot{M}c^2$, is radiated for a Schwarzschild hole, and $\sim$40% for an extreme Kerr hole.

According to the standard disk model, the accretion disk is geometrically thin and optically thick when the accretion rate $\dot{M}$ is sufficiently high (quantitative discussion follows). In such a disk, the thermal energy is radiated away in the form of blackbody radiation. The inner disk becomes as hot as 10^7 K, and the radiation is predominantly in the X-ray band. Depending on $\dot{M}$, the X-ray luminosity L_x of black-hole XBs can be as high as 10^{38} erg s^{-1}or may even go up to the Eddington limit, $L_{Edd} = 1.5 \times 10^{38}(M/M_\odot)$ erg s^{-1}.

In the case of neutron-star binaries, the accretion depends on the magnetic fields of the neutron stars. For strongly magnetized neutron stars, typically $\sim 10^{12}$ Gauss, the accretion disk stops at the magnetospheric boundary where the accretion pressure and the magnetic pressure are balanced, inside of which matter is funneled onto the magnetic poles. They manifest themselves as X-ray pulsars, an unambiguous signature of neutron stars.

On the other hand, most of neutron stars in LMXBs have much weaker magnetic fields of the order of 10^8 Gauss. Such weak magnetic fields do not disturb the accretion flow unless the accretion rate is extremely low, and the disk extends close to the neutron star surface. Since the radius of a canonical $1.4\,M_\odot$ neutron star is considered to be comparable with $3R_s$ ($\sim$12 km), this situation is not much different from a Schwarzschild black hole. Hence, the structure of the accretion disk is expected to be similar in both cases.

However, there is a fundamental difference between a neutron star and a black hole. It is the presence or the absence of a solid surface. If the compact object is a neutron star, accreting matter eventually hits the neutron star surface and releases the rest kinetic energy in the dense neutron star atmosphere. This energy will be emitted as additional blackbody radiation. On the other hand, for a black hole, matter simply disappears across the event horizon. This difference provides an important clue to distinguish between a black hole and a neutron star, as discussed in Sect. 16.4.3.

16.4.2 Soft X-Ray Transients

Transient outbursts have been discovered by various X-ray satellites, in particular those with large field-of-view detectors on board, e.g., Ariel V, Tenma, Ginga, Granat, CGRO and more recently RXTE, BeppoSAX.

Among ∼300 XBs known to date, roughly half of them are classified as transient sources. The transients are further divided into two different classes: the HMXB transients and the LMXB transients. Almost all HMXB transients are recurrent X-ray pulsars, in which a strongly magnetized neutron star in an eccentric orbit periodically encounters a stellar wind zone around a massive companion. They are not relevant to this chapter.

Of about 150 LMXBs known to date [34], one half are transients. These LMXB transients are characterized by episodic X-ray outbursts without a fixed periodicity. They are in a quiescent state for most of the time, and occasionally undergo dramatic X-ray outbursts. During an outburst, they show a soft X-ray spectrum (softer than the X-ray pulsars) characteristic of high-luminosity LMXBs, as explained in Sect. 16.4.3. These are called "soft X-ray transients," or sometimes "X-ray novae." The soft X-ray transients are a subset of LMXBs containing either a weakly-magnetized neutron star or a black hole. Many soft X-ray transients exhibit recurrent outbursts with intervals ranging from a few years to tens of years or even longer (e.g., Table 16.1). Perhaps all soft X-ray transients are recurrent. X-ray outbursts are accompanied by optical outbursts and sometimes radio jets that allow identification of the optical counterparts. For reviews, see [78] and [39].

As shown in the preceding section, 17 of the secure black-hole XBs are LMXBs and are all soft X-ray transients. In addition, ∼20 more soft X-ray transients are suspected to contain black holes (black hole candidates; see [39]) based on their X-ray and other properties, discussed in Sect. 16.4.3. This fact indicates that black-hole binaries occupy a major fraction of soft X-ray transients. The observed outburst rate allows a rough estimation of the total number of black-hole binaries existing in our Galaxy. Though the average recurrence period is largely uncertain, a modest estimate gives a few hundred, and could possibly be much more (e.g., [65, 78]).

Figure 16.1 shows X-ray light curves of the outbursts of four black-hole binaries, which exhibit monotonous exponential decays except for intermediate increases. However, many other outbursts show much more complex light curves (see [10, 39]. In most of the outbursts, the source brightens to an X-ray luminosity $L_X \sim 10^{38}$ erg s^{-1}or sometimes even $\sim 10^{39}$ erg s^{-1}at the outburst peak. The source returns to quiescence usually after a few months. There are exceptional cases: GRS 1915+105 has remained bright since the initial outburst in 1992. GX339–4 repeats frequent outbursts but has never decayed into the quiescent state. Radio outbursts, sometimes

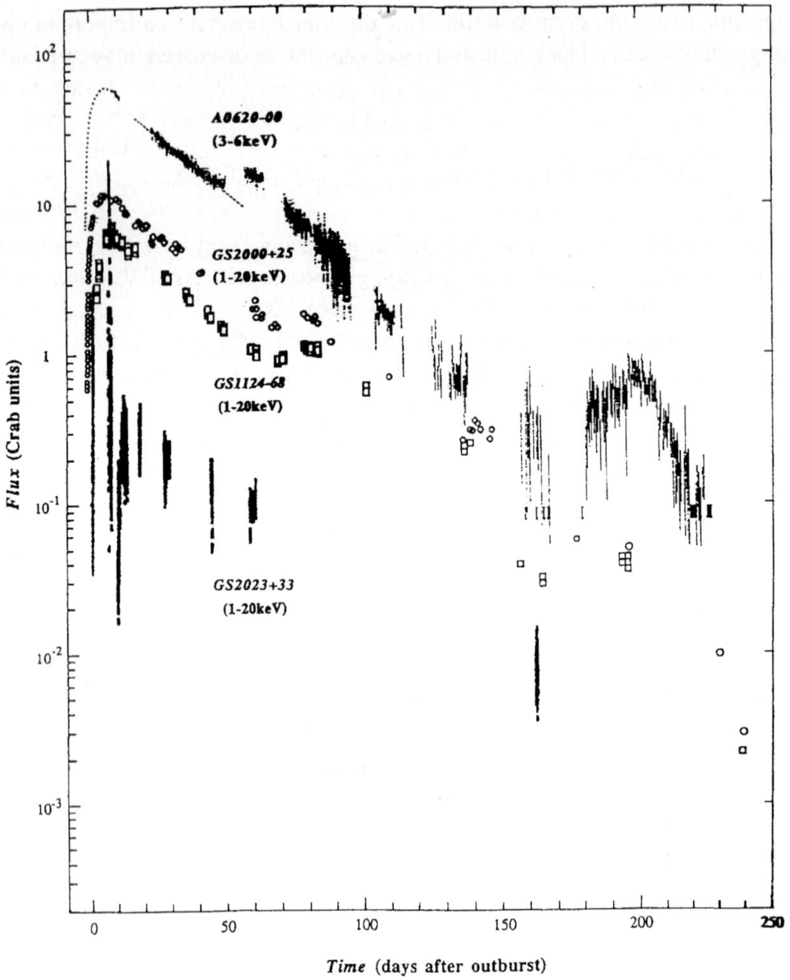

Fig. 16.1 X-ray light curves of the transient outbursts of four black-hole XBs [78]. The observed fluxes are shown in units of the Crab Nebula flux in an energy band indicated for each source

superluminal jets, are often observed at the X-ray outbreak, hence nicknamed "microquasars" (e.g., [48]).

An X-ray outburst is triggered by a sudden onset of the accretion flow onto the compact object, followed by the decay as the accretion rate gradually diminishes. Because the accretion rate changes over orders of magnitude through the decay, soft X-ray transients are extremely useful for studying the physics of mass accretion and X-ray properties as a function of the accretion rate. In fact, much of our current knowledge described in the following sections has been obtained from the studies of soft X-ray transients.

Whether a low-mass binary remains persistently X-ray active or not seems to depend on the accretion rate. Apparently, there is a minimum accretion rate $\dot{M}_{\rm min}$ below which X-ray emission is practically turned off. (Notice the abrupt fall in the last part of decay in Fig. 16.1) The available data indicate that $\dot{M}_{\rm min} \sim 10^{16}$ g s^{-1} corresponding to $L_{\rm X} \sim 10^{36}$ erg s^{-1}. In fact, there exist few persistent sources of $L_{\rm X} \ll 10^{36}$ erg s^{-1}. The long quiescence of soft X-ray transients between outbursts is probably because $\dot{M} < \dot{M}_{\rm min}$.

A soft X-ray transient outbursts are generally explained by the disk instability model originally developed for dwarf novae. The concept is as follows: An accretion disk has two stable states; a cool neutral state and a hot fully-ionized state. The quiescent state corresponds to the cool state where viscosity is too small to allow a stable accretion flow. As matter from the companion accumulates in the outer disk, the surface density and temperature increase gradually. When the surface density reaches a certain critical value, the disk jumps to the hot state because of a thermal instability. This triggers an accretion flow into the inner part of the disk, causing an X-ray outburst. When the surface density drops below another critical value, the disk returns to the cool quiescent state.

16.4.3 X-Ray Spectra

This section deals with the general characteristics of the X-ray spectra of black-hole XBs, and discusses the differences from those of neutron-star LMXBs of which neutron stars are weakly magnetized.

As shown later, there are at least three distinctly different states with respect to the spectral properties. These are a high-luminosity soft state (HS state), a low-luminosity hard state (LH state), and the quiescent state. These are believed to be related to intrinsic changes of the disk structure depending on the mass accretion rate $\dot{M}$ (hence $L_{\rm X}$). There appears to be a certain $L_{\rm X}$ value that divides the HS state and the LH state, which is around $L_{\rm X} \sim 10^{37}$ erg s^{-1}(or $\dot{M} \sim 10^{17}$ g s^{-1}). However, it may vary somewhat from source to source and from time to time. For previous reviews, see e.g., [39, 76, 78, 79].

16.4.3.1 X-Ray Spectrum at High Luminosities

In the HS state, black-hole XBs typically show a common characteristic spectral shape, consisting of a soft thermal component and a hard nonthermal tail, as shown in Fig. 16.2. Of the 20 secure black-hole XBs known so far (Table 16.1), 17 show such an X-ray spectral shape. (For three exceptions, see later.) In this state, the time variability of the soft component is generally small and slow. The hard component varies much more than the soft component.

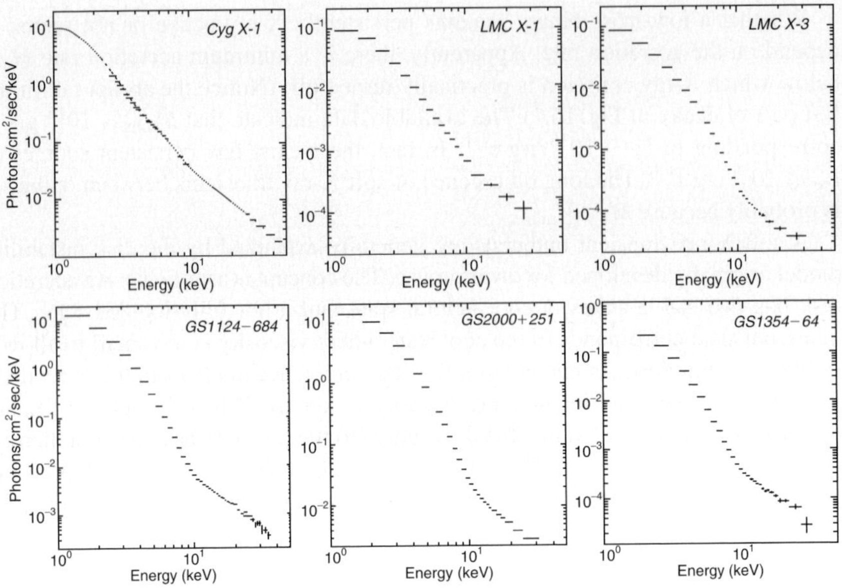

Fig. 16.2 X-ray photon spectra of secure black-hole XBs in the HS state obtained with Ginga, except for Cyg X-1, which is from the data of ASCA and RXTE [22]

The soft component shows the characteristics of thermal emission. It is interpreted as the emission from an optically-thick geometrically-thin accretion disk (hereafter abbreviated to "thin disk"). The observed soft component is well expressed by the "multicolor blackbody disk" (MCD) model developed by Mitsuda et al. [49], and later elaborated by Makishima et al. [36]. The disk emits blackbody radiation locally and the blackbody temperature increases toward the center, hence the emission is "multicolored." The MCD model is a formulation of such a spectrum based on the standard disk model. The simplest MCD model includes only two free parameters, i.e., R_{in} and kT_{in}, where R_{in} represents the innermost disk radius and kT_{in} is the temperature at R_{in}. The actual formula includes the source distance D and the disk inclination angle i in the form $R_{\text{in}}(cosi)^{1/2}/D$, which are assumed to be known here. Note that kT_{in} is the "color" temperature, and not the effective temperature.

The observed value of kT_{in} is typically ~1.0 keV at a soft X-ray luminosity $L_{\text{soft}} \sim 10^{38}$ erg s^{-1}, and becomes lower as the luminosity decreases. The color temperature is substantially higher than the effective temperature because of the electron scattering effect that dominates at such high temperatures [71].

The hard tail has a power-law form, and extends to well-over 100 keV, sometimes observed up to ~1 MeV without a cut-off, as shown for example in Fig. 16.6b. The luminosity of the hard component relative to the soft component varies irregularly by a large factor (see Fig. 16.6a). The photon index Γ of the power-law, for a photon number spectrum of the form $E^{-\Gamma}$, is ~2.5, and remains essentially constant against changes in intensity. This power-law spectrum has been considered to be produced

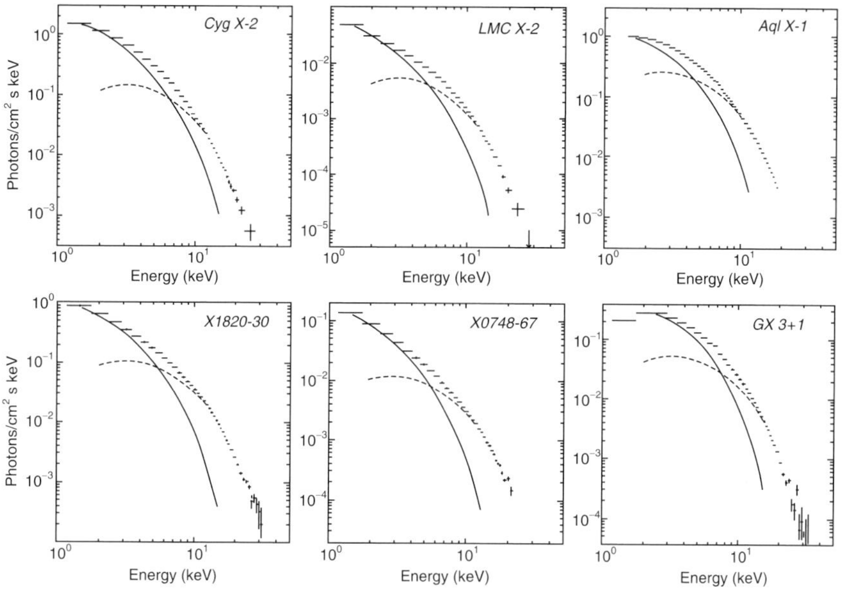

Fig. 16.3 X-ray photon spectra of neutron-star LMXBs in the HS state (Ginga), each consisting of a soft MCD component (*solid curve*) and a blackbody component (*dashed curve*)

as a result of multiple scattering of soft photons with high-energy electrons, gaining energy by the inverse Compton effect: the process called Comptonization [74]. Yet, the origin of the power-law tail is still unsettled, including the questions as to how the electrons are accelerated to $\gtrsim 1\,\text{MeV}$ and why Γ is commonly regulated at $\sim$2.5.

Turning to neutron-star LMXBs, they also reside in the HS state when $L_X \gtrsim 10^{37}$ erg s^{-1}. However, as shown in Fig. 16.3, their spectra are distinctly different from those of black-hole XBs. The spectrum of neutron-star LMXBs in the HS state actually consists of two separate components: a soft thermal component and another harder thermal component. It is to be emphasized that these two components are real, not artificially introduced for the purpose of reproducing the observed spectrum. Since these two components change in intensity with time independently of each other, they can be identified and their spectra can be determined separately (see [23, 49, 79]). An additional nonthermal hard tail is occasionally noticed, yet much less pronounced than in the case of the black-hole XBs.

This soft thermal component is also well fit with the MCD model, supporting that this is the emission from the standard thin disk. However, the observed color temperature kT_{in}, typically $\sim$1.5 keV for $L_X \sim 10^{38}$ erg s^{-1}, is significantly higher than that in the black-hole XBs at a similar L_{soft} (see below for the implication).

The harder thermal component shows a single blackbody spectrum with a color temperature of $\sim$2 keV. This component is most probably the emission from the neutron star surface where the kinetic energy of accreting matter is eventually deposited and thermalized. The intensity of this blackbody component varies irregularly without changing the shape, and its average luminosity is comparable to the soft component from the thin disk as expected. The fact that this spectrum is very similar to that of so-called Type I X-ray bursts further supports this interpretation. The Type I X-ray burst is a thermonuclear flash on the neutron star surface producing a blackbody radiation [32]. (More on the Type I X-ray burst in 16.4.3.2.) In fact, the Type I X-ray burst has never been observed from the secure black-hole XBs.

The above summarizes the X-ray spectral properties of the black-hole XBs and neutron-star LMXBs in the HS state. In what follows, the fundamental differences between them are discussed.

The first clear distinction is that none of the black-hole XBs show a "2-keV" blackbody component characteristic of the spectra of the neutron-star LMXBs in the HS state. For the reason mentioned earlier, the absence of this blackbody component and of the Type I X-ray bursts argue for the absence of a solid surface.

Furthermore, the lower $kT_{\rm in}$ values of the disk emission of black-hole XBs are understood in terms of the standard disk model. The radius of the innermost stable orbit is proportional to the mass of the compact object (see later). For a given luminosity, the larger the mass, the lower is the blackbody temperature ($kT_{\rm in} \propto M_{\rm X}^{-1/4}$). Hence, the X-ray spectrum of the disk around a black hole is expected to appear softer (lower temperature) than that around a neutron star. Moreover, the presence of the additional 2-keV blackbody component for the neutron-star LMXBs makes the thermal component of the black-hole XBs look much softer, hence sometimes called "ultrasoft" [83] (compare Figs. 16.2 and 16.3).

For at least three long-observed black-hole XBs, the observed value of $R_{\rm in}$ is found to remain essentially constant against large changes in the soft component luminosity $L_{\rm soft}$ [76], as shown in Fig. 16.4. This fact strongly supports an interpretation that $R_{\rm in}$ represents the radius of the innermost stable orbit. For a nonspinning black hole, the radius of the innermost stable orbit is $3R_{\rm s}$, hence proportional to the compact object mass. Therefore, the mass can be estimated from the $R_{\rm in}$-value, if the estimated source distance and the inclination angle are available. In fact, the observed values of $R_{\rm in}$ obtained for a set of black-hole XBs turned out to be larger than those for neutron-star LMXBs by a factor of 3 to 4 [79]. Thus estimated black hole masses are qualitatively consistent with those obtained from the mass functions (except for GRO J1655–40 and GRS 1915+105, as explained below). Similar results of constant $R_{\rm in}$ against changes in $L_{\rm soft}$ have later been found from more black-hole XBs (see [39]).

The above considerations strongly support that such a "soft + hard-tail" spectrum in the HS state is a signature of an accreting black hole. In addition to 17 among the 20 secure black-hole XBs, there are about 20 additional soft X-ray transients that exhibited this characteristic spectral shape. Also, they showed neither regular pulsation nor Type I X-ray burst (see 16.4.3.2). On the basis of these properties, they can be considered to be black hole candidates [39].

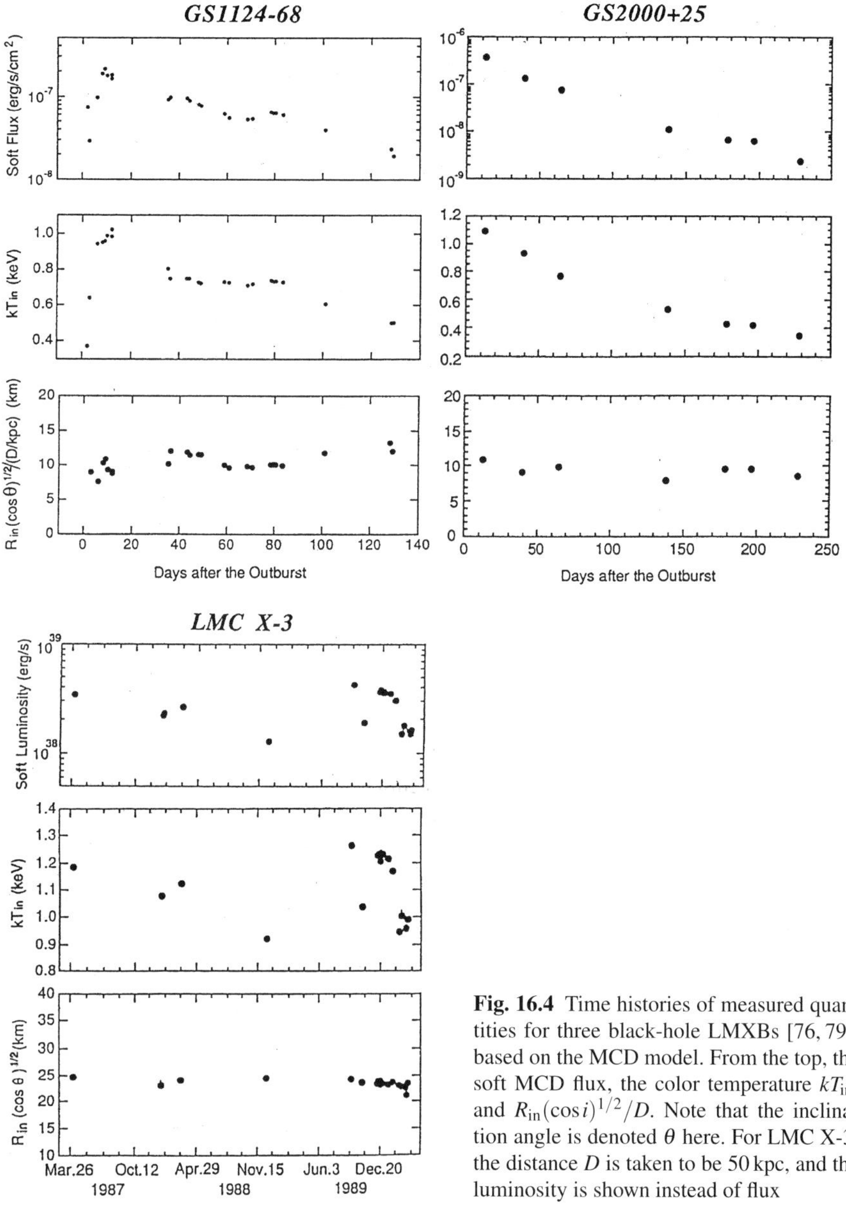

Fig. 16.4 Time histories of measured quantities for three black-hole LMXBs [76, 79], based on the MCD model. From the top, the soft MCD flux, the color temperature $kT_{\rm in}$, and $R_{\rm in}(\cos i)^{1/2}/D$. Note that the inclination angle is denoted θ here. For LMC X-3, the distance D is taken to be 50 kpc, and the luminosity is shown instead of flux

Among the 20 secure black-hole XBs, GS 2023+338, GRO J0422+32, XTE 1650–500, and XTE J1118+480 are exceptions. These four sources show an approximately single power-law spectrum typical for the LH state (see 16.4.3.2). It is possible that GRO J0422+32, XTE 1650–500, and XTE J1118+480 did not enter into the HS state, since their peak $L_{\rm X}$ were relatively low; $\sim 10^{37}$ erg s^{-1} for

GRO J0422+32 and XTE 1650–500, and $\sim 10^{36}$ erg s^{-1} for XTE J1118+480. On the other hand, GS 2023+338 was extremely luminous, $\sim 10^{39}$ erg s^{-1}, and probably reached the Eddington limit L_{Edd}. The reason why it did not show the soft + hard-tail spectrum is still unknown.

Two secure black-hole XBs GRO J1655–40 and GRS 1915+105, both exhibiting a "soft + hard-tail" spectrum and superluminal jets, show significantly higher kT_{in} than other black-hole XBs (hence smaller R_{in}). Zhang et al. [86] suggest that the black holes in these systems may be prograde Kerr holes, for which the maximum disk temperature can be substantially higher. The highly relativistic jets from these sources could be related to the high black-hole spin.

Occasionally, black-hole XBs become exceedingly luminous ($>0.2L_{Edd}$), and the hard power-law component dominates the X-ray spectrum, which looks significantly different from the typical HS state spectrum shown in Fig. 16.2. At the same time, (1) rapid and large variability associated with the intense hard component and (2) quasi-periodic oscillations (QPO) appear. Miyamoto et al. [50] first noted this behavior in GX 339-4, and named it the "very high (VH) state." The photon index Γ still remains in the typical range for the HS state or somewhat steeper, $\Gamma \gtrsim 2.5$, without showing a cut off. In this regard, the VH state may be called the "power-law dominated" HS (PLD-HS) state. This state has been observed for many black-hole XBs. Of particular importance is the appearance of high-frequency ($\gtrsim$ 100Hz) QPO in the PLD-HS state, as addressed in Sect. 16.5. (See [39] for more on the VH state.)

An additional difference from the standard HS properties is noticed in the PLD-HS state. The constancy of R_{in} (i.e., $L_{soft} \propto T_{in}^4$) breaks down. Compared with the normal HS state, T_{in} shows anomalously high values, and consequently R_{in} drops down. Such cases have been found from GRO J1655-40 [29, 72], XTE 1550–564 [30], and GRS 1915+105 [14]. Kubota et al. [29, 30] showed that the normal $L_{soft} - T_{in}$ relation is retained if a Comptonized MCD is included (see Fig. 16.5), and interpreted it to reveal an appearance of a hot corona that Comptonizes a part of the disk photons. This Comptonized component also contributes to the increase of the hard component in the PLD-HS state.

It has been theoretically predicted that the standard (gas pressure-dominated) thin disk is no longer stable when the radiation pressure becomes dominant (e.g., [33,70]). As $\dot{M}$ increases, the inner disk goes through an unstable regime, and shifts to another stable structure [1,25], which Abramowicz et al. [1] named a "slim disk." This instability predicts a possible limit-cycle variability between the thin disk and the slim disk. A unique behavior of GRS 1915+105 displaying repeated transitions between two flux levels [6] can be interpreted as such a case [84]. The highly-variable PLD-HS state probably corresponds to the intermediate unstable regime [30]. However, the limit-cycle behavior is rarely observed, which suggests that some effects suppress the onset. Energy disspation with the Comptonizing corona may be a possible one [14].

The concept of the slim disk applies when $\dot{M} \gg L_{Edd}/c^2$, where L_{Edd}/c^2 is called the critical accretion rate. The slim disk model has been elaborated later (e.g., [46, 56, 57]). The model predicts that L_x may well exceed L_{Edd} for extremely

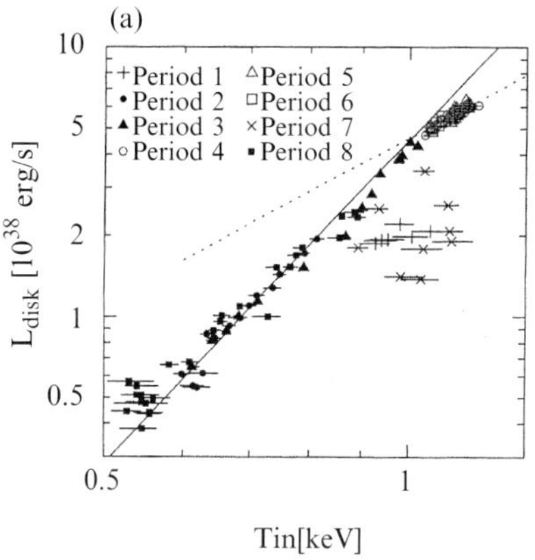

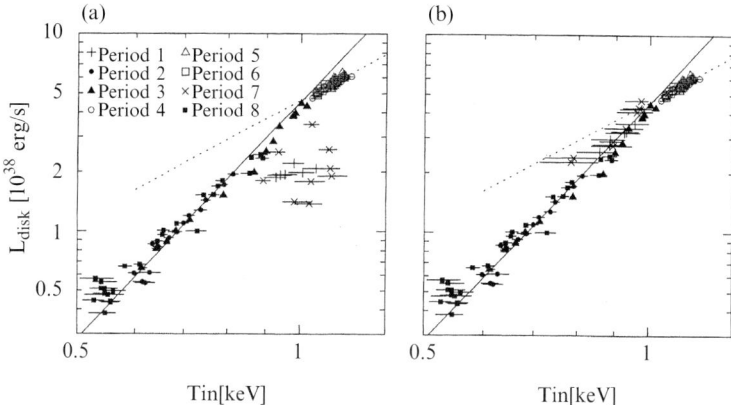

Fig. 16.5 $L_{disk}(= L_{soft}) - T_{in}$ relation of XTE 1550–564 [30]. (**a**) Fits with (MCD + power-law) model. (**b**) When Comptonized MCD is included in Period 1 and Period 7. The *solid* and *dashed lines* are for $L_{disk} \propto T_{in}^4$ and $L_{disk} \propto T_{in}^2$, respectively

high $\dot{M}$ [57]. The radiation efficiency of a slim disk is considered to be reduced because photons are trapped and carried away in the mass flow (advective energy transport). Hence, the $L_X - \dot{M}$ relation tends to become saturated [56, 57]. The spectrum is expected to be dominated by a MCD component, except that T_{in} is significantly higher than extrapolated from lower luminosity levels. Actually, such a feature has been observed in some black-hole XBs, e.g., GRS 1915+105 [14] and XTE 1550–564 [30] (see Fig. 16.5).

16.4.3.2 X-Ray Spectrum at Lower Luminosities

The X-ray spectrum at low luminosities is distinctly different from that at high luminosities. Because the spectrum is hard, much harder than in the HS state, it is called the low-luminosity hard (LH) state. Both neutron-star LMXBs and black-hole XBs exhibit a dramatic change, undergoing a transition of state, across $L_X \sim 10^{37}$ erg s^{-1}(or $\dot{M} \sim 10^{17}$ g s^{-1}), as mentioned in Sect. 16.4.3. The spectral shape in the LH state is a hard power-law form, as shown in Fig. 16.6a. Another outstanding difference is in the properties of time variability. When sources enter into the LH state, rapid large-amplitude intensity fluctuations (flickering) build up in all time scales down to milliseconds. Such a transition between the two spectral states has been observed in several soft X-ray transients during the decay, and also in Cyg X-1 and LMC X-3. Since this bimodal behavior is observed regardless of whether the compact object is a neutron star or a black hole, it is believed to be a fundamental property of an accretion disk, depending on the accretion rate. Once they go into the LH state, the spectral shape and variability are essentially the same for both

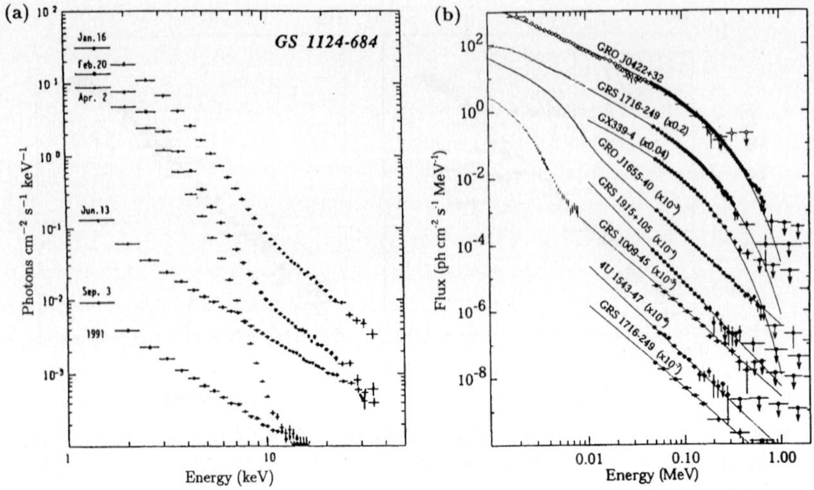

Fig. 16.6 (**a**) Changes in the spectral shape of the black-hole LMXB GS 1124–684 with luminosity [15]. The upper three are the spectra in the HS state, while the lower two are those after transition to the LH state. (**b**) Hard X-ray spectra of the black-hole soft X-ray transients [24]. Note that the lower five are those in the HS state, whereas the upper three are those in the LH state. The spectra are scaled by arbitrary factors (in *brackets*) for clarity

black-hole XBs and neutron-star LMXBs. Hence, these systems are no longer distinguishable [78].

The power-law spectrum in the hard state is clearly different from the hard tail of black-hole XBs in the HS state. It is substantially harder with the observed photon indices Γ in the range 1.7−1.9. In addition, unlike the hard tails of black holes in the HS state, the power-law spectrum shows a clear fall-off above several tens kiloelectronvolt (high-energy cut-off), as seen in a few examples in Fig. 16.6b (upper three). The power-law spectrum with a cut-off observed in the LH state can be reproduced by Comptonization of soft photons with thermal electrons of $kT \lesssim 100\,\mathrm{keV}$, a process called thermal Comptonization (see [74]).

The transition between the HS state and the LH state has been considered as due to a change in the disk structure. There is evidence that a tenuous hot plasma, the so-called "disk corona," builds up above the disk when a source goes into the LH state. For instance, when 4U 1608–522 (a neutron-star LMXB) was about to go into the LH state, the spectra of X-ray bursts (blackbody emission from the neutron star surface) began to show a hard tail [51], implying that a part of the burst photons were injected into the corona and Comptonized. Also, the disappearance of the 2 keV blackbody component of the neutron-star LMXBs in the LH state suggests that the neutron star surface is obscured by the disk corona. Whether the thin disk in the LH state still extends to the innermost stable orbit (embedded in the corona) or it recedes to a larger radius is an unsettled issue. The physical mechanism for the transition of state and other unique properties of the LH state are yet to be understood.

As mentioned erlier, the Type I X-ray burst is a thermonuclear flash of the matter accumulated on the surface of neutron stars (see [32] for a review), hence a unique signature of neutron-star LMXBs. Except for very luminous ones, such as Sco X-1 and LMC X-2, Type I X-ray bursts are frequently observed from neutron-star LMXBs. According to the LMXBs catalogue by Liu et al. [34], Type I X-ray bursts were detected from 63 out of total 150 LMXBs. Excluding the 17 secure black-hole LMXBs, it gives a high probability of $\sim$50%. The rest of LMXBs includes $\sim$20 transients that are selected as black hole candidates on the basis of a "soft + hard-tail" spectrum. Significantly, no Type I X-ray burst has been detected from them, despite a long watch before going into quiescence. Together with the absence of the 2-keV blackbody component in the HS state, this fact strongly supports that they are also black-hole LMXBs. (See also 4.3.3).

Another important characteristics of the LH state is the associated radio emission. While the sources are in the HS state, they are usually radio-quiet. When they go into the LH state, quasi-steady radio emission, sometimes resolved into jets, shows up.

Incidentally, the properties of XBs in the LH state are strikingly similar to those of many active galactic nuclei, i.e., similar power-law index, and high time-variability. These similarities suggest that despite huge differences in the system scale and power, the basic process of accretion is essentially the same in both systems. On this basis, many properties of active galactic nuclei are interpreted in the light of what is known about LMXBs. Most of active galactic nuclei are considered to be accreting supermassive black holes in the LH state.

16.4.3.3 X-Ray Spectrum in Quiescent State

Soft X-ray transients spend most of their life-times in a quiescent state. Since the X-ray luminosity in quiescence is below 10^{33} erg s^{-1}, observations of them require high sensitivity. Most of the observations so far have been made with ROSAT, ASCA and more recently with Chandra and XMM-Newton.

The observed results of the quiescent state show distinct differences between black-hole XBs and neutron-star XBs. Neutron-star XBs are systematically more luminous than black-hole XBs [40]. Moreover, there is an essential difference in the X-ray spectra in quiescence. Some examples are shown in Fig. 16.7. The available spectra of black-hole XBs show a hard power-law spectrum with a photon index $\Gamma \sim 2$ [3, 28, 54]. In contrast, neutron-star XBs commonly show a low-temperature blackbody-like component with $kT \lesssim 0.1$ keV in addition to a hard tail [3, 66–68]. This low-temperature component qualitatively accounts for higher luminosities of neutron-star XBs. It is most probably the thermal emission from the surface of a cooling neutron star. In fact, Rutledge et al. [66–68] showed that, if a proper atmosphere model is adopted, the estimated radius for the observed thermal component is of the order of 10 km, consistent with the canonical value for a neutron star. Here again, the lack of a thermal component in the spectra of black-hole XBs in quiescence provides further evidence for the absence of a visible surface.

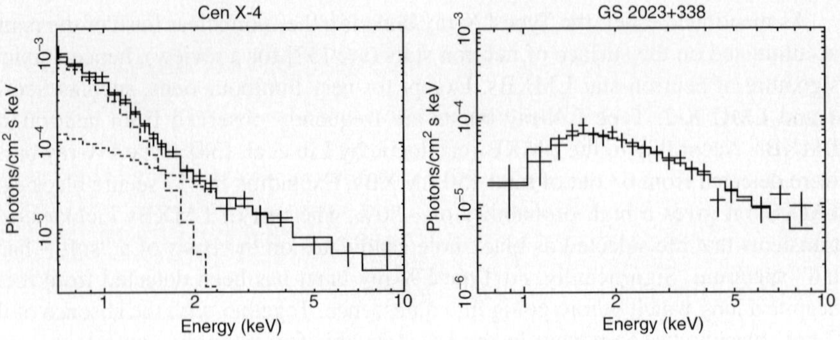

Fig. 16.7 X-ray spectra of two soft X-ray transients in quiescence: the neutron-star XB Cen X-4 and the black-hole XB GS 2023+338 [3]

The accretion flow in the quiescent state seems quite different from those in the HS and LH states. There are optical observations that prove continued mass transfer from the companion even during quiescence at a rate $\dot{M} \sim 10^{15}$ g s^{-1}(see [78]). On the other hand, the observed quiescent X-ray luminosities are 10^{30-32} erg s^{-1}, which indicates an extremely low efficiency of radiation of $\lesssim 0.01\%$.

Narayan, McClintock, and Yi [53] proposed a disk model consisting of two distinct zones: at large radii the disk is thin and cool in which a part of the transferred matter is stored, whereas the inner region is filled with a hot ($kT_e \sim 100$ keV) tenuous quasi-spherical plasma flow in which most of the energy released is carried away (advected) and only a tiny fraction is radiated. This advection-dominated accretion flow (ADAF) model also explains the observed power-law spectrum in quiescence. Variant models of low radiation-efficiency accretion flow were proposed later (see [39] for references).

16.4.4 Relativistic Iron Line

As shown in the preceding section, there is accumulating evidence for the absence of a material surface in the secure and candidate black-hole XBs. A further crucial step is to find strong relativistic effects; the unique features expected in the direct vicinity of the event horizon of black holes.

In some Seyfert galaxies, the iron K-lines are found to be relativistically broadened: highly asymmetric, extended to low energies (e.g., [19, 52, 77]). This feature strongly suggests that the line is emitted very close to the black hole where the relativistic Doppler effect and the gravitational redshift are quite significant. These effects produce such a broad, asymmetric iron K-line. The relativistically broad iron K-line from a Schwarzschild black hole was first investigated theoretically by Fabian et al. [18], and later generalized to that from a Kerr black hole [31] (see [19, 64] for a review).

Similar effects can be expected in the disk of the black-hole XBs. A generally accepted scenario is that the broad iron K-line is the fluorescent line (6.4 keV in the rest frame) generated in the innermost region of an optically-thick disk, which is irradiated by hard X-rays from an overlying corona. In this case, a reflection continuum with the iron K-absorption edge accompanies the iron K-line. The reflection component of XBs has recently been investigated extensively (see e.g., [85]). At high luminosities, the photoionization of the innermost disk may also produce a relativistic broad iron line [19, 64].

There have been a considerable number of reports on the detection of relativistically broad, asymmetric iron K-line from several black-hole XBs. The results obtained with proportional counters (e.g., on Ginga and RXTE) suffer from marginal energy resolution, and are to be taken with reservation (references in [39]). More recent results with higher energy resolution (ASCA, BeppoSAX, Chandra and XMM-Newton) include those from Cyg X-1 [20, 42], XTE J1650–500 [43, 47], XTE J1819.3–2525 [44] and GX 339-4 [45]. Most of them were observed in the HS state. Although the result depends on the model of the underlying continuum, particularly of the reflection component, the observed iron K-lines show a very broad redward-skewed profile, as shown for an example in Fig. 16.8. These results are often fitted to a Laor line [31] from a highly-ionized disk, and the authors suggest a Kerr black hole. However, the interpretation of these broad iron K-lines as a result of the genuine relativistic effects is still subject to further studies.

Comparing to the case of neutron-star LMXBs, where the disk may extend close to the neutron star surface comparable to $\sim 3R_s$, significant relativistic effects are also expected. Asai et al. [4] studied twenty NS-LMXBs of various luminosities with ASCA, and found iron K-lines from about half of them. Some of the lines appear slightly broad, but not as broad and asymmetric as those found in the black-hole XBs. Detections of iron K-lines were also reported from several neutron-star LMXBs in the HS state, e.g., Sco X-1, 4U 1608–52 [75], GX 17+2 [12], GX 349+2 [13]. These lines are not broad and located at $\sim$6.7 keV instead of 6.4 keV, corresponding to the line from highly-ionized iron. These differences are yet to be understood.

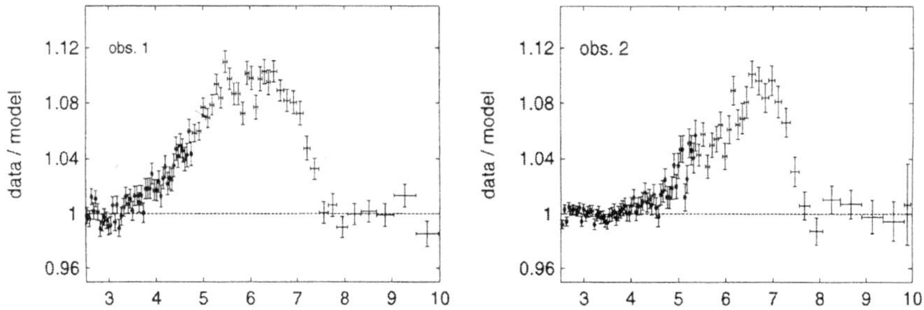

Fig. 16.8 Broad feature in XTE J1650-500 interpreted as a relativistic iron K-line [47]. Ratio of the data to a model: MCD + power-law with a smeared K-absorption edge. The 4–8 keV band is ignored in fitting

Note Added in Proof

Discoveries of relativistic broad iron lines in a few neutron star LMXBs have recently been reported from high-statistics observations with XMM-Newton [87] and Suzaku [88].

16.5 Quasiperiodic Oscillations

Quasiperiodic oscillations (QPOs) are generally observed in the PLD-HS state (see 16.4.3.1). Many black-hole transients exhibit low-frequency QPOs roughly in the range 1–20 Hz [39]. (See [80] for a general review of QPO.)

QPOs of higher frequencies (> several 10 Hz) may be directly related to the black-hole accretion disks. For instance, the orbital frequency at the innermost stable orbit around a Schwarzschild hole is $\sim 220(M/10M_{\odot})^{-1}$ Hz, and even higher for a Kerr hole. Several black-hole XBs do show high-frequency QPOs in the PLD-HS state in the range 40–450 Hz [39], suggesting that they originate in the inner disk near a black hole.

Remarkably, pairs of QPOs whose frequencies are commensurate in a 3:2 ratio were discovered from four black-hole XBs including one candidate. These are GRO J1655-40 at 450 and 300 Hz [61, 73], XTE J1550-564 at 276 and 184 Hz [41, 61] and a marginal QPO at 92 Hz (3:2:1 ratio) [61], GRS 1915+105 at 165 and 113 Hz [62], and H1743–322 at 242 and 166 Hz [63]. Importantly, these QPO frequencies do not shift with changes in luminosity, unlike the kHz QPOs in the neutron-star LMXBs. The frequencies seem to be fixed for each individual sources. For these black-hole XBs, the QPO frequency is found to be inversely proportional to the measured black hole mass ($\propto M_X^{-1}$) [39], except for H1743–322 whose mass is still unknown. These results prompted attempts for interpreting the high-frequency QPOs of black-hole XBs in the framework of general relativity.

There are three fundamental oscillation modes in the disk around a spinning black hole, which produce the orbital frequency, the radial epicyclic frequency and the polar epicyclic frequency. Abramowicz and Kluźniak [2] were the first to propose that the commensurate QPOs result from resonance between these modes. They later considered the resonance between the polar and radial modes for explaining the 3:2 ratio [27]. Remillard et al. [61] found that the 3:2 ratio is possible for each of the three couplings, i.e. orbital/radial, orbital/polar, or radial/polar, at respectively different orbital radii (where resonance occurs) and different angular momentum parameter a_* of a Kerr black hole, making use of the measured mass M_X.

Recently, Aschenbach [5] proposed a model that uniquely determines M_X and a_*. He reported that the coupling of the polar and the radial epicyclic modes is possible with a frequency ratio of not only 3:2 but also 3:1, and with the associated Kepler frequencies being commensurate at a 3:1 ratio. These conditions restrict the spin to just one value, i.e., $a_* = 0.99616$, which would be common to all 3:2 objects. The masses predicted from the measured QPO frequencies agree well with the dynamically determined masses (Table 16.1). Interestingly, at this high spin the topology of the particles' motion around the black hole unexpectedly changes such that their orbital velocity does not increase but decreases with decreasing distance in a region close to the innermost stable orbit [5]. This reversal of the velocity gradient might excite and maintain the radial epicyclic oscillations.

The commensurate high-frequency QPOs are unique to black holes and seem to be related to the black hole spin. Remarkably, these four sources with the commensurate QPOs all displayed jets, suggesting the role of the black hole spin in the jet production. Study of these QPOs is important as it may eventually lead to the determination of a_* of black holes in XBs.

16.6 Ultraluminous X-Ray Sources

The above-discussed Galactic black-holes in XBs have the masses around $10\,M_\odot$. Recently, there has been increasing attention to a possible presence of more massive black holes in external galaxies (e.g., [11, 16, 36]). They are ultraluminous X-ray sources (ULXs) with L_X in the range 10^{39}– $>10^{40}$ erg s^{-1}, located off-center of the galaxies (hence not AGN). Liu and Mirabel [35] published a catalogue of 229 ULXs. The observed spectra are either the MCD-type or of a power-law shape. Their luminosities are at least comparable to or well exceed the Eddington limit of a $\sim 10M_\odot$ black hole. The detected time variabilities argue that they are compact objects. This has led to the concept that ULXs are accreting black holes significantly more massive than $10\,M_\odot$, ranging to as high as $100\,M_\odot$ [17,36]. On the other hand, an alternative possibility has been considered that they are ordinary black-hole XBs radiating anisotropically [26]. In our Galaxy, no black-hole XB has been found to remain persistently super-Eddington, although a few of them temporarily exceeded the Eddington limit around the peak luminosity (see 16.4.3.1).

The slim disk concept allows for super-Eddington luminosities when $\dot{M} \gg L_\mathrm{Edd}/c^2$. Numerical simulations show that the luminosity may go up to several times L_Edd [56, 57]. Watarai et al. [81] analyzed the ULXs having the MCD-type spectra with the slim disk model, and found that they are mostly consistent with $\sim 10M_\odot$ black holes at extremely high accretion rates ($\dot{M} \gg L_\mathrm{Edd}/c^2$). Another effect at $\dot{M} \gg L_\mathrm{Edd}/c^2$ is that the radiation is expected to become anisotropic, i.e., stronger towards the disk axis [57]. Hence, if viewed at a small inclination angle, the luminosity assuming isotropy may be a considerable overestimate.

Even though most of ULXs may be explained as $\sim 10\,M_\odot$ black holes at very high accretion rates, the existence of intermediate-mass black holes exceeding several tens $M_\odot$ is by no means excluded at present. For an example, a variable off-center source in M 82 [37] can be a strong candidate for an intermediate-mass black hole. It brightened to $L_\mathrm{X} \sim 10^{41}$ erg s^{-1} corresponding to the Eddington limit for an object of several hundred $M_\odot$. If a class of intermediate-mass black holes were confirmed by future observations, the origin of them should be a challenging problem since such massive black holes could not be produced by supernovae.

References

1. Abramowicz, M.A., Czerny, B., Lasota, J.P., et al. 1988, Astrophys. J., 332, 646
2. Abramowicz, M.A., Kluźniak, W. 2001, Astron. Astrophys., 374, L19
3. Asai, K., Dotani, T., Hoshi, R., et al. 1998, Publ. Astron. Soc. Jpn., 50, 611
4. Asai, K., Dotani, T., Nagase, F., et al. 2000, Astrophys. J. Suppl., 131, 571
5. Aschenbach, B. 2004, Astron. Astrophys., 425, 1075
6. Belloni, T., Klein-Wolt, M., Méndez, M., et al. 2000, Astron. Astrophys., 355, 271
7. Bolton, C.T. 1972, Nature, 235, 271
8. Casares, J., Zurita, C., Shahbaz, T., et al. 2004, Astrophys. J., 613, L133
9. Casares, J. 2005, astro-ph/0503071
10. Chen, W., Shrader, C.R., Livio, M. 1997, Astrophys. J., 491, 312
11. Colbert, E.J.M., Mushotzky, R.F. 1999, Astrophys. J., 519, 89
12. Di Salvo, T., Stella, L., Robba, N.R., et al. 2000, Astrophys. J., 544, L119
13. Di Salvo, T., Robba, N.R., Iaria, R., et al. 2001, Astrophys. J., 554, 49
14. Done, C., Wardziński, G., Gierliński, M. 2004, Mon. Not. Roy. Astron. Soc., 349, 393
15. Ebisawa, K., Ogawa, M., Aoki, T., et al. 1994, Publ. Astron. Soc. Jpn., 46, 375
16. Fabbiano, G. 1989, ARAstron. Astrophys. 27, 87
17. Fabbiano, G., Zezas, A., Murray, S.S. 2001, Astrophys. J., 554, 1035
18. Fabian, A.C, Rees, M.J., Stella, L., et al. 1989, Mon. Not. Roy. Astron. Soc., 238, 729
19. Fabian, A.C., Iwasawa, K., Reynolds, C.S., et al. 2000, Publ. Astron. Soc. Pac., 112, 1145
20. Frontera, F., Palazzi, E., Zdziarski, A.A., et al. 2001, Astrophys. J., 546, 1027
21. Giacconi, R., Gursky, H., Paolini, F., et al. 1962, Phys. Rev. Lett., 9, 439
22. Gierliński, M., Zdziarski, A.A., Poutanen, J., et al. 1999, Mon. Not. Roy. Astron. Soc. 309, 496
23. Gilfanov, M., Revnivtsev, M., Molkov, S. 2003, Astron. Astrophys., 410, 217
24. Grove, J.E., Johnson, W.N., Kroeger, R.A., et al. 1998, Astrophys. J., 500, 899
25. Honma, F., Matsumoto, R., Kato, S. 1991, Publ. Astron. Soc. Jpn., 43, 147
26. King, A.R., Davies, M.B., Ward, M.J., et al. 2001, Astrophys. J., 552, L109
27. Kluźniak, W., Abramowicz, M.A. 2002, astro-ph/0203314
28. Kong, A.K.H., McClintock, J.E., Garcia, M.R., et al. 2002, Astrophys. J., 570, 277
29. Kubota, A., Makishima, K., Ebisawa, K. 2001, Astrophys. J., 560, L147
30. Kubota, A., Makishima, K. 2004, Astrophys. J., 601, 428
31. Laor, A. 1991, Astrophys. J., 376, 90
32. Lewin, W.H.G., van Paradijs, J., Taam, R.E. 1995, in W.H.G. Lewin, J. van Paradijs, E.P.J. van den Heuvel (eds.), X-Ray Binaries, Cambridge University Press, Cambridge, 175
33. Lightman, A.P., Eardley, D.M. 1974, Astrophys. J., 187, L1
34. Liu, Q.Z., van Paradijs, J., van den Heuvel, E.P.J. 2001, Astron. Astrophys., 368, 1021
35. Liu, Q.Z., Mirabel, I.F. 2005, Astron. Astrophys., 429, 1125
36. Makishima, K., Kubota, A., Mizuno, T., et al. 2000, Astrophys. J., 535, 632
37. Matsumoto, H., Tsuru, T.G., Koyama, K., et al. 2001, Astrophys. J., 547, L25
38. McClintock, J.E., Remillard, R.A. 1986, Astrophys. J., 308, 110
39. McClintock, J.E., Remillard, R.A. 2006, in W.H.G. Lewin, M. van der Klis (eds.), Compact Stellar X-Ray Sources, Cambridge University Press, Cambridge, 157
40. McClintock, J.E., Narayan, R., Rybicki, G.B. 2004, Astrophys. J., 615, 402
41. Miller, J.M., Wijnands, R., Homan, J., et al. 2001, Astrophys. J., 563, 928
42. Miller, J.M., Fabian, A.C., Wijnands, R., et al. 2002, Astrophys. J., 578, 348
43. Miller, J.M., Fabian, A.C., Wijnands, R., et al. 2002, Astrophys. J., 570, L69
44. Miller, J.M., Fabian, A.C., in't Zand, J.J.M., et al. 2002, Astrophys. J., 577, L15
45. Miller, J.M., Fabian, A.C., Reynolds, C.S., et al. 2004, Astrophys. J., 606, L131
46. Mineshige, S., Kawaguchi, T., Takeuchi, M., et al. 2000, Publ. Astron. Soc. Jpn., 52, 499
47. Miniutti, G., Fabian, A.C., Miller, J.M. 2004, Mon. Not. Roy. Astron. Soc., 351, 466
48. Mirabel, I.F., Rodríguez, L.F. 1999, ARAstron. Astrophys., 37, 409

49. Mitsuda, K., Inoue, H., Koyama, K., et al. 1984, Publ. Astron. Soc. Jpn. 36, 741
50. Miyamoto, S., Kimura, K., Kitamoto, S., et al. 1991, Astrophys. J., 383, 784
51. Nakamura, N., Dotani, T., Inoue, H., et al. 1989, Publ. Astron. Soc. Jpn., 41, 617
52. Nandra, K., George, I.M., Mushotzky, R.F., et al. 1997, Astrophys. J., 477, 602
53. Narayan, R., McClintock, J.E., Yi, I. 1996, Astrophys. J., 457, 821
54. Narayan, R., Barret, D., McClintock, J.E. 1997, Astrophys. J., 482, 448
55. Oppenheimer, J.R., Snyder, H. 1939, Phys. Rev., 56, 455
56. Ohsuga, K., Mineshige, S., Mori, M., et al. 2002, Astrophys. J., 574, 315
57. Ohsuga, K., Mori, M., Nakamoto, T., et al. 2005, Astrophys. J., 628, 368
58. Orosz, J.A., Groot, P.J., van der Klis, M., et al., 2002, Astrophys. J., 568, 845
59. Orosz, J.A., Polisensky, E.J., Bailyn, C.D., et al. 2002, BAAS, 201, 1511
60. Orosz, J.A., McClintock, J.E., Remillard, R.A., et al. 2004, Astrophys. J., 616, 376
61. Remillard, R.A., Muno, M.P., McClintock, J.E., et al. 2002, Astrophys. J., 580, 1030
62. Remillard, R.A., Muno, M.P., McClintock, J.E., et al. 2003, AAS/HEAD vol. 7
63. Remillard, R.A., McClintock, J.E., Orosz, J.A., et al. 2006, Astrophys. J., 637, 1002
64. Reynolds, C.S., Nowak, M.A. 2003, Phys. Rep., 377, 389
65. Romani, R.W. 1998, Astron. Astrophys., 333, 583
66. Rutledge, R.E., Bildsten, L., Brown, E.F., et al. 1999, Astrophys. J., 514, 945
67. Rutledge, R.E., Bildsten, L., Brown, E.F., et al. 2001, Astrophys. J., 551, 921
68. Rutledge, R.E., Bildsten, L., Brown, E.F., et al. 2001, Astrophys. J., 559, 1054
69. Shakura, N.I., Sunyaev, R.A. 1973, Astron. Astrophys., 24, 337
70. Shakura, N.I., Sunyaev, R.A. 1976, Mon. Not. Roy. Astron. Soc., 175, 613
71. Shimura, T., Takahara, F. 1995, Astrophys. J., 445, 780
72. Sobczak, G.J., McClintock, J.E., Remillard, R.A., et al. 1999, Astrophys. J., 520, 776
73. Strohmayer, T.E. 2001, Astrophys. J., 552, L49
74. Sunyaev, R.A., Titarchuk, L.G. 1980, Astron. Astrophys., 86, 121
75. Suzuki, K., Matsuoka, M., Inoue, H., et al. 1984, Publ. Astron. Soc. Jpn., 36, 761
76. Tanaka, Y., Lewin, W.H.G., 1995, in W.H.G. Lewin, J. van Paradijs, E.P.J. van den Heuvel (eds.), X-Ray Binaries, Cambridge University Press, Cambridge, 126
77. Tanaka, Y., Nandra, K., Fabian, A.C., et al. 1995, Nature, 375, 659
78. Tanaka, Y., Shibazaki, N., 1996, ARAstron. Astrophys., 34, 607
79. Tanaka, Y. 1997, in E. Meyer-Hofmeister, H. Spruit (eds.), Accretion Disk – New Aspects, Springer, Berlin Heidelberg New York, 1, Lecture Notes in Physics 487
80. van der Klis, M. 2006, in W.H.G. Lewin, M. van der Klis (eds.), Compact Stellar X-ray Sources, Cambridge University Press, Cambridge, 39
81. Watarai, K., Mizuno, T., Mineshige, S. 2001, Astrophys. J., 549, L77
82. Webster, B.L., Murdin, P. 1972, Nature, 235, 37
83. White, N.E., Marshall, F.E. 1984, Astrophys. J., 281, 354
84. Yamaoka, K., Ueda, Y., Inoue, H. 2002, in Ph. Durouchoux, Y. Fuchs, J. Rodriguez (eds.), New Views on Microquasars, Center for Space Physics, Kolkata (India), 326
85. Zdziarski, A.A. 2000, in P.C.H. Martens, S. Tsuruta, M.A. Weber (eds.), Highly Energetic Physical Processes and Mechanisms for Emission from Astrophysical Plasmas, IAU Symposium 195, 153
86. Zhang, S.N., Cui, W., Chen, W. 1997, Astrophys. J. Lett. 482, L155
87. Bhattacharyya, S., Strohmayer, T.D. 2007, Ap7664, L703
88. Cackett, E.M., Miller, J.M., Bhattacharyya, S. et al. 2007, arxiv: 0708.3615

17 X-Ray Studies of Supernovae and Supernova Remnants

R. Petre

17.1 Introduction

Supernova remnants (SNRs) hold an important place in the history of X-ray astronomy, being associated with many observational "firsts." The first radio/optical object associated with an X-ray source was the Crab Nebula. Cassiopeia A (Cas A) was the first X-ray source for which an X-ray spectrum showed evidence for line emission. SNRs were among the first extended X-ray sources to be imaged, and with their complex, filamentary structure, they are also among the most photogenic X-ray sources.

This review attempts to capture the most important insights gained into the physics of SNRs during the past few years of X-ray observations. Before doing so, the stage will be set by briefly describing SNR evolution, the significance of X-ray emission, and some of the key results from earlier X-ray observations. Interestingly, many of the fundamentally important recent results were hinted at by early observations (such as ionization nonequilibrium and nonthermal emission from shock accelerated electrons); bringing them to light required the enhanced capabilities of the most recent generation of observatories. This review focuses on the extended remnants, and includes little about either their compact central objects or pulsar wind nebulae. These objects merit an entire review in and of themselves.

This review also addresses X-ray observations of supernovae (SNe). This is relatively new area of X-ray studies, and for the most part, the X-ray emission mechanism of the SNe detected thus far is largely the same as that at work in their older cousins.

17.1.1 X-Ray Emission from SNRs

A SNR is the result of the interaction between the debris from an exploded star and its surrounding medium. The explosion can either arise as the detonation or deflagration of a white dwarf that has exceeded the Chandrasekhar limit as the result of mass accretion from a binary companion (Type Ia) or the gravitational collapse of the core of a massive star that has exhausted its nuclear fuel (Type II, Type Ib, and Type Ic). In the early stages of the remnant evolution, the differences between

J.E. Trümper and G. Hasinger (eds.), *The Universe in X-Rays.*

these explosions can lead to morphological differences, but these begin to blur as the remnant becomes increasingly dominated by the surrounding medium. In both cases the outer layers of the star are ejected with velocity of tens of thousands of kilometers per second. Interaction between the outer envelope and the ambient medium causes the development of a shock front that heats, compresses, and ionizes ambient gas. Undecelerated ejecta catching up with the shell of shocked material produce a reverse shock, so called because it propagates backwards in the ejecta frame of reference (though not in the observer's frame). A contact discontinuity separates the material swept up by the two shocks. For the first few hundred years, during the "free expansion" phase, the forward shock propagates essentially at constant velocity. Once the shock has swept up a mass comparable to that of the ejecta it starts to decelerate, signaling the onset of the so-called adiabatic evolutionary phase. In this phase the cooling time of swept up material exceeds the dynamic time scale, and so the remnant loses very little energy. Its idealized behavior is described by the Sedov–Taylor self-similarity equations. A recent, comprehensive analytic description of SNR evolution under various conditions can be found in [159].

During these two phases, X-ray emission is produced from the collisionally-ionized, shock heated plasma. Both shocks have sufficient velocity to heat the shocked material to high kinetic temperatures; subsequently, collisions with electrons ionizes the gas to predominantly their H-like and He-like states. X-rays are emitted as Bremsstrahlung continuum from the collisionally-heated electrons, plus a rich spectrum of lines from the materials composing the ejecta and the swept-up medium. During the initial phase, lines from the dense ejecta dominate, allowing X-ray observations to probe the properties of the progenitor star via the ejecta abundances and distribution, but as the material in the remnant becomes predominantly swept-up interstellar material, the line spectrum reflects the composition of the interstellar medium. Nonthermal components from electron synchrotron emission arise in central pulsar powered wind nebulae and at the forward shock from electrons accelerated to relativistic energies.

Once the radiative and dynamic timescales become comparable, the remnant enters the radiative or shell forming phase. Newly shocked material radiates its thermal energy and gets swept up into a thin, dense, cool shell. During this phase the shock velocity is too low to produce new X-ray emitting gas, but as described later, X-ray emission still arises from the interior.

17.1.2 Early SNR X-Ray Astrophysics

SNRs were observed by numerous early rocket experiments, largely because they are among the brightest objects in the sky at 1 keV. The spatial extent of these bright, soft objects facilitated the first structural studies via scanning observations, and so SNRs were the first cosmic X-ray sources whose morphology was investigated and compared with other bands. The Crab Nebula, though not a canonical SNR, figures prominently in early X-ray astronomy, in part because it is by far the brightest X-ray SNR. It was the first cosmic X-ray source identified with an optical counterpart, through a lunar occultation experiment involving a sounding rocket instrument [16]. Additionally, its pulsar was the first object seen to pulse in X-rays [35]. Soft X-ray rocket experiments detected and mapped in one and two-dimensions the emission from the large angular diameter Vela and Cygnus Loop remnants. Notable discoveries by early rocket experiments included the first detection of line emission from a cosmic X-ray source (Fe in Cas A; [141]); the first suggestion of a nonequilibrium plasma as the emission source [39]; and the first true SNR images (e.g., the Cygnus Loop; Fig. 17.1, [126]).

Early satellites provided higher quality spectra, as well as a few additional maps. The low energy concentrator on SAS-3 mapped the soft X-ray emission from large remnants such as Vela and Lupus, and discovered X-rays from SN 1006 (e.g., [50]). The HEAO-1 A2 proportional counters were used to study emission lines, especially Fe K, in the spectra of bright, young remnants (e.g., [123]), and made the first detection of hard spectral tails [122]. Prior to the launch of Einstein, about a dozen X-ray emitting SNRs were known.

As with many other classes of X-ray sources, the Einstein Observatory revolutionized the study of supernova remnants. Approximately 40 of the 150 Galactic X-ray remnants known at that time were detected by Einstein. Einstein images revealed that remnants have a variety of X-ray morphologies (centrally filled, composite, shell-like) as in the radio band, but that the X-ray and radio morphologies do not necessarily match [143]. One of Einstein's most notable contributions was the extension of X-ray study of SNRs beyond the galaxy, exemplified by the cataloging of remnants in the Magellanic Clouds (e.g., [100]).

While Einstein's imaging capability contributed substantially to our understanding of the structure of remnants, its spectral capability provided insight into the physics. The focal plane crystal spectrometer (FPCS), though it studied only a handful of objects, produced several important results. It resolved the lines in Cas A, providing the first X-ray detection of Doppler shifts and evidence of an asymmetric explosion [97]. The measurement of strong oxygen lines in Puppis A indicated that the progenitor star mass was larger than 25 $M_\odot$ [22]. The solid state spectrometer provided strong evidence of ejecta enhancement and nonequilibrium ionization conditions in the spectra of young SNRs (e.g., [9, 10]).

ROSAT, through its All Sky Survey, observed every SNR in the sky, making possible a flux-limited census of soft X-ray emission from SNRs. It detected 70 known Galactic SNRs and identified numerous additional SNR candidates [138]. It revealed distinctive new objects, such as the radio-dim remnants G165.1+5.7 and RX J1713.7-3946 and previously unknown features associated with known remnants (the Vela "shrapnel" [6] and G189.6+3.3, overlapping IC 443 [4]). It facilitated study of the largest solid angle remnants (e.g., Monogem [120]). Pointed observations with the PSPC and HRI provided high quality images of dozens of remnants, allowing spatially resolved spectroscopy, and facilitating in conjunction with older Einstein images the first X-ray expansion rate measurements.

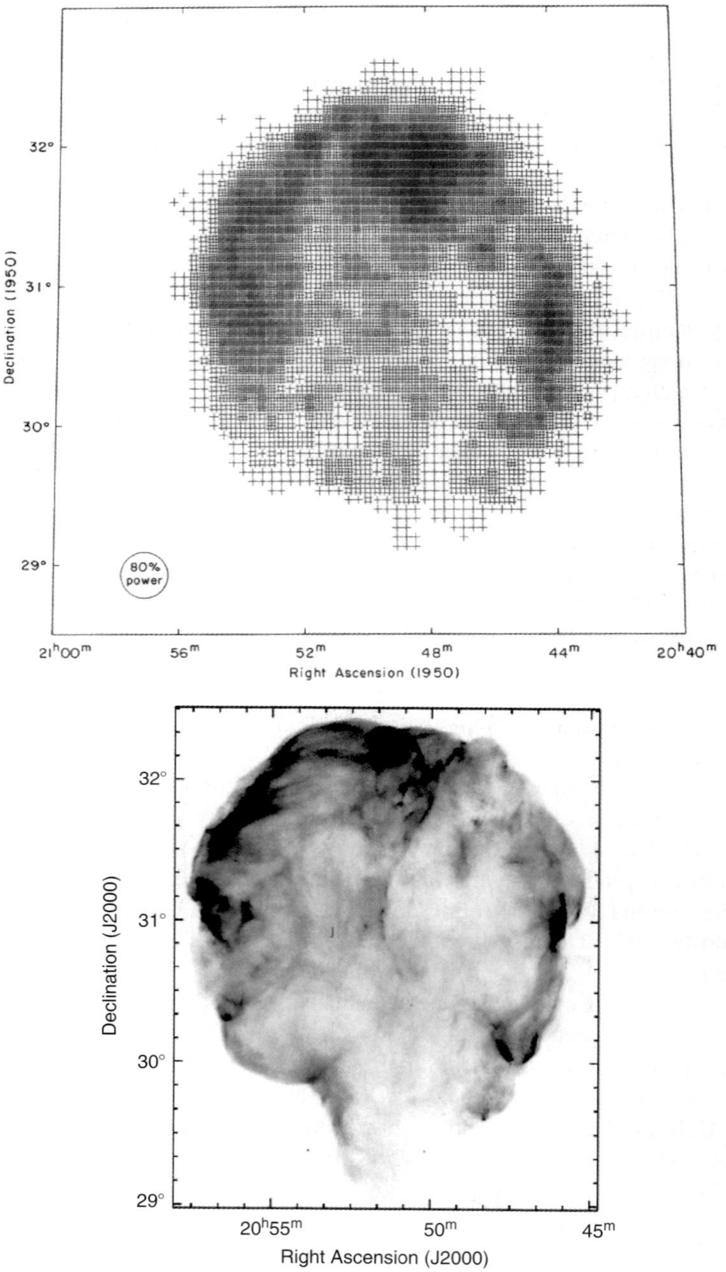

Fig. 17.1 Two images of the Cygnus Loop demonstrate the progress made in X-ray imaging capability. *Top*: A 300 s exposure from a sounding rocket instrument, the first true nonsolar X-ray image utilizing a Wolter-type mirror [126]. *Bottom*: A ROSAT HRI mosaic, with 5 arcsec resolution and an integrated exposure time of a million seconds (courtesy N. Levenson)

The numerous contributions of ROSAT, along with the subsequent contributions by ASCA, and the ongoing contributions by XMM-Newton and Chandra, are described in more detail further.

17.1.3 Supernovae

In contrast to the rich history of SNR observations through the Einstein era, observations of supernovae were sparse. Einstein's contribution was meager but important: it provided the first detection of a supernova soon after outburst. SN 1980K in the nearby star-forming galaxy NGC 6946 was detected 35 days after outburst but soon faded from view, becoming invisible 47 days after outburst [21]. The only other supernova to be detected in X-rays prior to the ROSAT era was SN 1987A. The mechanism responsible for the early X-ray emission in this object is fundamentally different from that of other X-ray detected SNe as explained later.

Before returning to supernovae, we first discuss the results of recent X-ray observations of supernova remnants.

17.2 Young SNRs

17.2.1 Ejecta Abundances, Distribution, and Ionization Structure in Young SNRs

The fact that the X-ray emission in young SNRs is heavily influenced by the presence of ejecta heated by a reverse shock has been known since the earliest spectroscopic measurements with moderate resolution spectrometers [9, 10]. It has also been known that their X-ray spectra are influenced by ionization conditions, specifically that the low density, shock-heated plasma is less ionized than one in collisional ionization equilibrium (CIE) at the same electron temperature. Models of so-called nonequilibrium ionization (NEI) plasmas preceded our ability to detect them in nature [44, 74].

As the fluxes from prominent X-ray lines are stronger in an NEI plasma than a CIE plasma with the same electron temperature, low and modest resolution spectrometers often yield ambiguous results between ionization state and abundance. Early measurements of SNR spectra were unable to distinguish between NEI and enhanced abundances (e.g., [9, 123]). The ability to accurately infer plasma conditions is further complicated by the fact that SNRs show a range of temperatures and ionization states (and abundance mixes) often along the same line of sight. This renders interpretation of spectra challenging. As a consequence, few, if any, X-ray measurements of abundances in SNRs are definitive, and all must be viewed critically. On the other hand, spectral measurements or narrow band equivalent width maps showing relative abundances can be instructive in understanding the distribution of ejecta in SNRs.

One desired use of X-ray abundances is for determining explosion type. The ejecta from core collapse SNe and deflagration/detonation (Type Ia) SNe have distinct abundance patterns. Specifically, the ejecta from core collapse SNe have high ratios compared with cosmic abundances of oxygen to iron; those from Type Ia's have the opposite. X-rays are the ideal band for seeking this signature, as broad band X-ray spectrometers cover the K lines from oxygen and iron, as well as the iron L band. The complications of abundance determination make SN typing challenging. Compounding the difficulty is our inability to know what fraction of the ejecta is visible in the X-ray band, and the fact that the oxygen abundance is difficult to determine in an unambiguous way as the oxygen line strength correlates with the column density in spectral fitting in moderate spectral resolution detectors.

Despite these difficulties, there have been some notable successes in measuring abundances, and deriving physical information from the measurement. The most successful was the use of Einstein FPCS observations of Puppis A to show an overabundance of oxygen with respect to iron, requiring the progenitor to have a mass of more than 25 $M_\odot$ [22]. Also of note are comprehensive analysis of Tycho data requiring ejecta [46] and the analysis of an EXOSAT observation of W49B requiring the presence of a substantial amount of shocked ejecta [151]. It is with the advent of spatially resolved spectroscopy through which shock structures can be isolated that most of the advances have taken place in our knowledge of the abundances of reverse shocked ejecta and forward shocked ISM. ASCA was the pathfinder mission in this regard, but XMM-Newton and Chandra excel in these studies. The recent studies are providing real insight into the ejecta masses and their degree of mixing, as well as the explosion mechanism.

17.2.1.1 Type Ia SNRs

There are only two confirmed Galactic Type Ia remnants, Tycho (Fig. 17.2) and SN 1006 (Fig. 17.10). In Tycho, recent spatially resolved observations have removed the ambiguity of the initial spectral results, from which the strong flux of lines from Si, S, and Ar could be interpreted as either due to enhanced abundances from ejecta or NEI [9, 123]. Circumstantial evidence for the presence of ejecta comes from the high mass of X-ray emitting material (5–15 $M_\odot$) required if solar abundance material is assumed [142]. Hamilton, et al. [46] analyzed all existing X-ray spectral data using a self-consistent model incorporating a forward shock encountering a uniform ISM and a reverse shock encountering stratified ejecta. They showed that satisfactory fits require the presence of ejecta, and that the total inferred ejecta mass is consistent with the 1.4 $M_\odot$ expected from a Type Ia explosion. Using Ginga data, Tsunemi et al. showed that the centroids of the brightest lines in Tycho (and Cas A) require NEI, and included NEI effects to infer abundances substantially higher than solar [161].

Broad band model fitting of the ASCA spectrum revealed the relative ejecta abundances, and spatially resolved spectroscopy revealed their distribution [64, 66]. The narrow band images in Mg, Si, S, Ar, Ca, Fe, and the continuum images all share a shell-like morphology, but each has distinct features. The Si and S ejecta have an average temperature of $(0.8\text{–}1.1)\times10^{7}$ K, and an average ionization age of $(0.8\text{–}1.3)\times10^{11}\ \mathrm{cm}^{-3}$ s. Azimuthal brightness variations in the two lines suggests a variation in the temperature of $\sim$1.5. The azimuthally averaged Fe K image peaks well within the other line images. The Fe K line radial surface brightness distribution and the centroid energy both indicate that the Fe ejecta have a temperature several times higher and an ionization age several times lower than the other ejecta. The non-Fe ejecta have abundances in good qualitative agreement with the predictions of standard SN Ia models. The apparently low abundance of Fe is consistent with the idea that the Fe ejecta are interior to the Si group ejecta, and largely unshocked.

Higher angular resolution views using Chandra and XMM-Newton refine this picture, but do not alter it in any fundamental way. The XMM-Newton image shows that the Fe K emission peaks at a smaller radius than the Fe L emission, verifying that the temperature with the ejecta increases toward the reverse shock [28]. The narrow band Si image corresponds well with the radio image, and probably marks the contact discontinuity, distorted by Rayleigh–Taylor instabilities. This latter conclusion is reinforced in dramatic fashion by the Chandra image of the Si emission in Tycho, which shows plume like structures throughout the interior [61, 171]. Some of the plumes viewed tangentially approach the outer shock.

The line emission in SN 1006 has been a secondary consideration to the nonthermal emission arising from the bright limbs (see Sect. 17.2.6 below). Line emission is clearly observed throughout the remnant, except in the bright nonthermal limbs. Along the northwestern rim, Chandra imaging spectroscopy shows a clear separation between the forward-shocked material and the ejecta [96]. The forward shock shows material at ordinary solar abundances, shock-heated to electron temperatures of $\sim$0.6–0.7 keV. Interior to both the nonthermal northeast shock and the thermal northwest shocks are clumpy structures similar to those observed in Tycho. Their presence invite the speculation that such structure is common in Type Ia remnants. Spectral analysis of these structures reveals enhanced O, Mg, Si, and Fe abundances. No quantitative X-ray based analysis of the ejecta mass has been performed for SN 1006. Abundance measurements in SN 1006 using X-rays are compromised by the known presence of a substantial amount of high-velocity, unshocked ejecta (Fe, Si, S, and O) interior to the reverse shock [179].

The dearth of Galactic Type Ia remnants is compensated by the large Magellanic Cloud (LMC), in which at least three others have been identified. Hughes et al. used ASCA spectra from three LMC SNRs to demonstrate how remnants can be typed using their broad-band X-ray spectra [57]. Two of these remnants, 0509-67.5 and 0519-69.0, were previously thought to be Type Ia remnants, based on their Balmer-dominated optical emission. The third, N103B, was thought to be a core collapse remnant, but its spectrum more closely resembles that of a Type Ia remnant. Subsequent studies of these remnants using Chandra and XMM-Newton have facilitated

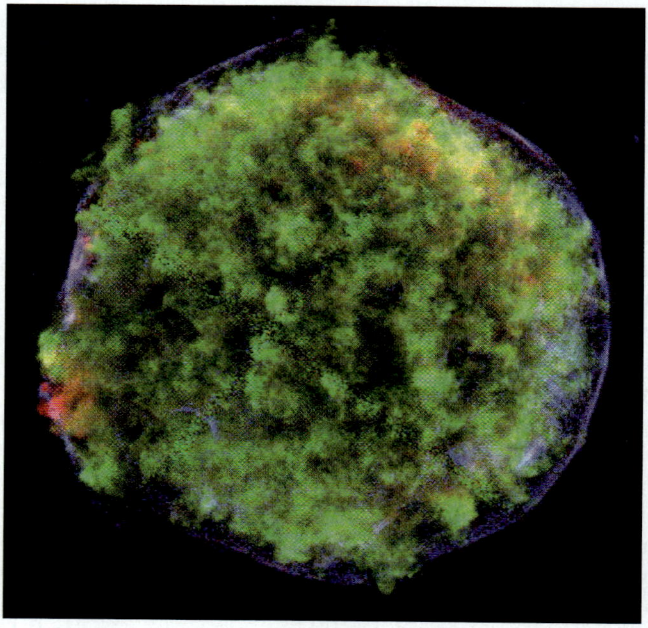

Fig. 17.2 Three-color composite Chandra image of Tychos SNR reveals a thin, hard outer rim that is the site of particle acceleration (Sect. 17.2.6) and plumes of ejecta-rich material in the projected interior, which are likely shaped by the Rayleigh–Taylor instability [171]. The *red*, *green*, and *blue* images correspond to photon energies in the 0.95–1.26, 1.63–2.26, and 4.1–6.1 keV bands. The angular diameter of Tycho's remnant is approximately 8 arcmin, corresponding to physical size of 5.8 pc at a distance of 2.5 kpc

spatially resolved spectroscopy of these distant objects with an equivalent spatial resolution to that provided by ASCA for Galactic remnants. Lewis et al. found that N103B has an ejecta distribution similar to Tycho: a bright shell dominated by Si and S, with an interior shell of hot Fe [95]. Interior to the hot Fe shell is a core of cooler Fe. The estimated masses of Si, S, Ar, Ca, and Fe do not match the predictions of any SN model, but are more consistent with a Type Ia model than a Type II. In contrast, the O, Si, and Fe mass estimates from an XMM-Newton RGS observation are markedly different and support a Type II origin [164]. Neither paper proposes a resolution to this discrepancy.

The LMC remnant 0509-67.5 also has a largely stratified structure, with ejecta confined to a shell interior to a shell of nonthermal emission [172]. The integrated spectrum shows clear evidence for enhanced metal abundances. Narrow band images reveal knots with Si and Fe abundances higher than the global average. Variations in the Fe and Si abundances from knot to knot suggest that they originated in the transition region between the Fe and Si layers of the progenitor. Inferred ejecta masses for the preferred model are 0.17 $M_{\odot}$ for O; 0.76 $M_{\odot}$ for Si; 0.37 $M_{\odot}$ for S; and 0.12 $M_{\odot}$ for Fe. These masses are shown to be more consistent with the yields

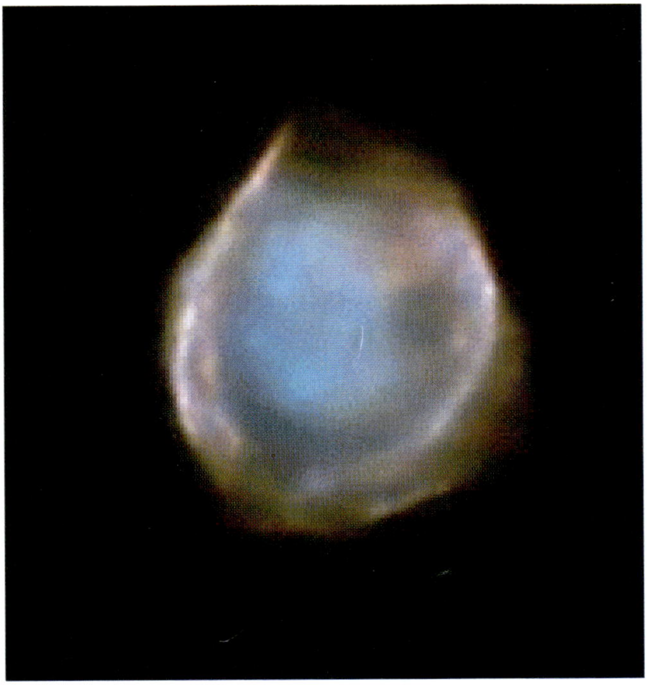

Fig. 17.3 The LMC remnant DEM L71 has an outer forward shock structure and an interior filled with Si and Fe rich material [56]. The color scale is: *red* 0.3–0.7 keV, *green* 0.7–1.1 keV, *blue* 1.1–4.2 keV. The angular extent of the remnant is approximately 1.4 by 1.2 arcmin, corresponding to 20 pc $\times$ 17 pc (Figure courtesy of Chandra X-ray Center and J. P. Hughes.)

for a delayed deflagration model for a Type Ia supernova than for a fast deflagration. The lack of Fe can be accounted for if the reverse shock has not propagated far into a cold, interior Fe layer.

DEM L71 is a much older LMC SNR; with an age $\geq$5 000 yrs, compared with less than 1 000 yrs for 0509-67.5 and $\sim$1 500 yrs for N103B. Nevertheless, it shows dramatic evidence of stratification [56]. In particular, its central core is markedly iron rich (Fig. 17.3). Despite its advanced age, the remnant preserves the two-shock structure observed in many young remnants. The X-ray spectra indicate a total ejecta mass of $\sim$1.5 $M_\odot$, most of which is composed of Fe.

17.2.1.2 Core Collapse Remnants

While Type Ia remnants universally seem to maintain stratification within their ejecta, the same cannot be said for core collapse remnants. They show a variety of ejecta structures, from highly stratified to nearly chaotic. Of course, Cas A, the most closely studied, shows the most anomalous structure.

Cassiopeia A: Narrow band ASCA images were the first to reveal the general ejecta distribution in Cas A [53]. The Si and S maps are virtually indistinguishable from each other or from the broad band map, but clearly different from the hard continuum. Narrow band images from the BeppoSAX MECS show clearly the difference between the line and hard continuum morphologies. Vink et al. used these data to estimate an ejecta mass of $4\,M_\odot$, and an X-ray emitting mass of $14\,M_\odot$, indicating the progenitor had a very high mass loss rate [170].

The first Chandra images revealed clear and unexpected differences in the distribution of metals (e.g., [65]). The Si, S, Ar, and Ca maps are similar to each other and to the distribution of fast optical knots. Equivalent width maps reveal the distribution of the prominent ejecta constituents. The structures in these maps contrast sharply with the 0.5–10.0 keV broadband map and the 4–6 keV continuum map. The northeastern jet, known from optical studies to contain Si group ejecta, shows up strongly in these maps. The Fe K emission has a very different morphology. In particular, in the southeast of the remnant, the Fe K emission is located at larger radii than the Si. This suggests that the inner Fe ejecta layers have been overturned and propelled beyond the Si group ejecta in this part of the remnant (Fig. 17.4). Such overturning is consistent with recent models of core collapse explosions (e.g., [17]).

Willingale et al. have used the XMM-Newton imaging data to infer the global metal abundance ratios and compare them with supernova models [178]. They show that the ratios with respect to Si of a large number of lines is most consistent with the theoretical nucleosynthesis yield for a $12\,M_\odot$ progenitor.

Detailed studies of individual knots show that they have a variety of compositions. Features with distinct composition can be found on the smallest size scales. While most knots show a mix of ejecta, some are dominated by Si group elements and others by Fe. At least one knot emits Fe lines exclusively, and apparently is devoid of lower mass material. The knots also show a variety of ionization conditions, which have been used in the context of analytic hydrodynamical models to constrain the ejecta density profile, the location of the knots in mass coordinates, and the degree of explosion asymmetry [67,90]. The ejecta show a range of density profiles, from very shallow ($\rho \propto r^{-n}$, where $n \leq 6$) to very steep, ($n \sim 30$–50). The ejecta close to the jet show the shallowest profile, possibly due to an asymmetric explosion in which more of the energy is directed along the jet than elsewhere. For a total ejecta mass of $2\,M_\odot$ expected from a $20\,M_\odot$ progenitor, the Fe-rich clumps are found to arise in a layer 0.7–0.8 $M_\odot$ from the center. The observed composition appears to be possible only if Si burning products are mixed with O burning products.

The overall appearance of Cas A contrasts starkly with the young Type Ia remnants. Cas A consists of small knots and thin filaments, not the emission plumes observed in Tycho. The similarity between the structures in Cas A and the prediction of models involving Fe bubbles has been noted. Laming and Hwang argue that the knots are not especially overdense compared with their surroundings, and their

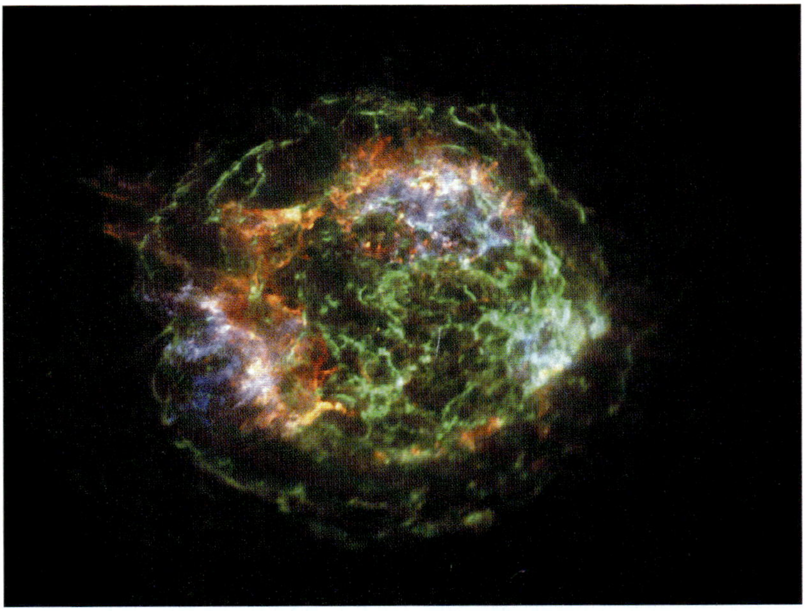

Fig. 17.4 Multicolor image of Cas A from a million-second Chandra observation reveals a variety of structures [62]. The *red* represents Si Heα (1.78–2.0 keV), the *blue* Fe K (6.52–6.95 keV), and the *green* 4.2–6.4 keV continuum. In the east, the Fe ejecta are external to the Si, indicating overturning of the inner Fe ejecta layers. The hard outer rim arises from shock-accelerated electrons; projected interior hard emission has a nonthermal bremsstrahlung component. The angular diameter is approximately 5 arcmin, corresponding to 5 pc at a distance of 3.4 kpc

high ionization ages and the proximity of some to the forward shock are the result of early passage through the reverse shock [90].

1E 0102.2-7219: The Small Magellanic Cloud remnant 1E 0102.2-7219 is another young core collapse remnant, with an estimated age of about 1 000 yrs. ASCA observations show that its emission is dominated by four lines (He-α and Lyα lines of O and Ne; [49]). Narrow band Chandra images show the remnant to be a nearly perfect ring, with a hotter, outer ring identified with the forward shock, and a cooler, inner ring of reverse-shocked ejecta [37]. Model fits to the ASCA and XMM-Newton composite spectra require multiple NEI components [49,137]. Observations using the grating instruments on Chandra and XMM-Newton have revealed the ionization structure and the distribution of the elements [34, 127]. The brightness distribution peaks of the Lyα lines from O, Ne, and Mg are exterior to those of the corresponding He-α lines, and there is a progression from smaller to larger radii of the brightness peaks of O, Ne, Mg, and Si. A simple model with constant electron temperature but variable ionization timescale accounts for the peak emission radius of all lines, suggesting that the radial structure is due to progressive ionization of a mixed plasma rather than stratification. From the Chandra grating observations,

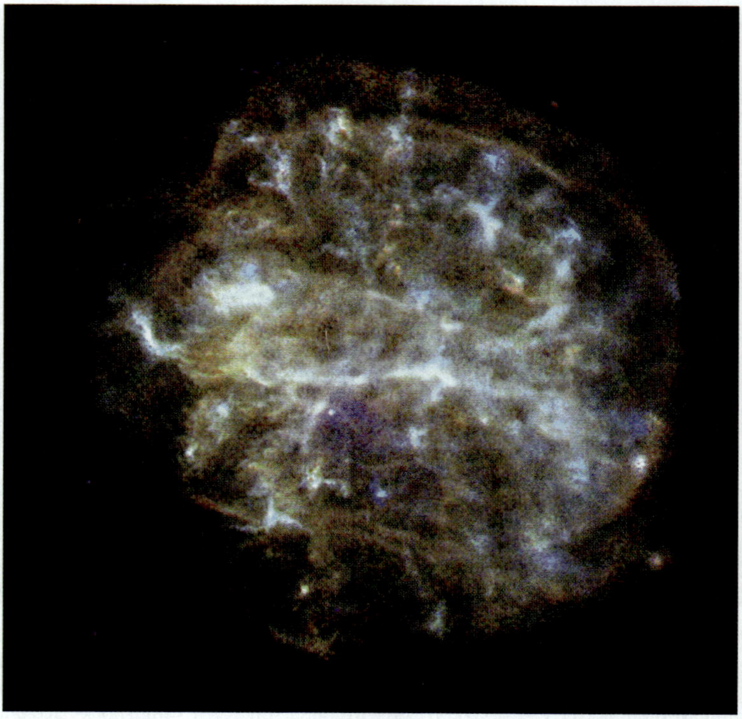

Fig. 17.5 Multicolor Chandra image of the O-rich SNR G292.0+1.8 has a filamentary appearance similar to that of Cas A, but has a very different atomic composition. In this image, enriched ejecta are *blue* and normal composition material is *yellow*. The central pulsar wind nebula also appears in *blue*. The angular diameter of the remnant is 6 arcmin, corresponding to a physical size of $\sim$11.5 pc at a distance of approximately 6 kpc (Figure courtesy of Chandra X-ray Center and J. P. Hughes.)

the mass of oxygen is estimated at $\sim 6\,M_\odot$, and the neon mass is estimated to be $\sim 2\,M_\odot$. The oxygen mass is consistent with a massive progenitor of $\sim 32\,M_\odot$.

G292.0+1.8: The young ($\leq$1 600 yrs) SNR G292.0+1.8 was thought to be a Type II remnant based on its O-rich optical spectrum. Its confirmation as such resulted from the discovery using ASCA of a central synchrotron nebula and X-ray pulsar near the center [60]. The Chandra image of the remnant, shown in Fig. 17.5, reveals a thin, nearly circular outer shell of hard emission filled with an array of knots and filaments rivaling Cas A in complexity and contrasting starkly with the Type Ia remnants [111]. The composition and distribution of the shocked ejecta are different from Cas A. The ejecta consist primarily of O, Ne, and Si, with less S and Ar, and very little Fe, and are distributed primarily around the remnant's periphery. An X-ray bright equatorial band has normal composition, and is thought to be associated with presupernova mass loss.

17.2.1.3 SNRs of Uncertain Type

Kepler: Possibly the most vexing puzzle concerning young SNRs is the nature of the progenitor of Kepler's supernova (SN 1604). Strong arguments have been made for either a Type Ia or a core collapse origin. Key evidence supporting a Type Ia origin is the object's large distance (600 pc) above the Galactic plane. The most compelling evidence of a Type II origin is the apparent presence of high-density material surrounding the object, thought to be the result of presupernova mass loss from the progenitor star. No compact source has been found.

The integrated ASCA spectrum shows strong emission from the K lines of Mg, Si, S, and Ca, and the L and K lines of Fe [82]. Iron appears to be in a low ionization state, with the bulk of the emission arising from Fe L lines. A model invoking a forward shock expanding freely into CSM and a reverse shock into ejecta was needed to satisfactorily fit the spectrum. High abundances of all metals are required, with the Fe abundance approximately ten times solar. Much of the Fe is thought to be still interior to the reverse shock, and so mixing of this material with exterior layers is needed to account for the Fe K line strength. A total ejecta mass of $\sim 1.4\,M_\odot$ is inferred. The relative abundances and total ejecta mass are consistent with a Type Ia origin.

Narrow band XMM-Newton images show a symmetric distribution of material, resembling Tycho more closely than Cas A [23]. The Fe L and Si images are generally consistent with each other, though the Si emission extends to a slightly larger radius. The Fe K emission peaks interior to the Fe L emission, suggesting an inward temperature gradient. There seems to be no substantial azimuthal asymmetry; azimuthal surface brightness variations are likely associated with density variations in the circumstellar medium. Detailed analysis of the ejecta abundances from Chandra and XMM-Newton observations have not yet been published.

W49B: W49B has one of the most impressive spectra of any cosmic X-ray source, with multiple strong Si, S, and Fe emission lines [151]. Early ASCA analysis suggested that the Si and S emission arose from a shell surrounding the centrally peaked Fe [36], but subsequent, more detailed analysis indicated similar morphologies for the three line complexes [68]. Nevertheless, each element requires a unique set of plasma conditions, although all are close to collisional ionization equilibrium. In fact, Kawasaki et al. suggest that the S line ratios require the plasma in the remnant to be recombining [78]. A high density is required; a minimum of $2\,\mathrm{cm}^{-3}$ is required, and the density is likely higher than this. Spectroscopy using XMM-Newton confirms the need for high density material but not a recombining plasma [102]. Both the ASCA and XMM-Newton spectra show features consistent with K lines from Cr and possibly Mn; this is the only remnant to show such lines.

The inferred abundances are high: the Si, S, Ca, Ar, Mg, and Fe all have abundance ratios to solar of 5–6. While the relative abundances support a Type Ia progenitor, a low mass Type II progenitor is consistent with the measured abundances, leaving the explosion type uncertain. On the other hand, the high-density environment in which W49B is expanding suggests a Type II progenitor. The high column to the remnant renders the O and Ne lines invisible, rendering impossible a direct O:Fe abundance comparison, and therefore a definitive determination. Kawasaki et al. argue that despite its apparent youth and anomalously high abundances, the morphology, thermodynamic state, and proximity to molecular clouds place W49B squarely in the class of mixed morphology remnants [78].

17.2.2 Identification of Shock Structures

In the young remnants, even the highest quality images prior to Chandra lacked the resolution or sensitivity to isolate shock structures. Many of these structures had been elusive in other bands as well.

The most prominent example is Cas A. The radio profile consists of a bright ring, thought to be associated with the reverse shock, and an outer plateau. The plateau emission shows no evidence of limb brightening, thus leaving the location of the forward shock unknown. The inner ring is so complicated that it is not possible to determine the location of the reverse shock. The highly filamentary optical emission was not useful for identifying shock structures. The initial deep Chandra image provided the first clear location of Cas A's forward and reverse shocks [41]. The 4–6 keV continuum image revealed a series of thin wisps at radii between 140 and 165 arcsec from the nominal center of the remnant. The correspondence of the outer edge of these X-ray filaments with a large jump in radio polarization indicates that they mark the previously unidentified forward shock. The reverse shock is marked by a correspondingly sharp rise of the emissivity of the X-ray line emission at 1.85 keV and the radio surface brightness, at an approximate radius of 1.8 arcmin. The approximate 3:2 diameter ratio of the forward and reverse shocks provides a means of assessing the evolutionary state of Cas A. Comparison with models (e.g., [159]) indicates that mass swept-up by the shock is comparable with that of the ejecta, and that Cas A is just entering the adiabatic phase.

Chandra and XMM-Newton images have also revealed quite dramatically the structure of the forward shocks in the other young SNRs. In Tycho, portions of the forward shock had been identified optically as an arc of nonradiative Hα filaments. The Chandra image reveals similarly thin X-ray filaments [61, 171]. As with Cas A, these appear most prominently in the 4–6 keV continuum band. Unlike those in Cas A, the emission is nearly continuous around the remnant, although its surface brightness does vary azimuthally by factors of several over scales of degrees. The radio rim corresponds closely with the X-ray rim, but the surface brightness of the two does not vary in any correlated way. As in Tycho, the X-ray emission from the bright NE and SW rims in SN 1006 corresponds closely with the radio and has a dramatically well-defined outer edge [96].

17.2.3 Equipartition of Ions and Electrons

While time dependent ionization has become a commonplace assumption in the fitting of the thermal spectra from SNRs, most results tacitly assume equipartition between the ion and electron temperatures.[1] Ions are instantaneously heated to a high kinetic temperature as a result of passage through the shock. Models of initial electron heating cover the full range between no heating ($T_e/T_i \sim m_e/m_i$) and instantaneous equipartition ($T_e = T_i$). The primary process by which electrons are thought to equilibrate is Coulomb collisions, but plasma processes can cause faster electron heating. In typical SNR shocks the collision timescale is long, and equipartition can require an appreciable fraction of the remnant life if it relies solely on Coulomb collisions. The bremsstrahlung continuum underlying the line emission in a SNR spectrum provides a direct measure of the electron temperature; since the electron temperature is expected to evolve behind the shock and vary with shock speed, an accurate measurement requires isolation of a small region of interest. Direct measurement of the ion temperature entails measurement of the thermal broadening of spectral lines. The ion temperature can also be inferred indirectly from the expansion velocity, provided there is not another channel into which shock energy can flow, such as particle acceleration.

While even the earliest time dependent ionization models of SNR spectra considered the effects of nonequipartition (e.g., [44]), X-ray measurements sensitive to the degree of equipartition came only with the high spectral and spatial resolution capabilities of Chandra and XMM-Newton. The primary reason is that previous observatories lacked the angular resolution to isolate regions near the forward shock, or the spectral resolution to sensitively measure the line ratios that allow inference of the ion temperatures. X-ray measurements are only now catching up to the better established optical or UV techniques. Rakowski summarizes all measurements of degree of equilibration [124]. Despite studies of several remnants, SN 1006 is the only one for which the degree of equilibration between particles has been well established. X-ray measurements have been made of SN 1006, DEM L71, Tycho, SN 1987A, and 1E 0102.2-7219. While each has large uncertainties, the measurements (Fig. 17.6) support a trend consistent with measurements in other bands of an inverse proportionality between shock velocity and degree of equilibration, with the youngest remnants (SN 1987A, Tycho, SN 1006) having a low electron-to-proton temperature ratio and the oldest (Cygnus Loop) having ratios consistent with equilibration [125].

Of special note are observations of two remnants: SN 1006 and DEM L71. In SN 1006, an XMM-Newton RGS spectrum was obtained from an isolated knot along the thermal northwestern limb [169]. The RGS measurement of the broadening of O V, O VI, and O VII lines implies a temperature for the O ions of 530 ± 150 keV. The simultaneous EPIC spectrum of the same region yields an electron temperature of 1.5 keV. An upper limit to $T_e/T_p < 0.05$ is obtained. This value is

[1] Fitting of moderate resolution X-ray spectra from the CCD detectors on Chandra and XMM-Newton is insensitive to the degree of equipartition.

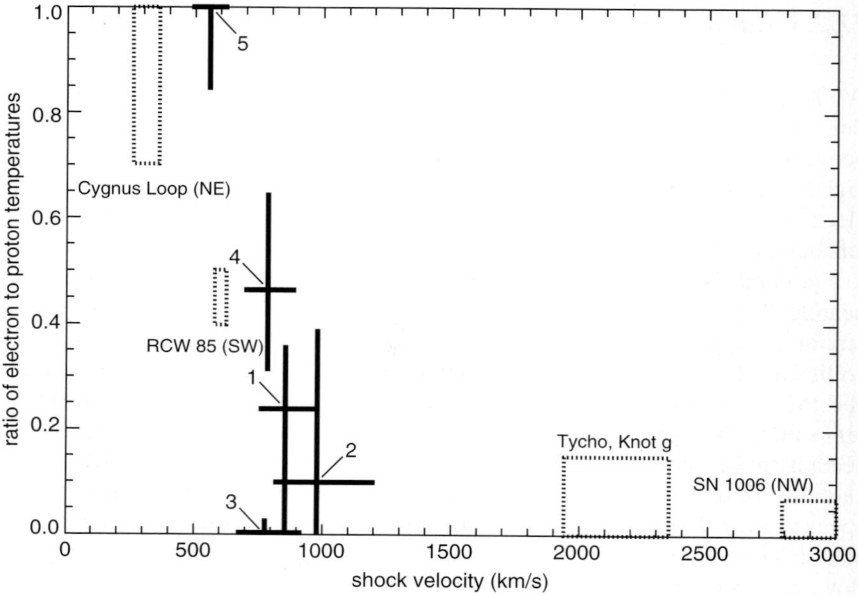

Fig. 17.6 Measurements to date indicate an inverse correlation between the degree of electron–proton equilibration and shock velocity [125]

consistent with optical and UV measurements of equilibration along the northwestern shock front in SN 1006. For DEM L71, the Chandra ACIS was used to extract spectra of three narrow strips at five positions along and immediately behind the forward shock. From these strips, the evolution of the electron temperature in different locations around the remnant was measured. A proton temperature measured optically was assumed. Three regions show results consistent with no equilibration. One region, with the slowest shock speed, is consistent with complete equilibration. The fifth region, with a shock speed similar to or slightly smaller than the regions with no equilibration, shows partial equilibration ($T_e/T_p \sim 0.5$).

17.2.4 Kinematics

17.2.4.1 Expansion Measurements

The availability of X-ray images of SNRs with angular resolution approaching those in the radio and optical made possible the first X-ray measurements of remnant expansion. For a remnant with an average forward shock velocity of 5 000 km s^{-1} at a distance of 3 kpc, the average angular expansion rate is 0.35 arcsec per year. Thus even with the few arc second angular resolution of the early high resolution imagers (Einstein and ROSAT) it became possible to measure global expansion rates of young, nearby remnants over the combined 10–15 yrs baseline spanned by these missions.

The first such measurement was facilitated by a deep ROSAT HRI image of Cas A taken in 1996, that could be used as a template. Koralesky et al. and Vink et al. independently used this observation together with archival ROSAT and Einstein HRI images to perform the first X-ray expansion rate measurements [87, 167] . They both found an average expansion rate of $(0.20 \pm 0.01)\%$ per year. The two studies differed on the presence of an azimuthal variation. The expansion rate corresponds to an expansion timescale of $\sim$500 yrs, which can be compared with the $\sim$340 yrs age of Cas A (in 1998). This difference is incompatible with the prediction of a model in which the remnant emission is dominated by ejecta interacting with a circumstellar shell created by a presupernova wind.

Early measurements of global expansion revealed a curious conundrum. In Cas A, Kepler and Tycho, the expansion rate measured in X-rays was roughly twice that measured in the radio [54, 55]. For example, the expansion rate of Kepler from the X-ray measurement of 0.239% per year is to be compared with the radio measurement of 0.125% per year. In all three remnants, representing different explosion types and ages, the radio and X-ray features are interspersed, and so a common explanation entailing coincidental surperposition of X-ray filaments passing through the radio filaments is highly unlikely. For Cas A, the problem was compounded by the fact that the cospatial, fast-moving optical knots have an expansion rate 50% still higher than the X-ray. The difference was ascribed to different bands measuring distinct hydrodynamical structures (e.g., [87]), but begged the question of what band measures the true expansion rate.

Detailed kinematic studies using Chandra have helped resolve the issue. DeLaney et al. compared ACIS images of Cas A taken just 2 yrs apart [29]. Their difference map reveals a wide variety of proper motion speeds and directions. They were able to measure the proper motion of hundreds of individual features in addition to measuring the overall expansion of the remnant. They found that the motions throughout the remnant are complex, and cannot be modeled with a homologous expansion (e.g., [159]). The continuum-dominated filaments at the forward shock show a range of expansion rates from 0.02 to 0.33% per year. The median expansion rate of 0.2% per year can be compared with the expected free expansion rate of 0.3% per year to demonstrate that most of the forward shock has undergone significant deceleration. While X-ray ejecta are moving twice as fast as the cospatial radio knots, matched individual X-ray/radio features have the same velocity. Their interpretation of this complex situation is that the small-scale X-ray and radio features in the bright ring (which dominates the emission and thus the global proper motion studies) represent ejecta at various stages of deceleration after passage through the reverse shock. This establishes a velocity gradient from the densest knots, which emit primarily in the optical band, to the radio features which are the most decelerated. Features in all three bands show brightness evolution, however, potentially complicating this straightforward interpretation [30].

In contrast, a similar, preliminary analysis of the X-ray proper motions in Kepler from a pair of Chandra images reveals a mean expansion rate of 0.15% per year, consistent with the radio proper motion [134].

The proper motion study of Cas A also revealed three different classes of X-ray features, separable by their spectral and/or kinematic properties [30]. The Si and Fe dominated ejecta and continuum dominated forward shock filaments have roughly the same proper motion (a mean expansion rate of 0.2% per year), but distinct spatial distributions. This suggests the two components are somehow dynamically coupled. The ejecta knots are associated spatially with the optical fast moving knots, but have been significantly decelerated relative to them, probably as a consequence of lower density. A slow-moving component (0.05% per year) with enhanced low energy emission is apparently associated with the optical quasi-stationary flocculi, and thus might represent a clumpy circumstellar component. Thus, the power of Chandra is beginning to allow us to reconstruct the details of the presupernova environment of Cas A.

Chandra observations have also made possible the first expansion measurement of an extragalactic SNR. By comparing Chandra and ROSAT HRI observations taken 15 yrs previously, Hughes et al. measured the expansion rate of 1E 0102.2-7219 in the SMC of 0.100 ± 0.025 percent per year, corresponding to a shock velocity of 6 000 km s^{-1} [58].

17.2.4.2 Doppler Measurements

Measurements of radial velocities via Doppler shifts of X-ray lines in SNRs are challenging for even the current generation of instrumentation. While dispersive spectrometers such as crystals and gratings have adequate resolution to resolve velocity broadening produced by motion of a few thousand kilometers per second expected in young SNRs, the motion gets blurred by the finite angular size of the remnants. Nondispersive imaging spectrometers generally lack the spectral resolution to detect the broadening or line centroid shift resulting from radial motion. In only three instances have Doppler shifts been detected: Cas A, as a result of its asymmetric ejecta distribution; 1E 0102.2-7219, because of small size and its well-defined ring shape; and SN 1987A, because of its very small angular size. SN 1987A is discussed in Sect. 17.5.1.

The first measurement of Doppler motions in Cas A was performed using the Einstein FPCS. Markert et al. detected broadening and asymmetry in the Si and S lines [97]. They detected a systematic redshift of the northwest half of the remnant with respect to the southeast, with a mean velocity difference of $1\,820 \pm 290$ km s^{-1}. Additionally, within each region a Doppler broadening of 5 000 km s^{-1} was detected. They suggested that the X-ray emission is concentrated in a ring inclined to the plane of the sky, with an expansion velocity greater than 2 000 km s^{-1}.

Subsequent results have largely substantiated these conclusions. Holt et al. constructed the first X-ray Doppler image of Cas A, using the ASCA SIS to measure spatially dependent shifts in the peak of the 1.85 keV Si Heα line [53]. Their map

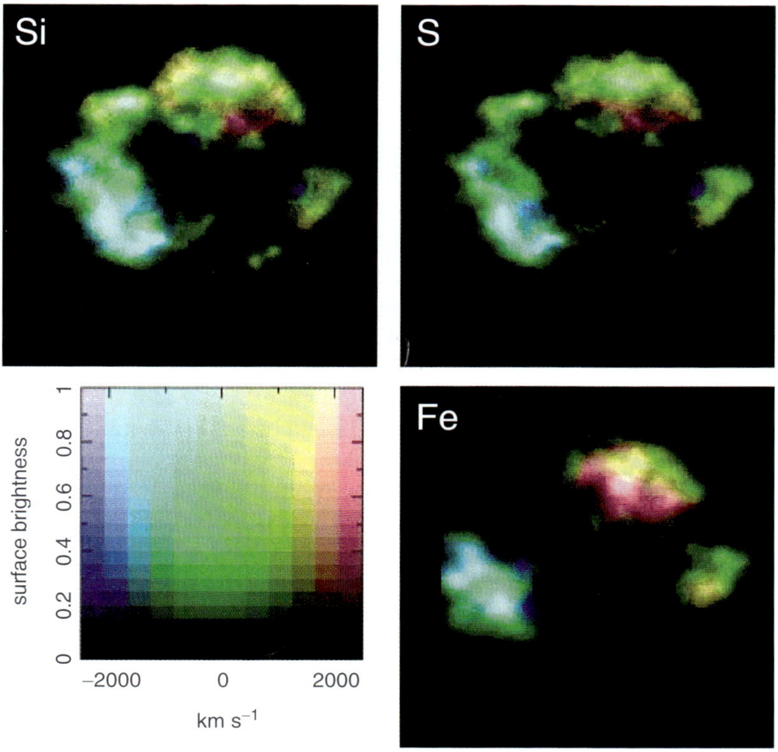

Fig. 17.7 Doppler maps of Si–K, S–K, and Fe–K emission lines in Cas A, extracted from XMM data [178]. The surface brightness of the line emission is color coded with the Doppler velocity using the scheme shown in the bottom left panel

confirmed the Markert et al. finding, but with substantially higher spatial detail. The X-ray emission geometry closely matches that observed in [S II]. Subsequent Chandra and XMM-Newton Doppler maps provided higher resolution views of the radial velocity structures [69, 178]. Willingale et al. used the XMM-Newton Doppler map to reconstruct a three-dimensional view of the Cas A X-ray emission (Fig. 17.7). They infer that the Si and S emission arises exclusively from ejecta and is confined to a narrow shell with radius 100–150 arc seconds. The Fe–K emission is confined in two large clumps, expanding faster and spanning a larger radius range of 110–170 arcsec. They liken these clumps to the ejecta bullets observed in Vela [6].

Flanagan et al. used the Chandra HETG to measure the radial velocities of the line emitting components in the SMC remnant 1E 0102.2-7219, along with their ionization structure (see Sect. 17.2.1.2) [34]. By measuring the different distortions among the dispersed line images, they found Doppler shifts consistent with bulk velocities of $\sim$1 000 km s^{-1}. As for Cas A, the bulk velocities suggest that the emission is confined to a ring, inclined to the line of sight.

In contrast to these results, spatially resolved spectroscopy of Tycho indicates the absence of a significant systematic Doppler shift, suggesting that the remnant has a spherical geometry [64].

17.2.5 Jets and Shrapnel

Cas A has long been known to have a "jet" of fast moving knots, with an extent of more than twice the outer shock radius. This knot was discovered optically; spectroscopy suggests that the knots composing it are rich in ejecta. Proper motion studies plus spectroscopy indicate that the motion of the jet is largely perpendicular to our line-of-sight. The optical spectra of the jet knots are dominated by [S II] emission, indicating they are ejecta. The jet is not prominent in the radio band. Early high quality X-ray images suggested low surface brightness emission coincident with the jet, but lacked the resolution or surface brightness contrast to allow detailed correlation study. The initial deep Chandra image reveals low surface brightness X-ray emission from the entirety of the jet with a striking resemblance to the optical image, especially in Si, S, Ar, and Ca [65].

The presence of a counterjet on the west side of the remnant was indicated by the presence of a large number of fast moving optical knots [33]. The initial deep Chandra image showed Si streamers in that direction, but lacked the sensitivity to unequivocally relate them to a counterjet. As shown in Fig. 17.8, the narrow band Si

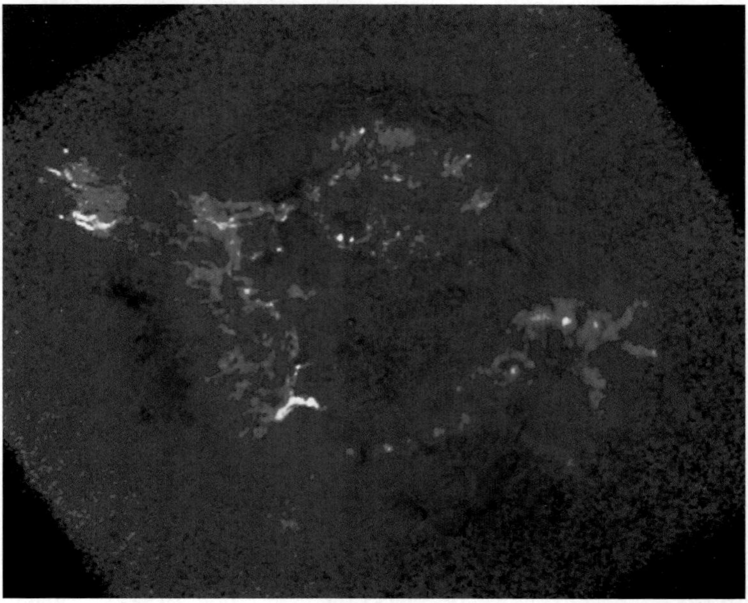

Fig. 17.8 The jet and counterjet extending to the east and west of Cas A, traced by the Si line emission in the million-second Chandra image [62]. The jet is revealed here via a ratio image between Si Heα (1.78–2.0 keV) and 1.3–1.6 keV (Mg Heα, Fe L)

map extracted using the recent million second Chandra exposure reveals the clear presence of the counterjet [62].

There is no clear explanation for the orientation of the jet. Models of jet-induced explosions generally predict a jet directed along the rotation axis of the progenitor star and a kick velocity imparted to the neutron star along that same axis (e.g., [80]). In Cas A, however, the jet axis and proper motion direction of the compact source are nearly perpendicular [157].

Fast moving ejecta knots are not found exclusively in young SNRs. One of the most spectacular results from the ROSAT all-sky survey was the discovery of half a dozen bright features well exterior to the shell of the nearby (250 kpc), evolved ($\sim 10^4$ yrs) large angular diameter ($\sim 8.3°$) Vela supernova remnant (Fig. 17.9) [6]. Each feature possesses morphology either like a truncated cone (fragments A and E) or an arc opening toward the remnant center (fragments B, C, D, and F). Simple geometrical arguments indicate a common origin, near the location of the Vela pulsar. These features were interpreted as dense fragments of the exploded star, or shrapnel, that have been decelerated less than the forward shock, and are now producing bow shock nebulae and Mach cones as they interact with ambient medium beyond the remnant boundary.

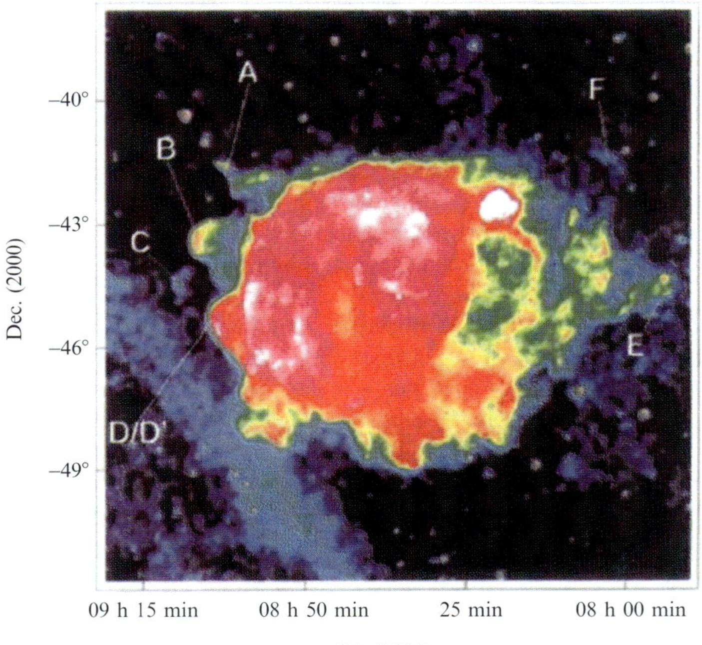

Fig. 17.9 The ROSAT All Sky Survey image of the Vela SNR showing the location of the shrapnel [6]

Detailed followup studies have largely confirmed this interpretation. Radio observations of fragment A show nonthermal emission along its leading edge, indicating particle acceleration at the shock front. Detailed X-ray observations of fragment A have been performed using ASCA and Chandra [104, 160]; XMM-Newton observations were performed on fragment D [77]. Fragment A was found to have a Si/O abundance ratio of 10 or more times the solar ratio, a mass on the order of 0.01 $M_\odot$, supporting an origin in the progenitor star. It has a velocity of about $500\,\mathrm{km\,s^{-1}}$, requiring substantial deceleration from its original velocity of $\sim 10^4\,\mathrm{km\,s^{-1}}$. In fragment D, a complex morphology is found, suggesting the onset of a Rayleigh–Taylor instability between the fragment and an interstellar cloud with which it is colliding. The X-ray spectrum suggests that the fragment is rich in O, Ne, and Mg, but not in Fe. The different composition suggests that fragment D originated in a different layer of the progenitor from fragment A.

Similar features have been found in at least one other SNR. A Chandra image of the evolved, core-collapse remnant N63A in the LMC reveals crescent-shaped plumes exterior to the remnant proper, similar to those associated with Vela [173]. Spectral analysis indicates that the plumes are not dominated by ejecta material, suggesting that if they are of similar origin as the Vela fragments, then they have undergone substantial mixing with ambient material.

17.2.6 Hard, Nonthermal Continua and Cosmic Ray Acceleration

It has long been thought that the shocks in supernova remnants are the primary sites of Galactic cosmic ray acceleration. This hypothesis is supported observationally by the presence of nonthermal radio emission from SNR shells arising from electrons accelerated to GeV energy. Theoretical models of diffusive shock acceleration (1st order Fermi acceleration) indicate that efficient acceleration can occur within SNR shocks. The signature of cosmic ray acceleration is subtle. Both electrons and protons have signatures in the TeV band, but this has been largely inaccessible, and is only now becoming open to sensitive observations. Synchrotron radiation from electrons accelerated to TeV energy falls in the X-ray band. Most bright SNRs have strong thermal emission from the candidate acceleration regions, and so finding the signature of TeV electrons required instruments with good angular resolution and a broad spectral band. It also required the existence of a fortuitous object, SN 1006.

SN 1006, the remnant of the brightest historical supernova, is located at high galactic latitude, in a region of apparently low ambient density. It is one of two (and possibly three) historical remnants of Type Ia SNe. The first broad band X-ray spectral observations showed a spectrum dominated by continuum, with no Fe K lines, in contrast to other young SNRs [11]. This led to competing interpretations, one requiring shock accelerated electrons [129], the other invoking continuum thermal emission from fully ionized carbon [45]. The first imaging data showed the remnant to have bright NE and SW rims, whose spectrum is considerably harder than elsewhere in the remnant. The hardness and brightness of the rims were problematic for a thermal model because one would expect brightness to correlate with density squared and temperature (hardness) to correlate inversely with density.

The required observational breakthrough came from ASCA, which showed that the emission from the bright rims was essentially free of line emission, while the interior radiated line emission from elements characteristic of Type Ia ejecta. The only plausible explanation for the hard continuum emission from the bright rims is synchrotron emission from electrons shock accelerated to TeV energy [89].

Chandra and XMM-Newton observations have provided clearer views of the SN 1006 rims (Fig. 17.10). The Chandra observations show that the hard, nonthermal rims follow the radio emission extremely closely indicating a single emission mechanism is responsible for both [96].

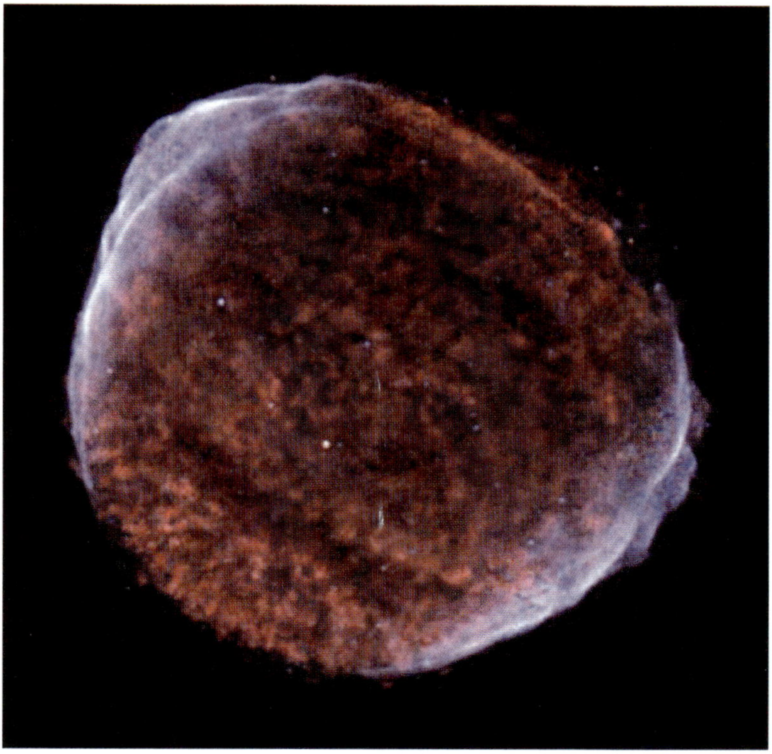

Fig. 17.10 Multicolor ACIS image of SN 1006 dramatically contrasts the hard, nonthermal emission from the NE and SW rims with the thermal emission from the bulk of the remnant. The nonthermal X-ray filaments correlate strongly with the radio filaments, indicating they arise from the same emission measure. The projected interior thermal emission shows plume-like structures like those in Tycho, the other known historical Type Ia remnant 17.2. The remnant is $\sim$30 arcmin across, correpsonding to a physical diameter of 18 pc at a distance of 2.1 kpc (Figure courtesy of Chandra X-ray Center and J. P. Hughes.)

Since the SN 1006 discovery, nonthermal emission has been discovered from a number of SNRs. These remnants fall into two groups with different emission characteristics. The first group comprises the majority of the historical remnants, including Cas A, Tycho, and Kepler. The X-ray emission is dominated by line and continuum from the forward and reverse shock-heated thermal plasma. Nonthermal emission has been detected in two ways. First, observations at energies above the band dominated by thermal emission has revealed hard tails. Hard tails extending to 10 keV can be identified in ASCA spectra, but the spectra do not have sufficient band pass or signal-to-noise to establish whether they are thermal or nonthermal. The RXTE spectra in the 10–20 keV band can be fit by power laws but are subject to multiple interpretations, as no morphological information is available [114]. Figure 17.11 shows a collection of RXTE spectra of young SNRs. Subsequently, Chandra and XMM-Newton resolved thin, hard outer rims, the spectra of which show evidence for a nonthermal continuum component. The isolation of the hard emission within these outer rims is a strong argument in favor of the shock acceleration interpretation.

In Cas A, there is a substantial contribution to the hard emission from the projected interior. This is likely to arise from nonthermal bremsstrahlung, as the interior magnetic fields are sufficiently high to reduce the maximum electron energy well

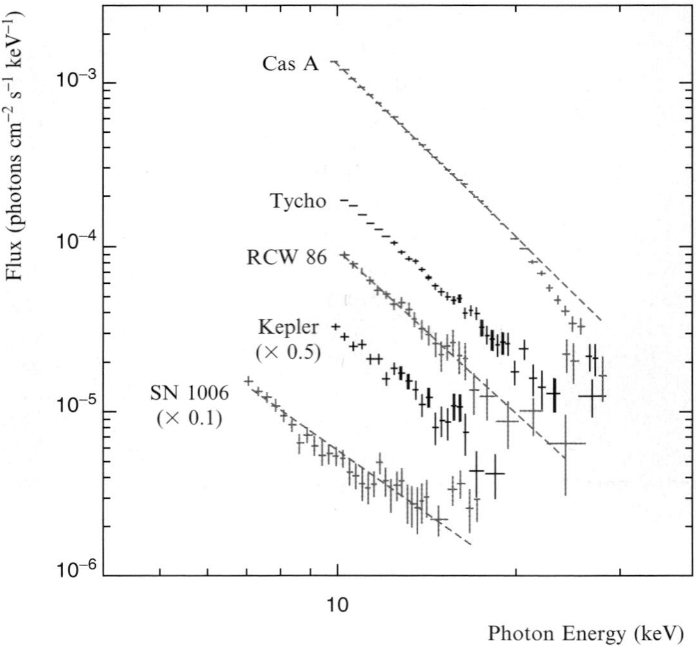

Fig. 17.11 RXTE PCA spectra of five historical SNRs with hard nonthermal tails probably produced by synchrotron emission from TeV electrons. Each spectrum is characterized by a power law with spectral index of approximately 2 [114]

below the limit for producing X-ray synchrotron emission [13, 168]. In contrast, the hard X-ray emission from the thin outer rim is likely to be synchrotron radiation from accelerated electrons. Vink and Laming use the width of the rim to estimate that the magnetic field near the shock front has a strength of 80–160 μG and maximum electron energies of ~40–57 TeV [168]. The magnetic field is weak compared with the average, but strong compared to the expectation for shocked ISM, suggesting either the progenitor wind had a high field or post shock field amplification has occurred.

The hard X-ray properties of RCW 86, thought to be the remnant of SN 185, contrast sharply with those of the other historical remnants. In the northeastern quadrant, images reveal a nearly continuous shock structure. Oddly, one segment is dominated by thermal emission and an adjacent one by nonthermal [166]. In the southwestern quadrant, where the remnant is interacting with an ISM structure, a hard emission region is located well behind the shock front, and is correlated with K line emission from low ionization states of Fe [132]. Rho et al. use Chandra data to show that the hard emission is associated with the reverse shock, and that it is likely synchrotron emission and not nonthermal bremsstrahlung. The presence of this emission requires the reverse shock to accelerate electrons to order 50 TeV.

The second group of remnants comprises evolved remnants with shell-like morphology, dominated by nonthermal emission. In addition to SN 1006, prominent members of this group are two remnants discovered in the ROSAT All-Sky Survey, RX J1713.7-3946 and RX J0852.0-4622. A possible fourth member is RXJ 04591+5147. RX J1713.7-3946 shows no evidence whatever for a thermal component [24], and the evidence in RX J0852.0-4622 is marginal [149]. These remnants share other important properties. They have low radio surface brightness, which correlates strongly with the X-ray morphology. TeV γ-ray emission from RX J1713.7-3946 (Fig. 17.12) and RX J0852.0-4622 by HESS shows a dramatic correlation with the X-ray surface brightness [1, 2].[2] The combination of X-ray and radio properties can be explained by expansion in a very low-density medium, in which little interstellar material is swept up. Consequently, the forward shock is not substantially decelerated and a strong reverse shock never forms. The combination of high forward shock velocity and low density results in a long synchrotron lifetime of the X-ray emitting electrons. Interestingly, both RX J1713.7-3946 and RX J0852.0-4649 have stellar remnants unlike the Ia remnant SN 1006, suggesting that the evolution into this class is not dependent on the type of the explosion.

In both groups of remnants, the similarity between the nonthermal X-ray features and the radio morphology indicates that the same emission process is responsible for both. The X-ray flux is substantially below that expected from an extrapolation of the radio spectrum, however. Thus if one spectrum fits both the radio and X-ray emitting particles, it must steepen between the radio and X-ray bands. There are ample mechanisms for producing a turnover: particle losses due to diffusion, the limitations on maximum particle energy from the age of the SNR, and synchrotron

[2] The detection of TeV emission from SN 1006, in contrast, is uncertain, with a claimed CANGAROO detection at odds with a more sensitive HESS upper limit [3, 156]

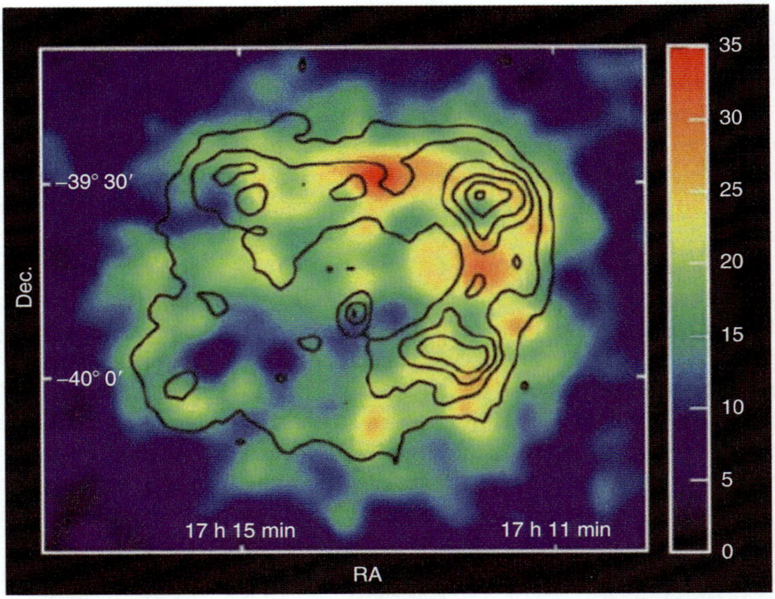

Fig. 17.12 In the nonthermal X-ray emission dominated remnant RX J1713.7-3946, there is a strong correlation between the TeV γ-ray (*colors*) and X-ray (*contours*) morphology [1]. The TeV γ-ray image is from HESS; the X-ray image is from ASCA

losses by the particles (as the more energetic particles radiate energy faster). Current hard spectra do not allow discrimination among the different mechanisms, but assuming any of them yields approximately the same turnover energy of the particle spectrum. For a sample of bright, young SNRs, the turnover photon energy is typically a few tenths of keV, which for reasonable magnetic fields corresponds to a turnover energy in the electron spectrum of a few TeV [130]. While the maximum electron energy falls far short of the "knee" in the cosmic ray spectrum generally associated with the highest energy Galactic cosmic rays, it is not clear how this energy relates to the maximum energy of protons, the dominant cosmic ray particle.

Indirect evidence for hadronic cosmic ray acceleration in SNR shocks in young remnants comes from comparison between theoretical models of shock structure and sensitive X-ray measurements, enabled by XMM-Newton and Chandra. Hydrodynamic modeling indicates that the channeling of shock energy into particle acceleration modifies shock structure in three observable ways: first, the postshock compression ratio can become higher than four; second, the distance between the forward shock, the contact discontinuity, and the reverse shock becomes compressed; and, third, the proton temperature is suppressed below would be expected from the shock velocity [27].

For the SMC remnant 1E 0102.2-7219, Hughes et al. found a significant discrepancy between the electron temperature inferred from spectral fitting of Chandra ACIS data (0.4–1.0 keV) and that inferred from their measurement of the shock velocity, assuming any plausible degree of electron heating (2.5–45 keV) [58]. From this they argued that a significant fraction of the shock energy is channeled elsewhere than heating the postshock ions and electrons. They concluded that this channel is the acceleration of cosmic rays. A similar analysis yielded the same conclusion for Tycho [61].

Also for Tycho, Warren et al. used Chandra data to measure the average distances between the forward and reverse shocks and the contact discontinuity [171]. The distance between the forward shock and contact discontinuity is smaller than hydrodynamic models predict, and can be explained if acceleration of ions is occurring in the forward shock. The nonthermal spectrum arising in the forward shock supports this conclusion. In contrast, both the distance between the contact discontinuity and the reverse shock and the thermal spectrum indicate that particle acceleration is not occurring in the reverse shock.

There is no firm evidence of hard, nonthermal emission from shock-accelerated electrons in older, thermal emission dominated SNRs. While neither Chandra nor XMM-Newton has fully mapped any of the prominent evolved Galactic remnants, the carefully studied regions of the shells of remnants such as Puppis A and the Cygnus Loop show no excess hard emission.

17.2.7 Stellar Remnants

The textbook picture of a core collapse supernova remnant includes an expanding blast wave sweeping up and shock heating ambient material and a central, rapidly spinning pulsar whose strong magnetic field interacts with its surroundings to produce a pulsar wind nebula (PWN). During the Einstein era, the most vexing mystery about SNRs was the dearth of such "composite" remnants. Even the closest Galactic match, the Vela remnant, deviated from the model in having no detected X-ray pulsations. The Crab Nebula, considered the prototype pulsar wind nebula, fails to satisfy the model because it lacks a thermally emitting shell. One of the great successes of the X-ray imaging missions, starting with Einstein, but brought to fruition by more recent missions, has been revealing how naïve this preconception was.

While a few pulsars with Crab-like properties were detected using Einstein, including fast pulsars in MSH 15-*52* [144] and the LMC remnant 0540-69.3 [146], most newly discovered central objects do not fit the paradigm. 1E 2259+586 in the center of the remnant CTB 109 was found to have an 8 s pulsation period [32]. The stellar remnants 1E 1613-509 in RCW 103 and 1E 1207.4-5209 in PKS 1209-52 were found to have modest luminosity, and no evidence for either pulsations or counterparts in other bands [51, 162]. None of these objects have an associated PWN. Of the roughly 50 Galactic remnants observed by Einstein, stellar remnants were known or found in 11 [143], leaving open the question of where (and what) the central stellar objects are.

The rich solution to this mystery came with ROSAT, ASCA, Chandra and XMM-Newton. ROSAT revealed a central stellar object in the ~4 000-year-old Puppis A remnant with a low X-ray luminosity, no radio or optical counterpart or PWN, and no pulsations[3] [115]. A discrete source was found in the LMC remnant N49, and associated with the bright γ-ray burst from that direction [98]. The Vela pulsar was found to pulse, but in soft X-rays only, suggesting that the pulsations arise from surface hot spots at the poles [107]. ROSAT discovered a shell around the pulsar wind nebula in the LMC remnant 0540-69.3 [145]; this remnant is now the closest known to the textbook picture. ASCA detection of variability in 1E 1613-509 in RCW 103 refuted the hypothesis that it is a quiescent neutron star emitting thermal X-rays from its surface [42]. The extended bandpass of ASCA allowed discovery of spectrally hard PWNe embedded in the soft thermal remnants W44 [47] and G11.2-0.3 [165]. Subsequently a 65 ms pulsar was found within G11.2-0.3 [158]. ROSAT and ASCA found stellar remnants were identified in the synchrotron-dominated shell-like remnants RX J1713.7-3946 and RX J0852.0-4622 [5, 150], indicating that environment is the fundamental determinant in whether a remnant is synchrotron dominated. The zoology of central stellar objects now includes pulsars, anomalous X-ray pulsars, soft gamma-ray repeaters, objects defying classification (such as SS 433), and neutron stars without a PWN, which are simply called "central compact objects."

XMM-Newton and Chandra have added to the wealth of observed phenomena. Chandra's high resolution revealed point sources previously hidden by their low luminosity, the most dramatic of which is the source in Cas A, revealed in the very first Chandra image. It is similar to the source in Puppis A, with no apparent pulsations, no counterparts, and no PWN [25]. Pulsations with period 0.42 s were discovered from 1E 1207.4-5209, the source in PKS 1209-52 [182]. Additionally, Chandra has been used to detect low luminosity pulsars hidden in PWNe. For instance, the location and detection of X-ray pulsations from the X-ray pulsar J0205+6449 in 3C 58 led to the subsequent discovery of radio pulsations [19]. Chandra and XMM-Newton have also revealed pulsar wind nebulae and bow shock nebulae embedded within thermal emission. Chandra revealed the presence of a PWN within the young, O-rich SNR G292.0+1.8 [59]. A region of hard emission detected using ASCA in the evolved remnant IC 443 was resolved into a compact source embedded in a bow shock nebula [108]. XMM-Newton observations show that the X-ray bow shock nebula is substantially larger than the radio nebula, a fact inconsistent with the expected inverse correlation between an electron's synchrotron lifetime and its energy [14].

On the other hand, the central stars in other SNRs remain elusive. A central hot spot in 3C 397 thought to house a compact stellar remnant showed no source to a limiting luminosity of 7×10^{33} erg s^{-1} [136]. Neither of the oxygen-rich Magellanic Cloud remnants, N132D or 1E 0102.2-72.2, shows evidence for a central source. A systematic optical/IR identification program of the unresolved Chandra

[3] Pulsations were reported based on ROSAT data, but not confirmed in Chandra data [113]

sources embedded in four SNRs revealed no stellar remnant candidates above an X-ray luminosity of 10^{31} erg s^{-1} [76].

Chandra and XMM-Newton have also found shells around Crab-like remnants. A soft thermal shell was found around 3C 58 [15]. A limb-brightened shell has also been found around the periphery of G21.5-0.9. This shell shows no evidence for thermal emission, and while dust scattering could contribute to its total flux, it is possible that the X-ray emission in the shell arises from shock accelerated electrons as in SN 1006 [99]. Ironically, all searches for thermal emission from the Crab have failed to find any, leaving the Crab as an exception rather than a true prototype.

17.3 Evolved SNRs

The majority of X-ray studies of Galactic remnants have concentrated on the early stages of evolution when the remnants are bright and shell-like with angular extent that can be covered in a modest number of observations. Evolved remnants are considerably larger than their younger counterparts; objects that fit in the field of view of most imaging instruments are distant, and therefore subject to absorption by intervening galactic material. If most of the emission is very soft, we get a distorted impression of them. Nearby evolved remnants, while not absorbed, subtend large solid angle and have low surface brightness, making observations challenging. ROSAT, with its soft response plus the unlimited field of view provided by the All-Sky Survey and the 2-degree field of view of the PSPC when used in pointed mode, was the first observatory to facilitate systematic study of cooler, evolved remnants. In contrast to the study of younger remnants, which has been revolutionized by XMM-Newton and Chandra, only limited advances have been made since ROSAT.

17.3.1 The Cygnus Loop

The Cygnus Loop is the best-studied evolved remnant. Its high surface brightness shell and modest (3°) angular diameter has made it accessible to most observatories. Its proximity (440 pc) and low column density make it accessible to soft X-ray (and UV) instruments. Its wealth and variety of shock structures make it an ideal laboratory for studying the interaction between a modest velocity shock and interstellar clouds.

The Cygnus Loop was the very first supernova remnant imaged, via a sounding rocket instrument [126]. That image showed the remnant as a nearly complete, limb brightened shell. Subsequent, higher resolution views have refined the overall impression provided by the first image, but have not substantially revised it. Complete mosaic images were constructed using the Einstein IPC, the ROSAT PSPC during the All-Sky Survey, the ASCA GIS, and most impressively, using a million seconds of ROSAT HRI exposure (Fig. 17.1). Chandra and XMM-Newton studies have concentrated on small regions of interest around the shell.

Global studies of the Cygnus Loop using the ROSAT PSPC and ASCA have revealed a number of interesting attributes. The ROSAT image reveals an abundance of features, including the complete south blowout region and a nearly circular bubble in the western interior [7]. The remnant is more strongly limb brightened at the lowest energy, and the interior is found to be hotter. A general correlation with radio surface brightness is found, but there are strong variations in the radio to X-ray surface brightness. Around the circumference, the displacement between the X-ray and radio edge is as much as ± 2 arcmin but averages to zero. The ASCA map shows that in the 1.5–5.0 keV band the remnant appears not limb-brightened, but centrally filled [106]. This is ascribed to a higher temperature in the interior than on the periphery.

17.3.2 Detailed Shock Physics in the Cygnus Loop and Puppis A

Chandra and XMM-Newton have made possible detailed spatial–spectral analysis of shock structures in nearby, evolved SNRs, allowing comparison with shock simulations, and in one case, ground-based experiments.

High-resolution Chandra observations of portions of the Cygnus Loop shell reveal enormous complexity, and multiple emission structures. Regions along the southeastern and western rims have been observed. In the west, the interaction is observed between the shock front and both a large interstellar cloud and the shell that surrounds the cavity formed by the progenitor's wind [94]. The X-rays uniquely trace the shock front within the dense cloud. The temperature where the shock front encounters the cloud is very low, $\sim$0.03 keV. Behind this shock front the temperature is higher, 0.2 keV, due to the reheating of material by a reflected shock. The interaction of the shock with the cavity boundary creates a thin (0.1 pc) X-ray structure showing clear radial temperature variation. The temperature at the outer shock is 0.12 keV; the slightly higher temperature material behind this region is also ascribed to a reflected shock. No evidence for turbulence associated with instabilities is detected along either shock structure, nor are anomalous metal abundances. The southeastern knot is thought to represent the early stage of an encounter between the blast wave and a large interstellar cloud. It appears to have an extremely complicated morphology, but the Chandra observations help demonstrate that the apparent complexity is due primarily to projection of multiple thin, fairly simple shock structures [93].

The Eastern Bright Knot (EBK) of the 4 000-year-old remnant Puppis A (at a distance of $\sim$2 kpc) was identified using the Einstein HRI as the site of the collision between the SNR shock and an isolated interstellar cloud. Subsequent radio and optical observations suggested that the EBK is related to a more extended interstellar feature, and that only a small portion has been encountered by the Puppis A

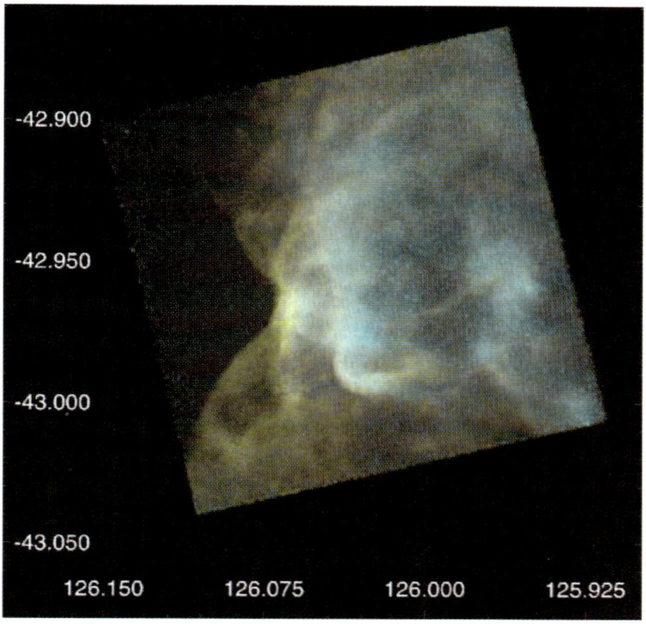

Fig. 17.13 The complex structure of the Eastern Bright Knot in Puppis A revealed by ACIS indicates it consists of two distinct features, a fully-shocked isolated cloud (*right*) and the result of more recent interaction with dense gas [63]. The soft band (O) is in *red*, medium band (Ne) is in *green*, and hard band (>1.5 keV) is in *blue*

shock. A spatial–spectral analysis using the Chandra ACIS indicates the EBK is two distinct features, with very different temperature structures [63]. These are apparent in Fig. 17.13. The inner of these is probably an isolated cloud; its morphology is strikingly similar to the results of laboratory simulation. Comparison with these simulations leads to inference of an interaction time of roughly three cloud-crushing timescales, which translates to 2 000–4 000 yrs. This is the first X-ray identified example of a cloud-shock interaction in this advanced phase. The X-ray emission of the compact knot, located approximately an arc minute closer to the edge of the remnant, implies a recent interaction with relatively denser gas, some of which is located in front of the remnant. Observations in other bands indicate that the gas cloud has a complicated morphology.

17.3.3 Mixed Morphology Remnants

ROSAT observations, and to a lesser extent observations using ASCA and subsequent missions, identified and placed into the evolutionary context a new class of SNRs. Most SNRs have similar radio and X-ray morphologies: either shell-like, which tend to be dominated by shock-heated, thermal emission; centrally filled, which tend to be synchrotron emission dominated pulsar wind nebulae; or composite, with a shell and a pulsar wind nebula. Einstein observations revealed that some SNRs have distinct X-ray and radio morphologies. Rho and Petre [133] carried out a systematic study of all X-ray emitting SNRs and found that a substantial fraction share this observational property. Using the ROSAT PSPC and non-X-ray data, they cataloged a number of additional properties. The remnants have either flat temperature profiles or temperatures gradually increasing towards the center. There is little if any evidence for a limb brightened shell in most. They tend to be found in complex regions of higher than normal ambient density, near molecular and H I clouds. Many are surrounded by complete or partial H I shells. PSPC spectra, and higher resolution ASCA spectra, suggested that the gas in these remnants is in or near collisional ionization equilibrium, indicating that the remnants are dynamically old. These remnants were termed mixed morphology (MM) remnants. Prominent examples include W44, 3C 391, MSH11-61A, W28, IC 443, and 3C 400.2.

Two models of these objects have gained acceptance. The first entails expansion of a SNR in a cloudy medium. Clouds overtaken by the passing shock front slowly evaporate in the remnant interior, and thus provide a reservoir of additional mass. The relative smoothness of the surface brightness profile sets constraints on the cloud size, while the overall brightness constrains their mass and filling factor. The second model assumes that the remnant has entered the radiative or shell forming stage. Thus the outer shock is moving too slowly to produce appreciable new X-ray emitting gas. Thermal conduction produces the smooth interior temperature and density profiles. This model accounts for the H I shell found around many MM remnants. The ROSAT data provided no strong discriminator between the two models.

Subsequent observations of these objects have helped clarify their nature. A ROSAT catalog of Magellanic Cloud SNRs revealed three MM candidates, N206, N120, and 0454-672 [177]. Detailed Chandra observations confirm the classification of N206 (and find an embedded bow shock nebula associated with a central point source) [176]. These are among the largest LMC remnants, and thus are likely to be among the most evolved. Spectral studies of various MM remnants using ASCA and Chandra show that they all are near coronal ionization equilibrium, and generally show evidence for a modest temperature gradient [78, 147]. The Chandra spatially resolved spectral study of 3C 391 shows variation of temperature on modest spatial scales that does not seem correlated with ISM structures [26]. Two remnants, IC 443 and W49B (considered by some to be a MM SNR), show evidence for overionization – the hydrogen-like to helium-like line ratio in abundant metals is too large to be consistent with collisional ionization equilibrium [78, 79]. This can happen in the interior of an SNR only if the gas cools faster than it recombines and requires thermal conduction.

An important new insight arose from the PSPC study of the large angular diameter remnant G65.2+5.7 [148]. This remnant is nearby and relatively unabsorbed. A mosaic produced from a collection of PSPC pointings showed that its 0.1–0.3 keV morphology is shell-like but that at higher energy the shell disappears and the remnant takes on morphology similar to other MM remnants. Temperature measurements show a gradual increase toward the center. This remnant provides the key to understanding the MM remnants: X-ray emission from newly shocked gas is too soft to be observed from most through the absorbing interstellar medium, and thus all that is observed is the warm central emission, the temperature profile of which is smoothed by thermal conduction.

While thermal conduction drives the evolution and appearance of MM remnants, it is not the only force at work. The complex temperature and density structure of 3C 391 suggests that cloud evaporation can play a role in MM remnants [26]. W28 also has a complex thermal structure, with a hot component not seen in other mixed morphology remnants, a strong temperature gradient across the remnant, and the presence of numerous X-ray knots and clumps [131]. While some of these features are environmentally induced (e.g., the temperature gradient arises from a density gradient in the ambient medium), others, such as the presence of the clumps, suggest departures from an idealized conduction model.

17.3.4 Ejecta in Evolved SNRs

Evidence of enhanced metal abundances is becoming commonplace in evolved remnants. ASCA observations revealed overabundances of Si, S, and Fe near the center of the Cygnus Loop [105]. The observed abundance ratios were shown to be consistent with a Type II supernova from a progenitor with mass of about 25 $M_\odot$.

At least four evolved, shell-like SNRs in the LMC show evidence for ejecta near their centers. DEM L71, thought to be 5 000-years-old, shows Fe-rich ejecta in its interior, presumably heated by the reverse shock [56]. The Fe:O ratio is larger than 5:1. Dynamical arguments suggest that the total ejecta mass is $\sim$1.5 $M_\odot$; the Fe mass based on the spectral fits is $\sim$0.8 $M_\odot$. Thus the X-ray measurements strongly support a Type Ia origin for this remnant. N49B (0525-66.0), with an estimated age of 10^4 yrs, shows enhanced Mg and Si in the center, but no evidence for either enhanced O or Fe [110]. Two other shell-like 10^4-year-old SNRs, 0548-70.4 and 0534-69.9, were found to have central enhancements of O, Mg, Si, S, and Fe [52]. The O:Fe ratios support a Type Ia origin for both remnants.

Some, but not all, MM remnants show evidence for residual ejecta. Spatially resolved spectroscopy along the long axis of the MM remnant W44 using Chandra reveals that the remnant's hot bright projected center is rich in Mg, Si, and Ne, and that the bright knots are regions of comparatively elevated elemental abundances [147]. The fact that Ne is among the elements with enhanced abundance indicates that ejecta contribute to the abundance trends. An underabundance of Fe from those same regions provides the first potential X-ray evidence for dust destruction in a supernova remnant. Additionally, Yokogawa et al. found evidence for abundance enhancements in three evolved (older than 10^4 yrs) SMC SNRs, all of which appear to be MM remnants [181]. W49B shows enhanced metal abundances throughout [68], but its evolutionary state and its membership in the MM class are a matter of debate. Rho and Petre suggest it is not, largely because of the presence of ejecta [133]; Kawasaki et al. argue it is because of plasma properties [78]. The evolved, small-diameter remnant G347.7+0.2, which has all the attributes of MM remnants except a centrally filled X-ray morphology, shows slightly enhanced Si throughout, and a S enhancement on its western side [91].

17.3.5 The Monogem Ring

The Monogem Ring, a large angular diameter nearby old SNR, provides an excellent example of how the RASS provided new insights into low surface brightness objects [120]. This object, also known as the Gemini-Monoceros X-ray enhancement, is a 25° diameter bright region in the 0.1–0.5 keV band. It was imaged for the first time by the RASS. It is shown to have an average temperature of $10^{6.15}$ K, and a column density of approximately $5 \times 10^{19}\,\mathrm{cm}^{-2}$. The PSPC observations confirm that the ring is most likely a SNR. Models suggest it is still in the adiabatic stage. If a distance of 300 pc is assumed, then the age is 86 000 yrs, and remnant is expanding in a cavity of very low ambient density, $5 \times 10^{-3}\,\mathrm{cm}^{-2}$. The overall pressure of the shocked, X-ray emitting gas is only 2–4 times that of the ambient medium. The remnant is likely to reach pressure equilibrium with the surrounding medium before it reaches the radiative evolutionary stage. This has potentially important implications for SNR evolution models in suggesting that SNRs in low-density media may end their lives differently from remnants in higher density media.

17.3.6 Newly Discovered Evolved SNRs

One of the most significant contributions of the ROSAT All Sky Survey is the revelation of X-ray emission from many previously undetected or unknown SNRs, especially remnants with large angular diameters and low surface brightness. Two of the best known of these, RX J1713.3-3946 and RX J0852.0-4622, have already been discussed. Two others worth mention are G189.6+3.3 and G156.2+5.7.

G189.6+3.3 lies partially in the line of sight to IC 443. The ROSAT All Sky Survey revealed the remnant to have a complete shell that partially overlaps IC 443 [4]. It is located in front of IC 443 and the thin molecular cloud with which IC 443 is colliding. Spectral analysis yields an age of $\sim 10^5$ yrs, more than an order of magnitude older than IC 443. Where the G189.6+3.3 shell overlaps the bright northeastern region of IC 443, an increased amount of absorption is measured in the IC 443 spectrum.

G156.2+5.7, also known as RX J04591+5147, was the first new SNR discovered in the RASS [117]. It is a circular SNR with a diameter of 108 arcmin. One of the brightest X-ray SNRs, it is by contrast one of the lowest surface brightness remnants in the radio. This suggests the remnant is expanding into a very low-density medium. Unlike other low radio surface brightness remnants, the soft X-ray spectrum of G156.2+5.7 is dominated by a thermal component [180], and its X-ray and radio morphologies do not correspond closely. The thermal emission is described using a NEI model with a temperature of about 0.4 keV, and an ionization timescale of $1\text{–}3 \times 10^{11}$ cm^{-3} s. The ambient density is $\sim$0.2 cm^{-3}, higher than that found in the synchrotron-dominated shell-like remnants. The age is estimated to be $\sim 1.5 \times 10^4$ yrs. The remnant also has a hard tail. ASCA observations indicate that the hard component arises in the remnant interior as well as along the rim, and were unable to discriminate between a thermal and a nonthermal origin for the hard component [180].

17.4 Extragalactic SNRs

X-ray studies of extragalactic SNRs were first facilitated by the imaging capabilities of the Einstein Observatory. While Einstein detected (and discovered) remnants in the Magellanic Clouds, it did not have sufficient sensitivity to detect a substantial number of SNRs in more distant galaxies. ROSAT and ASCA made major contributions to the study of Magellanic Cloud remnants. Additionally, ROSAT also carried out detailed studies of the source populations of nearby galaxies and in so doing was the first observatory to detect a significant number of SNRs. ROSAT was unable to spatially resolve remnants detected beyond the Magellanic Clouds and those it detected were the most luminous. ROSAT studies were thus restricted to spatial and luminosity distributions. Nevertheless, ROSAT identified 16 SNRs in M31 with $L_X > 4 \times 10^{35}$ erg s^{-1} [154]. In M 33, it found 16 SNRs with $L_X > 7 \times 10^{35}$ erg s^{-1}, 12 of which had radio and/or optical counterparts [43]. In the more distant NGC 300, ROSAT detected six SNRs with $L_X > 5 \times 10^{36}$ erg s^{-1}, four of which are detected in other bands [109, 128]. In NGC 7793, only two SNRs of 33 were detected in X-rays in a survey sensitive to $L_X \sim 2 \times 10^{37}$ erg s^{-1} [109].

In general, identifying SNRs in other galaxies has relied on searches for one or more of three properties: radio spectrum, high [S II]/H-α in the optical, and X-ray brightness/color. The most reliable searches combine all three properties. One major result from studies utilizing ROSAT is the small overlap among the SNR candidates in the three bands (e.g., [109]). Thus, all three bands must be searched if a complete SNR census is to be obtained.

XMM-Newton and Chandra have advanced the X-ray identification of SNRs by facilitating searches to lower flux. They thereby increase the potential for detecting in X-rays SNRs found in other bands. It is possible to identify SNR candidates on the basis of X-ray spectral information using an X-ray color–color diagram in which SNRs have a nearly unique location (a technique originally applied using ROSAT e.g., [128]). Additionally, the availability from Chandra of high-resolution imaging allows identification by extent as well, at least for the largest remnants in the nearest galaxies. Using these observatories, it is possible for the first time to carry out X-ray imaging and spectral study of at least a few individual remnants beyond the Magellanic Clouds.

In M 31 an XMM-Newton survey has dramatically lowered the limiting flux and doubled the number of SNR candidates [118]. With a limiting 0.2–4.5 keV luminosity of $\sim 4.4\times10^{37}$ erg s^{-1}, 21 SNRs were found based on correlations with radio or optical catalogs, as well as 23 SNR candidates, 22 of which are new, based on X-ray selection criteria alone. Another object may be the first pulsar wind nebula discovered beyond the Magellanic Clouds.

In M 33, XMM-Newton and Chandra observations raise the total number of identified SNRs in all bands to 100. Of these, 37 have X-ray counterparts with $L_X > 7\times10^{35}$ erg s^{-1} [38, 119]. At least four appear extended in the Chandra data. Chandra observations indicate that while the number of SNRs in the LMC and M 33 with $L_X > 10^{35}$ erg s^{-1} are comparable, M33 has substantially fewer with $L_X > 10^{36}$ erg s^{-1}. There is no clear reason for this difference; it cannot be fully explained by uncertainty in spectral model parameters, nor can it be accounted for by the abundance differences between the two galaxies.

In NGC 300, XMM-Newton observations have a factor of 10 lower limiting luminosity than the ROSAT observations. The number of SNRs and candidates is increased only from six to nine [20].

A Chandra survey of 11 nearby (4–13 Mpc), nearly face-on (inclination $<33°$) spiral galaxies yields about 150 SNRs [81]. The vast majority of these are found in the galaxies with high star-forming rates: M 51, M 83, and M 94.

It is also now possible to study in some detail the X-ray properties of individual extragalactic remnants. In particular, it is possible using Chandra to spatially resolve the physically largest local group galaxies.

The remnant CXOM31 J004327.2+411829 is the first remnant beyond the Magellanic Clouds to be spatially resolved in X-rays [84]. It appears ring-like, with a diameter of 42 pc. The age is estimated to be 3 210–22 300 yrs and the ambient density 0.003–0.3 cm^{-3}. Subsequent study of the time history of the object using XMM-Newton and Chandra suggest that a variable source is embedded in the northwestern portion of the remnant [174]. The temporal properties suggest the source is a low-mass X-ray binary (LMXB). If so, then it is the first direct association between an LMXB and an SNR.

Four other M 31 SNRs have been resolved using Chandra and associated with radio and/or optical shells [86, 175]. They have diameters ranging from 12 to 40 pc, and luminosities between 7×10^{35} and 4×10^{37} erg s^{-1} erg s^{-1}. All are likely propagating through low density ISM.

The remnant Ho 12 in local group dwarf irregular galaxy NGC 6822 is resolved using Chandra. The remnant appears shell-like, with a diameter of $\sim$24 pc. Its morphology is consistent in radio, optical ([O III], H-alpha, and [S II]), and X-ray. Spectral analysis yields $kT \sim 2.8$ keV, an age 1 700–5 800 yrs, and a very low ambient medium density. Optical emission suggests the remnant expanded into a pre-existing cavity, making it an extragalactic analog to the Cygnus Loop [85].

17.5 X-Ray Supernovae

The discussion of extragalactic SNRs leads naturally into a discussion of X-ray observations of SNe. No supernova has been observed in our Galaxy for 400 yrs. Thus, observations in all bands must be performed on extragalactic supernovae. Thus only the most luminous such objects (L_X typically larger than 10^{39} erg s^{-1}) are detectable. The detection of X-ray emission from SNe is a rarity. Compared with the hundreds of SNe detected optically over the past three decades, only 30–40 SNe have been detected in X-rays. Most initial detections have been made using ROSAT, XMM-Newton, or Chandra.[4]

The paucity of the detections can be traced to the circumstances under which substantial X-ray flux is produced in a supernova. Several mechanisms might produce X-ray emission, but one is primarily responsible for detected X-rays. Possible mechanisms include a thermal flash associated with breakout of the shock from the stellar surface; Compton scattered γ-rays from metals synthesized in the explosion; and thermal emission from shock-heated circumstellar matter. Only the latter has been observed more than once.

The X-ray observations of SNe are usually interpreted in the context of a model describing the interaction between the SN shock and a circumstellar medium (CSM). A detectable X-ray flux is produced only in the rare circumstance when the SNR shock encounters high density material, either within the star's neighborhood or as the consequence of substantial pre-supernova mass loss by the progenitor. In models assuming a smooth medium, the interaction between the outflowing and ambient material causes the development of a pair of shocks: a forward shock that heats, compresses, and ionizes the ambient medium; and behind it a reverse shock into which fast-moving ejecta collide. Material interacting with either shock becomes heated to X-ray emitting temperatures. It is expected that the material interacting with the forward shock will be heated to a higher temperature than the denser ejecta interacting with the slightly slower moving reverse shock. Depending on the medium density and the degree of compression of the ambient material (and thus the cooling time) a neutral shell might form between the two shocks. If an X-ray spectrum shows two components, the hotter of the two is interpreted as arising from the forward shock and the cooler from the reverse shock. The cooler component is sometimes associated with a larger absorption column, which is interpreted as shock compressed material. As the remnant ages and encounters ever less dense medium, we expect the temperatures of the two components to decrease and the relative flux of hotter component to decrease. Any intrinsic column density

[4] An up-to-date listing of the SNe detected in X-rays, their properties, and relevant references is maintained by Immler at http://lheawww.gsfc.nasa.gov/users/immler/supernovae list.html.

of the cooler component will also decrease as the density decreases. The two-shock structure is analogous to that in young supernova remnants: in effect, a high density medium accelerates the evolution of a supernova into a supernova remnant.

An alternative set of models involves the propagation of a shock through a structured medium, consisting of dense clumps or filaments These structures arise from an unstable presupernova wind. The hot component arises from the propagation of the shock through the interclump medium, and the cool component arises from shocks driven into clouds. The clouds are compressed but not accelerated.

For a supernova expanding into a CSM from a progenitor wind, the X-ray luminosity can be written as

$$L_X = 4/(\pi m^2)\Lambda(T)(\dot{M}/v_w)^2(v_s t)^{-1}\,erg\,s^{-1}$$

Here $\Lambda(T)$ is the temperature-dependent plasma emissivity; $\dot{M}$ is the progenitor mass loss rate; v_w is the wind velocity; and v_s is the shock velocity. If v_s is known (generally from measuring the width of the broad Hα lines), then $(\dot{M}/v_w)$, the ratio between the mass loss rate and the wind velocity, can be determined at the radius r $(= v_s t)$. If one assumes that for a typical supernova the progenitor wind velocity is $10\,\mathrm{km\,s^{-1}}$ and the average shock velocity is $10^4\,\mathrm{km\,s^{-1}}$, then every year over which X-ray observations span samples 10 000 yrs of presupernova evolution. Thus, the SN X-ray light curve traces the mass loss history of the star. For the oldest X-ray supernovae, it has been possible to trace the pre-supernova mass loss history for $\geq$25 000 yrs before the explosion.

The density profile of the circumstellar medium is generally characterized using a power law, $\rho_{csm} \propto r^{-s}$. If the progenitor has a constant $\dot{M}/v_w$, then $s = 2$, and the supernova light curve will have the form $L_X \propto t^{-1}$. If the light curve is steeper, then we can infer that $s > 2$ and thus $\dot{M}/v_w$ increased as the star neared core collapse; alternatively, a flatter light curve suggests $s < 2$ and a decrease in $\dot{M}/v_w$ prior to the explosion. Deviation from a smooth light curve can be interpreted as evidence for either mass loss rate changes, or for a clumped or filamentary CSM due to wind instabilities.

A wide variety of SNe have been detected, including Type II, Type IIL, Type IIP, Type IIn, Type IIb, Type Ib/c, and Type Ic. They have been detected at a broad range of distances, between 3.2 Mpc (SN 2004dj) and 100 Mpc (SN 1988Z). A broad range of luminosities has been observed from $10^{38}\,\mathrm{erg\,s^{-1}}$ (SN 1970G) to $10^{41}\,\mathrm{erg\,s^{-1}}$ (SN 1988Z). The limited number of detections prevents the discernment of any relation between luminosity and type, although the Type IIn SNe tend to be among the most luminous.

Figure 17.14 shows an assemblage of X-ray SN light curves. The curves have a range of slopes, but with the exception of SN 1987A, they all show either constant or decreasing flux. The sampling of most light curves is sparse. Multiple observations exist for only a handful of supernovae; these include SN 1987A, SN 1993J, SN 1978K, and SN 1979C, which we discuss later.

Applying the relationship between luminosity and density described above to the light curves of some of these supernovae has led to density profile estimates. Some

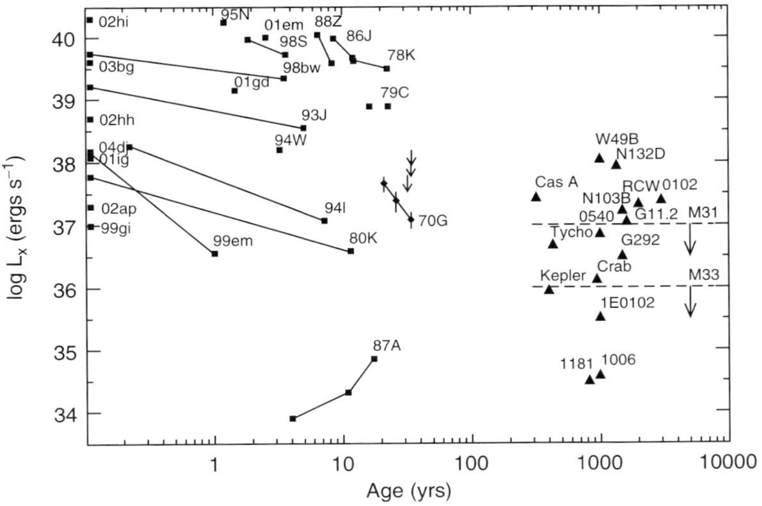

Fig. 17.14 X-ray light curves (0.32 keV) band of all SNe detected to date (*filled squares*). A range of behaviors is observed, from flat to dropping steeply. Only SN 1987A shows increasing luminosity [73]. The figure suggests that the luminosities of historical SNRs (*filled triangles* on the right) can be extrapolated from the SN luminosities

of these are shown in Fig. 17.15. All of the profiles can be described by power laws of the form $\rho_{\rm csm} \propto \rho_0^{-s}$, where $1.0 \leq s < 2.0$. This plot suggests that none of the supernovae are expanding into the remnants of a wind that is constant on a scale of thousands of years, but that $\dot{M}/v_{\rm w}$ decreased as the progenitor star approached its cataclysm. Furthermore, light curves from SN 1993J and SN 1998S with many samples show nonmonotonic behavior, signaling the presence of a clumpy or filamentary CSM.

From the inferred densities, it has been possible to estimate a mass loss rate for many supernovae. The inferred values range between 10^{-6} and $10^{-4}\,{\rm M}_{\odot}\,{\rm yrs}^{-1}$. There is no correlation between mass loss rate and supernova type among the detected objects.

Independent confirmation of the existence of a dense medium surrounding X-ray supernovae comes from radio observations. Radio emission arises from the strong interaction between the forward shock and dense material. This interaction produces nonthermal electrons, which are accelerated in the compressed magnetic field behind the shock. Every X-ray supernova is accompanied by radio emission although the converse is not true. Nevertheless, the detection of radio emission from a supernova serves as a trigger for an X-ray search. The relationship between the radio and X-ray light curves is not always simple.

X-ray spectra, taken for a few of the highest flux SNe, including SN 1978K, SN 1986J, SN 1993J, and SN 1988S, provide informative plasma diagnostics. These spectra have revealed the presence of two temperature components, with the hotter one generally ascribed to the forward shock and the cooler to the reverse. In

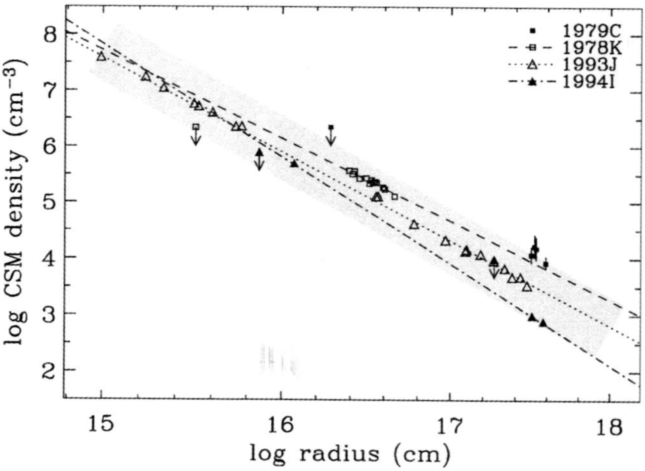

Fig. 17.15 Circumstellar matter density profile as a function of shell expansion radius for several SNe, including 1994I (*filled triangles*), 1993J (*open triangles*), 1978K (*open boxes*), and 1979C (*filled boxes*) [72]. The profiles can be expressed as power laws with indices between 1 and 2. A uniform density medium gives a power law index of 2

SNe in which spectra at various times are available these components evolve at different rates, the hotter one cooling and fading more rapidly (e.g., [139]). In some spectra, the cooler component also has a higher column density, indicating the presence of neutral matter between the two shocks. Also, evidence for enhanced metal abundances has been detected in the cool components of SN 1978K [139] and SN 1998S [121], indicating they arise in shocked ejecta (Fig. 17.16). Finally, the spectra can be compared with supernova models to constrain the ejecta profile index n where ($\rho_{\text{ejecta}} \propto r^{-n}$). For SN 1978K, n is found to be approximately 5 [139] and for SN 1999em n $\sim$ 7.

Below we discuss some interesting examples of X-ray supernovae.

17.5.1 SN 1987A

SN 1987A is distinctive among X-ray detected supernovae. It is the only detected Type IIL supernova, and has by far the lowest luminosity of any X-ray detected SN. The detection can be ascribed to its proximity, at a distance of 50 kpc in the LMC. As can be seen in Fig. 17.14, its low flux demonstrates why similar, more distant supernovae have not been detected. As the nearest supernova in 400 yrs, it has been the subject of intensive observation in all wavelengths. While several comprehensive reviews provide detailed descriptions of the phenomenology and theoretical interpretation (e.g., [101]), the object continues to evolve rapidly. In the X-ray band, the

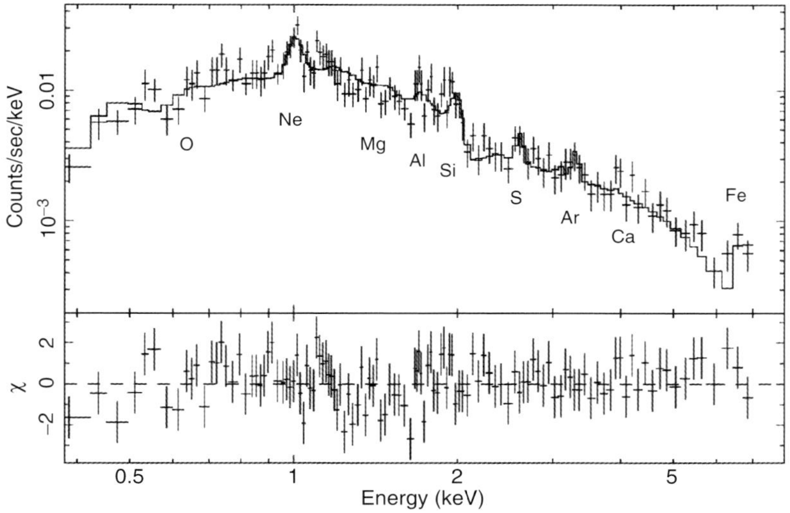

Fig. 17.16 The ACIS spectrum of SN 1998S shows that the spectrum arises in shock-heated gas and shows evidence of ejecta [121]

object has changed dramatically since the most recent publications, and it continues to be observed intensively by Chandra and XMM-Newton.

SN 1987A was discovered optically on January 27, 1987. At the time, the only orbiting X-ray observatories were the HEXE instrument on the Mir space station and Ginga. Observations over the first 5 months after the explosion detected no prompt signal [31]. Models suggested that X-rays downscattered from the γ-radiation from decaying Ni thought to power the declining optical light curve would eventually be detected as the outer expanding envelope decreased in density. Nevertheless, it was a surprise when hard X-rays were independently detected by HEXE and Ginga 6 months after the explosion [31, 153]. This surprising result was interpreted as evidence that the ejecta were mixed early during the expansion. Ginga also detected an iron line from the end product of the radioactive Ni decay. Ginga and HEXE followed the hard light curve for several months until SN 1987A became undetectable.

While initial ROSAT observations in June of 1990 yielded only an upper limit, SN 1987A was detected 8 months later, in February 1991 [12, 40]. The supernova was initially detected at a 0.2–2.4 keV band luminosity of $\sim 10^{34}$ erg s^{-1}, several orders of magnitude lower than any previous or subsequent supernova. This flux was interpreted as arising from the interaction of the supernova ejecta with circumstellar material from the blue giant wind.

ROSAT and ASCA monitored SN 1987A throughout their lifetime. From 1991 through 1995, ROSAT detected a nearly linear flux increase, by a factor of 5–6 from the initial detection [48]. Contemporaneous ASCA observations also showed a linear brightness increase. The average 0.5–10.0 keV spectrum during this interval was consistent with a 1.8 keV plasma with sub-solar metal abundances [75].

The object's light curve steepened markedly around the time of the launch of Chandra and XMM-Newton, allowing the full power of these observatories to be brought to bear. The initial Chandra observation resolved the emission into a ring consistent with the inner bright optical ring thought to be a equatorial band of circumstellar material, indicating that the shock was now encountering this inner ring [18]. An early grating spectrum revealed lines of N, O, Ne, Mg, and Si. Broadening of the composite line leads to an inferred shock velocity of $(3\,400 \pm 700)\,\mathrm{km\,s^{-1}}$ [103].

Subsequent observations have followed the spectacular evolution of the supernova [112]. As the interaction between the forward shock and the band has evolved, Chandra has observed the ring to grow in diameter and overall brightness, and has resolved individual features along the ring brightening and dimming. The flux has increased exponentially; as of day 6 500 the supernova was approaching a 0.5–2.0 keV flux 50 times higher than the initial ROSAT detection. The most recent observations suggest the development of a new emission component, presumably associated with circumstellar material not in the equatorial band. At the same time, a decrease in the overall expansion rate by a factor of two is observed, suggesting strong interaction with the dense ring material. No evidence for a central compact object has yet been found.

A deep Chandra grating observation reveals the presence of numerous lines from multiply-ionized O, Ne, Mg, S, Si, and Fe [183]. Many lines are broadened, and inferred expansion velocities range from 300 to $1\,700\,\mathrm{km\,s^{-1}}$, much smaller than the global radial expansion measured from images. The lower velocities are interpreted as arising in shocks produced as the supernova blast wave encounters dense protrusions on the ring.

17.5.2 SN 1993J

SN 1993J, in the nearby spiral M81 (3.4 Mpc), is the most intensively monitored X-ray SN after SN 1987A. It was initially detected 6 days after its discovery in the visible band, at the time the earliest X-ray detection of a SN [185]. Its early evolution was monitored extensively using ROSAT [70] and ASCA [163]. More recently, it has been observed using both Chandra and XMM-Newton [155, 184]. It is the only SN besides SN 1987A for which a substantial number of X-ray spectral observations have been performed.

The early detection indicated the presence of dense circumstellar material in close proximity to the progenitor star. For the first month after the explosion, the X-ray spectrum was too hot to be characterized by ASCA ($kT > 30$ keV; [83]), and the detection by OSSE instrument on the Compton Gamma Ray Observatory suggests an initial temperature on the order of 10^9 K [92].

ASCA observations between days 8 and 572 show dramatic spectral evolution [163]. A second, cooler component emerges ($kT \sim 1$ keV), with a higher column density than the hot component. The two components are consistent with the expected emission from the forward and reverse shocks. The spectra soften dramatically with time as temperatures of both components decrease. Additionally, the column density to the reverse shock decreases, making the reverse shock the dominant contributor to the soft flux. Contemporaneous ROSAT PSPC data, which cover a narrower band, reflect this softening, with the 0.2–2.4 keV spectra prior to day 60, too hot to be constrained, and the spectra between days 200 and 400 consistent with a 1 keV temperature [70]. The most recent high quality spectra from XMM-Newton and Chandra show that this trend continues. The Chandra spectrum (taken on day 2 594) requires three components, two with low temperatures (0.35 ± 0.06 and 1.01 ± 0.05 keV) absorbed by a relatively high column density (4×10^{21} cm^{-2}), and a high temperature component ($kT = 6.0 \pm 0.9$ keV) with Galactic absorption (5×10^{20} cm^{-2}). The low temperature components have some metal abundances deviant from solar. The XMM-Newton spectrum is fit by a two-component model with temperatures 0.34 ± 0.04 keV and 6.54 ± 4 keV. The column density for both components is consistent with the larger of the two Chandra values (4×10^{21} cm^{-2}). Metal abundances are consistent with solar.

The 0.3–2.4 X-ray light curve between 1993 and 2001 shows remarkable structure, with a constant decay proportional to $t^{-0.3}$ for the first 200 days, followed by dips and rises relative to this constant decay. The $t^{-0.3}$ slope suggests an average CSM density profile ρ^{-s}, where $s \sim 1.65$. The variations about this average have been interpreted as the result of mild density variations. From both the average slope and the variations it can be inferred that the mass loss rate of progenitor star was not constant, decreasing slowly on average as the explosion approached, but with brief episodes of lower and higher mass loss rates a few thousand years before the explosion.

17.5.3 SN 1978K

SN 1978K is one of the most remarkable X-ray supernovae. It is located in the nearby late-type SB spiral galaxy NGC 1313 (at a distance of 4.1 Mpc). A 1980, Einstein IPC image showed two bright sources in NGC 1313 with $L_X > 10^{39}$ erg s^{-1}, one near the nucleus and one 6 arcmin to the south. A 1992 ROSAT PSPC image revealed a third source to the east-southeast of the nucleus, nearly as bright as the other two, and nearly forming an equilateral triangle with them. An optical counterpart had been discovered 2 yrs earlier and reported as a nova. Also, it had been detected as a bright radio source in 1982. Coupled with the X-ray luminosity, these observations led to the conclusion that this object is a supernova [135]. SN 1978K is thus the first supernova identified as such from X-ray observations. Studies of archival plates indicated an explosion date around June 1, 1978, approximately 19 months before the IPC observation.

SN 1978K has been monitored extensively, initially using ROSAT and ASCA, and subsequently XMM-Newton and Chandra. Its 0.5–2.0 keV flux since discovery has remained remarkably constant, while its 2–10 keV light curve shows a slight decrease starting after 2000 (22 years post-explosion). The 0.5–10 keV ASCA spectra taken in 1993 were reasonably well fit by a single thermal spectrum with $kT \sim 3.0$ keV, with low metal abundances [116]. Recent ACIS and EPIC spectra require two thermal components, with temperatures of ~0.6 keV and ~3.0 keV [139]. These spectra show a hint of enhanced Si. Schlegel et al. interpret the warmer and cooler components as emanating from the forward and reverse shocks [139]. The decline of the flux from the hotter component is interpreted as the first hint of increasing transparency of the matter surrounding the supernova. The flat light curve is consistent with an unusually flat CSM density profile ($\rho \propto r^{-1}$), indicating an unusual mass loss history [72].

17.5.4 SN 1998bw

While not especially noteworthy as an X-ray supernova, the importance of SN 1998bw comes from its possible association with the gamma ray burst GRB 980425. This was a highly unusual gamma ray burst, with a luminosity four orders of magnitude lower than a typical one, and an X-ray afterglow with a substantially slower rate of decline. BeppoSAX observations revealed two possible counterparts to the GRB in the host galaxy ESO 184-G82, but a Chandra observation showed that only one of these had shown a steady luminosity decline. This source is spatially coincident with the radio position of SN 1998bw [88]. The X-ray light curve shows a gentle decline over the first 200 days, followed by an order of magnitude decline over the next 1 000 days. This later behavior is typical of other X-ray SN light curves, suggesting that while the supernova explosion itself may have been unusual, the subsequent interaction with the interstellar medium is not. SN 1998bw has been classified as a supernova Type Ib/c. Its late time X-ray luminosity is somewhat higher than other Ib/c or Ic SNe, suggesting a higher pre-supernova mass loss rate.

17.5.5 SN 1970G

A deep Chandra image of M101 allowed the detection of SN 1970G 35 yrs after its explosion, making it the oldest supernova detected in X-rays [73]. The deep observation allowed detection of the object with a relatively low luminosity, $L_X = (1.1 \pm 0.2) \times 10^{37}\,\mathrm{erg\,s^{-1}}$. This luminosity is consistent with a mass loss rate of $(2.6 \pm 0.4) \times 10^{-5} (v_w / 10\,\mathrm{km\,s^{-1}})\,M_\odot\,\mathrm{yrs^{-1}}$. Its long term X-ray light curve, reconstructed from archival observations, is steep compared with those of most X-ray SNe, with a profile $L_X \sim t^{(2.7 \pm 0.9)}$. The soft Chandra spectrum ($kT \sim 2$ keV) is interpreted as arising in the reverse shock region. As the oldest supernova detected in X-rays, the SN 1970G detection is important, because it begins to bridge the observational gap between supernovae and supernova remnants (the youngest of which, Cas A, is 340-years-old), and offers hope that other old SNe can be detected.

17.5.6 Type Ia SNe

The explosion mechanism of Type Ia supernovae is the detonation or deflagration of a white dwarf star that exceeds the Chandrasekhar limit, and thus is very different from the core collapse of a massive star leading to a Type Ib, Ic, or II explosion. The exploding star is thought to need a binary companion from which to accrete. The mass transfer rate is thought to be on the order of $10^{-6}\,M_\odot\,yrs^{-1}$, considerably lower than the mass loss rates of $10^{-4-5}\,M_\odot\,yrs^{-1}$ responsible for the dense CSM in X-ray supernovae. It is possible that an optically thick wind stabilizes the mass transfer within the binary system. The time history of the wind is nonetheless likely to be complex, depending on the mass loss rate of the companion, the binary system orbital parameters, as well as the means of accretion by the white dwarf star. Residual CSM will be compressed and heated to X-ray temperatures as the shock passes through it. Gaining insight into the surroundings of Type Ia supernovae can potentially provide insight into the precise nature of the explosion mechanism, which is still the topic of debate,[5] and could lead to important information about their usefulness as standard candles for cosmological studies.

The most sensitive published observations of Type Ia SNe near optical maximum come from ROSAT and Swift, with upper limits of a few $10^{-14}\,erg\,cm^{-2}\,s^{-1}$. A ROSAT observation of SN 1992A, observed 16 days after maximum, yielded an upper limit to the mass loss rate from the secondary of a few times $10^{-6}\,M_\odot\,yrs^{-1}$ [140]. A Swift survey of eight Ia events yields luminosity upper limits as small as $\sim 1\times10^{39}\,erg\,s^{-1}$, and mass loss rates as low as $7\times10^{-6}\,M_\odot\,yrs^{-1}$ [71]. These upper limits only begin to test models of Type Ia progrenitor systems. One object, SN 2006ke, showed a peculiar UV light curve, suggesting interaction with circumstellar material and was marginally detected in X-rays at a flux of $4 \pm 1\times10^{-15}\,erg\,cm^{-2}\,s^{-1}$, corresponding to a progenitor mass loss rate of $3\pm1\times 10^{-6}\,M_\odot\,yrs^{-1}$. Even if this detection is treated as an upper limit, it is the most constraining upper limit on mass loss rate from a Type Ia progenitor.

Limits on the flux from older Type Ia supernovae are routinely obtained using deep observations of their host galaxy. For instance, a 3σ upper limit to the luminosity of the Ia SN 1983N of $L_X < 1.7\times10^{36}\,erg\,s^{-1}$ (0.3–8 keV) was obtained from a deep Chandra observation of M83 [152]. This corresponds to a mass loss rate $\dot{M} \leq 10^{-8}\,M_\odot\,yrs^{-1}(v_w/10\,km\,s^{-1})$, approximately 20 000 years before the explosion, but provides no insight into the nature of the material near the progenitor.

[5] Comparison of models with the integrated Chandra spectra of the young Type Ia remnants Tycho and 0509-67.5 suggest delayed detonation as the most likely explosion mechanism [8, 172]

17.6 Conclusion

The unprecedented capabilities of XMM-Newton and Chandra (and ROSAT and ASCA before them) have led to important new insights into the astrophysics of supernova remnants. As these missions continue we can expect further new developments. Examples include direct expansion measurements in additional young SNRs and sensitive searches for trace elements. The new Suzaku Observatory with its sensitive Hard X-ray Detector will measure the spectra of the hard component in many SNRs and crudely map it for large remnants. High spectral resolution spatially resolved spectroscopy will someday allow detailed plasma diagnostics.

Much can be done for X-ray supernovae as well. The high sensitivity of XMM-Newton and Chandra allow the light curves of the known SNe to be tracked to much lower flux. Benefits include mapping of the pre-SN wind density to ever longer times, and further narrowing the observational gap between SNe and the historical SNRs. SN research in the X-rays, as in other bands, still relies primarily on serendipity, however. Fortunately, the Swift Observatory, with its rapid response to new phenomena, allows SNe to be observed in X-rays within hours of discovery and frequently monitored thereafter. Of course, it is the hope of all researchers in this field that before the end of these missions the opportunity arises to observe a new Galactic supernova.

References

1. Aharonian, F. A. & et al. 2004, Nature, 432, 75
2. Aharonian, F. A. & et al. 2005a, A&A, 437, L7
3. Aharonian, F. A. & et al. 2005b, A&A, 437, 135
4. Asaoka, I. & Aschenbach, B. 1994, A&A, 284, 284
5. Aschenbach, B. 1998, Nature, 373, 587
6. Aschenbach, B., Egger, R., & Trümper, J. 1995, Nature, 373, 587
7. Aschenbach, B. & Leahy, D. A. 1999, A&A, 341, 602
8. Badenes, C., Bravo, E., Borkowski, K. J., & Domínguez, I. 2003, ApJ, 593, 358
9. Becker, R. H., Holt, S. S., Smith, B. W., et al. 1980a, ApJ, 235, L5
10. Becker, R. H., Smith, B. W., White, N. E., et al. 1979, ApJ, 234, L73
11. Becker, R. H., Szymkowiak, A. E., Boldt, E. A., Holt, S. S., & Serlemitsos, P. J. 1980b, ApJ, 240, L33
12. Beuermann, K., Brandt, S., & Pietsch, W. 1994, A&A, 281, L45
13. Bleeker, J. A. M., Willingale, R., van der Heyden, K., et al. 2001, A&A, 365, L225
14. Bocchino, F. & Bykov, A. M. 2001, A&A, 376, 248
15. Bocchino, F., Warwick, R. S., Marty, P., et al. 2001, A&A, 369, 1078
16. Bowyer, S., Byram, E., Chubb, T. A., & Friedman, H. 1964, Sci, 146, 912
17. Burrows, A., Hayes, J., & Fryxell, B. 1995, ApJ, 450, 830
18. Burrows, D. N., Michael, E., Hwang, U., et al. 2000, ApJ, 543, L149
19. Camilo, F. & et al. 2002, ApJ, 571, L41
20. Campano, S., Wilms, J., Schirmer, M., & Kendziorra, E. 2005, A&A, 443, 103
21. Canizares, C. R., Kriss, G. A., & Feigelson, E. D. 1982, ApJ, 258, 225
22. Canizares, C. R. & Winkler, P. F. 1981, ApJ, 246, L33
23. Cassam-Chenaï, G., Decourchelle, A., Ballet, J., et al. 2004a, A&A, 414, 545

24. Cassam-Chenaï, G., Decourchelle, A., Ballet, J., et al. 2004b, A&A, 427, 199
25. Chakrabarty, D., Pivovaroff, M. J., Hernquist, L., Heyl, J. S., & Narayan, R. 2001, ApJ, 548, 800
26. Chen, Y., Su, Y., Slane, P. O., & Wang, Q. D. 2004, ApJ, 6162, 885
27. Decourchelle, A., Ellison, D. C., & Ballet, J. 2000, ApJ, 543, L57
28. Decourchelle, A., Sauvageot, J. L., Audard, M., et al. 2001, A&A, 365, L218
29. DeLaney, T. & Rudnick, L. 2003, ApJ, 589, 818
30. DeLaney, T., Rudnick, L., Fesen, R. A., et al. 2004, ApJ, 613, 343
31. Dotani, T., Hayashida, K., Inoue, H., Itoh, M., & Koyama, K. 1987, Nature, 330, 230
32. Fahlman, G. G. & Gregory, P. C. 1982, ApJ, 261, L1
33. Fesen, R. A. 2001, ApJS, 133, 161
34. Flanagan, K. A., Canizares, C. R., Dewey, D., et al. 2004, ApJ, 605, 230
35. Fritz, G., Henry, R. C., Meekins, J. F., Chubb, T. A., & Friedman, H. 1969, Sci, 164, 709
36. Fujimoto, R. & et al. 1995, PASJ, 47, L31
37. Gaetz, T. A., Butt, Y. M., Edgar, R. J., et al. 2000, ApJ, 534, L47
38. Ghavamian, P., Blair, W. P., Long, K. S., et al. 2005, AJ, 130, 539
39. Gorenstein, P., Harnden, F. R., & Tucker, W. H. 1974, ApJ, 192, 661
40. Gorenstein, P., Hughes, J. P., & Tucker, W. H. 1994, ApJ, 420, L25
41. Gotthelf, E. V., Koralesky, B., L.Rudnick, et al. 2001, ApJ, 552, L39
42. Gotthelf, E. V., Petre, R., & Vasisht, G. 1999, ApJ, 514, L107
43. Haberl, F. & Pietsch, W. 2001, A&A, 373, 438
44. Hamilton, A. J. S., Chevalier, R. A., & Sarazin, C. L. 1983, ApJS, 51, 115
45. Hamilton, A. J. S., Sarazin, C. L., & Szymkowiak, A. E. 1986a, ApJ, 300, 698
46. Hamilton, A. J. S., Sarazin, C. L., & Szymkowiak, A. E. 1986b, ApJ, 300, 713
47. Harrus, I. M., Hughes, J. P., & Helfand, D. J. 1996, ApJ, 464, L161
48. Hasinger, G., Aschenbach, B., & Truemper, J. 1996, A&A, 312, L9
49. Hayashi, I., Koyama, K., Ozaki, M., et al. 1994, PASJ, 46, L121
50. Hearn, D. A., Larsen, S. E., & Richardson, J. A. 1980, ApJ, 235, L67
51. Helfand, D. J. & Becker, R. H. 1984, Nature, 307, 215
52. Hendrick, S. P., Borkowski, K. J., & Reynolds, S. P. 2003, ApJ, 593, 370
53. Holt, S. S., Gotthelf, E. V., Tsunemi, H., & Negoro, H. 1994, PASJ, 46, L151
54. Hughes, J. P. 1999, ApJ, 528, 298
55. Hughes, J. P. 2000, ApJ, 545, L53
56. Hughes, J. P., Ghavamian, P., Rakowski, C. E., & Slane, P. O. 2003a, ApJ, 582, L95
57. Hughes, J. P., Hayashi, I., Helfand, D., et al. 1995, ApJ, 444, L81
58. Hughes, J. P., Rakowski, C. E., & Decourchelle, A. 2000, ApJ, 543, L61
59. Hughes, J. P., Slane, P. O., Burrows, D. N., et al. 2001, ApJ, 559, L153
60. Hughes, J. P., Slane, P. O., Park, S., Roming, P. W. A., & Burrows, D. N. 2003b, ApJ, 591, L139
61. Hwang, U., Decourchelle, A., Holt, S. S., & Petre, R. 2002, ApJ, 581, 1101
62. Hwang, U. & et al. 2004, ApJ, 615, L117
63. Hwang, U., Flanagan, K. A., & Petre, R. 2005, ApJ, 635, 355
64. Hwang, U. & Gotthelf, E. V. 1997, ApJ, 475, 665
65. Hwang, U., Holt, S. S., & Petre, R. 2000a, ApJ, 537, L119
66. Hwang, U., Hughes, J. P., & Petre, R. 1998, ApJ, 497, 833
67. Hwang, U. & Laming, J. M. 2003, ApJ, 597, 362
68. Hwang, U., Petre, R., & Hughes, J. P. 2000b, ApJ, 532, 970
69. Hwang, U., Szymkowiak, A. E., Petre, R., & Holt, S. S. 2001, ApJ, 560, L175
70. Immler, S., Aschenbach, B., & Wang, W. D. 2001, ApJ, 561, L107
71. Immler, S., Brown, P. J., Milne, P., et al. 2006, ApJ, 648, L119
72. Immler, S. & et al. 2005, ApJ, 632, 283
73. Immler, S. & Kuntz, K. D. 2005, ApJ, 632, L99
74. Itoh, H. 1977, PASJ, 29, 813
75. Itoh, M., Asai, K., Uno, S., et al. 1999, Astronomische Nachrichten, 320, 333

76. Kaplan, D. L., Frail, D. A., Gaensler, B. M., et al. 2005, ApJS, 153, 269
77. Katsuda, S. & Tsunemi, H. 2005, PASJ, 57, 621
78. Kawasaki, M., Ozaki, M., Nagase, F., Inoue, H., & Petre, R. 2005, ApJ, 631, 935
79. Kawasaki, M., Ozaki, M., Nagase, F., et al. 2002, ApJ, 572, 897
80. Khokhlov, A. M., Höflich, P. A., Oran, E. S., et al. 1999, ApJ, 524, L107
81. Kilgard, R. E. & et al. 2005, ApJS, 159, 214
82. Kinugasa, K. & Tsunemi, H. 1999, PASJ, 51, 239
83. Kohmura, Y. & et al. 1994, PASJ, 46, L157
84. Kong, A. K. H., Garcia, M. R., Primini, F. A., & Murray, S. S. 2002, ApJ, 580, L125
85. Kong, A. K. H., Sjouwerman, L. O., & Williams, B. F. 2004, AJ, 128, 2783
86. Kong, A. K. H., Sjouwerman, L. O., Williams, B. F., Garcia, M. R., & Dickel, J. R. 2003, ApJ, 590, L21
87. Koralesky, B., Rudnick, L., Gotthelf, E. V., & Keohane, J. W. 1998, ApJ, 505, L27
88. Kouveliotou, C. & et al. 2004, ApJ, 608, 872
89. Koyama, K., Petre, R., Gotthelf, E. V., et al. 1995, Nature, 378, 255
90. Laming, J. M. & Hwang, U. 2003, ApJ, 597, 347
91. Lazendic, J. S., Slane, P. O., Hughes, J. P., Chen, Y., & Dame, T. M. 2005, ApJ, 618, 733
92. Leising, M. & et al. 1994, ApJ, 431, L95
93. Levenson, N. A. & Graham, J. R. 2005, ApJ, 622, 366
94. Levenson, N. A., Graham, J. R., & Walters, J. L. 2002, ApJ, 576, 798
95. Lewis, K. L., Burrows, D. N., Hughes, J. P., et al. 2003, ApJ, 582, 770
96. Long, K. S., Reynolds, S. P., Raymond, J. C., et al. 2003, ApJ, 586, 1162
97. Markert, T. H., Canizares, C. R., Clark, G. W., & Winkler, P. F. 1983, ApJ, 268, 134
98. Marsden, D., Rothschild, R. E., Lingenfelter, R. E., & Puetter, R. C. 1996, ApJ, 470, 513
99. Matheson, H. & Safi-Harb, S. 2005, JRASC, 99, 140
100. Mathewson, D. S., Ford, V. L., Dopita, M. A., et al. 1983, ApJS, 51, 345
101. McCray, R. 2005, in Cosmic Explosions, ed. J. M. Marcaide & K. W. Weiler, 77
102. Miceli, M., Decourchelle, A., Ballet, J., et al. 2006, ArXiv Astrophysics e-prints
103. Michael, E. & et al. 2002, ApJ, 574, 166
104. Miyata, E., Tsunemi, H., Aschenbach, B., & Mori, K. 2001, ApJ, 559, L45
105. Miyata, E., Tsunemi, H., Kohmura, T., Suzuki, S., & Kumagai, S. 1998, PASJ, 50, 257
106. Miyata, E., Tsunemi, H., Koyama, K., & Ishisaki, Y. 2000, Advances in Space Research, 25, 555
107. Ögelman, H., Finley, J. P., & Zimmermann, H. U. 1993, Nature, 361, 136
108. Olbert, C. M., Clearfield, C. R., Williams, N. E., & Keohane, J. W. 2001, ApJ, 554, L205
109. Pannuti, T. G., Duric, N., Lacey, C. K., et al. 2002, ApJ, 565, 966
110. Park, S., Hughes, J. P., Slane, P. O., et al. 2003, ApJ, 592, L41
111. Park, S., Roming, P. W. A., Hughes, J. P., et al. 2002, ApJ, 564, 39
112. Park, S., Zhekov, S. A., Burrows, D. N., et al. 2005, ArXiv Astrophysics e-prints
113. Pavlov, G. G., Sanwal, D., Garmire, G. P., & Zavlin, V. E. 2002, in ASP Conf. Ser. 271: Neutron Stars in Supernova Remnants, ed. P. O. Slane & B. M. Gaensler, 247
114. Petre, R., Allen, G. E., & Hwang, U. 1999, Astron. Nachr., 320, 199
115. Petre, R., Becker, C. M., & Winkler, P. F. 1996, ApJ, 432, L43
116. Petre, R., Okada, K., Mihara, T., Makishima, K., & Colbert, E. J. M. 1994, PASJ, 46, L115
117. Pfeffermann, E., Aschenbach, B., & Predehl, P. 1991, A&A, 246, L28
118. Pietsch, W., Freyberg, M., & Haberl, F. 2005, A&A, 434, 483
119. Pietsch, W., Misanovic, Z., Haberl, F., et al. 2004, A&A, 426, 11
120. Plucinsky, P. P., Egger, R., Edgar, R. E., & McCammon, D. 1996, ApJ, 463, 224
121. Pooley, D. & et al. 2002, ApJ, 572, 932
122. Pravdo, S. H. & Smith, B. W. 1979, ApJ, 234, L195
123. Pravdo, S. H., Smith, B. W., Charles, P. A., & Tuohy, I. R. 1980, ApJ, 235, L9
124. Rakowski, C. E. 2005, in X-Ray Diagnostics of Astrophysical Plasmas: Theory, Experiment, and Observation, ed. R. Smith, 203
125. Rakowski, C. E., Ghavamian, P., & Hughes, J. P. 2003, ApJ, 590, 846

126. Rappaport, S., Petre, R., Kayat, M. A., et al. 1979, ApJ, 227, 285
127. Rasmussen, A. P., Behar, E., Kahn, S. M., den Herder, J. W., & van den Heyden, K. 2001, A&A, 365, L231
128. Read, A. M. & Pietsch, W. 2001, A&A, 373, 473
129. Reynolds, S. P. & Chevalier, R. A. 1981, ApJ, 245, 912
130. Reynolds, S. P. & Keohane, J. W. 1999, ApJ, 525, 368
131. Rho, J. & Borkowski, K. 2002, ApJ, 575, 201
132. Rho, J., Dyer, K. K., Borkowski, K., & Reynolds, S. P. 2002, ApJ, 581, 1116
133. Rho, J. & Petre, R. 1997, ApJ, 503, L167
134. Robinson, P. E., Ennis, J. A., Rudnick, L., DeLaney, T., & Petre, R. 2005, American Astronomical Society Meeting Abstracts, 206
135. Ryder, S., Staveley-Smith, L., Dopita, M., et al. 1993, ApJ, 416, 167
136. Safi-Harb, S., Dubner, G., Petre, R., Holt, S. S., & Durouchoux, P. 2005, ApJ, 618, 321
137. Sasaki, M., Stadlbauer, T. F. X., Haberl, F., Filipović, M. D., & Bennie, P. J. 2001, A&A, 365, L237
138. Schaudel, D., Becker, W., Aschenbach, B., et al. 2002, in Neutron Stars, Pulsars, and Supernova Remnants, ed. W. Becker, H. Lesch, & J. Trümper, 26
139. Schlegel, E. M., Kong, A., Kaaret, P., DiStefano, R., & Murray, S. 2004, ApJ, 604, 644
140. Schlegel, E. M. & Petre, R. 1993, ApJ, 412, L29
141. Serlemitsos, P. J., Boldt, E. A., Holt, S. S., & Ramaty, R. 1973, ApJ, 184, L1
142. Seward, F., Gorenstein, P., & Tucker, W. 1983, ApJ, 266, 287
143. Seward, F. D. 1990, ApJS, 73, 781
144. Seward, F. D. & Harnden, F. R. 1982, ApJ, 256, L45
145. Seward, F. D. & Harnden, F. R. 1994, ApJ, 421, 581
146. Seward, F. D., Harnden, F. R., & Helfand, D. J. 1984, ApJ, 287, L19
147. Shelton, R. L., Kuntz, K. D., & Petre, R. 2004a, ApJ, 611, 906
148. Shelton, R. L., Kuntz, K. D., & Petre, R. 2004b, ApJ, 615, 275
149. Slane, P., Hughes, J. P., Edgar, R. J., et al. 2001, ApJ, 548, 814
150. Slane, P. O., Gaensler, B. M., Dame, T. M., et al. 1999, ApJ, 525, 357
151. Smith, A., Jones, L. R., Peacock, A., & Pye, J. P. 1985, ApJ, 296, 469
152. Stockdale, C. J., Maddox, L. A., Cowan, J. J., et al. 2006, AJ, 131, 889
153. Sunyaev, R. & et al. 1987, Nature, 330, 227
154. Supper, R. & et al. 2001, A&A, 373, 63
155. Swartz, D. A., Ghosh, K. K., McCollough, M. L., et al. 2003, ApJS, 144, 213
156. Tanimori, T. & et al. 1998, ApJ, 497, L25
157. Thorstensen, J. R., Fesen, R. A., & van den Bergh, S. 2001, AJ, 122, 297
158. Torii, T., Tsunemi, H., Dotani, T., & Mitsuda, K. 1997, ApJ, 489, L145
159. Truelove, J. K. & McKee, C. F. 1999, ApJS, 120, 299
160. Tsunemi, H., Miyata, E., & Aschenbach, B. 1999, PASJ, 51, 711
161. Tsunemi, H., Yamashita, K., Masai, K., Hayakawa, S., & Koyama, K. 1986, ApJ, 306, 248
162. Tuohy, I. R. & Garmire, G. P. 1980, ApJ, 239, L107
163. Uno, S., Mitsuda, K., Inoue, H., et al. 2002, ApJ, 565, 419
164. van den Heyden, K. J., Behar, E., Vink, J., et al. 2002, A&A, 392, 955
165. Vasisht, G., Aoki, T., Dotani, T., Kulkarni, S. R., & Nagase, F. 1996, ApJ, 456, L59
166. Vink, J. 2005, in AIP Conf. Proc. 745: High Energy Gamma-Ray Astronomy, ed. F. A. Aharonian, H. J. Völk, & D. Horns, 160
167. Vink, J., Bloemen, H., Kaastra, J. S., & Bleeker, J. A. M. 1998, A&A, 339, 201
168. Vink, J. & Laming, J. M. 2003, ApJ, 584, 758
169. Vink, J., Laming, J. M., Gu, M. F., Rasmussen, A., & Kaastra, J. S. 2003, ApJ, 587, L31
170. Vink, J., Maccarone, M. C., Kaastra, J. S., et al. 1999, A&A, 334, 289
171. Warren, J. S. & et al. 2005, ApJ, 634, 376
172. Warren, J. S. & Hughes, J. P. 2004, ApJ, 608, 261
173. Warren, J. S., Hughes, J. P., & Slane, P. O. 2003, ApJ, 583, 260
174. Williams, B. F., Barnard, R., Garcia, M. R., et al. 2005, ApJ, 634, 365

175. Williams, B. F., Sjouwerman, L. O., A. K, . H. K., et al. 2004, ApJ, 615, 720
176. Williams, R. M., Chu, Y.-H., Dickel, J. R., et al. 2005, ApJ, 628, 704
177. Williams, R. M., Chu, Y.-H., Dickel, J. R., et al. 1999, ApJS, 123, 467
178. Willingale, R., Bleeker, J. A. M., van der Heyden, K. J., Kaastra, J. S., & Vink, J. 2002, A&A, 381, 1039
179. Wu, C.-C., Leventhal, M., Sarazin, C., & Gull, T. R. 1983, ApJ, 269, L5
180. Yamauchi, S., Koyama, K., Tomida, H., Yokogawa, J., & Tamura, K. 1999, PASJ, 51, 13
181. Yokogawa, J., Imanishi, K., Koyama, K., Nishiuchi, M., & Mizuno, N. 2002, PASJ, 54, 53
182. Zavlin, V. E., Pavlov, G. G., Sanwal, D., & Trümper, J. 2000, ApJ, 540, L25
183. Zhekov, S. A., McCray, R., Borkowski, K. J., Burrows, D. N., & Park, S. 2005, ApJ, 628, L127
184. Zimmermann, H.-U. & Aschenbach, B. 2003, A&A, 406, 969
185. Zimmermann, H. U., Lewin, W., Predehl, P., et al. 1994, Nature, 367, 621

18 The Interstellar Medium

D. Breitschwerdt, M. Freyberg, and P. Predehl

18.1 Introduction

18.1.1 Gas

Looking at the sky on a clear and dark night, the first and most impressive view is the myriad of stars twinkling on the firmament and emitting light from inconceivable distances. The space between them seems to be empty. Apart from other, apparently moving objects, such as comets and planets, it was thought for a long time that the universe was infinite and static, and Newton considered the stars to represent an absolute frame of reference. Progress in astronomy is often driven by technology. The increasing use of telescopes in the seventeenth and eighteenth centuries revealed structures hitherto unseen. Messier, hunting for comets, discovered similarly looking patches on the sky that were not moving and decided therefore to make a compilation (published in several catalogues of star clusters and nebulae until 1781), to avoid further confusion. Among them Andromeda (Messier 31, or M 31 for short) and Orion (M 42) are the most famous ones.[1] The physical nature of the nebulae became only clear with the advent of *spectroscopy* in the nineteenth century, its main proponents being Fraunhofer, who discovered absorption lines in the spectrum of the sun, and Bunsen and Kirchhoff who associated spectral lines in the laboratory with light emitted by specific atoms. By then it had become evident that the space between the stars, the so-called interstellar medium (ISM), is by no means empty.

The history of the ISM is generally thought to commence with the discovery of stationary CaII lines in the spectrum of the spectroscopic binary system δ Orionis [24]. There remained, however, a shadow of doubt, until it could be demonstrated unambiguously that these lines are *interstellar*, because they increase in strength with distance, and their velocity is only half of the value expected from Galactic rotation [43]. Further highlights in ISM research are the discovery of interstellar extinction by Trumpler (1930), who found that open star clusters become brighter with increasing distance, and, therefore, concluded an overestimate of the distance due to the interstellar extinction of starlight by about 1 magnitude per kiloparsec [70].

[1] Incidentally, M 1 is the famous Crab Nebula, a supernova remnant – and a typical object of the hot Interstellar Medium.

J.E. Trümper and G. Hasinger (eds.), *The Universe in X-Rays.*

Following a suggestion of Oort, the hyperfine structure transition of H I – because of the scarcity of atomic collisions in the ISM with an average density of $\sim 1\,\mathrm{cm}^{-3}$ – had been predicted to be observable in the 21 cm wavelength [72]. This was observationally confirmed by Oort and collaborators a few years later. Thus in the first half of the twentieth century the ISM consisted of cold neutral hydrogen, and warm ionized gas surrounding O and B stars in so-called H II regions with sizes calculated on the basis of ionization equilibrium by [65].

The birth of the hot ISM (HIM) had to await the launch of rockets and satellites, equipped with sensitive high energy radiation detectors. Remarkably, the first diffuse (apparently all-sky) background radiation ever discovered, i.e., before the cosmic microwave background, was in X-rays in 1962 [22], and its soft component (below 2 keV) in 1968 [5]. The interpretation of the measurements remained ambiguous, in particular the possible sources and their spatial distribution as we shall see below. Therefore, initially, the hallmark of the HIM became the O VI resonance line in the UV, discovered in absorption toward background stars with the Copernicus satellite, launched in 1972 as a result of Spitzer's insight into astrophysical plasmas and his promotion for mounting telescopes on satellites. It turned out that the widespread O VI line was not attributable to the circumstellar environment, because the column density increased with distance, thus leading to the establishment of the hot phase of the ISM [29, 49, 78]. Attempts to link this to the soft X-ray background (SXRB) mentioned above were doomed to fail, because the excitation temperatures were too different: collisionally ionized O VI traces gas of $\sim 3\times 10^5$ K, whereas the SXRB monitors $\sim 10^6$ K gas. In collisional ionization equilibrium (CIE) this difference is impossible to reconcile, since at SXRB temperatures the ion fraction of O VII is dominating O VI by more than two orders of magnitude.

It became soon clear that a new "phase"[2] of the ISM had been discovered, which should dominate by volume, and due to the implied temperature range ($2\times 10^5 \le T \le 2\times 10^6$ K) must be generated by shock heating in supernova remnants (SNRs). It was shown that a supernova rate of 1 per 50 years is sufficient to maintain a large scale Galactic tunnel network [13]. However, as we shall see below, the O VI line primarily traces older SNRs or superbubbles, the latter being generated by successive explosions in OB associations. It is, therefore, the *emission in soft X-rays* that gives us direct information on the structure and evolution of the hot ISM. To monitor the soft X-ray sky, a program launching sounding rockets, equipped with proportional counters and narrow bandpass filters, was initiated at the University of Madison, Wisconsin. The Wisconsin Survey (1983) [37] showed an anisotropic patchy distribution in the low energy, so-called C-band (0.16 – 0.284 keV), increasing in flux from disk to pole by about a factor of 2–3, not much different from the B-band (0.13 – 0.188 keV) and the Be-Band (0.07 – 0.111 keV). This is remarkable, as soft X-ray photons are very sensitive to photoelectric absorption from intervening neutral ISM (proportional to energy E^{-3}), with the cross-section between the latter two bands differing by a factor of about 6. These findings led to the so-called displacement [52] and Local Hot Bubble model [67]. The idea is that the solar system is

[2] Somehow loosely defined as the regions of stability in the pressure–temperature diagram.

surrounded by a bubble, filled with hot plasma with a temperature of about $\sim 10^6$ K and a density of $n \sim 5 \times 10^{-3}\,\mathrm{cm}^{-3}$ [58], *displacing* HI. Such a local HI void does indeed exist [20], and has similar extensions to the Local Bubble (LB; for recent observations cf. [32]).

A major break-through in the hot ISM research came with the launch of the ROSAT satellite [69], with its fast X-ray resolving optics [1] and the low noise PSPC detector (Position Sensitive Proportional Counter) [44]. It is fair to say that the ROSAT PSPC All-Sky Survey is still today an invaluable source of data showing the distribution of hot gas in the Galaxy. The data base is unrivalled in completeness and still complements the large observatories Chandra and XMM-Newton in the energy range $0.1 - 0.3$ keV, in which they are not very sensitive.

18.1.2 Dust

At optical and UV wavelengths our view of the structure of the Galaxy is severely limited by ISM absorption; at radio, infrared, and X-ray wavelengths (above energies of 1 keV), however, the interstellar medium becomes sufficiently transparent to allow direct observations of a significant fraction of the Galaxy. Specifically, at soft X-ray energies between 0.1 and 3 keV, the absorption cross section of the interstellar medium changes by more than three orders of magnitude, and therefore observations in the keV-range are excellently suited for probing the absorption of the interstellar medium toward distant galactic X-ray sources. Assuming a given set of element abundances, the absorption column density (usually expressed as an equivalent hydrogen column density) along a given line of sight can be measured. In this context, it is worthwhile noting that the X-ray absorption (at least at energies above 0.5 keV) is primarily produced by heavier elements, mostly by oxygen and iron [40]; further, at these energies the absorption cross sections exclusively arise from K or L shell absorption, and therefore the absorption at X-ray wavelengths is independent from the physical or chemical constitution of the absorbing matter. This situation must be contrasted with the case of optical extinction which arises primarily from scattering (rather than genuine absorption) off grains of dust (i.e., quite complex aggregates of atoms).

Naturally, dust grains also affect the propagation of X-rays in the interstellar medium. The interaction of X-rays with cosmic dust grains leads not only to mere absorption but also to scattering, which is strongly biased into forward direction. Therefore, X-ray sources behind sufficiently dense dust clouds are expected to be surrounded by halos of faint and diffuse X-ray emission. As a consequence, X-ray observations offer an unique advantage over observations at all other wavelength ranges: with one single measurement one can determine, first, from the observed X-ray cutoff the total extinction (which is mostly due to photoelectric absorption) along the direct line of sight toward a given X-ray source, and second, from the surrounding diffuse X-ray halo the total scattering (which is exclusively due to dust grains). Since the dust grains are also made from heavy elements, the correlation between the simultaneously measured effects of X-ray absorption and X-ray scattering should depend on the depletion of heavy elements in dust grains in the interstellar medium.

18.1.3 Outline

In the following, we will discuss the most important observations of the X-ray emitting gas in our Galaxy, mainly ROSAT and XMM-Newton, characterizing the HIM, and outline the history and state of the art of modelling (as of 2005) of the ISM. In Sect. 18.4 we summarize the knowledge on dust scattering halos, and show how they can be used to study the cold ISM using X-rays.

18.2 Observations of the Hot Interstellar Medium

18.2.1 The Pre-ROSAT Era

Since its discovery in 1968 [5], the soft X-ray background (SXRB) below 2 keV has been understood only rudimentarily because of its complexity. The main observational fact was an approximate 1:3 variation in intensity below 0.28 keV from galactic plane to galactic pole. Moreover, at higher energies a few large-scale emission features could be identified, such as the *North Polar Spur* (NPS) or the *Vela supernova remnant*, besides an otherwise quite homogeneous X-ray sky.

The general observed anticorrelation of X-ray intensity with galactic neutral hydrogen column density favored an "absorption model," which placed the emitting regions behind the absorbing material. This could not explain a nonvanishing flux in the galactic plane where the large absorbing column density would not allow the distant flux to be observable. "Interspersed" models mixed the emitting and absorbing material, however, had to clump the absorbing gas by more than allowed from 21 cm observations [8]. A "displacement model" suggests that the emitting X-ray flux originates in front of the absorbing material and that larger intensities are related to longer lines-of-sight through the emitting plasma [58]. This model could best explain the observations in the pre-ROSAT era. Not much soft X-ray flux from distant regions was expected. This model was supported by the fact that the solar system seemed to be embedded in a cavity almost void of neutral matter; and only at distances larger than $40-100$ pc an almost sudden increase in interstellar absorption (of the order of 10^{20} cm^{-2}) was observed in stellar spectra (e.g., [21]).

18.2.2 The Contributions by ROSAT

After the launch of ROSAT, the picture changed significantly. The ROSAT PSPC All-Sky Survey (RASS) provided a new large-scale data base in the soft X-ray band with higher spatial and spectral resolution than available before, accompanied by an extensive follow-up programme of pointed observations.

The following major components of the SXRB have been identified since the early days, mainly with ROSAT data: (i) unresolved Galactic point sources (e.g., stars, accreting compact objects), (ii) unresolved extragalactic point sources, comprising about 80% of the total emission (e.g., AGN), (iii) diffuse *local* emission (Local Bubble, Loop I), (iv) diffuse *distant Galactic* emission (hot ISM and Galactic halo), (v) a diffuse warm-hot intergalactic medium (WHIM), and (vi) a yet unquantified very local emission component due to charge exchange reaction of solar wind ions with geocoronal and heliospheric gas.

Maps of the diffuse soft X-ray background (after point source removal) in several energy bands have been constructed (see Fig. 18.1). The morphology of the $1/4$ and $3/4$ keV band maps reveals large differences: while the $3/4$ keV (and also 1.5 keV) emission show generally a homogeneous emission except for the *North Polar Spur* (NPS) and the Sco-Cen region (in the general direction of the galactic center) the softer $1/4$ keV emission is full of small-scale and large-scale variations. It is much more intense at high latitudes, and still of the order of $1/3$ in the galactic plane [63].

These maps can be combined to band ratio maps or to composite color images (see Fig. 18.2). Red color indicates soft emission while blue regions illustrate either hard emission or presence of strong absorption (e.g., in the galactic plane). Green indicates plasma emission related to supernova remnants and superbubbles.

The efficient background rejection of the ROSAT PSPC allowed for the first time a detailed imaging of the distribution of faint diffuse soft X-ray emission and allowed also the discovery of X-ray shadows. One of the first milestone exposures has been the observation of the crescent Moon, where the dark side casts a shadow onto distant X-ray emission, proving that the bulk of the soft X-rays were originating beyond the Moon [54]. Furthermore, in a pointed observation as well as in the ROSAT survey a shadow by the Draco cloud (distance $\sim$600 pc) has been detected in the $1/4$ keV band [9, 59]. The detected X-ray flux showed an anticorrelation with the cloud content as traced by the 21 cm HI line and IRAS 100 μm maps. For the first time clear evidence had been obtained that a major fraction ($>$50%) of the SXRB is emitted at large distances, and that the distribution of material responsible for the SXRB may be more complicated.

The ROSAT survey, however, suffered from the only moderate spectral resolution of proportional counters. A detailed spectroscopy was, therefore, not possible because of ambiguities in the spectral modelling. Various approaches were used to disentangle local and distant plasma components.

Looking toward opaque absorbers (e.g., dense molecular clouds) gives the foreground emission, off-cloud measurements sample both local and distant emission components: a comparison of both yields information on the distribution along the

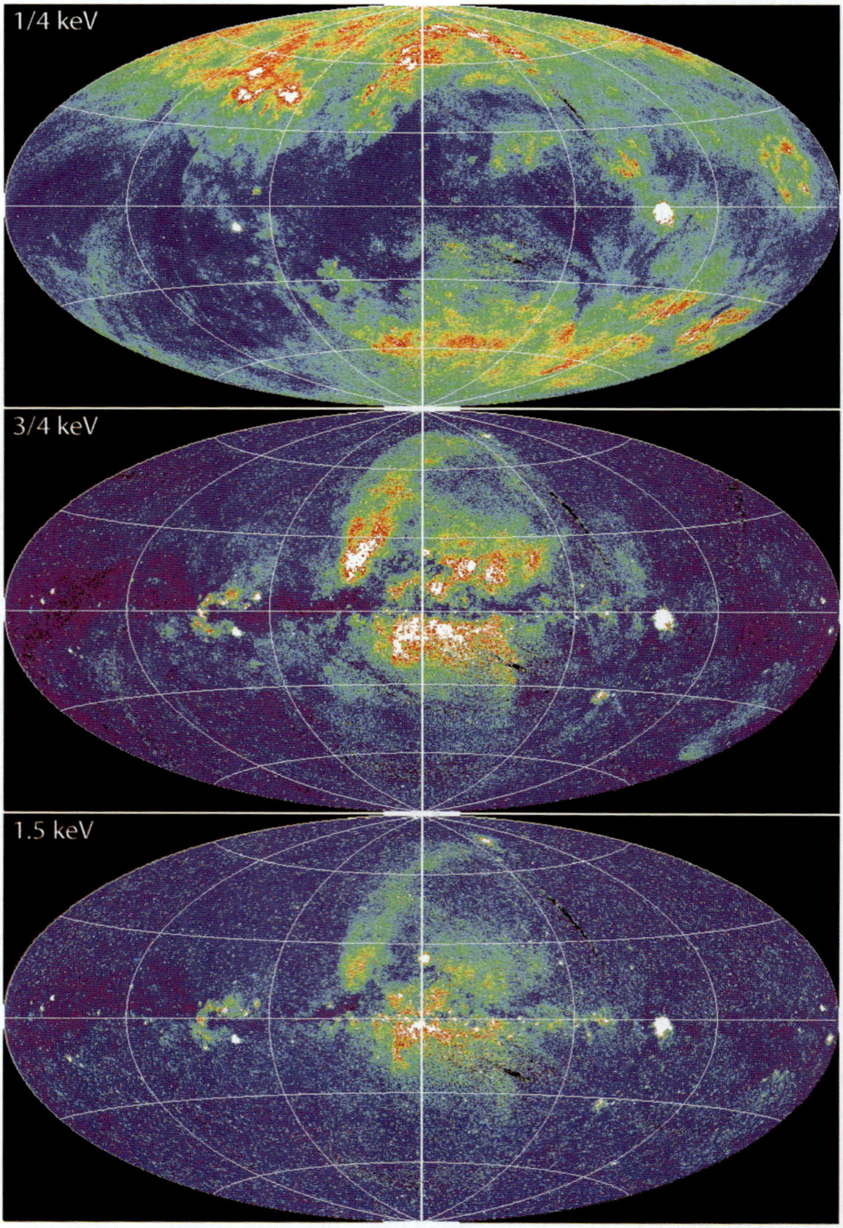

Fig. 18.1 *Top panel*: ROSAT PSPC All-Sky Survey maps of the SXRB in three energy bands (top to bottom): 1/4, 3/4, 1.5 keV. These maps have been used to construct the composite image in Fig. 18.2. Spatial resolution here is 20 arcmin; point sources have been excluded. The galactic centre is at $l = 0°$ with longitudes increasing to the left. The map production is described in [57], the details of the survey completion and these final maps can be obtained from [17]

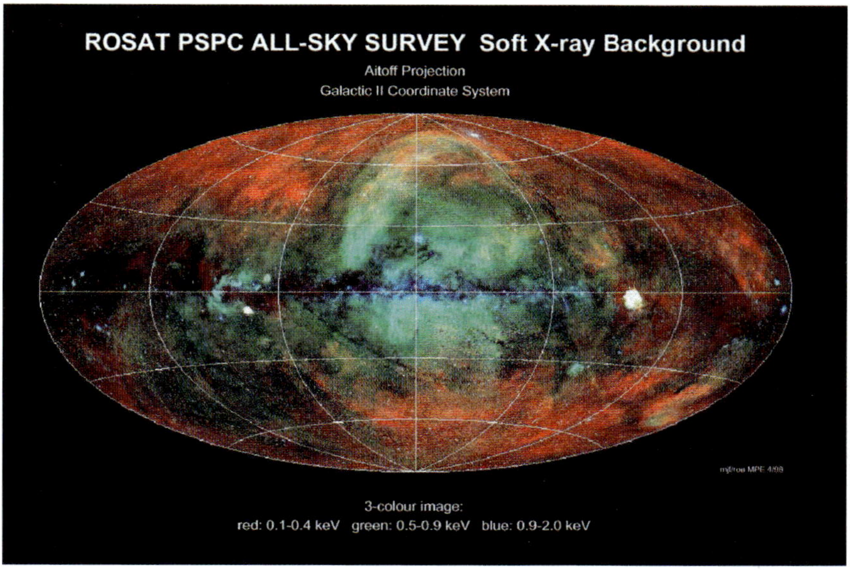

Fig. 18.2 Multispectral X-ray view of the soft X-ray background as seen in the ROSAT PSPC All-Sky Survey. This RGB image covers three energy bands (red 1/4 keV, green 3/4 keV, and blue 1.5 keV, respectively). The galactic centre is at $l = 0°$ with longitudes increasing to the left ([17], and references therein)

line of sight. Another approach is by means of band ratio maps, the most sensitive one is the ratio of the two softest ROSAT bands.[3] Variations of this band ratio away from known distinct emission features (e.g., supernova remnants, superbubbles) can be interpreted as spectral variations of foreground emission as optical depth of unity is already reached below $N_{\mathrm{H}} < 10^{20}\,\mathrm{cm}^{-2}$, at the Local Bubble boundary [64].

Further observations of high-latitude molecular clouds at known distances (e.g., MBM 12, [60]) and relations of other soft X-ray shadows indicated that the distant emission is distributed very inhomogeneously in the 1/4 keV band (cf. [63]). MBM 12 is one of the closest molecular clouds, at a distance similar or slightly larger than the extent of the Local Bubble.

One difficulty arises from the fact that the real zero level of the SXRB has to be known. "Noncosmic" contributions can be discriminated by means of temporal variations. In the RASS the PSPC (114 arcmin field-of-view diameter) scanned the sky in great circles along constant ecliptic longitude with a survey progression of about 4 arcmin in the ecliptic plane per revolution. Therefore, each part of the sky has been observed multiple times in these largely overlapping scans. Time variations within a measured sky portion in spectrum and intensity could thus be discriminated and assigned to "noncosmic" origins. The most prominent ones were solar X-rays scattered off Earth's residual atmosphere along the line of sight, and "long-term

[3] ROSAT bands R1 and R2, which together form the 1/4 keV band, band ratio defined as R2/R1.

enhancements" lasting for several $(4-9)$ h. The latter could be related to changes in solar wind parameters [15]. They are probably due to X-rays generated from charge exchange processes by the solar wind with neutral exospheric material (similar to X-rays produced at comets) [16]. Variations on time-scales of months or years (e.g., solar cycle) could not be determined by the only 6 months long RASS; these were proposed if heliospheric material were responsible. The measured zero-level of the SXRB may therefore only be an upper limit of the "true" zero-level. The future mission eROSITA [48] will scan the sky many times within 4 yrs and will, therefore, provide a much better database.

18.2.3 The CCD Era and The Future

X-ray CCDs like the ones used in the EPIC-pn CCD camera onboard XMM-Newton [66] provide higher spectral and also spatial resolution compared to proportional counters. It became possible to produce maps in narrower separated energy bands and to resolve some ambiguities in spectral modelling. In Fig. 18.3 we show the first X-ray shadow detected by XMM-Newton. The images have been obtained from three partly overlapping exposures across a part of the Ophiuchus molecular cloud with a strong gradient, so that the series covers the range from *on-cloud* to *off-cloud*, in the $0.5-0.9$ keV band (cf., [18]).

The very dense Bok globule *Barnard 68* with an absorbing hydrogen column density exceeding 10^{23} cm^{-2} allowed to determine the foreground emission to even higher energies above 1 keV [19].

Higher spectral resolution than possible with CCDs has been obtained using crystal spectrometers aboard the *DXS* instrument [53], and microcalorimeters flown with rockets [38]. However, these observations suffer from low quantum efficiency and deteriorated spatial or temporal resolution, as the high resolution data have to be rebinned due to low statistics. Therefore, the observations cover larger portions of the sky. Nevertheless, these SXRB spectra show clearly that they are dominated by spectral lines of ionized carbon and oxygen, e.g., O VII and O VIII), and are thus of thermal origin, though it is hardly possible to describe them by a single isothermal plasma in collisional ionization equilibrium. The already mentioned eROSITA mission will provide a major increase in collecting area at these low energies, a much better spectral response at soft energies compared with ROSAT or XMM-Newton, and a much longer time basis for screening against variable components.

18.3 Models of the Interstellar Medium

McKee and Ostriker have tried to put all the wisdom of the ISM at the time (1977) into a coherent picture, coined the three-phase ISM model [39]. One of the most interesting features is the transformation of the "phases" into one another by

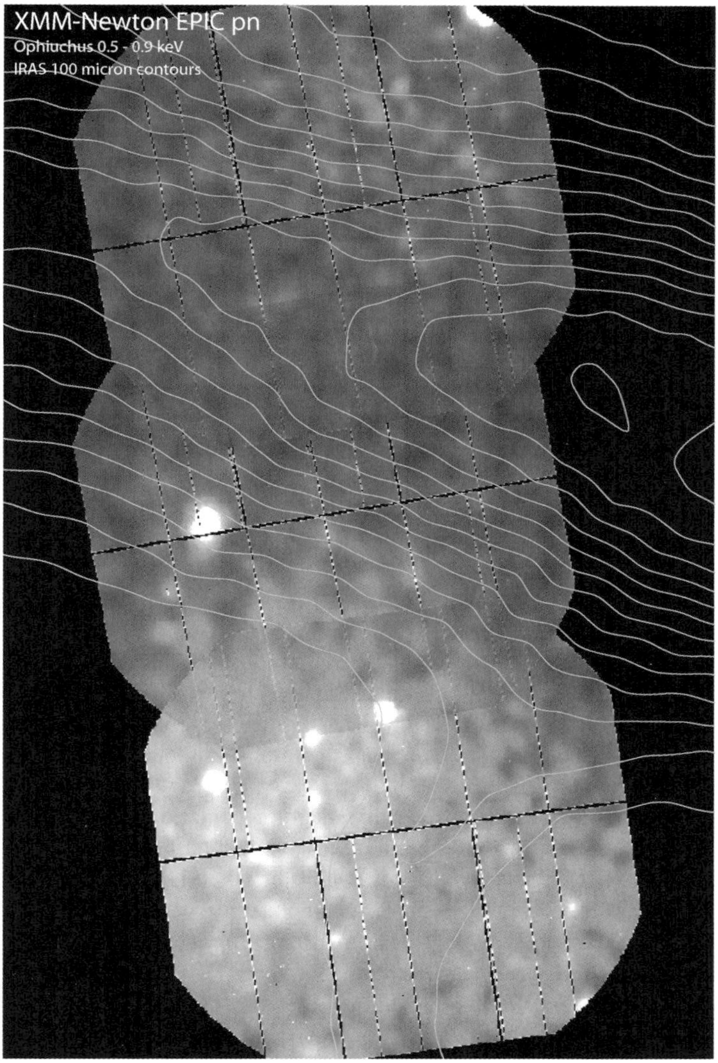

Fig. 18.3 Merged image of three XMM-Newton EPIC-pn observations of the Ophiuchus molecular cloud in the $0.5 - 0.9$ keV band with IRAS 100 μm contours overlaid. The cloud casts a shadow onto the X-ray background; the X-ray intensity is anticorrelated with the absorption by the molecular cloud (taken from [18])

different physical processes, such as e.g., radiative cooling, heat conduction, photoionization, and evaporation. The underlying assumption of the model is that the ISM is in global pressure equilibrium according to Spitzer's principle [64]. This might be expected with some optimism, if self-regulation is the dominant feature: on the one hand supernova explosions, and to a minor extent stellar winds and HII regions, inject energy locally. The expansion of remnants and bubbles will distribute it across the disk. A simple estimate shows that practically every point in the ISM within the star-forming Galactic disk will be overrun by a hot bubble within about 10^7 yrs or less. On the other hand, both radiative and adiabatic cooling will prevent the disk from disruption – a boiling pot needs pressure release valves to prevent explosion. One of these has been ignored in the original three-phase model, viz. the Galactic fountain, proposed by [55] to reconcile observations of the $1/4$ keV SXRB with the existence of highly ionized species such as O VI. It is, therefore, not too surprising that the three-phase model in its original form fails, e.g., in the predicted volume filling factors of the diffuse ionized gas (DIG) and especially of the HIM; the latter was predicted at least a factor of 2 too high. The fountain model gained wider attraction, when observations of supershells [26] and superbubbles [10] argued for about half of the supernova explosions to occur in associations (spatially and temporally correlated). This could lead to locally highly overpressured regions, which were hard to confine to the Galactic disk, although strong disk parallel magnetic fields were sometimes invoked to prevent break-out into the halo. These findings are summarized in another analytical model, the so-called "chimney" model [41]. Here break-out and establishment of the fountain flow was accounted for, but the physics of break-out was rather simplistic.

A heavily nonlinear system, such as the ISM, is however very sensitive to changes in the *initial conditions*, as it is known from deterministic chaos theory. Therefore, the richness of phenomena cannot be covered in an analytical model, and detailed numerical simulations are required. A comparison to the weather forecast might be appropriate. Short term predictions are generally very reliable, but as soon as nonlinear terms in the underlying equations of the system become important, the system undergoes bifurcations and breaks out into chaotic behavior, with e.g., Lorenz attractors being just a very simple representation.

The ISM is a multicomponent (containing gas, dust particles, cosmic rays, magnetic fields), and with respect to the gas, a multiphase (cold neutral medium, warm neutral medium, warm ionized medium, hot ionized medium) system. The fact that at least locally the energy density in the major components, gas (and here separately for kinetic, thermal and turbulent energy density), magnetic field and cosmic rays, amongst others, is comparable and of the order of about $1\,\mathrm{eV\,cm^{-3}}$, suggests that interaction between these components should be strong. As we shall see this is indeed the case. Having stressed this, it is clear that further progress in our understanding of the ISM must rely on numerical simulations, supplemented by the tools of nonlinear dynamics. Earlier work (e.g., [30, 51, 71, 74]) clearly showed the multiphase structure of the ISM and its turbulent nature, but remains exploratory, because of severe restrictions such as e.g., limited grid size, simulation in only two dimensions, too short evolution time scales. Therefore, in the remainder of this section most recent 3D high resolution simulations on large grids, based on adaptive mesh refinement, will be discussed [3, 4].

The magnetohydrodynamical (MHD) equations are solved on a Cartesian grid of $0 \le (x,y) \le 1$ kpc size in the Galactic plane and extending $-10 \le z \le 10$ kpc into the halo. This will ensure the full tracking of the Galactic fountain flow. The grid is centered on a typical ISM patch on the orbit of the Sun around the Galactic center, with a finest adaptive mesh refinement resolution of 1.25 pc in the layer $-1 \leq z \leq 1$ kpc. Following the observations, basic physical ingredients of the model are: (i) massive star formation in regions of converging flows ($\nabla \cdot v < 0$, where v is the gas velocity) where density and temperature are $n \geq 10\,\mathrm{cm}^{-3}$ and T ≤ 100 K, respectively; (ii) the distribution of masses and individual life times of stars are derived from a Galactic initial mass function; (iii) supernova explosions occur at a Galactic rate and with a canonical energy of 10^{51} erg (including the scale height distributions of types Ia, Ib, and II); (iv) the gas is immersed in a gravitational field provided by the stellar disk following [31]; (v) radiative cooling, assuming an optically thin gas in collisional ionization equilibrium, with a temperature cut off at 10 K, and uniform heating due to starlight varying with z [77] *; (vi) an initially disk parallel magnetic field composed of random (B_r) and uniform (B_u) components, with total field strength of 4.5 μG ($B_u = 3.1$ and $B_r = 3.3\,\mu$G). The establishment of the fountain flow up to heights of $5-10$ kpc in the vertical direction takes about $100-200$ Myr, so that a simulation time of 400 Myr seems adequate to recover the main features. Boundary conditions are periodic on the side walls of the computational box and outflow on its top and bottom.

The results are quite remarkable. Figure 18.4 shows the vertical distribution of the magnetic field (left panel) and the corresponding density (right panel) after 330 Myr of evolution. The former shows a detailed loop structure of the field, with large scale outflow in between. Therefore, an important result is that break-out of the disk cannot be inhibited, but only delayed. Once holes have been punched into the thick disk, the flow is channelled through them and out into the halo. The density distribution in z-direction exhibits a thin dense gas disk, with clouds ascending into and descending from the halo, i.e., the disk-halo-disk circulation is in full swing. Inspection of corresponding cuts parallel to the disk (see Fig. 18.5) reveals a filamentary structure of the magnetic field (left panel). Most interesting is the temperature distribution (right panel). First, hot bubbles with temperatures in excess of 10^6 K are clearly visible, and in case they are more evolved, i.e., larger in size, they are also elongated. This must be an effect of the magnetic field. Suppose the initial field was parallel to the disk. Then, as the bubble radius increases magnetic tension forces become stronger, and expansion along the field will prevail. It may even be possible that toward the end of expansion, provided that the bubble is not disrupted by nearby explosions, Maxwell stresses will cause the bubble to shrink again. Second, the cold medium exhibits a filamentary structure with lots of small-scale wiggles arising from turbulent motions.

Gas associated with bubbles should be clearly visible in soft X-rays, namely in O VII, whereas the ubiquitous O VI absorption line mainly traces gas in old remnants. Our long term simulations show that the volume filling factors in the disk are quite different from the three-phase model, in particular for the hot gas. Irrespective of the presence of a magnetic field, it is always less than 25%, in agreement with the disk

* It has been pointed out that the plasma is out of ionization equilibrium in many regions [6,7]; the implementation of non-equilibrium ionization (NEI) simulations is underway.

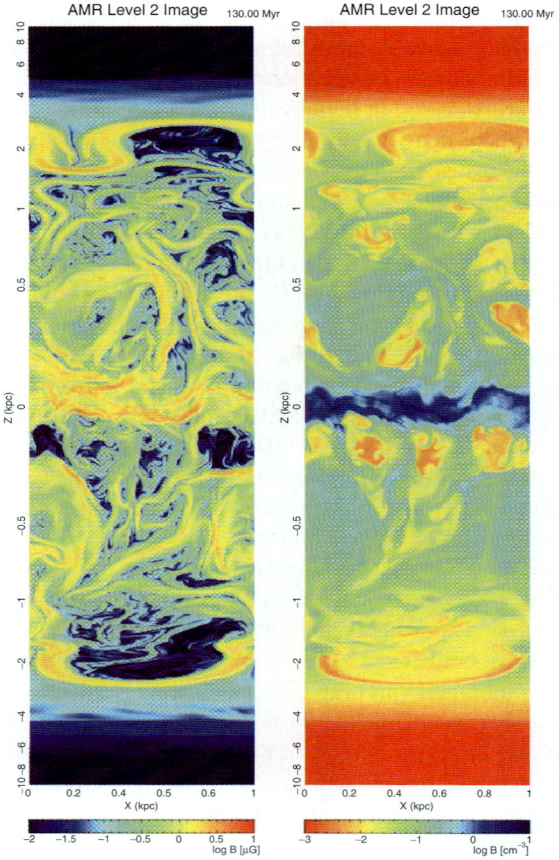

Fig. 18.4 Large scale numerical hydro- and magnetohydrodynamic simulations of a supernova driven ISM performed on parallel computers with typically 1 million hours of CPU time per run. The vertical distribution (perpendicular to the midplane) of the magnetic field (*left panel*) and the density (*right panel*) are shown at time $t = 330$ Myr for a mean field of 3 μG. The disk stretches horizontally from $z = 0$ (*midplane*), and the extension into the halo reaches $z = \pm 10$ kpc (abscissa range 1 kpc, ordinate range ± 10 kpc). The color coding is *red/blue* for low/high field (logarithmically $10^{-2} - 10^{1}$ μG) and density (logarithmically $10^{-3} - 10^{1}$ cm^{-3}), respectively. The break-out of superbubbles through the disk, and the generation of magnetic loops in the lower halo is striking

coverage of supershells in our and in external galaxies. The main reason for these low values is the pressure release in the disk because of the set-up of the fountain flow. It should be stressed at this point that numerical simulations can be rather arbitrary if they are resolution dependent. Therefore, it is reassuring that the filling factors shown in Table 18.1 do not change when the resolutions is increased by a factor of 2 (i.e., to 0.625 pc).

Another result that could not be derived from the analytical model is the large amount of turbulence in the ISM. This is not completely unexpected in a supernova

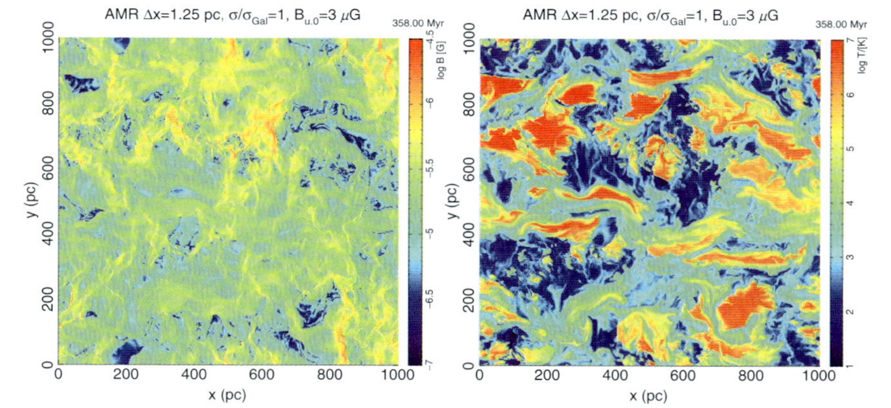

Fig. 18.5 The distribution of magnetic field (*left panel*) and the temperature (*right panel*) in the Galactic disk at time $t = 358$ Myr for a mean field of 3 μG. The color coding is the same as in Fig. 18.4. Note the elongated bubbles in the disk, and the filamentary structure of the cold (∼100 K) gas (from [4])

Table 18.1 Summary of the average values of volume filling factors, mass fractions, and root mean square velocities of the disk gas at the different thermal regimes for pure hydrodynamical (HD) and MHD runs, taken from Avillez and Breitschwerdt (2005)

T	$\langle f_V \rangle^a$ [%]		$\langle f_M \rangle^b$ [%]		$\langle v_{rms} \rangle^c$	
[K]	HD	MHD	HD	MHD	HD	MHD
<200 K	5	6	44.2	39.9	7	10
$200 - 10^{3.9}$	46	29	49.0	43.7	15	15
$10^{3.9} - 10^{4.2}$	10	11	4.4	8.5	25	21
$10^{4.2} - 10^{5.5}$	22	33	2.0	7.4	39	28
$>10^{5.5}$	17	21	0.3	0.5	70	55

[a] Occupation fraction (volume filling factor)
[b] Mass fraction
[c] Root mean square velocity in units of km s^{-1}

driven ISM, since expanding bubbles and colliding streams of gas generate lots of shear and vorticity. As the simulations show, turbulent mixing between hot and cold gas is fairly efficient, and there seems to be no need for heat conduction to promote phase transitions as in the three-phase model. A rather surprising result of these simulations, obtained by massively parallel computing is the amount of gas in classical *thermally unstable* regimes. Most of the disk mass is found in the $T \leq 10^{3.9}$ K gas, with the cold ($T \leq 200$ K) and thermally unstable gases ($200 < T \leq 10^{3.9}$ K) harboring on average 80 and 90% of the disk mass in the MHD and HD runs. About 55–60% of the thermally unstable gas ($200 < T \leq 10^{3.9}$ K) has $500 \leq T \leq 5000$ K in both runs, consistent with recent observations of the warm neutral medium [27].

An explanation for such an unexpected behavior, contradicting all analytical models, in particular the failure of the [14] criterion, may be found in the rôle turbulence (both supersonic and superalfvénic) plays. Turbulence can have a stabilizing effect thereby inhibiting local condensation modes. The situation is reminiscent of the existence of the solar chromosphere, consisting of gas at around 10^5 K in the thermally unstable regime. Here, heat conduction can prevent thermal runaway on small scales. In other words, diffusion processes may have a stabilizing effect. In our case, it is *turbulent diffusion* that replaces the rôle of conduction, again, most efficient for large wavenumbers. The turbulent viscosity $\nu_{\rm turb} \sim Re\,\nu_{\rm mol}$ can be orders of magnitude larger than the molecular viscosity, $\nu_{\rm mol}$; here Re is the Reynolds number of the highly turbulent flow. Therefore, with increasing eddy wavenumber, the eddy crossing time becomes shorter than the cooling time.

These new and exciting findings have provided new insight into the complex behavior of the ISM, and have opened up new paths to follow. However, it is fair to state that both on the observational and theoretical side, there are still more questions open than have been solved.

18.4 Dust Scattering Halos

18.4.1 History

Overbeck (1965) [42] was the first to point out the existence of dust scattering halos and their use as a powerful diagnostic tool. Further studies by many other authors developed a worked-out theory of the physics of X-ray scattering, and provided a detailed description of the observational signatures of X-ray scattering halos, e.g., [25,33]. While most of the early papers on scattering halos were of rather theoretical nature, the first detection of an X-ray halo around GX 339-4 was published by Rolf (1983) [50], using the Einstein Observatory.

A new era in the observational study of dust scattering halos began in 1990 with the launch of ROSAT [69]. While all previous studies of dust scattering halos relied on a careful subtraction of an instrumental halo (caused by scattering off the X-ray optics) from the observed brightness distribution, the superb performance of the ROSAT X-ray telescope (which is characterized by an extremely low level of mirror scattering) allowed, for the first time, a direct mapping of X-ray halos around many of the brighter galactic bulge sources. Another method for observing dust-scattering halos is using lunar occultations (Fig. 18.8). It was performed for the first time in 1992 with ROSAT [46]: The X-ray halo around GX 5-1 was directly visible while the compact X-ray source was actually behind the lunar disk. In an up to now most comprehensive study, 28 halos around galactic X-ray sources were detected and studied (e.g., Fig. 18.7) [45]. This led to fundamental relationships between X-ray scattering, absorption, and visual extinction. Also some information about the physical constitution of interstellar dust grains could be achieved.

Because of their good energy resolution, the current two large observatories, i.e., the Chandra Observatory and XMM-Newton are excellently suited for more detailed studies of X-ray scattering halos. Consequently, further progress is being made, particularly in studying the chemistry of interstellar grains.

18.4.2 X-Ray Scattering on Dust Grains

The physical situation we consider is the small-angle scattering of X-rays emitted by a bright X-ray source from dust grains in the ISM near the line of sight, which will create an X-ray halo around the point source. The scattered angles are small enough that, for astronomical purposes, we can assume that the path traveled by the unscattered primary X-rays is nearly identical to the paths of scattered photons, in terms of their distances and environments.

Traditionally, the scattering is treated by the so called Rayleigh-Gans theory, a simplification of the general Mie theory. This approximation makes two assumptions: First, the reflection from the surface of the grain is negligible, i.e., that $|m-1| \ll 1$ with m is the complex index of refraction of the grain. Second, the phase of the incident wave is not shifted, i.e., that $k_o a|m-1| \ll 1$, where k_o is the wave number of the X-ray and a is the radius of the grain. Here, we restrict ourselves to the Rayleigh-Gans approximation, although it has been shown that it holds for X-ray energies above 1 keV only [56]. The differential cross section for a single grain is given by [36]

$$\frac{d\sigma_{\rm sca}(\Theta_{\rm sca})}{d\Omega} = c_1 (\frac{2Z}{M})^2 (\frac{\rho}{3\,{\rm g\,cm^{-3}}})^2 a(\mu{\rm m})^6 [\frac{F(E)}{Z}]^2 \, \Phi^2(\Theta_{\rm sca}) \tag{18.1}$$

where $c_1 = 1.1\,{\rm cm^2\,sr^{-1}}$, Z is the mean atomic charge, M the mean molecular weight (in amu) ρ the mass density, E the energy in keV, $F(E)$ the atomic scattering factor as, for instance, given by [28], $\theta_{\rm sca}$ the scattering angle, and $\Phi(\Theta_{\rm sca})$ a "form factor," which depends on the shape of the particle. For spherical particles, the form factor is well approximated by a Gaussian, which yields, simplified, for the differential cross section [35]

$$\frac{d\sigma_{\rm sca}(\Theta_{\rm sca})}{d\Omega} \sim \frac{a^6}{E^2} \exp(-0.4575 E^2 a^2 \frac{\alpha^2}{(1-x)^2}) \tag{18.2}$$

with α the radial distance and x the fractional distance between grain and observer (see Fig. 18.6). With integrating the differential cross section over all solid angles we obtain the *total* scattering cross section, again simplified:

$$\sigma_{\rm sca} \sim a(\mu{\rm m})^4 E({\rm keV})^{-2}\,{\rm cm}^2. \tag{18.3}$$

The total scattering cross section depends roughly on E^{-2} and on the grain size with a^4. Polycyclic hydrogen carbonates (PAHs), which are considered to be ultra

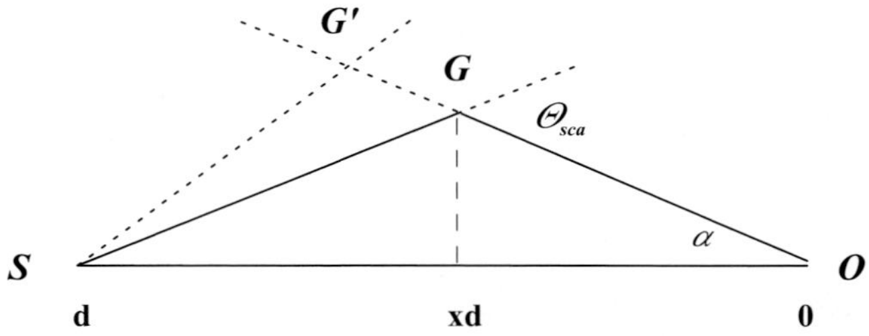

Fig. 18.6 Scattering geometry with X-ray source *S*, observer *O*, and grains *G*. Typical values for the angular distance α from the direct line of sight toward the source are of the order of 10 arcmin

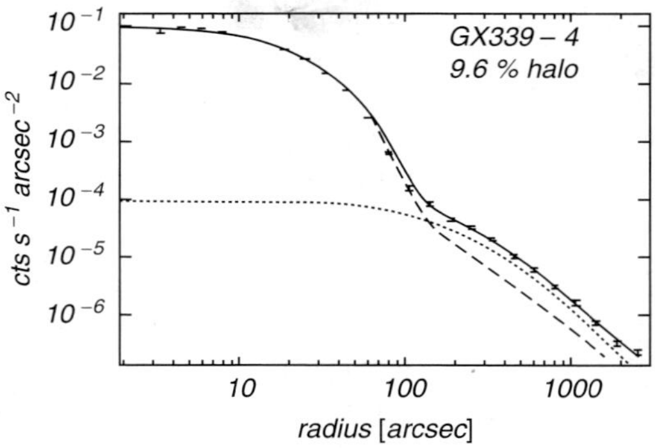

Fig. 18.7 The radial brightness distribution of an X-ray source with bright halo (from [46]). The brightness distribution (*solid line*) is a composite of a dust scattering halo (*dotted line*) and the instrument's point response function (*dashed line*)

small particles [23] play no role in small angle X-ray scattering. On the other hand, the grain size distribution follows approximately a powerlaw $\sim a^{-3.5}$, which means that over a size range around the mean value $a = 0.1\,\mu\mathrm{m}$ all grain sizes contribute similarly to the scattering halo.

The observed radial brightness profile around a source is derived by integration of the differential cross section over all grains along the line of sight and a large solid angle, the size distribution of grains and the photon spectrum. According to (18.2) large grains, high energies, and dust in the vicinity of the source produce narrow brightness profiles, while small grains, low energies, and dust close to the observer produce wide profiles. The scattering cross section is also a strong function of the scattering angle. Since scattering close to the observer occurs under smaller scattering angles (at given angular distance), it contributes much more than scattering in the vicinity of the source. The fractional scattered intensity is $I_{\mathrm{halo}} = I_{\mathrm{total}}(1 - \exp(-\tau_{\mathrm{sca}}))$, with the optical depth in scattering $\tau_{\mathrm{sca}} = \sigma_{\mathrm{sca}} N$, with N the dust column density between source and observer. Multiple scattering can be neglected in most cases because the absorption cross section is much greater than the scattering cross section.

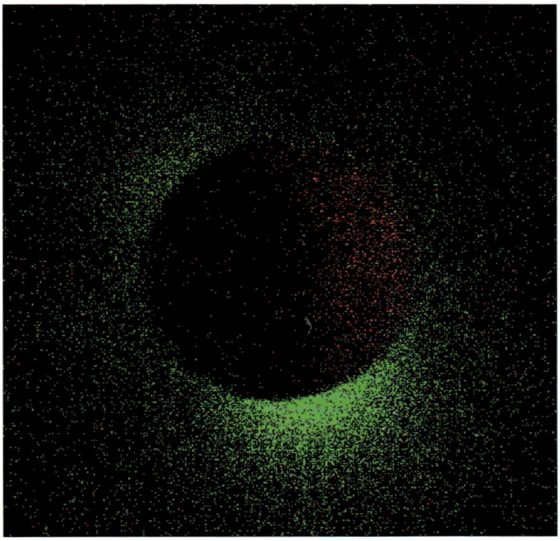

Fig. 18.8 In rare cases using the moon as a "shutter," a halo can be made directly visible without disturbing instrumental effects. This halo around Sco X-1 was observed with ROSAT in 1998. The image is centered with respect to the moon while Sco X-1 is moving; therefore, the halo appears to be asymmetric. The soft photons from the moon are coded in *red*, the harder photons from Sco X-1 in *green*

Scattered radiation has to travel along a slightly longer path than the direct, unscattered light. Any intensity variation of the source, therefore, occurs somewhat delayed in the halo. Although proposed as method to measure the distance of variable X-ray sources already in 1973 [68], it took almost 30 yrs, before this effect could be measured for the first time directly. The attractiveness of the "halo-method" is that it yields a *geometrical* rather than an *physical* distance as other methods like the 21 cm absorption, visual extinction, or X-ray absorption. The time delay is given by

$$\mathrm{d}t = \frac{d}{2c}\frac{x\alpha^2}{1-x} = 1.15\ d\ \frac{x\alpha^2}{1-x} \qquad (18.4)$$

if d is given in kpc and α in arcs (c is the speed of light).

18.4.3 Observations and Results

18.4.3.1 Physical Constitution of Dust Grains

The shape of the halo profile is primarily determined by the grain size or size distribution, respectively, and the dust distribution along the line of sight. Various studies show that the derived grain size distributions are consistent with common dust

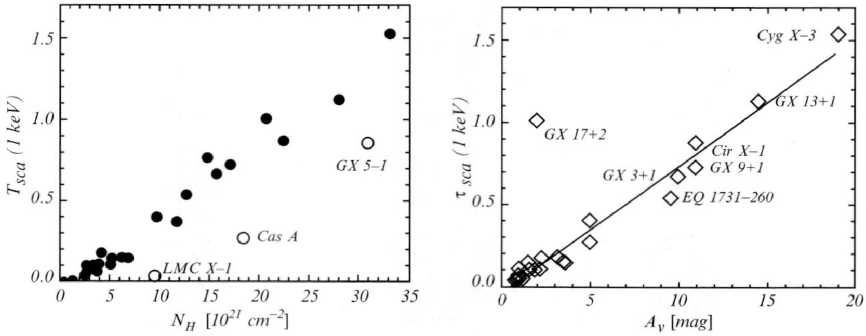

Fig. 18.9 Optical depth in X-ray scattering vs. X-ray absorption (expressed as equivalent hydrogen column density (*left*) and vs. visual extinction of the optical counterpart (*right*). These diagrams led to the conclusion that LMC X-1, Cas A, and GX 5-1 are, in addition to the interstellar absorption, locally absorbed. GX 17+1 was optically misidentified

models [34, 75]. In some cases, the halo profile reflects also the inhomogeneous structure of the milky way, e.g., [11].

From the clear correlation between dust scattering and visual extinction (Fig. 18.9), one derives that both visual extinction as well as X-ray scattering must essentially be due to the same dust grains.

Since absorption is a measure for the total amount of heavy elements in the interstellar medium, and scattering a measure for the amount of those elements locked in grains, a correlation between both quantities should reveal the degree of depletion of metals in the ISM. A simple calculation shows in first place that either the depletion is much less than already supposed or dust is not composed of solid grains but more of fluffy grains [45]. This result has later been debated in the light of a revised scattering theory [12], and updated elemental abundances [76].

18.4.3.2 Chemistry of Dust Grains

Making use of the intrinsically good energy resolution of imaging semiconductor X-ray detectors, the *pure* scattered radiation, i.e., after eliminating external contribution (instrument and source spectrum) can be studied. The moderate ISM absorption toward Cyg X-2 for instance allowed a precise study of the halo spectrum down to 0.4 keV [12] (Fig. 18.10). Scattering features from oxygen, magnesium, and silicon could be fitted with dust model containing a major contribution from olivines ($Fe_{2-x}Mg_xSiO_4$) and pyroxenes ($Fe_{1-x}Mg_xSiO_3$) [75].

18.4.3.3 Distance Determination of Variable X-Ray Sources

Although the idea is fascinating, its realization is difficult to perform; distances of only a few sources could be measured using this method. Since usually the scattering

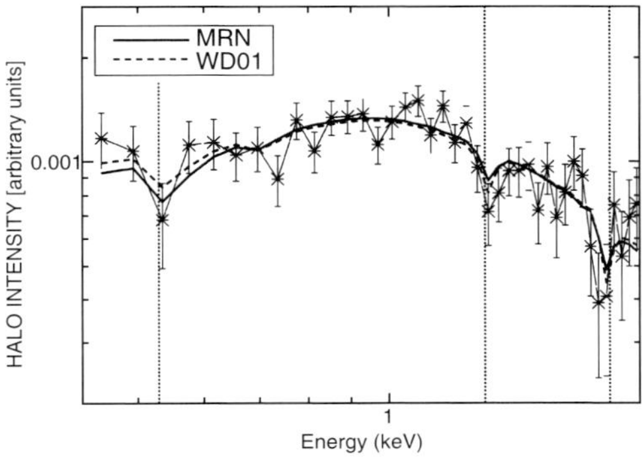

Fig. 18.10 Spectrum of the halo around Cyg X-2 between 2.1′ and 4′ angular distance [12]. Data taken with the XMM-NewtonEPIC-pn camera compared with the best fit dust models WD01 ([75], *dashed*) containing 27% pyroxenes and 73% olivines, and the MRN model ([34], *solid line*)

does not happen at one particular point between source and observer, the time delay is accompanied by a damping of the lightcurve variation, which becomes more pronounced at larger angular distances. Best suited are sources with a sharp drop in intensity (eclipses) seen by a telescope with very good angular resolution. A few attempts have been made even without an imaging telescope by modeling the lightcurve at different energies: the lightcurve at higher energies is intrinsically produced, at lower energies (with halo) the time delay and damping effects become dominant. The direct, imaging method worked for the first time with Cyg X-3 [47], recently also with other sources [2].

An interesting consequence arise by reversing the method: if the source is at "infinite" or known distance, the distances toward dust clouds in the foreground can be measured [11]. Recently, an expanding halo could be detected around a GRB. This is due to the afterglow scattered on dust in our galaxy. Thus, the distance to two different dust clouds could be well determined [73].

References

1. Aschenbach, B. 1988, Appl. Opt. 27, 1404
2. Audley, M.D., Nagase, F., Mitsuda, K., Angelini, L., Kelley, R.L. 2006, Mon. Not. Roy. Astron. Soc., 367, 1147
3. Avillez, M.A., Breitschwerdt, D. 2004, Astron. Astrophys., 425, 899
4. Avillez, M.A., Breitschwerdt, D. 2005, Astron. Astrophys., 436, 585
5. Bowyer, C.S., Field, G.B., Mack, J.E. 1968, Nature, 217, 32
6. Breitschwerdt, D., Schmutzler, T. 1994, Nature, 371, 774
7. Breitschwerdt, D., Schmutzler, T. 1999, Astron. Astrophys. 347, 650
8. Burrows, D.N. 1989, Astrophys. J., 340, 775
9. Burrows, D.N., Mendenhall, J.A. 1991, Nature, 351, 629
10. Cash, W., Charles, P., Bowyer, S., et al. 1980, Astrophys. J., 238, L71
11. Clark, G. 2004, Astrophys. J., 610, 956

12. Costantini, E. 2004, Absorption and scattering of interstellar dust. PhD Thesis, Ludwig-Maximilians Universität, München
13. Cox, D.P., Smith, B.W. 1974, Astrophys. J., 189, L105
14. Field, G.B. 1965, Astrophys. J., 142, 531
15. Freyberg, M.J. 1994, Untersuchung der kosmischen und nichtkosmischen Komponenten der Röntgenhintergrundstrahlung mit ROSAT, PhD Thesis (in German), Ludwig-Maximilians-Universität, München
16. Freyberg, M.J. 1998, in D., Breitschwerdt, M.J., Freyberg, J., Trümper (eds.), The Local Bubble and Beyond. Lyman-Spitzer Colloquium, Proceedings of the IAU Colloquium No. 166, LNP 506, 311
17. Freyberg, M.J., Egger, R. 1999, in B., Aschenbach, M.J., Freyberg (eds.), Highlights in X-ray Astronomy – International Symposium in honour of Joachim Trümper's 65th birthday, MPE Report 272, 278
18. Freyberg, M.J. 2004, Astrophys. Space Sci., 289, 229
19. Freyberg, M.J., Breitschwerdt, D., Alves, J. 2004, Memorie della Societa Astronomical Italiana (MmSAI), 75, 509
20. Frisch, P.C., York, D.G. 1983, Astrophys. J., 271, L59
21. Fruscione, A., Haukins, I., Jelinsky, P., Wiercigroch, A. 1994, Astrophys. J., 94, 127
22. Giacconi, R., Gurski, H., Paolini, F., Rossi, B.B. 1962, Phys. Rev. Lett. 9, 439
23. Greenberg, J.M., Hage, J.I. 1990, Astrophys. J., 361, 260
24. Hartmann, J. 1904, Sitzb. Kgl. Akad. Wiss. 527 (translated in Astrophys. J., 19, 268)
25. Hayakawa, S. 1970, Prog. Theor. Phys. 43, 1224
26. Heiles, C. 1980, Astrophys. J., 235, 833
27. Heiles, C., Troland, T.H., Astrophys. J., 586, 1067
28. Henke, B.L. 1981, in D.T., Attwood, B.L., Henke (eds.), Low Energy X-ray Diagnostics, AIP, New York, 146
29. Jenkins, E.B., Meloy, D.A. 1974, Astrophys. J., 193, L121
30. Korpi, M.J., Brandenburg, A., Shukurov, A., Tuominen, I., Nordlund, Å. 1999, Astrophys. J., 514, L99
31. Kuijken, K., Gilmore, G. 1989, Mon. Not. Roy. Astron. Soc., 239, 651
32. Lallement, R., Welsh, B.Y., Vergely, J.L., Crifo, F., Sfeir, D. 2003, Astron. Astrophys. 411, 447
33. Martin, P.G., 1970, Mon. Not. Roy. Astron. Soc., 149, 221
34. Mathis, J.S., Rumpl, W., Nordsieck, K.M. 1977, Astrophys. J., 217, 425
35. Mathis, J.S., Lee, C.-W. 1991, Astrophys. J., 376, 490
36. Mauche, C.W., Gorenstein, P. 1986, Astrophys. J., 302, 371
37. McCammon, D., Burrows, D.N., Sanders, W.T., Kraushaar, W.L. 1983, Astrophys. J., 269, 107
38. McCammon, D., Almy, R., Apodaca, E., et al. 2002, Astrophys. J., 576, 188
39. McKee, C.F., Ostriker, J.P. 1977 Astrophys. J., 218, 148
40. Morrison, R., McCammon, D. 1983, Astrophys. J., 270, 119
41. Norman, C.A., Ikeuchi, S. 1989, Astrophys. J., 345, 372
42. Overbeck, J.W. 1965, Astrophys. J., 141, 864
43. Plaskett, J.S., Pearce, J.A. 1933, Publ. Dom. Astr. Obs. Victoria, B.C. 5, 167
44. Pfeffermann, E., Briel, U.G., Hippmann, H., et al. 1986, Proc. SPIE 733, 519
45. Predehl, P., Schmitt, J.H.M.M. 1995, Astron. Astrophys., 293, 889
46. Predehl, P., Schmitt, J.H.M.M., Snowden, S.L., Trümper, J. 1992, Science, 257, 935
47. Predehl, P., Burwitz, V., Paerels, F., Trümper, J. 2000, Astron. Astrophys., 357, L25
48. Predehl, P., Friedrich, P., Hasinger, G., Pietsch, W. 2003, Astron. Nachr., 324, 128
49. Rogerson, J.J., York, D.G., Drake, J.F., et al. 1973, Astrophys. J., 181, L110
50. Rolf, D.P. 1983, Nature, 302, 46
51. Rosen, A., Bregman, J.N. 1995, Astrophys. J., 440, 634
52. Sanders, W.T., Kraushaar, W.L., Nousek, J.A., Fried, P.M. 1977, Astrophys. J., 217, L87

53. Sanders, W.T., Edgar, R.J., Kraushaar, W.L., McCammon, D., Morgenthaler, J.P. 2001, Astrophys. J., 554, 694
54. Schmitt, J.H.M.M., Snowden, S.L., Aschenbach, B., et al. 1991, Nature, 349, 583
55. Shapiro, P.R., Field, G.B. 1976, Astrophys. J., 205, 762
56. Smith, R.K., Dwek, E. 1998, Astrophys. J., 503, 831
57. Snowden, S.L., Egger, R., Freyberg, M.J., et al. 1997, Astrophys. J., 485, 125
58. Snowden, S.L., Cox, D.P., McCammon, D., Sanders, W.T. 1990, Astrophys. J., 354, 211
59. Snowden, S.L., Mebold, U., Herbstmeier, U., Hirth, W., Schmitt, J.H.M.M. 1991, Science, 252, 1529
60. Snowden, S.L., McCammon, D., Verter, F. 1993, Astrophys. J., 409, L21
61. Snowden, S.L., Freyberg, M.J., Plucinsky, P.P., et al. 1995, Astrophys. J., 454, 643
62. Snowden, S.L., Egger, R., Finkbeiner, D.P., Freyberg, M.J., Plucinsky, P.P. 1998, Astrophys. J., 493, 715,
63. Snowden, S.L., Freyberg, M.J., Kuntz, K.D., Sanders, W.T. 2000, Astrophys. J., 128, 171
64. Spitzer L., Jr. 1956, Astrophys. J., 124, 20
65. Stroemgren, B. 1939, Astrophys. J., 89, 526
66. Strüder, L., Briel, U.G., Dennerl, K., et al. 2001, Astron. Astrophys. 365, L18
67. Tanaka, Y., Bleeker, J.A.M. 1977, Sp. Sci. Rev. 20, 815
68. Trümper, J., Schönfelder, V. 1973, Astron. Astrophys., 25, 445
69. Trümper, J. 1983, Adv. Space Res., 2, 241
70. Trumpler, R.J. 1930, Publ. Astron. Soc. Pac., 42, 267
71. Vázquez-Semadeni, E., Passot, T., Pouquet, A. 1995, Astrophys. J., 441, 702
72. van de Hulst, H.C. 1945, Ned. Tijdschr. Natuurk. 11, 201
73. Vaughan, S., Willingale, R., O'Brien, P.T., et al. 2004, Astrophys. J., 603, 5
74. Wada, K., Norman, C.A. 2001, Astrophys. J., 547, 172
75. Weingartner, J.C., Draine, B.T. 2001, Astrophys. J., 548, 296
76. Wilms, J., Allen, A., McCray, R. 2000, Astrophys. J., 542, 914
77. Wolfire, M.G., McKee, C.F., Hollenbach, D., Tielens, A.G.G.M. 1995, Astrophys. J., 453, 673
78. York, D.G. 1974, Astrophys. J., 193, L127

19 The Galactic Center

P. Predehl

19.1 Introduction

The center of our Galaxy lies, within the constellation Sagittarius at a distance of $\sim$8 kpc, behind a large column density of gas and dust. This prevents observations over a wide band ranging from optical to soft X-ray wavelengths. The exploration of this region had to await the instrumental developments of, first, radio, later also of submillimeter and infrared, and most recently, X-ray- and γ-ray astronomy. The Galactic Center is one of the scientifically most interesting regions for astrophysics. It is the closest galactic nucleus, a factor of 100 closer than the next external nucleus, M31. Thus, the relative proximity of our Galactic Center, combined with high angular resolution instruments now available, provides the opportunity to differentiate and study a wealth of highly unusual objects, possibly unique, but probably standard occupants of a normal galactic core.

19.1.1 Morphology of the Galactic Center

The first detection of the Galactic Center was made at radio frequencies ("Sgr A," Fig. 19.1). With observations at increasing resolution it turned out that Sgr A is a rather dense and complex region: at the dynamical center of the Galaxy, we have the nonthermal compact radio source Sgr A*. In recent years, observations made in the near-infrared revealed orbital motions of stars in the immediate vicinity of Sgr A*. A detailed study of these motions confirmed that Sgr A* is most probably a supermassive ($\sim 3 \times 10^6\, M_\odot$) black hole, e.g., [23]. Sgr A* is located at the center of the thermal radio source Sgr A West, which consists of a spiral-shaped group of thermal gas clouds. Sgr A West is surrounded by the circumnuclear disk (CND), a shell or ring of molecular matter. The radius of the CND is about 30". The CND is clumpy and rotating around IRS 16, a dense cluster of hot stars. The nonthermal shell-like radio source Sgr A East is surrounding Sgr A West in projection, but its center is offset by about 50". The nonthermal (and expanding) shell, in turn, is surrounded by a dust ring or molecular ridge. The molecular cloud M-0.02+0.07 (or "+50 km s^{-1} cloud") is located north-east of Sgr A East. At larger distances up

J.E. Trümper and G. Hasinger (eds.), *The Universe in X-Rays.*

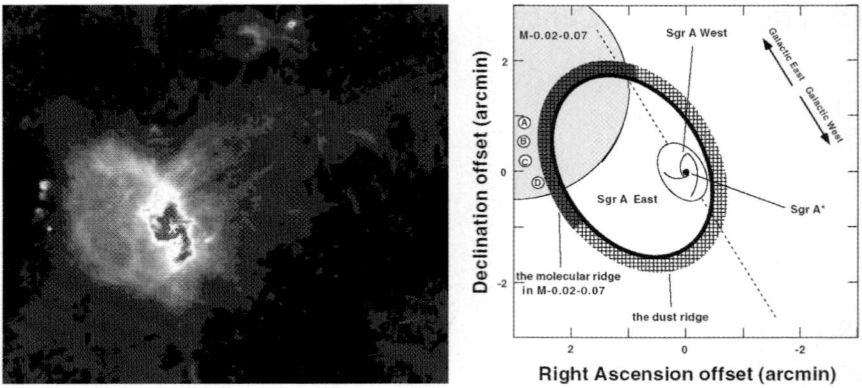

Fig. 19.1 VLA radio continuum image of the Galactic Center showing the shell-like structure of nonthermal Sgr A East and the spiral-shaped structure of Sgr A West at $\lambda = 6\,\mathrm{cm}$ with a resolution of $3.4'' \times 2.9''$ (*left*, [28]). Schematic diagram (*right*, [9]) showing the sky locations and rough sizes and shapes of Galactic Center sources. The coordinates are relative with respect to the compact nonthermal radio source Sgr A*. One arcminute corresponds to about 2.3 pc at the distance of 8 kpc

to 100 pc, there is a variety of clusters of bright stars, synchrotron-bright filaments, supernova remnants, and giant molecular clouds.

The nonthermal radio emission from Sgr A East in the direction of Sgr A West is heavily absorbed by Sgr A West [28], a convincing indication of the fact that, along the line of sight, Sgr A West lies in front of the Sgr A East shell, although at an uncertain distance. Possibly, the front edge of the expanding shell has reached and passed through Sgr A*, at the center of Sgr A West. Whether the outward shock pushed gas toward the central black hole causing a bright flash about 1 000 years ago, is under discussion. Once passed Sgr A*, this shock might have swept away the gas thereby being responsible for the currently dim nature of the black hole.

19.1.2 Early X-Ray Observations

Early X-ray observation in this region with rocket and satellite instrumentation have revealed many variable, but few steady, sources. Uhuru observations showed the first evidence for an extended source, although not distinguishable from a source complex. Subsequent observations with other satellite instruments have led to the detection of several new sources, including bursters. But not earlier than with the Einstein Observatory in 1979, X-ray imaging of the Galactic Center could be performed [27] (Fig. 19.2).

In addition to a large, diffuse emission extending over more than $30'$ in NE-SW direction (i.e., along the galactic plane), enhanced flux could be measured from the direction of Sgr A. Although the angular resolution was only of the order of $1'$

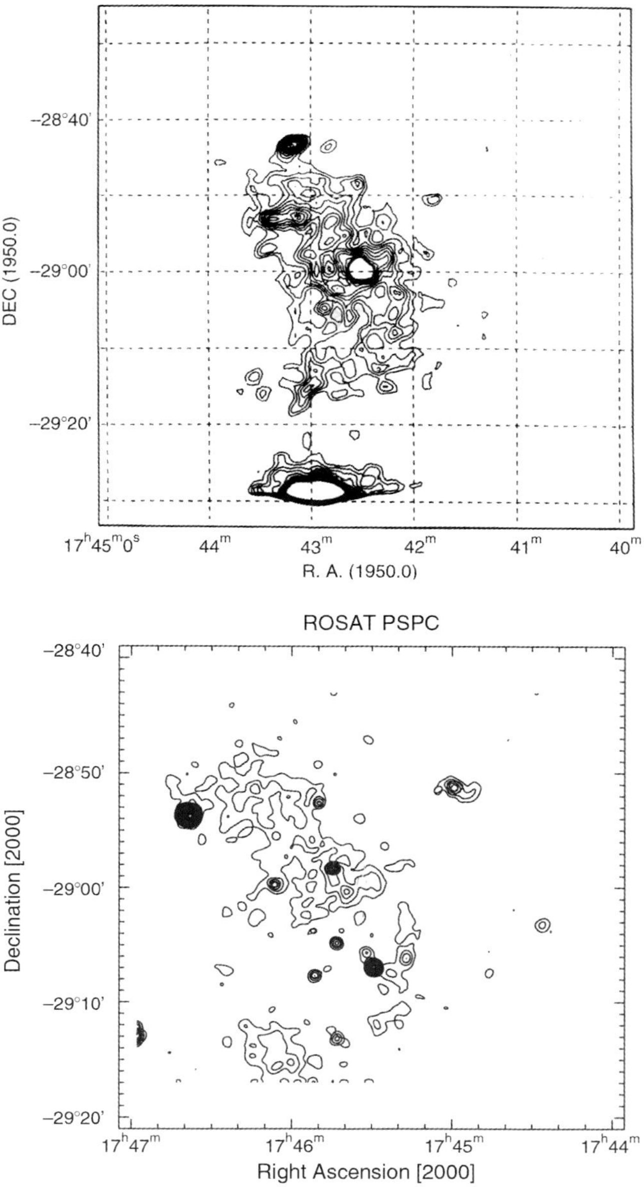

Fig. 19.2 Early X-ray images of the Galactic Center region observed with the Einstein Observatory (*top*, [27]) and ROSAT (*bottom*, [19])

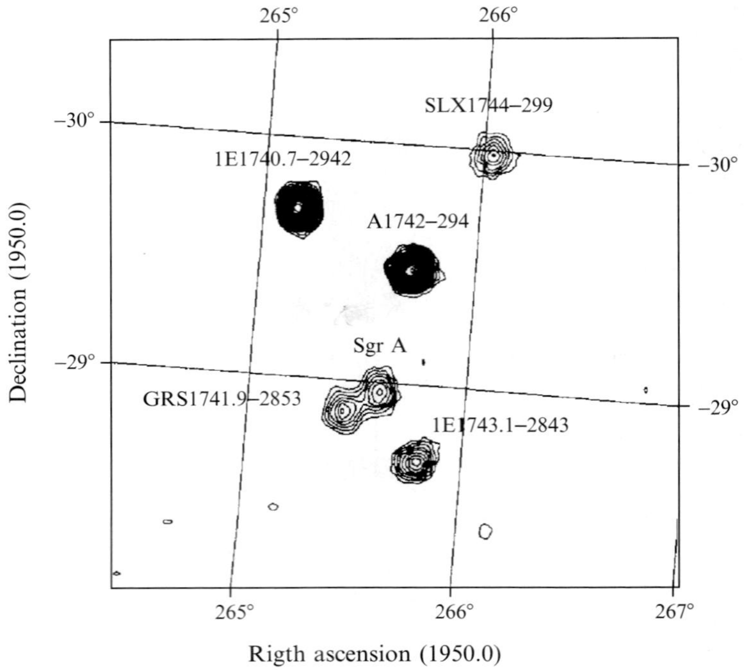

Fig. 19.3 *ART-P* observations, spring 1990 [17]. Image of a $2.3^\circ \times 2.3^\circ$ region near the Galactic Center (3-17 keV). Map contours correspond to a statistical significance of source detection at the 3.5, 5, 6.5 standard deviation levels. Each pixel is $1.3'$ on a side

with the imaging proportional counter (IPC), 12 point sources could be resolved, among them also one coincident with the Center itself. Its nature as well as the nature of the apparently diffuse emission remained unclear. Twelve years later, the same region was observed with ROSAT, and a few more sources could be resolved, although the energy bandpass of ROSAT was a bit too soft compared with the large column density toward the Galactic Center and the resulting low energy cut-off due to the interstellar absorption. The source at the Center found by Einstein (1E1742.5-2859) could be resolved into three individual objects (Fig. 19.2), one out of which is coincident with Sgr A* within $8''$ [19].

The luminosity of this source (or source complex given the prosumed density in this region) was derived to be $L_x = 1.1 \times 10^{35}\,\mathrm{erg\,s^{-1}}$. At higher energies, instruments like *ART-P* on the Russian Granat satellite could detect a few point sources, most of them variable but also some diffuse emission on a larger scale [17] (Fig. 19.3). Brightest source in this field is a burster (A1742-294) and the "Einstein-source" 1E1740.7-2942, a source that exhibits a 511 keV line and is, therefore, often referred as the "Great Annihilator." Another bright source is the double source SLX1744-299/300 slightly south of the Galactic Center.

Fig. 19.4 Chandra "true color" X-ray mosaic of the central $2° \times 0.8°$ band of the GC region [26]. The three energy bands are 1–3 keV (shown in *red*), 3–5 keV (*green*), and 5–8 keV (*blue*). The spatial resolution ranges from $0.5''$ on-axis to $5''$ at the near edge of the CCD and to $10''$ at the diagonal edge. This image is adaptively smoothed with a signal-to-noise ratio of 3 and is plotted logarithmically to emphasize low surface brightness emission. The saw-shaped boundaries of the map, plotted in Galactic coordinates, results from a specific roll angle of the observations

Fig. 19.5 Galactic Center region seen with XMM-Newton (*left*, [20]). *Red color* represents photon energies below 2.5 keV, *dark blue color* energies above 6 keV, *green color* is for intermediate energies between 2.5 and 6 keV. The bright supernova remnant Sgr A East is at the center of the image. There is diffuse emission along the galactic plane into NE direction and filaments of 6.4 keV emission (neutral iron K_α line, *light blue color*). The corresponding Chandra Observatory image (*right*, smaller field of view, higher resolution) shows the detailed structure of Sgr A East and, across to the galactic plane, various lobes of hot gas, probably remnants of past explosions at the center [13]

At higher energies, the center of the Galaxy itself (Sgr A*) belongs to the less striking sources in the field. These results made it clear that our Galactic Center is a rather weak representative of galactic nuclei (Fig. 19.4).

Meanwhile, with the use of high resolution and high sensitivity X-ray telescopes (Chandra Observatory and XMM-Newton) Sgr A* could be established as an X-ray source, primarily by its flaring behavior. At the same time also in near infrared flares could be detected from this compact radio source. Thus the large variety of theoretical models for the Galactic Center source could be somewhat constrained.

19.2 Sgr A East and its Environment

19.2.1 The Nature of Sgr A East

Its oval shell-like, nonthermal structure in the radio continuum supports the view that Sgr A East is a supernova remnant (SNR), although other explanations have also been discussed, e.g., multiple explosions based on the, originally derived total kinetic energy of 6×10^{52} erg [12]. However, measurements with the Chandra Observatory and XMM-Newton have revealed that the total energy of the hot plasma is much smaller ($\approx 1.5 \times 10^{49}$ erg), a value even smaller than the nominal energy for a single SNR. Thus one SNR can easily account for Sgr A East [9]. Since the plasma has already reached thermal equilibrium, the estimate does apply to the full thermal energy in the observable X-ray band. With an age of about 8 000 y [12], the mass of the plasma may originate as either the ejecta or swept-up interstellar material. The X-ray spectrum is enriched by heavy elements (see later), which suggests that the plasma is dominated by the supernova ejecta. SNRs showing centrally concentrated thermal X-rays lying within a shell-like nonthermal radio shell are defined as a relatively new class of SNRs and are called "mixed morphology" (MM) supernova remnants. About 20 members of this class are identified up to now, including Sgr A East itself [9].

19.2.2 X-Ray Imaging and Spectroscopy of Sgr A East

Most of the emission within the central few arcminutes comes from a region within the radio shell of Sgr A East. The brightest spot corresponds to the location of Sgr A*. The X-ray region around Sgr A* is found to be extended given the angular resolution of XMM-Newton, whereas the Chandra Observatory with its $0.5''$ resolution could resolve a few more sources within the innermost $10''$. With the CCD detectors onboard of XMM-Newton and Chandra Observatory, it became possible for the first time to combine high resolution imaging with good spectroscopic performance. Since the Galactic Center region is known to be full of molecular clouds producing different amounts of absorption, a spectral analysis of an extended region requires a model of patchy absorption. The average fitted temperature of Sgr A East is about 2–4 keV. The various element abundance across the SNR is remarkable: it seems that iron is overabundant ($Z \approx 3.5$) at the center of the SNR and less abundant in the outer regions ($Z \approx 0.5$), whereas the other metals do not significantly vary within Sgr A East from solar abundances [22]. At the rim of the remnant, a 6.4 keV iron fluorescent line is detected. This outer region is nearly coincident with the observed dust ring [12]. Hence, the detection of the neutral iron line supports the idea that Sgr A East is surrounded by the dust shell, or Sgr A East is interacting with the molecular ridge.

19.2.3 Bipolar Lobes

There is further X-ray emission to the north-east along the Galactic plane, emission that was also seen before by the Einstein Observatory, ROSAT, and ASCA (Fig. 19.5).

A number of diffuse, extended structures are present in the image, but the most prominent of them are aligned along a line passing through the center of Sgr A East (or better: Sgr A*) and oriented perpendicular to the Galactic plane. These "bipolar lobes" [13] apparently are filled structures with discernible sharp edges. Thermal fits to the blobs (i.e., the brightest regions within the lobes) give temperatures around 2 keV on average. A compelling interpretation for the bipolar lobes straddling Sgr A* is that they result from energetic mass ejections or explosions from the immediate environment of Sgr A*, presumably from an accretion disk.

19.3 Sgr A*

*19.3.1 X-Ray Detection of Sgr A**

After 20 y searching for high energy emission from Sgr A* [19, 27], it has come to a turning point when Chandra Observatory with its $0.5''$ resolution could detect weak emission from the radio source position [2]. Surprisingly, Sgr A* is much fainter than expected from accretion onto a super-massive black hole. In particular, in the 2–10 keV energy band its (quiescent) X-ray luminosity is only about $2.2 \times 10^{33}\,\mathrm{erg\,s^{-1}}$ within a radius of $1.5''$ [3]. This value may in fact be considered as an upper limit since this region contains other components such as stars, hot gas, etc. Thus, Sgr A* radiates in X-rays at about 11 orders of magnitude less than its corresponding Eddington luminosity.

*19.3.2 Flaring Sgr A**

Then, during a Chandra observation in October 2000, the same source was seen to flare up by a factor of $\approx$50 within 3 h [2]. The flare had a duration of about 10 ks, with L(2–10 keV) = $1.0 \pm 0.1 \times 10^{35}\,\mathrm{erg\,s^{-1}}$ for the flare peak. The discovery

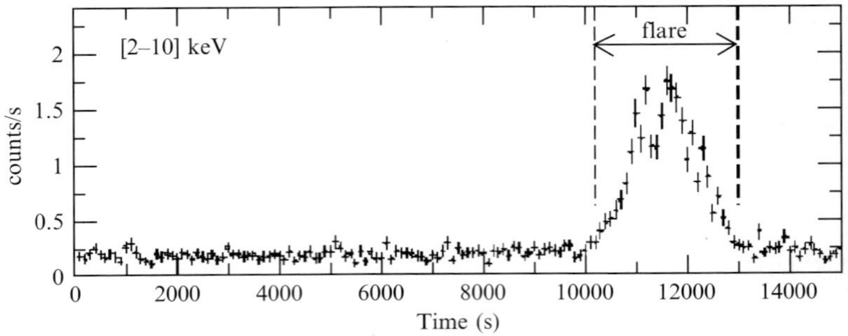

Fig. 19.6 XMM-Newton X-ray light curve (EPIC MOS 1+MOS2+PN) within a radius of 10″ around the Sgr A* position. The time binning is 100 s, and the error bars indicate 1σ uncertainties. The 2–10 keV light curve shows the quiescent and the flare periods [18]

of X-ray flares from Sgr A* has provided new exciting perspectives for the understanding of the processes at work in the galactic nucleus.

A second significant flare was detected with XMM-Newton [7]. It showed a monotonic flux rise up to a factor of about 20–30 in the last 900 s of the observation. As in the Chandra Observatory observation, the spectrum was rather hard with a power law photon index $\Gamma = 0.9$ and a maximum observed 2–10 keV luminosity of about 6×10^{34} erg s^{-1}. On the basis of and further flare detections, the flare rate could be estimated to be slightly more than on flare per day.

Then, in October 2002, the brightest flare so far was detected with XMM-Newton. With a peak luminosity of almost 4×10^{35} erg s^{-1}, it was 160 times brighter than at quiescent level [18], Fig. 19.6. Remarkably for this flare was its extremely short duration of less than 1 h and its rather soft spectrum with $\Gamma \approx 2.5$.

*19.3.3 The Nature of Sgr A**

A detailed analysis of this and the preceding flares revealed quasi-periodic oscillations with several distinct periods between ≈100 and 2 000 s, which were assumed to be characteristic cyclic modes of the accretion disk. From these modes, both the mass and the angular momentum of the central black hole could be derived [1].

Since also in the near-infrared regime Sgr A* could be detected first by its flaring behavior, it was quite natural to search for simultaneous flares both in NIR and X-rays. Success came finally in 2004 using the NACO adaptive optics instrument at the Very Large Telescope of the European Southern Observatory and the Chandra Observatory [6].

Current models that explain the Sgr A* spectral energy distribution invoke radiatively inefficient accretion flow models, RIAFs [16] including advection dominated accretion flows, ADAF [15], convection-dominated accretion flows, ADIOS [4] or Bondi-Hoyle accretion [11] and jet-models [10]. The recent simultaneous X-ray/NIR detection of the Sgr A* counterpart suggests that at least for the observed flare it is the same population of electrons that is responsible for both the IR and the X-ray emission, regardless of the emission mechanism. While it is not yet possible to completely rule out any of the proposed models, it is found that an attractive mechanism to explain the observed simultaneous NIR/X-ray flare is the synchrotron self-Compton (SSC) process. In this model, the X-ray photons are produced by up-scattering of millimeter or submillimeter photons [6].

Doppler boosting could quite conveniently explain the flaring behavior of Sgr A*. Doppler boosting will occur in models that involve relativistic outflows or jets pointing toward the observer at a small angle to the line of sight [10]. In the context of this jet model, the emitting component would be located close to the jet base and would have a size of a few Schwarzschild radii or less.

19.4 X-Ray Luminous Molecular Clouds

19.4.1 X-Ray Reflection Nebulae

With the imaging capability in the wide energy band up to ten kiloelectronvolt, ASCA found 6.4 keV line emission in the Galactic Center region [8]. This emission is largely coincident with giant molecular clouds, e.g., Sgr B2, the first cloud from which the 6.4 keV line was detected (Fig. 19.7). Because the 6.4 keV line is a characteristic radiation from neutral iron, it is natural to believe that X-rays are emitted by these molecular clouds. However, clouds are cold and cannot emit X-rays by themselves. Therefore, it has been proposed that these clouds are irradiated by an external X-ray source, and emit fluorescent X-rays. To explain the strength of the 6.4 keV line from the Sgr B2 molecular cloud, a strong X-ray source with a luminosity of $\approx 10^{39}\,\mathrm{erg\,s^{-1}}$ is required. Since there is not such a bright source in the vicinity, it has been argued that the nucleus of our Galaxy was much brighter than today some hundred years ago [24]. This corresponds to a light travel time between Sgr A* and Sgr B2 which is located about 100 pc away from the Galactic Center but probably at an only slightly shorter distance from us (Fig. 19.8).

Using archival data of ASCA, BeppoSAX, Chandra Observatory, and XMM-Newton, no significant variability of the line flux during the period 1993–2001 is found. This excludes alternative explanations with a transient source inside of Sgr B2. Recently, a hard X-ray source associated with Sgr B2 has been found with the *IBIS* instrument onboard of INTEGRAL (IGR J17475-2822). The broad band (3–200 keV) spectrum of the source constructed from data of different observatories strongly supports the idea that the X-ray emission of Sgr B2 is Compton scattered and reprocessed radiation emitted in the past by Sgr A* [21]. Radiation 6.4 keV was also found from other molecular clouds, e.g., Sgr C and clouds within the radio arc region. From their brightnesses and their respective distances from Sgr A*, its putative luminosity in the past could be estimated. It turns out from these studies that the luminosity of Sgr A* decreases continously over the last several hundreds of years [14].

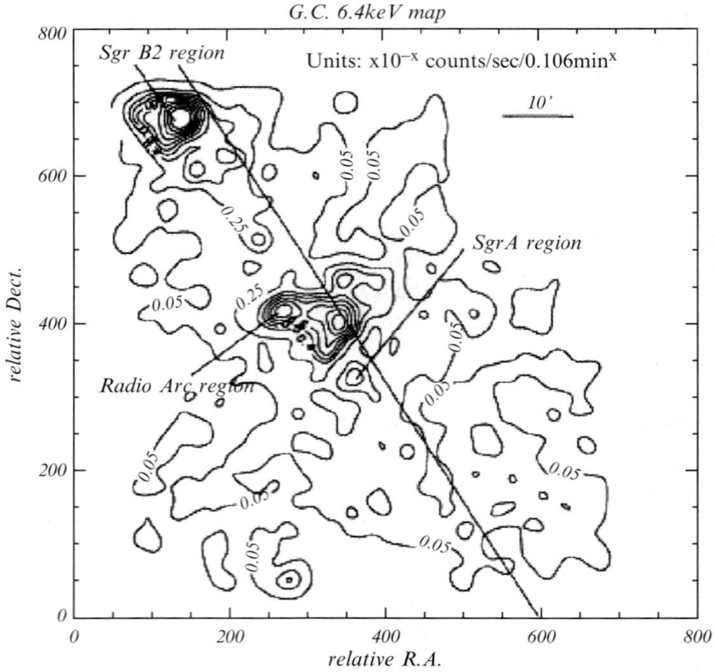

Fig. 19.7 The distribution of the 6.4-keV line intensity in the GC region observed with ASCA [14]

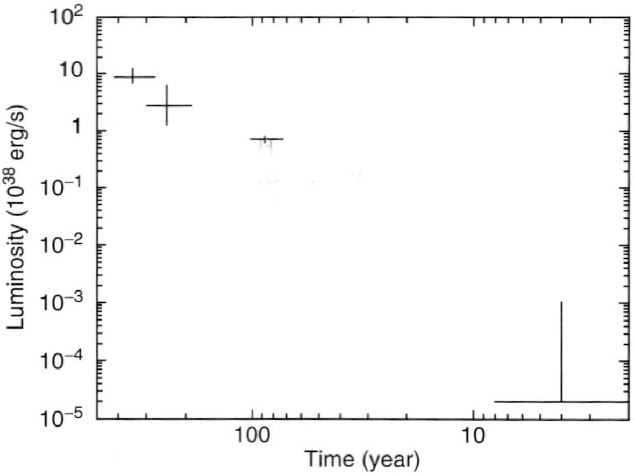

Fig. 19.8 The past luminosity of Sgr A* estimated from the luminosity of 6.4 keV line of clouds at different distances from Sgr A* (Sgr B2, Sgr C, M0.11-0.08, from left to right; the right-most data point is the current luminosity of Sgr A*) [14]

19.4.2 X-Ray Tubes?

However, there are a few arguments allowing also different explanations for the radiation from molecular clouds in the vicinity of the Galactic Center. Fluorescence is a natural consequence of absorption. Compton scattered radiation plus a fluorescence line should be accompanied by an absorption edge, in this case by the iron K-edge at 7.1 keV. Given the iron K fluorescence yield, about three times more photons must be absorbed than reradiated in the K_α line. However, the filamentary structures within the radio arc do not exhibit such an edge [20]. Furthermore despite their different distances from Sgr A*, the filaments have more or less comparable surface brightnesses that is not necessarily a consequence of the continously decreasing luminosity of Sgr A*. The spectra of the reflection nebulae are reasonably fit by bremsstrahlung and lines from neutral iron. Those spectra could also be created by the impact of electrons (or protons at higher energies) as in laboratory X-ray tubes. There are sufficient hints on the existence of those electrons: one of the great mysteries is the Galactic ridge X-ray emission [5,25,29], which has a scale height of $\approx$100 pc and is greatly enhanced toward the Galactic Center. The overall spectrum of the ridge emission suggests a characteristic thermal temperature of 7–10 keV. Diffuse hot gas with such temperatures would escape fast into the galactic halo, demanding a very efficient heating source, which is not found in the Galactic Center. As an alternative, a nonthermal origin of the galactic X-ray background has been proposed, e.g., electrons that could be also responsible for the 6.4 keV fluorescence line–similar to an ordinary X-ray tube in laboratory.

References

1. Aschenbach, B., Grosso, N., Porquet, D., Predehl, P. 2004, Astron. Astrophys., 417, 71
2. Baganoff, F.K., Bautz, M.W., Brandt, W.N., et al. 2001, Nature, 413, 45
3. Baganoff, F.K., Maeda, Y., Morris, M., et al. 2003, Astrophys. J., 591, 891
4. Blandford, R., Begelman, M. 1999, Mon. Not. Roy. Astron. Soc., 303, L1
5. Dogiel, V.A., Ichimura, A., Inoue, H., Masai, K. 1998, Publ. Astron. Soc. Jpn., 50, 567
6. Eckart, A., Baganoff, F.K., Morris, M., et al. 2004, Astron. Astrophys., 427,1
7. Goldwurm, A., Brion, E., Goldoni, P., et al. 2003, Astrophys. J., 584, 751
8. Koyama, K., Maeda, Y., Sonobe, T., et al. 1996, Publ. Astron. Soc. Jpn., 48, 249
9. Maeda, Y., Baganoff, F.K., Feigelson, E.D., et al. 2002, Astrophys. J., 570,671
10. Markoff, S., Falcke, H., Yuan, F., et al. 2001, Astron. Astrophys., 379, L13
11. Melia, F., Falcke, H. 2001, Annu. Rev. Astron. Astrophys., 39, 309
12. Mezger, P.G., Zylka, R., Salter, C.J., et al. 1989, Astron. Astrophys., 209,337
13. Morris, M., Baganoff, F.K., Muno, M.P., et al. 2003, Astron. Nachr., 324, 167
14. Murakami, H., Senda, A., Maeda, Y., Koyama, K. 2003, Astron. Nachr., 324, 117
15. Narayan, R., Yi, I., Mahadevan, R. 1995, Nature, 374, 623
16. Quataert, E. 2003, Astron. Nachr., 324, 435
17. Pavlinskii, M.N., Grebenev, S.A., Sunyaev, R.A. 1992, Sov. Astron. Lett 18, 116
18. Porquet, D., Predehl, P., Aschenbach, B., et al. 2003, Astron. Astrophys., 407, L17
19. Predehl, P., Trümper, J. 1994, Astron. Astrophys., 290, L29
20. Predehl, P., Costantini, E., Hasinger, G., Tanaka, Y. 2003, Astron. Nachr., 324, 73

21. Revnivtsev, M.G., Churazov, E.M., Sazonov, S.Yu., et al. 2004, Astron. Astrophys., 425, L49
22. Sakano, M., Warwick, R.S., Decourchelle, A. 2003 Astron. Nachr., 324, 197
23. Schödel, R., Ott, T., Genzel, R., et al. 2002, Nature, 419, 694
24. Sunyaev, R.A., Markevitch, M., Pavlinsky, M. 1993, Astrophys. J., 407, 606
25. Valinia, A., Tatische, V., Arnaud, K., et al. 2000, Astrophys. J., 543, 733
26. Wang, Q.D., Gotthelf, E., Lang, C. 2002, Nature, 415, 148
27. Watson, M.G., Willingale, R., Grindlay, J.E., Hertz, P. 1981 Astrophys. J., 250, 142
28. Yusef-Zadeh, F., Melia, F., Wardle, M. 2000, Science 287, 85
29. Yusef-Zadeh, F., Law, C., Wardle, M. 2002, Astrophys. J., 568, L121

Part III

Extragalactic X-Ray Astronomy

20 X-Rays from Nearby Galaxies

W. Pietsch

20.1 Introduction

In part II several different classes of X-ray sources have been introduced, which can be observed within the Milky Way. One of the largest uncertainties in their study is the distance of the more luminous sources, which are often known to not better than a factor of two. This results in an order of magnitude uncertainty in the X-ray luminosity and prevents the determination of the luminosity function of the various types of sources. In addition to these problems, the absorption within the plane of the Galaxy strongly suppresses soft X-ray emission. Therefore, supersoft X-ray sources (SSS) and the hot interstellar medium (ISM) can only be observed in the solar neighborhood.

Observations of the X-ray source population and the ISM in nearby galaxies help us to overcome these difficulties. All X-ray sources in a galaxy are – to first order – seen at the same distance. For the study of SSS and the hot ISM, galaxies are of specific interest which are observed in directions with low Galactic foreground absorption; late type galaxies seen face-on are best suited for studies of the source population and the ISM in the galaxy disk; galaxies seen edge-on allow us to resolve emission from the hot ISM in the galaxy halo and the interaction region from point sources and the ISM in the disk.

In this chapter we will discuss the different X-ray emission components of nearby galaxies leaving aside the emission from active galactic nuclei (AGN), which are present in some of the galaxies and which will be dealt with in Chap. 22.

20.2 History of X-Ray Observations of Galaxies

X-ray observations of galaxies were started with the first X-ray satellite Uhuru, which carried a collimated proportional counter instrument sensitive for photons in the energy range 2–10 keV and was launched in 1970. Uhuru discovered intense X-ray emission from compact stellar remnants (white dwarfs, neutron stars, black holes) in binary systems in the Milky Way, from AGN and from hot plasma in clusters of galaxies. Also X-ray emission from the Local Group galaxies M 31 and the Magellanic Clouds (MC) were detected; however, emission from other normal

J.E. Trümper and G. Hasinger (eds.), *The Universe in X-Rays.*

galaxies was below the Uhuru detection threshold (see [27] for a review of Uhuru results). The first satellite with a focusing X-ray telescope was the Einstein Observatory launched in 1978. Its much higher sensitivity and spatial resolution resulted in the detection and study of a significant number of normal galaxies. For a review of the Einstein observations of galaxies see e.g. [17] and the catalogs of Einstein images and spectra of galaxies [20, 42, 43].

More than a decade later, the focal plane instruments aboard the ROSAT observatory provided much higher sensitivity at low X-ray energies and slightly better angular resolution, its telescope had a wider field of view, and the mission lasted significantly longer compared to the Einstein Observatory. All these points have led to a better understanding of the spatial morphology of the X-ray emission of galaxies, and to the detection and study of the soft X-ray-emitting ISM of spiral galaxies.

ASCA and BeppoSAX extended the observable energy band to 10 keV and above. In addition, the CCD detectors as focal plane instruments aboard ASCA provided improved energy resolution by factors of 5–10 compared to previous instruments. This allowed the study of different spectral components of the X-ray emission. However, the significantly inferior angular resolution of these satellites compared to Einstein and ROSAT prohibited in most cases detailed studies of individual galaxies.

The situation drastically chanced with the launch of the Chandra and XMM-Newton observatories in June and December 1999. These satellites represent a big step forward in sensitivity, imaging capabilities, and spectral resolution compared to the previous generation of X-ray observatories and provide a new and much deeper look at galaxies in X-rays. For a more detailed discussion of the history of the X-ray observations of galaxies see [19].

With the increased sensitivity of the X-ray instrumentation, galaxy science has developed to an important topic of X-ray astronomy. This is best demonstrated by the fact that observations of fields in the Large Magellanic Cloud (LMC) were selected as first light targets for ROSAT and XMM-Newton (see Fig. 20.1), and pointings

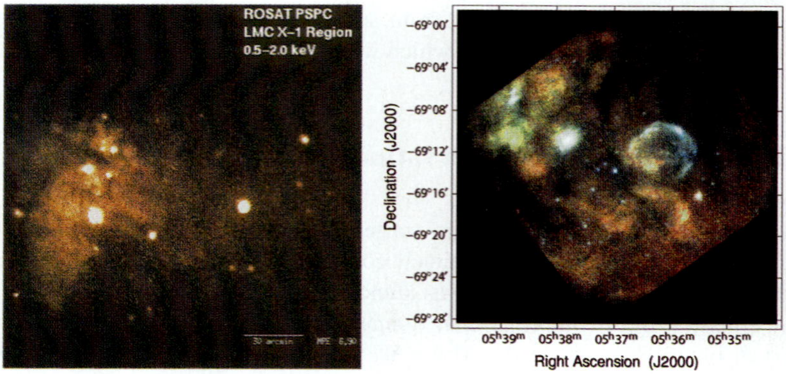

Fig. 20.1 ROSAT PSPC (*left*) and XMM-Newton EPIC pn (*right*) first light images pointing at a LMC area slightly west of 30 Dor. The XMM-Newton image [14] is a zoom-in by about a factor of six compared to the ROSAT image

to nearby galaxies (e.g., LMC, Small Magellanic Cloud SMC, M 31, NGC 253) were used as calibration and performance verification targets for XMM-Newton and Chandra.

20.3 Point-Like Emission Components

The bright X-ray sources in galaxies discussed in this section include X-ray binaries (XRBs), supersoft sources (SSS), supernovae (SN) and supernova remnants (SNRs), nuclear sources, and a number of very luminous sources, so-called ultra-luminous X-ray sources (ULXs). ULXs are radiating at luminosities above the Eddington limit for a 1 $M_{\odot}$ object and their nature is not fully understood. But at least part of them could belong to the classes of objects mentioned above.

Often, the X-ray source population of a target galaxy is confused by foreground stars in the Milky Way and background objects (galaxies, galaxy clusters, and AGN) observed in the same field. For fainter X-ray sources, the fraction of these spurious objects rises in number and in Local Group galaxies it can dominate the detected sources. One therefore has to find ways to separate the different galactic and extragalactic source classes. This can be achieved by identifying detected X-ray sources with the help of their position and/or time variability with counterparts that were categorized in surveys in other wavelength regimes (e.g., as SNR, globular clusters, SN). Because of the high source density in the fields, good X-ray positions are of utmost importance for this approach. Classification can also be based on X-ray properties like time variability, extent, hardness ratios, or energy spectra. Classification schemes including X-ray hardness and extent were successfully applied to ROSAT PSPC sources in the LMC (758 sources cataloged), SMC (517) (see Fig. 20.2) and

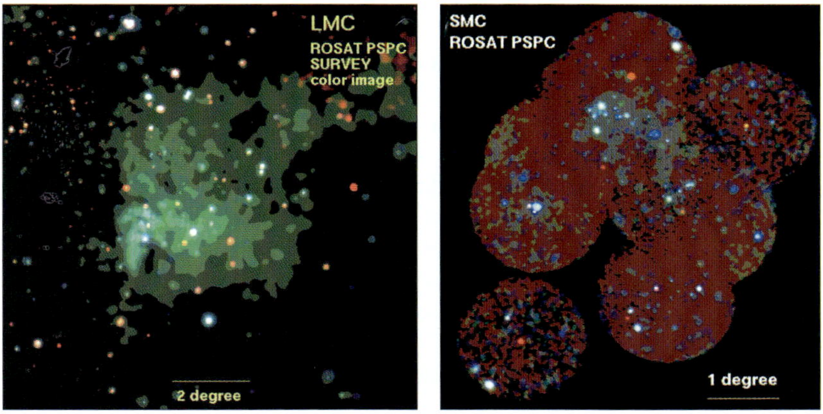

Fig. 20.2 ROSAT PSPC images of the LMC observed in the ROSAT all sky survey (*left*) and of the SMC (*right* [31]) in the energy bands 0.1–0.4, 0.5–0.9, and 0.9–2 keV combined in red-green-blue color coded images

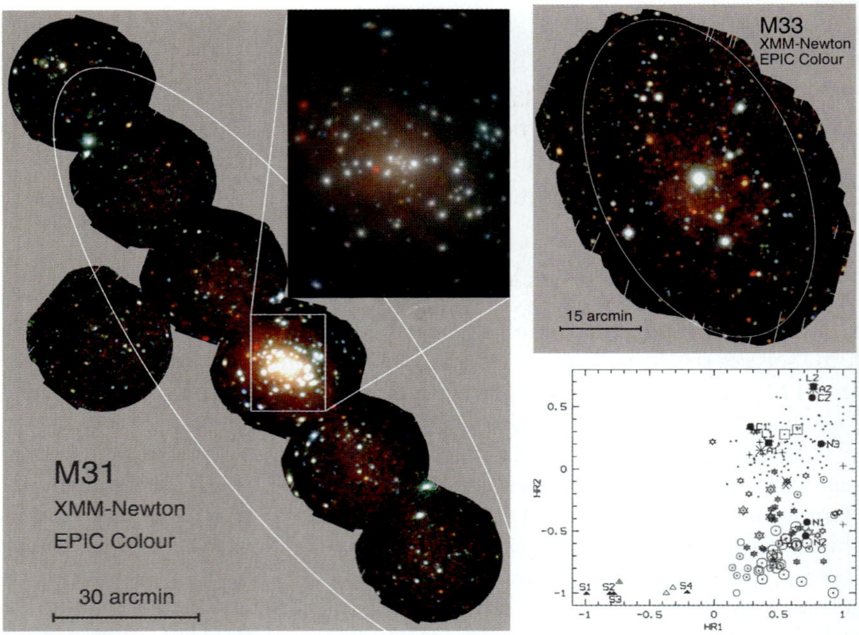

Fig. 20.3 Three color XMM-Newton EPIC images of the Andromeda galaxy M 31 (*left*: together with a zoom-in on the M 31 bulge area, [52]) and M 33 (*upper right* [55]). Red, green, and blue show, respectively, the 0.2–1.0, 1–2, and 2–12 keV bands. The ellipses indicate the optical extent of the galaxies. The X-ray color/color plot (*lower right*) [55] using energy bands 0.2–0.5, 0.5–1, and 1–2 keV can be used – together with informations from other wavelength regimes – to classify the detected X-ray sources as SSS, SNRs, foreground stars, or hard sources (XRBs, plerions, background sources)

M 33 (184) fields [31–33]. Figure 20.3 shows X-ray color images of deep XMM-Newton EPIC surveys of the bright Local Group spirals M 31 and M 33 together with a hardness ratio diagram for M 33. More than 850 respectively 400 point-like sources were detected in these surveys [52, 55]. The X-ray sources have been classified using similar schemes as developed for the ROSAT observations. Many foreground stars, SSS, SNRs, and candidates can be identified by these procedures. However, even with the broader energy band of XMM-Newton and the better energy resolution, Crab-like SNRs, XRBs, and AGN do not separate in the hardness ratio diagrams. XRBs may be separated if they show strong time variability, or as sources correlating with an optically identified globular cluster in the galaxy. A source may be classified as AGN or Crab-like SNR if additional optical or radio indicators are available. The results of the identifications will be discussed later and compared to results for source populations in other galaxies.

20.3.1 X-Ray Binaries

In the absence of an AGN or a large amount of hot gas, XRBs contribute the major fraction to the host galaxy's X-ray luminosity, as is the case for the Milky Way or the Andromeda galaxy (see, e.g., a recent review [22]). XRBs are subdivided into low-mass and high-mass systems depending on the mass of the donor star (LMXB, $M_{\rm opt} \lesssim 1\,{\rm M}_\odot$; HMXB, $M_{\rm opt} \gtrsim 8\,{\rm M}_\odot$, respectively) and have very different evolutionary time-scales. The lifetime of HMXBs is limited by the nuclear time-scale of the massive donor star to less than 10^6–10^7 yrs, i.e., it is comparable to the duration of a star formation event. The onset of the X-ray active phase of a LMXB after the formation of the compact object is determined by the nuclear evolution time-scale of the donor star and/or the binary orbit decay time-scale to about 10^9–10^{10} yrs. The active phase then may last for a similar time. Therefore, HMXBs radiate during or shortly after a star formation event while LMXB lifetimes are comparable to that of the host galaxy. The population of LMXBs may be proportional to the total stellar mass of a galaxy (see e.g. [29]).

Pointed archival ROSAT PSPC and HRI observations covered large parts of the LMC. The surveys [32, 62] revealed 17 XRBs and candidates. Besides the LMXB LMC X-2 and one LMXB candidate in the LMC bar the other sources are HMXBs or candidates. Ten have been identified as Be-type X-ray binaries, one is a wind fed HMXB, one is fed by Roche lobe overflow, and two are black hole HMXBs [47]. The observed ratio of LMXBs and HMXBs of about 1/7 is as expected from the low LMC mass and moderate star formation rate [65].

X-ray surveys of the SMC with ASCA, BeppoSAX, ROSAT, RXTE, and XMM-Newton revealed a large number of XRBs (67). Most of them are transient and were discovered through the detection of X-ray pulsations, indicating the spin period of a neutron star. Optical follow-up observations revealed in all cases Be stars as optical companions in HMXBs (with the only secure exception of the supergiant system SMC X-1), making the SMC very different to the Milky Way. These Be-type XRB systems have a quiescent luminosity of $\sim 10^{34}$ erg s^{-1} and are brighter by about a factor of 100 during outbursts, which generally occur close to the time of periastron passage of the neutron star. They sometimes in addition show giant outbursts ($L_{\rm X} \gtrsim 10^{37}$ erg s^{-1}) lasting for several weeks or even months, which are connected to Be star activity. Only during such outbursts pulsations from Be-type XRBs in the MCs were detectable with satellites before XMM-Newton [9, 34–36].

M 33 is an Sc spiral in the Local Group at a distance of 795 kpc, about 15 times the distance of the MCs that makes the detection and identification of XRBs much more difficult. In M 33 only two candidates for XRBs could be identified up to now. One is the brightest source in the Local Group, M 33 X-8 with an X-ray luminosity of 1.5×10^{39} erg s^{-1} in the 0.5–10 keV band, which – based on a Chandra analysis of its X-ray spectrum, time variability, and correlation with a radio source – most likely is an ultraluminous black hole XRB system at the M 33 nucleus similar to the galactic microquasar GRS 1915+105 [15]. Even with Chandra positioning it is not clear yet if the system correlates with an optical star bright enough for an high mass optical counterpart. The second system is the HMXB M 33 X-7 with 3.45 d orbital period, the first eclipsing XRB outside the Milky Way and the MCs. In 1989 it was suggested as an eclipsing XRB with 1.7 d orbital period based on Einstein observations. Adding ROSAT observations revealed that the orbital period is twice as long and allowed to refine the shape of the eclipsing light curve and binary ephemeris. The position of X–7 was covered in several observations of the XMM-Newton survey of M 33. The collecting power of the instruments not only led to a much clearer orbital light curve (resolution of 1 000 s) but also allowed detailed modeling of the source spectrum with a disk blackbody or bremsstrahlung spectrum. With the improved X-ray position of the source (including Chandra observations) a B0I to O7I star could be identified as optical counterpart that showed an optical heating light curve of a HMXB with the M 33 X-7 period. The X-ray spectrum together with the lack of an X-ray pulsation period proposed M 33 X-7 as the first eclipsing black hole XRB (see Fig. 20.4 and detailed references in [56]). Further XRB candidates are expected from a more detailed analysis of the time variability of sources in the XMM-Newton raster observations of M 33.

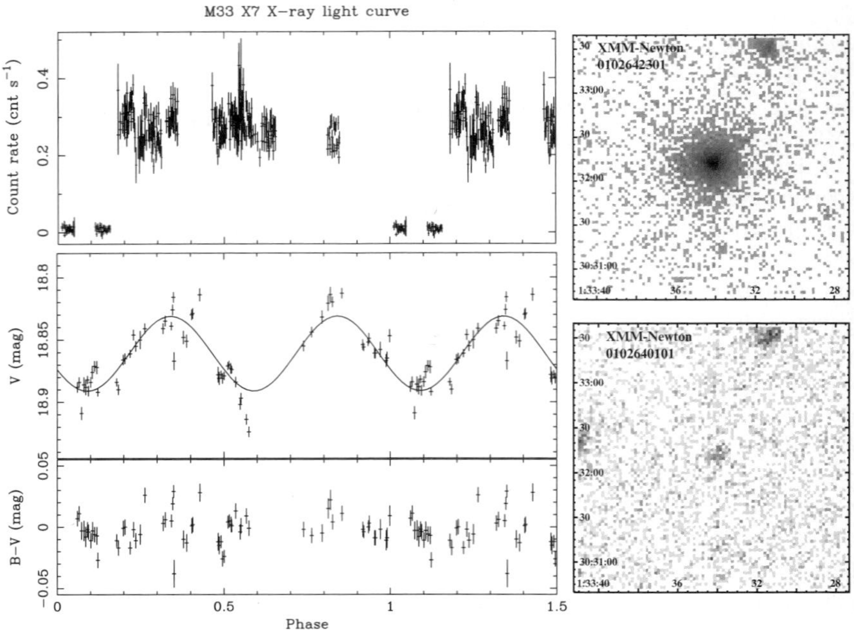

Fig. 20.4 Light curve of the XRB M 33 X–7 in the 0.5–3.0 keV band and in optical V and B–V folded over the 3.45 d orbital period (*left*). XMM-Newton EPIC images during the on-state (*above*) and eclipse (*below*) demonstrate the strong intensity change of the XRB compared to a constant nearby source (*right*) [56]

The Andromeda galaxy M 31, a massive SA(s)b galaxy in the Local Group similar to the Milky Way, is located at about the same distance as M 33. As in the Milky Way, a significant part of the luminous X-ray sources are found in globular clusters. Most of the globular cluster sources in the Galaxy show bursts and therefore are low mass neutron star XRBs (see 3.4). There have been extensive surveys for globular clusters in M 31 in the optical and infrared bands. The XMM-Newton catalog of M 31 sources lists 37 globular cluster sources and candidates covering an absorbed luminosity range from 4.5×10^{35} to 2.4×10^{38} erg s^{-1} in the 0.2–4.5 keV band well in the range allowed for neutron star LMXBs. One of the sources shows intensity dips every 2.78 h that indicate the orbital period of a neutron star LMXB [74]. A search for X-ray bursts in the archival XMM-Newton observations at globular cluster X-ray source positions detected bursts from two sources that can be interpreted as type I radius expansion bursts from sources in M 31 radiating at maximum with a 1 keV black body spectrum with 3.8×10^{38} erg s^{-1}. The bursts identify the sources as neutron star LMXBs in M 31. These type I X-ray bursts are the first detected outside the Milky Way and show that with the help of XMM-Newton X-ray bursts can be used to classify neutron star LMXBs in Local Group galaxies (Fig. 20.5 [53]). In addition to these LMXB sources, detailed work on individual M 31 sources using XMM-Newton and/or Chandra Observatory data, has identified four black hole XRBs, three neutron star LMXBs, an XRB pulsar, and several transients (for references see [52]). While in general transient behavior of bright X-ray sources indicates an XRB nature, with such a selection one may pick up also some highly variable background objects. The XRBs in M 31 selected in this way cover an absorbed luminosity range from 8.4×10^{35} to 2.8×10^{38} erg s^{-1} in the 0.2–4.5 keV band. Up to now no HMXBs have been identified in M 31. However, several emission line objects correlating with XMM-Newton catalog sources may be good candidates for Be-type HMXBs.

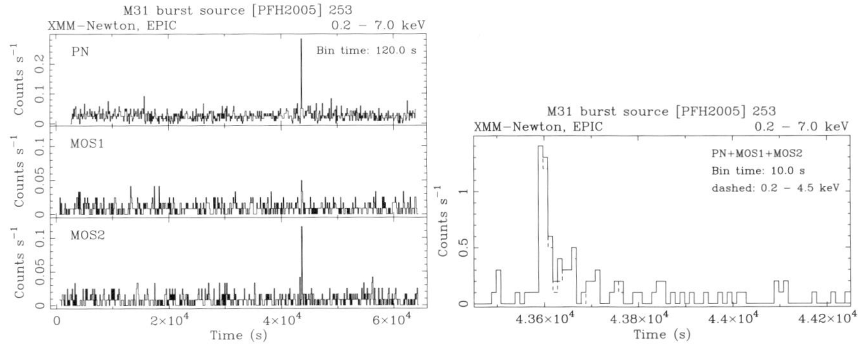

Fig. 20.5 XMM-Newton EPIC light curves of source [PFH2005] 253 on January 6/7, 2002 for the individual cameras integrated over 120 s (*left*) and for the time of the burst integrated over 10 s for all cameras added (*right*) [53]

Several late type galaxies outside the Local Group have been searched for X-ray point-like sources with ROSAT, XMM-Newton, and Chandra. This resulted in source catalogues of up to 100 sources in galaxies with distances of 10 Mpc and more (e.g. M 81, M 83, and M 101). However, at these larger distances identifying XRBs gets more and more difficult. [54] report the detection of an eclipsing XRB in the starburst galaxy NGC 253 (distance 2.58 Mpc) based on two changes from low to high state in Chandra Observatory and XMM-Newton observations separated by about a year. They use additional XMM-Newton, Chandra, ROSAT, and Einstein observations to further constrain the orbital parameters and determine an X-ray luminosity during the high state, which is close to the Eddington limit for a $1.4\,M_{\odot}$ neutron star. In the other galaxies XRB candidates may be identified from globular cluster correlations or their location in the spiral arms together with spectral and time variability arguments. Often, X-ray luminosity functions (XLF) are the only tools to get a hint on the XRB population.

In galaxies with more intense star formation, observations show flatter XLF slopes, which indicate the presence of very luminous sources. The best example is the galaxy merger system NGC 4038/39 (The Antennae, distance 19 Mpc), where nine ULXs were discovered with Chandra Observatory [23]. [30] suggest that the XLFs of star forming galaxies scale with the star formation rate (SFR) and propose HMXBs as SFR indicators in galaxies. They demonstrate that for starburst galaxies the total X-ray luminosity of a galaxy is linearly correlated to the SFR and suggest a universal luminosity function for HMXBs, a proposal that seems to contradict the results of the XLF slopes determined for a minisurvey of starburst galaxies with Chandra Observatory and the prediction of theoretical models. Independent of these discrepancies, comparing XLFs determined for different galaxies with XLFs determined via evolution models of the parent galaxies will help to understand the nature of the X-ray sources and the parent galaxy stellar population (see [22]).

In E and S0 galaxies, XRBs could not be detected directly with pre-Chandra Observatory telescopes, because of the distance of these galaxies and the limited angular resolution of the telescopes. Already [72] predicted the presence of XRBs in E and S0 galaxies based on an analogy with the bulge of M 31. The detection of hard spectral components in early type galaxies supported this prediction. Chandra Observatory images now resolve many point-like sources in E and S0 galaxies, which often correlate with globular clusters and therefore can be identified as LMXBs (see e.g. [1,61]).

20.3.2 Supersoft Sources, Optical Novae

Luminous SSSs were recognized as an important new class of intrinsically bright X-ray sources in the LMC by [75]. Their spectra and luminosities are in the range of $(10^5–10^6)$ K and $10^{36}–10^{38}$ erg s^{-1}, respectively. The first two sources were detected with the Einstein Observatory in the LMC and optical identifications established the binary nature of the sources. ROSAT detected many more SSSs in the MCs (see red sources in the images of Fig. 20.2), some in the Galaxy and several in nearby galaxies, especially also in M 31 and M 33. Their observed characteristics are consistent with those of white dwarfs, which are steadily or cyclically burning hydrogen-rich matter accreted onto the surface at a rate of order $10^{-7}\,M_{\odot}\,yrs^{-1}$ for timescales of (10^6–10^7) yrs. Steady burning can also occur in a post-nova stage for shorter time scales and SSS have been observed in a few classical novae and symbiotic novae in the galaxy, LMC and one in ROSAT observations of M 31. With the improved sensitivity of XMM-Newton and Chandra many more SSS were detected in nearby galaxies (e.g. 18 in M 31 and 5 in M 33 in the XMM-Newton surveys, with Chandra 8 in M 81 (distance 3.6 Mpc), 10 in M 101 (7.2 Mpc)). [61] discovered with Chandra the first three SSS in an elliptical galaxy, NGC 4697 (16 Mpc). Several of the SSS are transients. In addition the first supersoft pulsator (865 s) has been discovered with XMM-Newton as a transient source in M 31 [48, 73]. For reviews see [40, 41].

As mentioned above, it was known that optical novae could contribute to the class of SSS. [51] therefore correlated optical novae detected in M 31 and M 33 with XMM-Newton, Chandra, and ROSAT data from catalogs and archives. It turned out that the majority of SSS in M 31 are novae in their SSS state (21) and even M 33 where the optical nova catalogues are much less complete than for M 31 revealed two X-ray source/optical nova correlations (see Fig. 20.6). This work more than tripled the number of known optical novae with SSS phase. For many of the novae, X-ray light curves could be determined. From the delay of the onset of the SSS phase after the outburst of two of the novae, one could estimate the ejected hydrogen masses to 10^{-5} and $10^{-6}\,M_{\odot}$. As many novae at different time after outburst can be observed in each observation of the M 31 bulge, monitoring of this area will lead to a better understanding of the X-ray emission of novae and nova outburst properties.

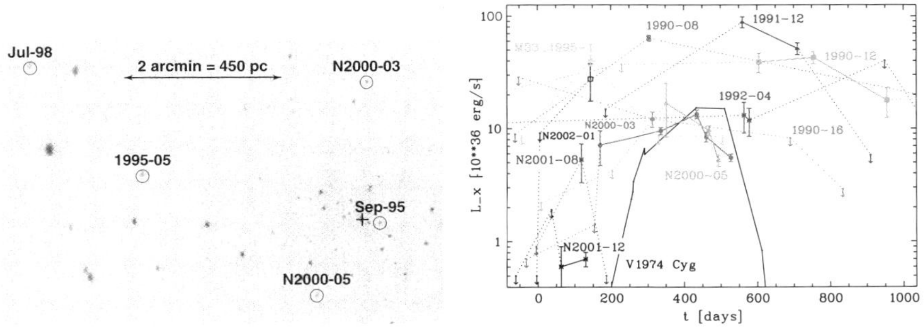

Fig. 20.6 Chandra HRC I image of the center area of M 31 on Oct 31, 2001. Circles with 5″ radius indicate nova positions. The cross indicates the M 31 center, the aim point of the observation (*left*). Soft X-ray light curves of novae in M 31 and M 33 that were detected within 1 000 d after the optical outburst. The light curve of the Galactic nova V1974 Cyg that was monitored by ROSAT after its outburst in 1992, is shown for comparison (*right*) [51]

20.3.3 Supernova Remnants and Supernovae

Observations in our own Galaxy established SNRs as bright sources of both thermal and nonthermal X-rays (see Chap. 17). After a few hundred to thousands of years they mark the location of a supernova explosion heating the interstellar medium. Samples of SNRs have been obtained for different galaxies and studied to probe their evolution and gain information on the ISM of the host galaxy. In X-rays, sources have been identified with SNRs not only in Local Group galaxies but also in galaxies as distant as $\sim$10 Mpc. However, in many galaxies the overlap of X-ray detected SNRs with optically and radio selected SNRs is small. This may be caused by distance dependant effects in the sensitivity to detect SNRs in different wavelength regimes [49].

With Einstein and ROSAT, SNRs have been identified mostly in Local Group galaxies. XMM-Newton and Chandra Observatory now resolve details in SNRs of the MCs. In both, M 33 and M 31, XMM-Newton observations identified 21 SNRs and classified an additional 23 using X-ray hardness ratio criteria and correlations with optical or radio SNR candidates [52, 55]. The SNRs cover a luminosity range from 4×10^{34} to 5×10^{36} erg s^{-1} in the 0.2–4.5 keV band. With the superior point spread function (PSF) of the Chandra Observatory mirror/detector system five SNR could be spatially resolved in M 31 [79, and references therein] and four (plus three marginally) in M 33 [26], respectively. In addition, two X-ray selected M 33 SNR candidates from [55] were identified as SNRs in optical images [26] proving the X-ray selection strategy. With the number of SNRs known it is now possible to compare X-ray luminosity functions of SNRs in different Local Group galaxies. As a result LMC SNRs on average show higher luminosities than SNRs in M 31 and M 33. This effect may be caused by differences in interstellar abundances between the galaxies.

But one does not have to wait for hundreds of years to see X-ray emission from the location of supernova (SN) explosions. About 25 SN were observed in X-rays within days to years after the outburst in nearby galaxies (see [38] for a recent review and Chap. 7). SN are classified as type II or I based on the presence or absence of hydrogen lines in their optical spectra. While most of the X-ray detected SN are of type II, six are of type Ib/c, Ic, or Ic/pec. No type Ia SN – believed to be nuclear detonations of carbon+oxygen white dwarfs when exceeding the Chandrasekhar limit through accretion – has been detected to date in X-rays. Also some gamma-ray bursts and their X-ray afterglow have been connected to SN explosions in distant galaxies (see Chap. 6). Here I only shortly want to mention some results of the very close by SN 1987A (in the LMC, 50 kpc) and SN 1993J (in M 81, 3.6 Mpc).

SN 1987A in the LMC was the closest SN exploding during the X-ray era. For the first time, hard X-rays (5×10^{37} erg s^{-1} in 45–100 keV band) due to emission from radioactive decay of the debris could be observed for a SN starting about half a year after the explosion [71] till it dropped below the sensitivity limit of the instruments 2 yrs later. Only in 1991, the SN was redetected as a faint soft X-ray source in ROSAT observations ($\sim 10^{34}$ erg s^{-1} in 0.5–2 keV band [3, 28]). Its X-ray flux,

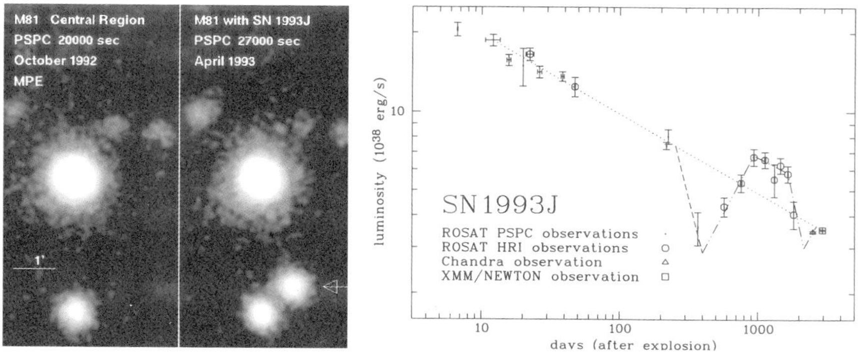

Fig. 20.7 X-ray emission from the supernova SN 1993J in the galaxy M81 was detected by ROSAT only six days after the explosion at the end of March 1993. The supernova location, south of the center of M81, is indicated by the arrow (right half of the image). In an earlier observation in October 1992, no X-ray emission was detected at the supernova position (left half of the image) (*left*). The light curve of the X-ray luminosity of SN 1993J gives informations on the mass loss rate and wind velocity of the supernova progenitor star (*right* [80])

now due to circumstellar interactions, has been continuously rising to soft X-ray luminosities of 10^{36} erg s^{-1} till mid 2005 [50].

SN 1993J in M 81 exploded very early in the ROSAT mission live time. Soft emission with an absorption corrected luminosity of 3×10^{39} erg s^{-1} in the 0.1–2.4 keV band was detected within 6 days (see Fig. 20.7, [81]). According to standard SN models, the SN X-ray light curve is determined by the interaction of the SN ejecta with the stellar wind of the progenitor star. The well sampled X-ray light curve of SN 1993J in this way revealed the pre-supernova evolution of the progenitor (see e.g. [37,80]).

20.3.4 Ultra-Luminous X-Ray Sources

Ultra-luminous X-ray sources (ULX) are normally defined as sources with luminosities in the X-ray band above 10^{39} erg s^{-1}. They were first detected with the Einstein Observatory and at that time named super-Eddington sources, because their luminosity was well above the Eddington limit for a neutron star ($\sim 2\times 10^{38}$ erg s^{-1}), suggesting accreting objects with masses of 100 $M_\odot$ or more (see, e.g., review [17]). Such masses exceed those of stellar mass black holes in XRBs and may indicate that ULXs represent a new class of astrophysical objects. Certainly several ULXs can be explained as bright background objects or even as X-ray luminous SN in the host galaxy (see, e.g., SN 1993J in M 81), but several of them are recurrent and/or connected to star forming regions in galaxies excluding such models. They could represent sources that fill the gap in the black hole mass distribution between stellar mass black holes detected in X-ray binaries and supermassive black holes found in the nuclei of many galaxies. These sources have been called intermediate mass black holes (IMBHs). The high spatial resolution of Chandra allows to resolve these sources also in more distant galaxies and with the collecting power of XMM-Newton detailed X-ray spectra were collected. Several of the X-ray spectra of the brightest ULX sources showed emission components that have been modeled by cool accretion disks. As the temperature is inversely related to black hole mass, this has been taken as an argument for them containing an IMBH (see e.g. [46]).

20.3.5 Galactic Nuclei

HST photometry and ground-based kinematics of nearby galaxies revealed that most galaxies have to harbor a central massive black hole (SMBH) [44]. However, only few galactic nuclei show up as AGN in the optical, radio, or X-ray wave bands. This can be either explained by an extremely low accretion rate or by an accretion process, which is radiatively inefficient.

The closest example of such an inactive SMBH is the black hole in the center of the Milky Way, Sgr A*, with a quiescent X-ray luminosity in the 0.5–7 keV band of $\sim 2\times 10^{33}$ erg s^{-1}, which sometimes shows flares (see, e.g., [2] and Chap. 19.3). Also M 31 harbors a SMBH in the center, but no bright X-ray source coincides with the M 31 center. Chandra images show a 2.5σ evidence for a faint ($\sim 10^{36}$ erg s^{-1}) discrete source consistent with the position of M 31* [25], indicating again a radiatively inefficient accretion flow. These and further deep X-ray observations of SMBHs in nearby galaxies will place severe constraints on the radiative processes in such a flow.

20.4 Hot Plasma Components

Fabbiano and collaborators have systematically analyzed all Einstein galaxy observations (see [17, 20, 42, 43]). They find normal galaxies of all morphological types as spatially extended sources of X-ray emission with luminosities in the 0.2–3.4 keV band in the range of 10^{38} to 10^{42} erg s^{-1}. Spiral galaxies only reach a few 10^{41} erg s^{-1}. On average the X-ray spectra for spiral galaxies are harder than for ellipticals. This is explained by the commonly accepted view that X-rays from elliptical galaxies originate from a hot interstellar medium while the emission of spiral galaxies is dominated by point-like sources with a harder spectrum (XRBs and SNRs). Also nuclear sources may be present and contribute significantly to the total X-ray emission. These sources may either be connected to star forming or Seyfert-like activity. Some spiral galaxies showed in addition to the expected hard component a soft component in their spectra, indicating that an extended gaseous component can also be present in spirals.

Many of these galaxies have been investigated in greater detail in the 0.1–2.4 keV band with ROSAT and are targets of XMM-Newton and Chandra observations. The good spatial and spectral resolution of present X-ray observatories allows us to separate the emission of distinct sources from that arising from surrounding gas within the galaxies. In the interstellar space, stars are born out of the densest regions and massive stars transfer matter back to the ISM via stellar winds and supernova explosions. Therefore, the diffuse component of the X-ray emission from galaxies will help us to better understand the interaction between stars and the ISM as well as the matter cycle within galaxies.

20.4.1 Hot Interstellar Medium and Gaseous Outflows in Spiral and Starburst Galaxies

The existence of a hot gaseous component in the ISM of late type galaxies with temperatures around 10^6 K was proposed from theoretical considerations. It should originate from SNRs in the galaxies and via fountains also partly fill the halo of galaxies (see e.g. [4, 5, 10, 12, 68]). The pre-ROSAT knowledge of the hot ISM was rather sparse. For a review of the local ISM in the Milky Way see [11]. While a hot ISM was detected with Einstein in the LMC and less convincingly in the SMC [77, 78], studies of edge-on galaxies [6] and the large face-on galaxy M101 [45] could only derive upper limits for diffuse emission from hot gas. The only Einstein detections of hot ISM in late type galaxies outside the Local Group were reported for the starburst galaxies M 82, NGC 253, and NGC 3628 [16, 18, 21, 79].

Thanks to the improved sensitivity of ROSAT, X-ray emission from the hot ISM was resolved in many late type galaxies. If fitted with thin thermal plasma models, temperatures were in the million K range. Examples range from the plane and halo of the Milky Way [66] to galaxies at 20 Mpc distance and more (see e.g. [13, 60]). For M 101 the detected diffuse emission most likely not only originates from the hot ISM in the disk but also from the halo of the galaxy. In addition the ROSAT measurements for the first time showed evidence for shadowing of the soft X-ray background at about 0.25 keV by a M101 spiral arm [67]. Many of the ROSAT results on the hot ISM and gaseous outflows in late type galaxies have been confirmed and investigated in more detail with Chandra Observatory and XMM-Newton (see, e.g., [69, 70, 76]). In the following we will discuss three typical examples for the study of the hot ISM in galaxies with ROSAT from – with increasing distance – the MCs (Fig. 20.8), the prototypical starburst galaxy NGC 253 (Fig. 20.9), and the LINER galaxy NGC 3079 (Fig. 20.11).

Diffuse X-ray emission from the ISM of the LMC was clearly detected in the ROSAT All Sky Survey and in merged images of all pointed observations (see Fig. 20.2). The diffuse X-ray emission of the MCs was systematically studied in all archival pointed ROSAT PSPC observations [64]. Contributions from the X-ray point-like sources in the ROSAT PSPC and HRI catalogues [31, 32, 62, 63] were

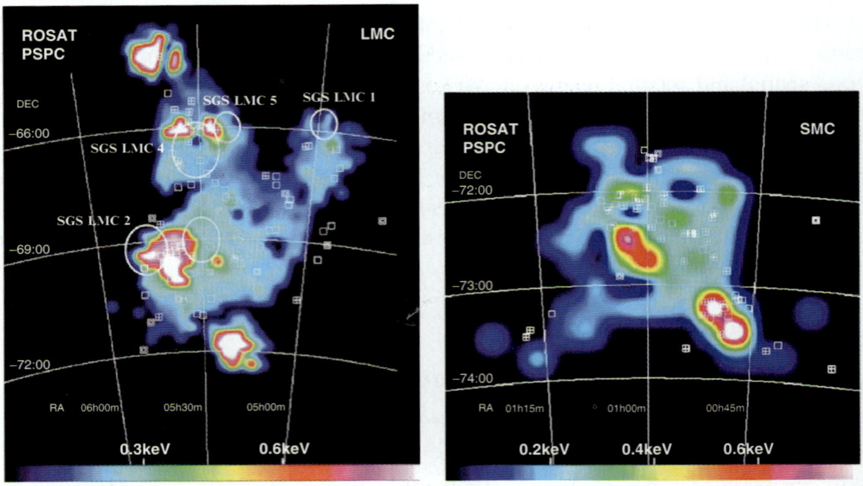

Fig. 20.8 Temperature distribution image of the LMC (*left*) and SMC (*right*). Positions of SNRs observed by ROSAT are shown by *squares*, of XRBs by *crossed squares*, and for SSSs by *double squares*. For the LMC the position of the supergiant shells SGS LMC 1–5 are marked as ellipses. Hot ISM with ($T = 10^6$–10^7 K) extends over the whole LMC and SMC [64]

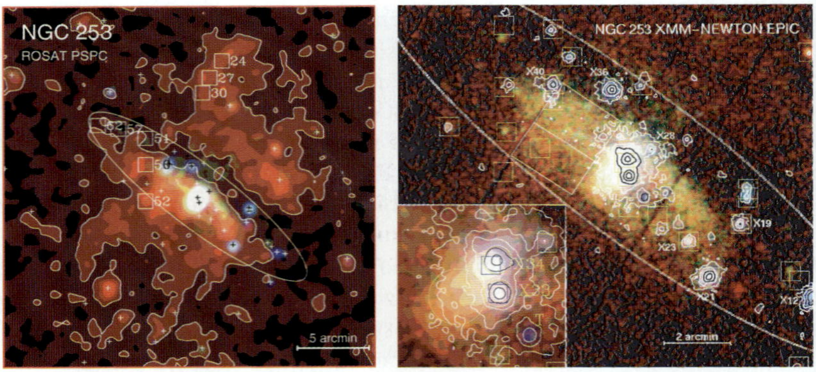

Fig. 20.9 ROSAT PSPC (*left* [59]) and XMM-Newton EPIC (*right* [57]) images of the starburst galaxy NGC 253. The images are color coded with red representing lower energy X-ray emission (0.1–0.5 keV) and blue higher energy X-ray emission (1–2 keV). Hard emission (2–10 keV) is shown superimposed in the EPIC image as contours. The ellipse indicates the optical extent of the galaxy. The ROSAT image clearly shows the extended soft emission from the galaxy halo, which is absorbed in the northwest by the ISM of the galaxy disk. The XMM-Newton image shows many point-like sources resolved as well as emission of the hot ISM in the NGC 253 disk. The plume emanating from the nuclear area to the southeast is interpreted as manifestation of an outflow of hot plasma from the starburst nucleus (see factor 3 zoom-in in inset)

cut out. The spectral analysis yielded characteristic temperatures of $(10^6–10^7)$ K for the hot thin plasma of the ISM, which extends over the whole LMC and SMC (see Fig. 20.8). The total unabsorbed luminosity in the 0.1–2.4 keV band within the observed area amounts to 3.2×10^{38} erg s^{-1} in the LMC and 1.1×10^{37} erg s^{-1} in the SMC, respectively. The X-ray luminosity in the LMC is comparable to that of other nearby galaxies with pronounced star formation. In the LMC, hot regions were found especially around the supergiant shell (SGS) LMC 4 and in the field covering SGS LMC 2 and LMC 3. The highest temperatures for the SMC are located in the southwestern part of the galaxy. The diffuse emission is most likely a superposition of the emission from hot gas in the interior of the shells and super-shells as well as from the halo of these galaxies.

As mentioned above NGC 253 was one of the few starburst galaxies in which extended X-ray emission was detected by the Einstein Observatory [16, 21]. Because of the low Galactic foreground absorption, its big optical extent, and the edge-on viewing geometry, the prototypical starburst galaxy NGC 253 is ideally suited for a detailed analysis of the X-ray emission from disk and halo. ROSAT PSPC and HRI observations revealed diffuse soft X-ray emission from NGC 253, which contributes 80% to its total X-ray luminosity (5×10^{39} erg s^{-1}, corrected for foreground absorption). The nuclear area, disk, and halo contribution to the luminosity is about equal. The starburst nucleus itself is highly absorbed and not visible in the ROSAT band. The hollow-cone shaped plume traces the outflow of the nuclear starburst and interactions with the ISM to an extent of $\sim$700 pc along the SE minor axis. The diffuse emission in the disk follows the spiral arm structure and can be separated in a bright inner and a fainter outer component along the major axis with extents of $\pm$3.4 and $\pm$7.5 kpc, respectively. The coronal halo emission (scale height 1 kpc) is absorbed in the NW by the intervening ISM of the disk. The outer halo can be traced to projected distances from the disk of 9 kpc and shows a horn-like structure with a harder spectrum in the NW halo than in the SE. The emission in the corona and the outer halo is most likely caused by the strong starburst wind. An additional contribution to the coronal emission may come from hot gas fueled by galactic fountains originating in the boiling star-forming disk. A two temperature thermal plasma model ($kT = 0.13$ and 0.62 keV) or a nonequilibrium cooling model of the halo plasma (see, e.g., [7]) are needed to explain the X-ray spectra (see left panel of Fig. 20.9, [59]).

XMM-Newton EPIC observations of NGC 253 [57] allowed to better constrain the diffuse emission in the nuclear area, plume, and disk. The unresolved emission of the two disk regions can be modeled by two thin thermal plasma components ($kT = 0.13$ and 0.4 keV) plus residual harder emission, with the low temperature component originating from above the disk. The nuclear spectrum can be modeled by a three temperature plasma (0.6, 0.9, and 6 keV) with the higher temperatures increasingly absorbed. The high temperature component most likely originates from the starburst nucleus. The combination of EPIC and RGS also sheds new light on the emission of the complex nuclear region and plume (see right panel of Fig. 20.9 and Fig. 20.10).

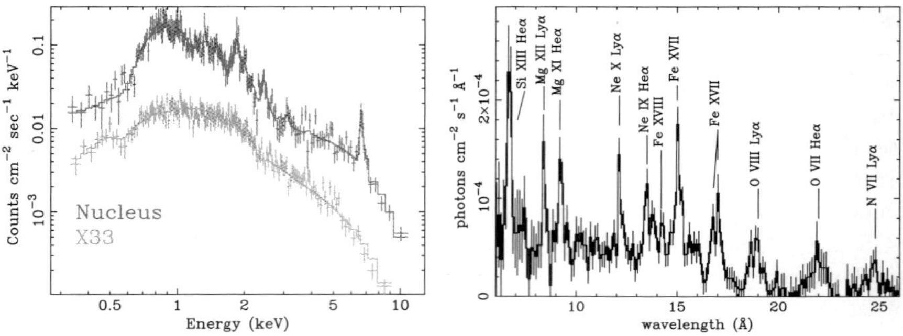

Fig. 20.10 The XMM-Newton EPIC pn spectrum of the nuclear area (*left*, upper spectrum) clearly shows emission lines indicative of emission from a hot thin plasma and can be compared to the featureless disk-blackbody model which best describes the X-ray spectrum of the black hole XRB NGC 253 X33 (*left*, spectrum below). The XMM-Newton RGS spectrum of the bright nuclear area of NGC 253 (*right*) is dominated by bright emission lines [57]

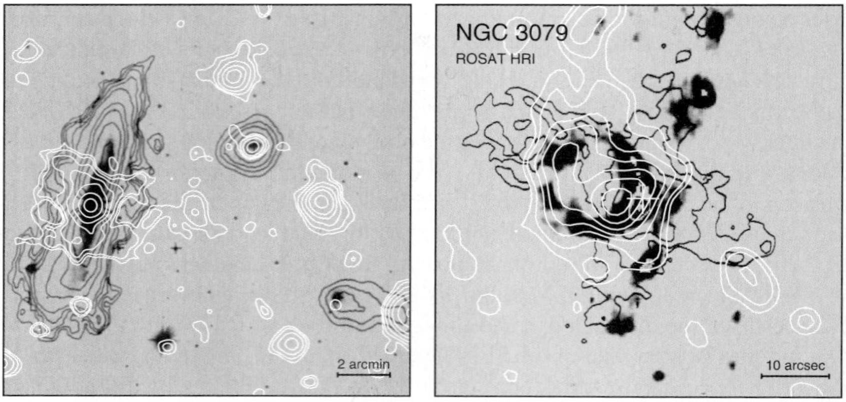

Fig. 20.11 *Left:* Contour plot of the broad band PSPC X-ray emission of the region of NGC 3079 and its companions overlaid in white on an reproduction of the POSS red plate. H I-contours are superposed in black. *Right:* Contour plot of the central emission region of NGC 3079 for ROSAT HRI overlaid on the continuum-subtracted distribution of H_α + [N II λλ6548, 6583 line emission (grey-scale) and the 20 cm continuum distribution (black contours) [58]

ROSAT PSPC and HRI observations resolved complex emission from the inner 5′ around the LINER galaxy NGC 3079 which is seen edge-on. The extended emission in the innermost region is extended and has a luminosity of 1×10^{40} erg s^{-1}. It coincides with a super-bubble seen in optical and the east lobe of bipolar radio emission originating from the nucleus. The active nucleus of the galaxy known from radio observations may contribute to the X-ray emission as a point source. In addition there is emission from the disk of the galaxy (7×10^{39} erg s^{-1}) that can partly be resolved by the HRI in three point-like sources with luminosities of $\sim 6 \times 10^{38}$ erg s^{-1} each. The PSPC resolves very soft X-shaped emission from the halo, with $L_X = 6 \times 10^{39}$ erg s^{-1}, extending to a diameter of 27 kpc. The X-ray luminosity of NGC 3079 is higher by a factor of 10 compared to other galaxies of similar optical luminosity. This may be caused by the presence of an AGN rather than by starburst activity (Fig. 20.11, [58]). Chandra observations confirmed the ROSAT findings for the NGC 3079 nuclear super-bubble and large-scale superwind [8, 69].

20.4.2 Hot Gaseous Emission in Early Type Galaxies

Einstein observations demonstrated that E and S0 galaxies may have large hot gaseous halos, dominating their X-ray emission. Central cooling flows similar to some galaxy clusters have been suggested and the X-rays have been used to trace the gravitational potential of these galaxies (see [17] and references therein). However, not all E and S0 galaxies are able to retain a large amount of hot ISM. For some galaxies an extensive halo is clearly displaced from the stellar component and in some cases extending to very large radii. This has been shown by ROSAT for several galaxies and studied in detail for NGC 4406 with XMM-Newton [24]. Other galaxies show much fainter X-ray emission compared to the optical, which typically is coincident with the stellar body seen in the optical. From ROSAT hardness ratio arguments this emission has been attributed to LMXBs [39]. This proposal was nicely confirmed by high resolution Chandra and HST observations that show X-ray point sources correlating with globular clusters in the galaxies (see, e.g., [61]).

20.5 Future Prospects

The Einstein Observatory opened the field of X-ray observations of normal galaxies. With ROSAT we were able to investigate many galaxies in greater detail, thanks to longer observing times. Of specific importance was the much higher low energy sensitivity, which allowed us to detect diffuse emission from disk and halo of several galaxies. However, the moderate point spread function (several to tens of arcseconds) and effective area of the instruments restricted detailed investigations to a few nearby galaxies. XMM-Newton and Chandra observations of nearby galaxies have demonstrated the importance of high collecting power for time variability and high spectral resolution investigations and of sub-arcsecond spatial resolution to separate different emission components, respectively. To gain further understanding of the X-ray emission properties of galaxies next generation X-ray missions should combine high collecting power with good spatial and spectral resolution and also allow for long observing times. With such observations not only detailed source population studies in the individual galaxies will be possible but one will also be able to study several X-ray source classes (like novae, Be-type XRBs, ULXs, and SN) in external galaxies with much better efficiency than – if at all possible – in the Milky Way.

References

1. Angelini, L., Loewenstein, M., & Mushotzky, R. F. 2001, ApJ, 557, L35
2. Baganoff, F. K., Maeda, Y., Morris, M., et al. 2003, ApJ, 591, 891
3. Beuermann, K., Brandt, S., & Pietsch, W. 1994, A&A, 281, L45
4. Bregman, J. N. 1980a, ApJ, 236, 577
5. Bregman, J. N. 1980b, ApJ, 237, 681
6. Bregman, J. N. & Glassgold, A. E. 1982, ApJ, 263, 564
7. Breitschwerdt, D. & Schmutzler, T. 1999, A&A, 347, 650
8. Cecil, G., Bland-Hawthorn, J., & Veilleux, S. 2002, ApJ, 576, 745
9. Coe, M. J., Edge, W. R. T., Galache, J. L., & McBride, V. A. 2005, MNRAS, 356, 502
10. Corbelli, E. & Salpeter, E. E. 1988, ApJ, 326, 551
11. Cox, D. P. & Reynolds, R. J. 1987, ARA&A, 25, 303
12. Cox, D. P. & Smith, B. W. 1974, ApJ, 189, L105
13. Dahlem, M., Weaver, K. A., & Heckman, T. M. 1998, ApJS, 118, 401
14. Dennerl, K., Haberl, F., Aschenbach, B., et al. 2001, A&A, 365, L202
15. Dubus, G. & Rutledge, R. E. 2002, MNRAS, 336, 901
16. Fabbiano, G. 1988, ApJ, 330, 672
17. Fabbiano, G. 1989, ARA&A, 27, 87
18. Fabbiano, G., Heckman, T., & Keel, W. C. 1990, ApJ, 355, 442
19. Fabbiano, G. & Kessler, M. F. 2001, in: The Century of Space Science, eds. J. A. M. Bleeker, J. Geiss & M. C. E. Huber (Dortrecht: Kluwer), Vol. 1, p. 561
20. Fabbiano, G., Kim, D.-W., & Trinchieri, G. 1992, ApJS, 80, 531
21. Fabbiano, G. & Trinchieri, G. 1984, ApJ, 286, 491
22. Fabbiano, G. & White, N. E. 2006, in: Compact Stellar X-ray Sources, eds. W. Lewin & M. van der Kliss (Cambridge: Cambridge Univ. Press), 475
23. Fabbiano, G., Zezas, A., & Murray, S. S. 2001, ApJ, 554, 1035
24. Finoguenov, A., Pietsch, W., Aschenbach, B., & Miniati, F. 2004, A&A, 415, 415
25. Garcia, M. R., Williams, B. F., Yuan, F., et al. 2005, ApJ, 632, 1042
26. Ghavamian, P., Blair, W. P., Long, K. S., et al. 2005, AJ, 130, 539
27. Giacconi, R. & Gursky, H., eds. 1974, X-Ray Astronomy (Dortrecht: Reidel, ASSL Vol. 43)
28. Gorenstein, P., Hughes, J. P., & Tucker, W. H. 1994, ApJ, 420, L25
29. Grimm, H.-J., Gilfanov, M., & Sunyaev, R. 2002, A&A, 391, 923
30. Grimm, H.-J., Gilfanov, M., & Sunyaev, R. 2003, MNRAS, 339, 793
31. Haberl, F., Filipović, M. D., Pietsch, W., & Kahabka, P. 2000, A&AS, 142, 41
32. Haberl, F. & Pietsch, W. 1999, A&AS, 139, 277
33. Haberl, F. & Pietsch, W. 2001, A&A, 373, 438
34. Haberl, F. & Pietsch, W. 2004, A&A, 414, 667
35. Haberl, F. & Pietsch, W. 2005, A&A, 438, 211
36. Haberl, F., Pietsch, W., Schartel, N., Rodriguez, P., & Corbet, R. H. D. 2004, A&A, 420, L19
37. Immler, S., Aschenbach, B., & Wang, Q. D. 2001, ApJ, 561, L107
38. Immler, S. & Lewin, W. H. G. 2003, LNP Vol. 598: Supernovae and Gamma-Ray Bursters, 598, 91
39. Irwin, J. A. & Sarazin, C. L. 1998, ApJ, 499, 650
40. Kahabka, P. & van den Heuvel, E. P. J. 2006, in: Compact Stellar X-ray Sources, eds. W. Lewin & M. van der Kliss (Cambridge: Cambridge Univ. Press), 461
41. Kahabka, P. & van den Heuvel, E. P. J. 1997, ARA&A, 35, 69
42. Kim, D.-W., Fabbiano, G., & Trinchieri, G. 1992a, ApJS, 80, 645
43. Kim, D.-W., Fabbiano, G., & Trinchieri, G. 1992b, ApJ, 393, 134
44. Magorrian, J., Tremaine, S., Richstone, D., et al. 1998, AJ, 115, 2285
45. McCammon, D. & Sanders, W. T. 1984, ApJ, 287, 167
46. Miller, J. M., Fabian, A. C., & Miller, M. C. 2004, ApJ, 614, L117
47. Negueruela, I. & Coe, M. J. 2002, A&A, 385, 517

48. Osborne, J. P., Borozdin, K. N., Trudolyubov, S. P., et al. 2001, A&A, 378, 800
49. Pannuti, T. G., Duric, N., Lacey, C. K., et al. 2002, ApJ, 565, 966
50. Park, S., Zhekov, S. A., Burrows, D. N., & McCray, R. 2005, ApJ, 634, L73
51. Pietsch, W., Fliri, J., Freyberg, M. J., et al. 2005a, A&A, 442, 879
52. Pietsch, W., Freyberg, M., & Haberl, F. 2005b, A&A, 434, 483
53. Pietsch, W. & Haberl, F. 2005, A&A, 430, L45
54. Pietsch, W., Haberl, F., & Vogler, A. 2003, A&A, 402, 457
55. Pietsch, W., Misanovic, Z., Haberl, F., et al. 2004a, A&A, 426, 11
56. Pietsch, W., Mochejska, B. J., Misanovic, Z., et al. 2004b, A&A, 413, 879
57. Pietsch, W., Roberts, T. P., Sako, M., et al. 2001, A&A, 365, L174
58. Pietsch, W., Trinchieri, G., & Vogler, A. 1998, A&A, 340, 351
59. Pietsch, W., Vogler, A., Klein, U., & Zinnecker, H. 2000, A&A, 360, 24
60. Read, A. M., Ponman, T. J., & Strickland, D. K. 1997, MNRAS, 286, 626
61. Sarazin, C. L., Irwin, J. A., & Bregman, J. N. 2000, ApJ, 544, L101
62. Sasaki, M., Haberl, F., & Pietsch, W. 2000a, A&AS, 143, 391
63. Sasaki, M., Haberl, F., & Pietsch, W. 2000b, A&AS, 147, 75
64. Sasaki, M., Haberl, F., & Pietsch, W. 2002, A&A, 392, 103
65. Shtykovskiy, P. & Gilfanov, M. 2005, A&A, 431, 597
66. Snowden, S. L., Hasinger, G., Jahoda, K., et al. 1994, ApJ, 430, 601
67. Snowden, S. L. & Pietsch, W. 1995, ApJ, 452, 627
68. Spitzer, L. J. 1956, ApJ, 124, 20
69. Strickland, D. K., Heckman, T. M., Colbert, E. J. M., Hoopes, C. G., & Weaver, K. A. 2004a, ApJS, 151, 193
70. Strickland, D. K., Heckman, T. M., Colbert, E. J. M., Hoopes, C. G., & Weaver, K. A. 2004b, ApJ, 606, 829
71. Sunyaev, R., Kaniovsky, A., Efremov, V., et al. 1987, Nature, 330, 227
72. Trinchieri, G. & Fabbiano, G. 1985, ApJ, 296, 447
73. Trudolyubov, S. P., Borozdin, K. N., & Priedhorsky, W. C. 2001, ApJ, 563, L119
74. Trudolyubov, S. P., Borozdin, K. N., Priedhorsky, W. C., et al. 2002, ApJ, 581, L27
75. Trümper, J., Hasinger, G., Aschenbach, B., et al. 1991, Nature, 349, 579
76. Tüllmann, R., Pietsch, W., Rossa, J., Breitschwerdt, D., & Dettmar, R.-J. 2006, A&A, 448, 43
77. Wang, Q. 1991, ApJ, 377, L85
78. Wang, Q., Hamilton, T., Helfand, D. J., & Wu, X. 1991, ApJ, 374, 475
79. Williams, B. F., Sjouwerman, L. O., Kong, A. K. H., et al. 2004, ApJ, 615, 720
80. Zimmermann, H.-U. & Aschenbach, B. 2003, A&A, 406, 969
81. Zimmermann, H. U., Lewin, W., Predehl, P., et al. 1994, Nature, 367, 621

21 X-Ray Flares in the Cores of Galaxies

S. Komossa

It has long been suggested that supermassive black holes in nonactive galaxies can be tracked down by occasional tidal disruptions of stars approaching the black holes. A tidal disruption event would reveal itself by a luminous flare of electromagnetic radiation. This chapter describes the X-ray detection of the highest amplitudes of variability ever recorded among galaxies and their interpretation in terms of the long-sought tidal disruption events.

21.1 Introduction: Tidal Disruption of Stars by Supermassive Black Holes

There is strong evidence for the presence of massive black holes at the centers of many galaxies (see [16] for a review). Does this hold for *all* galaxies? Questions of particular interest in the context of galaxy evolution are what fraction of galaxies have passed through an active phase, and how many now have nonaccreting and hence unseen supermassive black holes (SMBHs) at their centers (e.g., [19, 20])? How do SMBHs *grow* to masses of 10^6–$10^9 M_\odot$?

It has been pointed out that an *unavoidable consequence*, and one of the best diagnostics of the presence of supermassive black holes at the centers of nonactive galaxies would be occasional tidal disruptions of stars approaching the supermassive black hole [18, 20]. The flare, produced when the star is tidally disrupted and subsequently accreted, can be used to find and study these supermassive black holes and their immediate vicinity.

A star on a radial "loss-cone" orbit gets tidally disrupted once the tidal force excerted by the black hole exceeds the self-gravity of the star (e.g., [10]). This happens at the tidal radius, r_t, given by

$$r_t \simeq 7\,10^{12}\,(\frac{M_{BH}}{10^6 M_\odot})^{\frac{1}{3}}\,(\frac{M_*}{M_\odot})^{-\frac{1}{3}}\,\frac{r_*}{r_\odot}\ \mathrm{cm}\,, \qquad (21.1)$$

where M_{BH} is the black hole mass, M_* is the mass of the star in units of solar mass $M_\odot$, and r_* is the radius of the star in units of solar radius $r_\odot$. About 50–90% of the gaseous debris becomes unbound and is lost from the system (e.g., [1]). The

J.E. Trümper and G. Hasinger (eds.), *The Universe in X-Rays.*

rest will eventually be accreted by the black hole. This produces a flare of radiation, lasting on the order of months to years [21].

21.2 X-Ray Flares from Inactive Galaxies

Given its broad astrophysical relevance and the intense theoretical attention (e.g. [2,4,10,20,24] and Sect. 3 of [11] and references therein]), a key question is: do these flares occur in nature, how frequent is tidal disruption of stars, and what are its properties?

All-sky X-ray surveys, similar to the one performed with the ROSAT satellite [23], are the ideal experiment to detect such flares, since hundredths of thousands of galaxies are sampled in the soft X-ray band.

With ROSAT, several giant-amplitude, nonrecurrent X-ray outbursts from a handful of "normal" galaxies – NGC 5905 [12], RX J1242−1119 [13], RX J1624+7554 [8], RX J1420+5334 [6] – were detected (see [11] for a review). All of them share similar properties:

- Huge soft X-ray peak luminosities (up to $L_{sx} \sim 10^{44}$ erg s^{-1})
- Large amplitudes of variability (up to a factor ∼200)
- Ultra-soft X-ray spectra ($kT_{bb} \simeq 0.04$-0.1 keV, where T_{bb} is the black body temperature)
- Complete absence of any signs of ongoing Seyfert activity in ground-based optical spectra (confirmed with the Hubble Space Telescope, except for NGC 5905 that shows a faint high-excitation core in its nucleus [5])

The observed events were interpreted as excellent candidates for the long-sought tidal disruption events. To perform key tests of the favored outburst scenario, follow-up observations with the new generation of X-ray observatories, Chandra and XMM-Newton, were obtained.

21.3 Chandra and XMM-Newton Follow-Up Observations

Chandra and XMM-Newton allow us to measure the long-term evolution of the light curves and spectra of the flares. In particular, Chandra with its high spatial resolution enables us to pinpoint the counterpart of the flare precisely. In case of tidal disruption, the emission should come directly from the center of each galaxy.

RX J1242–1119 [13] was the first target of choice for follow-up X-ray observations because it flared most recently, so the probability of catching the source in the declining phase–before it had faded away completely–was highest.

Chandra detected a huge drop in X-ray flux by a factor ∼200 [14], compared with the high-state ROSAT observation. The point-like late-phase flare emission

Fig. 21.1 *Upper panel*: Artist's sketch of the tidal disruption of a star. The star is ripped apart by the tidal forces of a massive black hole. Part of the stellar debris is then accreted. This causes a luminous flare of radiation that fades away as more and more of the matter disappears into the black hole. *Lower panel*: optical image of the galaxy (pair) RX J1242-1119A, B (*right*) and "afterglow" of the X-ray flare from its center (*left*) [image credit: NASA/CXC/M.Weiss/ESO/MPE; Komossa et al. 2004]

coincides with the center of the galaxy RX J1242-1119A (Fig. 21.1). With XMM-Newton, for the first time a good-quality X-ray spectrum of one of the few flaring galaxies was obtained [14]. The spectrum is well fit by a power law (photon index $\Gamma_x = -2.5$), harder than during outburst. This spectral shape is typical for the emission spetrum of matter in the immediate vicinity of a black hole.

A further decline of the X-ray emission is expected as more and more of the stellar debris is accreted by the black hole. Indeed, reobservation of RX J1242–1119 with Chandra in 2004 showed a further fading of the X-ray source by a factor of several (Komossa et al. 2007, in prep.) such that the total amplitude of variability exceeds a factor of 1 000.

Two more flares were followed up with Chandra. Only few, if any, photons from the galaxies' centers were detected, making their total amplitude of variability extremely large, a factor >1 000 (NGC 5905, [9]) and >6 000 (RX J1624+75, [9, 22]).

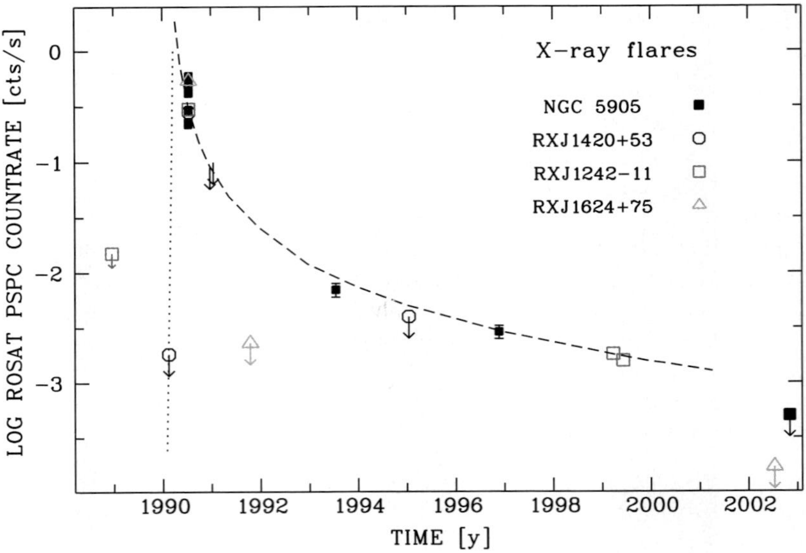

Fig. 21.2 Collective X-ray light curves of the four flaring galaxies, all shifted to the same peak time. The *dashed line* follows a $t^{-5/3}$ law matched to the high- and first low-state data point of RX J1242-1119. The *dotted line* corresponds to the least fastest rise to high-state consistent with the upper limit of RX J1420+5334. The last data point of NGC 5905 corresponds to the central point source only, while previous X-ray measurements of NGC 5905 plotted in this figure include both extended and core emission not spatially resolved prior to Chandra [15]

Figure 21.2 shows the collective X-ray light curve of the flare events, shifted in time to the same high-state as NGC 5905. The light curves of all events are relatively similar, and consistent with a faster rise, and a slower decline on the timescale of months–years. Most of the data points are consistent with a $t^{-5/3}$ decline law, which is expected for the "fall-back" phase of tidal disruption (e.g. [21,17]).[1]

In *summary*, the observed events match basic predictions from models of the tidal disruption of stars by supermassive black holes at the centers of the flaring galaxies (see [12,9,14] for a more detailed discussion). The observed X-ray flares have the highest amplitudes of variability–up to a factor >6 000 – ever recorded among galaxies, including active galaxies. The rate of tidal disruption events, estimated on the basis of ROSAT observations, is $\sim$1 event per galaxy per 10^4 yrs [3], roughly consistent with theoretical predictions [24].

[1] This law only holds if the bolometric correction is constant, i.e., if most of the emission remains in the X-ray band, as time evolves. It is approximate for other reasons as well.

21.4 Future Observations and Applications

X-ray outbursts from optically nonactive galaxies provide important information on the presence of SMBHs in these galaxies, and the link between active and normal galaxies. In particular, the capture and disruption of stars by SMBHs is one of the three major processes studied in the context of black hole growth, together with accretion of interstellar matter and black hole - black hole merging. The relative importance of these three processes in feeding black holes is still under investigation. Different processes may dominate in different stages of the evolution of galaxies.

Future X-ray all-sky surveys will be valuable in finding more of these outstanding events [e.g., 7]. Rapid follow-up observations at all wavelengths will then be important. We estimate that the planned all-sky survey mission eROSITA will detect a few hundred flares during the first years of its operation, which will allow us to study these events in detail.

References

1. Ayal, S., Livio, M., Piran, T. 2000, Astrophys. J., 545, 772
2. Carter, B., Luminet, J.P. 1982, Nature, 296, 211
3. Donley, J., et al. 2002, Astron. J., 124, 1308
4. Gurzadyan, V.G., Ozernoi, L.M. 1979, Nature, 280, 214
5. Gezari, S., Halpern, J., Komossa, S., Grupe, D., Leighly, K. 2003, Astrophys. J., 592, 42
6. Greiner, J., Schwarz, R., Zharikov, S., Orio M. 2000, Astron. Astrophys., 362, L25
7. Grindlay, J.E. 2004, in P. Kaaret, et al. (eds.), X-ray Timing 2003: Rossi and Beyond, AIP Conference Proceedings, 714, American Institute of Physics, Melville, NY, 413
8. Grupe, D., Leighly, K., Thomas, H. 1999, Astron. Astrophys., 351, L30
9. Halpern, J., Gezari, S., Komossa, S. 2004, Astrophys. J., 604, 572
10. Hills, J.G. 1975, Nature, 254, 295
11. Komossa, S. 2002, in R.E. Schielicke (eds.), Reviews in Modern Astronomy, Wiley, Weinheim 15, 27
12. Komossa, S., Bade N. 1999, Astron. Astrophys., 343,775
13. Komossa, S., Greiner J. 1999, Astron. Astrophys., 349, L45
14. Komossa, S., Halpern J., Schartel S., et al. 2004, Astrophys. J., 603, L17
15. Komossa, S. 2005, in A. Merloni, et al. (eds.), Growing Black Holes: Accretion in a Cosmological Context, ESO Astrophysics Symposia, Springer, Berlin Heidelberg New York, 159
16. Kormendy, J., Richstone, D.O. 1995, Astron. Astrophys., 33, 581
17. Li, L.-X., Narayan, R., Menou, K. 2002, Astrophys. J., 576, 753
18. Lidskii, V.V., Ozernoi, L.M. 1979, AZh Pis'ma, 5, L28
19. Lynden-Bell, D. 1969, Nature, 223, 690
20. Rees, M. 1988, Nature, 333, 523
21. Rees, M. 1990, Science, 247, 817
22. Vaughan, S., Edelson R., Warwick R.S. 2004, Mon. Not. Roy. Astron. Soc., 349, 1
23. Voges, W., et al. 1999, Astron. Astrophys., 349, 389
24. Wang, J., Merritt, D. 2004, Astrophys. J., 600, 147

22 Active Galactic Nuclei

T. Boller

22.1 General Introduction to Active Galaxies

X-ray observations of active galactic nuclei (AGN) are of wide astrophysical interest. The amount and efficiency of the energy release within the immediate neighbourhood of black holes belong to the most extreme physical processes observed to date. The most probable explanation for the huge amount of energy output (and for other observational parameters, such as the width of the optical emission lines and the strength of the radio emission) is given by the transformation of potential energy into radiation by accretion of matter onto a supermassive central black hole [43]. The velocities of the accreting matter reach values of about one third of the velocity of light, deduced by relativistically broadened line profiles (centred at about 6.4 keV), [49]. The emission from the matter around the black hole may vary on very short time scales of only a few hundred seconds. The corresponding changes in luminosity reach values of about 10^{10} solar luminosities. All this is further suggestive for the presence of supermassive black holes. Energy production processes and radiation mechanisms for the innermost regions of AGN are an important research field in X-ray astronomy and open a unique possibility to study matter under extreme gravity. Other astrophysical important aspects include the detection and study of binary black holes, expected to lead to strong gravitational wave emission and tidal disruption events of stars in the dense core around the central black holes. At larger distances from the black hole, the emission from optically thin plasma can be studied. Imprinted absorption and emission lines give information on the chemical composition of the gas and of infall and outflow velocities. The log N−log S distribution of AGN and resulting luminosity functions allow to study the density and luminosity evolution of AGN in dependence of redshift. The first compact obscured objects in the universe can be best studied with X-rays.

The early rocket flights in the sixties detected about 50 X-ray sources, including three AGN. The first AGN discovered was M 87 in the Virgo cluster [13], the other two were the nearby active galaxy NGC 1275 and the quasar 3C 273. Ariel V established the Seyfert 1 galaxies as a class of X-ray emitters [14]. HEAO-1 provided luminosity functions of AGN [32] and numerous broad band spectra of AGN [35, 57]. Further progress in the field was made with EXOSAT and Ginga. Observations with the Einstein satellite further improved our knowledge on accreting sources and enabled the study of distant quasars, previously detected in other wavebands. The first

J.E. Trümper and G. Hasinger (eds.), *The Universe in X-Rays.*

X-ray all sky survey with an imaging X-ray telescope carried out with ROSAT provided large complete samples of AGN. The ROSAT pointed observations lead to many discoveries and answered in particular one of the oldest questions of X-ray astronomy by resolving the X-ray background into discrete sources, mainly AGN. Presently the advanced telescopes on Chandra, XMM-Newton and Suzaku contribute many new results to the field.

22.1.1 Nuclear Components of Active Galaxies

In the following the most relevant nuclear components of active galaxies are described. Figure 22.1 gives a schematic illustration. Within the centre, the hypothetical black hole is located, surrounded by an accretion disk. At larger distances from the central black hole are located, the region where the broad emission lines (Broad Line Region, BLR) originate and the region of the narrow emission lines (Narrow Line Region, NLR). Another component detected in some active galaxies is the molecular torus, which is probably located within the NLR. The molecular torus leads to strong absorption of the emission from the nuclei of active galaxies. It is

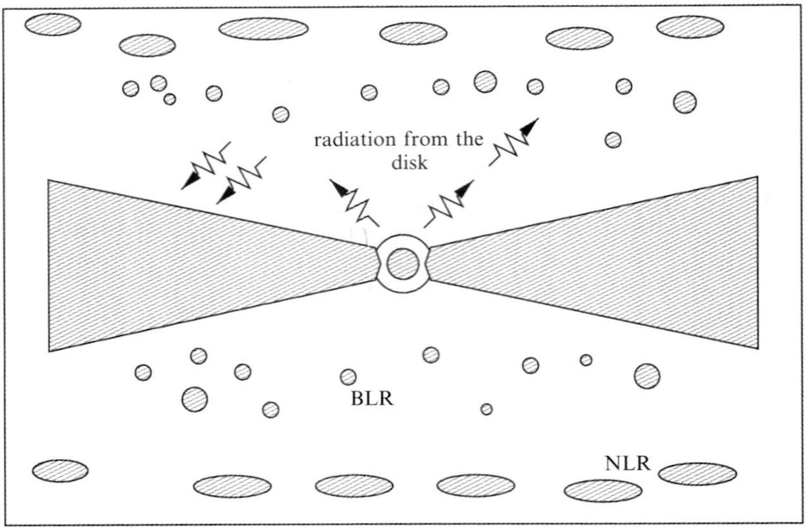

Fig. 22.1 Schematic illustration of the principal components in the nuclei of active galaxies. The black hole is surrounded by an accretion disk. The radiation from the accretion disk, which is considered as the primary energy source for the excitation of the optical emission lines in active galaxies, is marked by the arrows. The accretion disk is surrounded by the Broad Line Region (BLR). A system of relatively cool (about 10^4 K) clouds (marked by the dashed circles), thought to emit the broad emission lines in active galaxies, is probably embedded in a hot (about 10^8 K) ionized gas. Both components are in pressure equilibrium. Further out is seen the region of the primary origin of the narrow emission lines (NLR), marked by the dashed ellipses

assumed, that the main difference between Seyfert 1 and Seyfert 2 galaxies is due to orientation effects. The Full-Width at Half Maximum (FWHM) of the Balmer lines in Seyfert 1 galaxies reaches values up to about 100 000 km s^{-1}, in contrast to Seyfert 2 galaxies where the typical FWHM values are of the oder of only a few hundred km s^{-1}. Absorption along the line of sight, caused by the molecular torus can explain the absence of broad lines in Seyfert 2 galaxies. Part of the emission of the central source is scattered in the observers direction. This leads to polarization and the polarized part of the emission of Seyfert 2 galaxies shows the typical broad emission lines as observed in Seyfert 1 galaxies [1].

22.1.2 The Black Hole

Black holes are solutions of the general relativistic field equations. The solutions for motions of test particles for a non-rotating, spherical symmetric black hole are given by equ. 1 and 2 (cf. Sect. 9.1 of [19]).

$$\left(\frac{dr}{ds}\right)^2 + V^2(r) = E^2$$

$$\frac{dt}{ds} = E\left(1 - \frac{2GM}{rc^2}\right)^{-1}$$

Here $V^2(r) = (1 - 2GM/rc^2)(1 + h^2/r^2)$ is the effective potential, h is the angular momentum of a test particle in the gravitational potential, t is the time measured by a distant observer (the distance r to the black hole approaches infinity), M is the mass of the black hole and s gives the time dependent on the strength of the gravitational field. The term E describes the total energy of a test particle per unit mass, G is the gravitational constant and c is the velocity of light.

$$\text{For} \quad r \longrightarrow \frac{2GM}{c^2} = R_S$$

$$\text{follows} \quad V \longrightarrow 0, \frac{dt}{ds} \longrightarrow \infty$$

and a singularity results. The term R_S is therefore called the event horizon or Schwarzschild radius. Below R_S radiation and particles cannot leave the black hole, as R_S is the radius of the maximum allowed escape velocity $v_{escape} = (2GM/R_S)^{1/2} = c$.

22.1.3 The Accretion Disk

The possible existence of accretion disks was originally predicted from theoretical considerations. An accretion disk occurs, when gas with non-zero angular momentum accretes onto a compact central object. The gas particles, mainly electrons and ions, interact by collisions. In addition electromagnetic forces transport angular momentum outwards. A mean free path length λ is used to describe these interactions. The gas particles move along the path length before they change their direction or velocity due to a collision with another particle. For a distant observer ($r >> \lambda$), the accretion flow can be described by a fluid motion, characterized by the parameters velocity v, temperature T and density ρ at a certain position. The calculating of the accretion process onto a compact central object and the emitted radiation spectrum is complicated. First, solutions for the movement of particles in the gravitational potential of the central black hole have to be found. Such solutions result from the continuity equation, the equation of state for an ideal gas, and the Euler equation (Chap. 2, [19]). Assuming spherical symmetric accretion, and neglecting forces other than the gravitational force, solutions are shown in Fig. 22.1 of [19]. In the calculation of the emitted spectrum, cooling and heating processes have to be considered. In addition, the influence of magnetic fields can be of importance. A detailed discussion of all these effects is given in Chap. 4 of [44].

Accretion onto a compact object is an effective process in the release of radiation. To first order, this radiation release can be approximated by the change in potential energy $\Delta E_{accretion}$ of a test particle with mass m, in the gravitational field of a black hole, as

$$\Delta E_{accretion} = \frac{GMm}{R_S}$$

The accretion luminosity $L_{accretion}$, which represents the emitted amount of energy per time unit, can be derived from $\Delta E_{accretion}$ as

$$L_{accretion} = \zeta \cdot \frac{GMdm/dt}{R_S} = \zeta \cdot \frac{GM\dot{M}}{R_S} = \frac{\zeta}{2} \cdot \dot{M}c^2 = \eta \cdot \dot{M}c^2$$

$\dot{M}$ is the accretion rate and $\eta = \frac{\zeta}{2}$ is the term which describes the efficiency of the accretion process. The efficiency of converting matter into radiation in the accretion process onto a non-rotating black hole is $\eta = 0.057$ [44]. For accretion onto a rotating black hole the corresponding value is $\eta = 0.29$ [52]. With an accretion rate of only one solar mass per year, a luminosity of 10^{46} erg s^{-1} is obtained (assuming $\eta = 0.1$). This is typical of values measured in high- luminosity active galaxies.

22.1.4 Signatures of Activity

The energy release of active galactic nuclei (reaching values above 10^{10} solar luminosities) is emitted within only a few Schwarzschild radii, deduced from the rapid X-ray variability. A more detailed definition is given in terms of the following signatures of activity. One observed signature is sufficient to classify a galaxy as an AGN.

22.1.4.1 Emission Line Widths (500 – 10 000 km s^{-1})

Assuming Keplerian motions, the width of the emission lines is determined by the mass of the central black hole M and the distance of the line emitting clouds. The maximum values in non-active galaxies are less than about 500 km s^{-1}. The motions of line emitting clouds close (a few light days) to central black holes with masses of a few 10^6 solar masses up to about 10^9 solar masses, result on the other hand, in much larger line widths, up to about 10 000 km s^{-1}. Observational evidence for this assumption is seen through reverberation mapping observations. Within these observations, the response of the optical line widths and intensities to changes in the ionizing continuum, are investigated. The time difference between the changes of the ionizing radiation and the change in the intensity of the emission lines is interpreted as the light travelling time between the inner parts of the accretion disk and the BLR, and hence serves as a measure for the distance r. Characteristic values for r are of the order of a few light days. At these distances from the central black hole, the high measured line widths are achieved [38].

22.1.4.2 Emission Line Ratios (Diagnostic Diagrams)

The galaxy classification is based on the work of [54]. The following emission line ratios are used: $[\mathrm{NII}]/\mathrm{H}\alpha$, $[\mathrm{SII}]/\mathrm{H}\alpha$, $[\mathrm{OI}]/\mathrm{H}\alpha$ versus $[\mathrm{OIII}]/\mathrm{H}\beta$ (cf. Fig. 22.2). Non-active galaxies are separated from active galaxies in terms of their positions in the diagnostic diagrams. The main reason is the stronger ionization of elements by a non-stellar continuum, arising from the nuclei of active galaxies, compared with the ionization which can be obtained by the emission of stars. A further sub-division between Seyfert 1 and Seyfert 2 galaxies is given by the ratio $[\mathrm{OIII}]/\mathrm{H}\beta$ [45].

22.1.4.3 Fe II Lines

Based on the papers of [3] and [56] the detection of Fe II multiplet emission is indicative of emission from high-density regions $>10^6$ particles per cm^3. Such high densities are assumed to exist in the circumnuclear environments of active galaxies, probably in the accretion disk and in the line emitting clouds of the BLR. Fe II exhibits a complicated energy level diagram, and the calculation of the transition probabilities by [3], which was originally used for stellar astrophysics, serves presently as an important criterion for the classification of active and non-active galaxies. The Fe II spectrum consists of thousands of lines. The most intense lines form the so-called Fe II multiplet line bands between 2200–2600 Å, 3000–3400 Å, 4500–4600 Å, und 5250–5350 Å. It is thought that X-ray photons are required, to heat and to partially ionize these dense regions. The temperature in the Fe II emitting regions is smaller than 40 000 K (at higher temperatures Fe III would be formed, which is not observed), and the ratio of neutral to ionized hydrogen N_{H^0}/N_{H^+} is

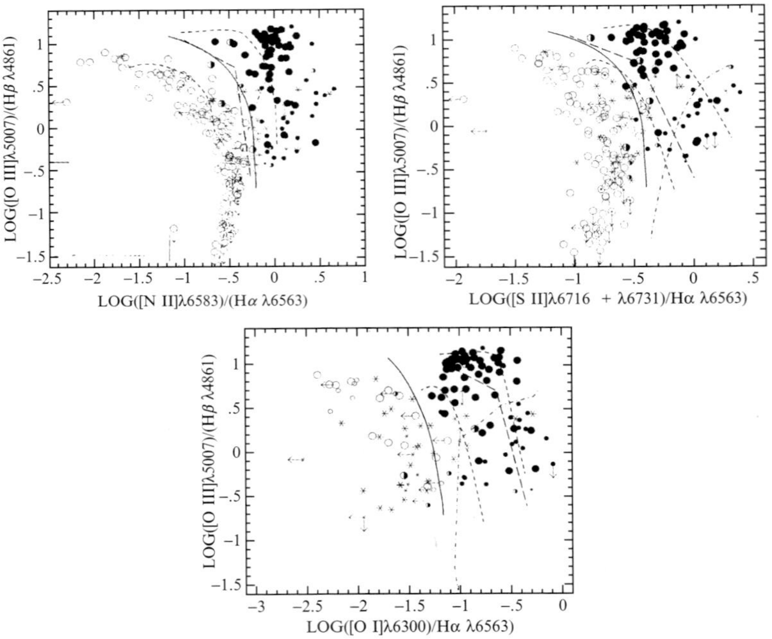

Fig. 22.2 Emission line ratios of [OIII]/Hα versus [NII]/Hα (left panel), [SII]/Hα (right panel) and [OI]/Hα (lower panel) from Veilleux and Osterbrock (1987). Active galaxies are marked by the filled circles, open circles mark the location of non-active galaxies in the classification diagrams. The dotted and solid lines represent different models for the emission from non-active galaxies (see Chap. 3 of [54] for a detailed discussion). Narrow-line Seyfert 1 galaxies form an exception with respect to the classification of Seyfert galaxies in the diagnostic diagrams, as they can be located in all regions of the diagnostic diagrams discussed above

about 10. The strength of the Fe II emission is referred to in the literature as the 'Fe II problem'. The observed intensity of the Fe II lines reaches the intensity of the strongest emission line of hydrogen, that of the Lyα line. The Fe II line width derived from photoionization models only reaches values between about 1/3 and 1/2 of the observed Lyα line intensity. The ratio of the optical Fe II multiplet emission centred at 4500–4600 Å to the Hβ line width, reaches values of up to about 10 times the predicted value. Additionally, there are problems in explaining the ratio of the optical to the ultraviolet Fe II multiplet emission with the above mentioned photoionization models. Fe II emission is observed in Seyfert 1 galaxies, narrow-line Seyfert 1 galaxies and quasars, but not in Seyfert 2 galaxies, and not in non-active galaxies. In particular, the presence of Fe II line emission in narrow-line Seyfert 1 galaxies serves as an important diagnostic tool in classifying these objects as Seyfert 1 type objects.

22.1.4.4 Rapid Variability

From the time scales of the intensity changes in the continuum emission of active galaxies, the size of the emitting regions can be estimated, assuming isotropic emission: $r < c \cdot \Delta t$. Especially in X-rays, significant changes on short time scales of the continuum emission are often observed. The shortest time scales are of the order of a few hundred seconds, with changes in the luminosity of up to about 10^{44} erg s^{-1}. Such sizes of the emitting regions are comparable to the volume of the inner parts of our solar system. The emitted luminosity within that volume corresponds to the emission of up to about 10^{10} solar luminosities.

22.1.4.5 Relativistic Jet Emission

Another signature for AGN activity has been detected in radio-loud objects. Relativistic jets are emitted, most probably perpendicular to the accretion disc, up to distances comparable to the diameter of the host galaxy. Prominent examples for active galaxies with X-ray jets are Centaurus A [51], 3C 273 [2] and M 87 [26]. For Centaurus A, it was found that the X-ray data are dominated by a point source and that an X-ray jet exists which is aligned with the optical jet. Recent papers on 3C 273 discuss the underlying physical mechanisms for the X-ray jet emission, mostly synchrotron versus inverse Compton emission. According to the Kellerman formalism [29], the ratio of the 4.76 GHz to the monochromatic B magnitude is used for the classification into ratio-loud ($f_R/f_B > 1$) and radio-quiet objects. It is assumed that the inner parts of the accretion disc are rejected perpendicular to the disc up to relativistic velocities. Observational evidence is given by the anticorrelation between the increasing radio flux and the decreasing X-ray flux during the jet emission. Even for moderate relativistic velocities of 0.97 c, the flux in the observers frame in increased up to a factor of about 1000. The jet emission is anisotropic and increases with the velocity of the jet emission. Typical apparent separations of blobs emitted from the disc at a distance of 1 Gpc are of the order of some arcsec.

22.2 Introduction to Narrow-Line Seyfert 1 Galaxies

ROSAT, ASCA and BeppoSAX observations have shown many NLS1s to have characteristic, unique and extreme X-ray properties. These include strong soft X-ray excess emission, extremely rapid and large-amplitude X-ray variability, steep 2–10 keV power-law continua, and unusual discrete spectral features. The discovered relations between the slopes of the soft and hard X-ray continua and the optical line widths in NLS1s may allow us to make statements on the physical conditions in the accretion disc and the accretion disc corona. X-ray observations of NLS1s are therefore a new and effective tool in the study of the innermost parts of active galactic nuclei. NLS1s do not appear to form a distinct class, but are rather connected to the 'standard' broad-line Seyfert 1s through a continuum of properties. More generally, soft X-ray excess strength is directly connected with the strong set of optical emission line correlations studied by [8]. These correlations emerged as the 'primary eigenvector' of the Principal Component Analysis (PCA) performed by Boroson & Green, and they relate optical line width, Fe II emission and forbidden-line emission. The primary eigenvector represents the strongest set of optical emission line correlations found among type 1 Seyferts/quasars, and ultrasoft NLS1s fall toward the narrow-Hβ/strong-Fe II/weak-[OIII] extreme of the primary eigenvector. A proper understanding of the Boroson & Green eigenvector would be a major step forward towards determination of the structure and kinematics of Seyfert/quasar emission line regions.

22.3 The X-ray Slope - Optical Line Widths Relation

NLS1s exhibit a larger than previously thought dispersion in their spectral continuum slopes in the X-ray energy range. This large diversity is found in the strength of the soft excess in the soft energy range, as well as in the underlying power-law in the hard energy range. The steepness of the slope in both energy bands correlates with the width of the optical emission lines, especially with the FWHM of the Hβ line. In Fig. 22.3 (left panel), the correlation between the steepness of the spectral energy distribution in the soft energy range and the FWHM of the Hβ line is shown. Fig. 22.3 (right panel) gives the relation between the slope of the power-law in the hard energy range and the FWHM Hβ line width.

22.3.1 Correlation in the Soft Energy Range

A significant correlation between the slope of the 0.1−2.4 keV X-ray continua and the FWHM of the Hβ line was found for Seyfert 1 galaxies [4]. This work was based on a systematic investigation of 46 NLS1s with the X-ray satellite ROSAT. The 46 NLS1s were selected from available data in the literature (in 1996). 32 of these 46 NLS1s are located within the field of view of public ROSAT pointed observations. 31 of these 32 narrow-line Seyfert 1 galaxies show significant X-ray emission in ROSAT pointed observations, and only one object does not show detectable X-ray emission. For these 31 X-ray-detected NLS1s, power-law fits were obtained. The photon index Γ is compared with the corresponding value found in broad-line Seyfert 1 galaxies, analyzed using ROSAT All-Sky Survey data [55]. The relation between the photon index Γ and the Hβ line widths is presented in Fig. 22.3. The results of power-law fits to the ROSAT data of narrow- and broad-line Seyfert 1 galaxies can be summarized as follows:

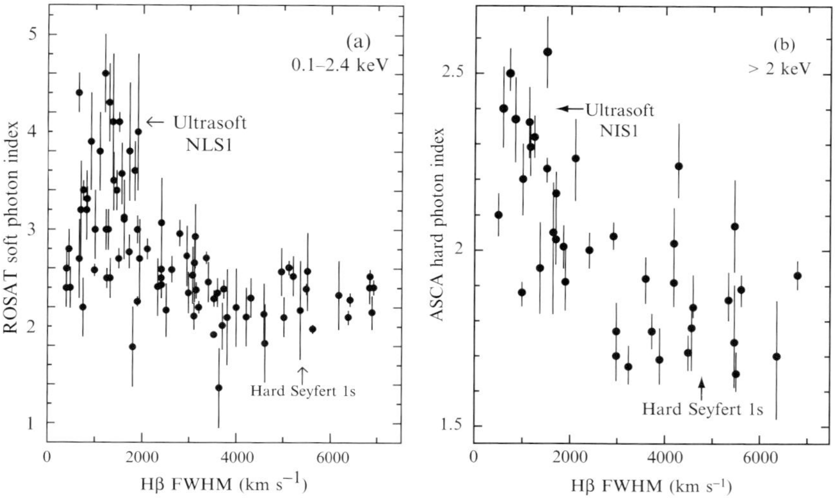

Fig. 22.3 Left: Photon index in the 0.1–2.4 keV energy range as a function of the FWHM line width of the Hβ line. The photon index serves as a measurement of the steepness of the X-ray continuum. All objects in the diagram are Seyfert 1 galaxies. The narrow-line Seyfert 1 galaxies Hβ line widths smaller than 2000 km s^{-1} are taken from [4]. The broad-line Seyfert 1 galaxies were investigated by [55]. A significant correlation between the slope of the X-ray continuum and the FWHM Hβ line width is found. Seyfert 1 galaxies with large line widths (FWHM Hβ greater than 2000 km s^{-1}) show a relatively small dispersion in their values of the photon index with a mean of about 2.3. NLS1s show a large dispersion in their values of the photon index. Some of these objects exhibit values of the photon index up to about 5. A region of the diagram, with values of the photon index larger than 3 and optical line widths larger than about 3000 km s^{-1} is not occupied by Seyfert galaxies. This region is sometimes called as 'zone of avoidance'. There is probably an underlying physical parameter, which causes the absence of objects in this region. One possible explanation is, that this physical parameter is the ratio of the accretion rate to the black hole mass. **Right:** Photon index in the 2–10 keV energy range as a function of the FWHM Hβ line width discovered by [10]. The measurements were obtained with the Japanese X-ray satellite ASCA. Also in the hard X-ray band NLS1s show a stronger dispersion in their photon indices than previously thought. Typical values of the photon index in the hard X-ray energy band range between about 1.7 and 2.6. In this energy range, the spectrum is dominated by the power law component. The correlation between the slope of the X-ray spectral energy distribution and the optical line width is expected to allow statements on the physical conditions in the accretion disc corona

- Seyfert1 galaxies with Hβ line widths greater then 3000 km s^{-1} display flat X-ray continua with a mean value of about $\Gamma = 2.3$. The dispersion from the mean value is low. The minimum value of Γ is 1.3 and the maximum value is 2.6.
- NLS1s show a significantly stronger dispersion in the values of the photon index compared with broad-line Seyfert 1 galaxies. The photon index reaches values of up to about 5. Steep X-ray continua with values of the photon index greater than 3 are found only in NLS1s. As a class, NLS1s show therefore steeper X-ray continua than broad-line Seyfert 1 galaxies.
- A region in the Γ – FWHM Hβ plane with $\Gamma > 3$ and FWHM Hβ greater than 3000 km s^{-1} is not occupied by Seyfert 1 galaxies. This region is sometimes

called the 'zone of avoidance'. There are no obvious selection effects which could cause the absence of objects in this region.

- The distribution of the values Γ and FWHM Hβ show a continuous increase in the slope of the spectral continuum distribution with decreasing FWHM Hβ line width. This suggests that narrow- and broad-line Seyfert 1 galaxies form essentially the same class of objects and that there might be an underlying physical parameter which controls the distribution of objects shown in Fig. 22.3.
- In the 0.1−2.4 keV energy range, the soft excess component as well as the power-law component contribute to the measured spectral energy distribution (in contrast to the 2−10 keV energy range, where only the power-law component is contributing to the continuum emission). Therefore, in the 0.1−2.4 keV energy range, the photon index serves as a measure of the relative ratio of the emission of the soft excess- to the power-law component. As the soft X-ray excess component is thought to arise primarily in the accretion disc, the distribution of photon indices at soft X-rays should allow us to make statements regarding the physical conditions in the accretion disc. In addition, as the soft excess component is probably the high energy tail of the big blue bump, the photon index in the soft energy range is probably a measure of the strength of the ionizing radiation from the innermost regions of active galactic nuclei. The distribution of the data points in Fig. 22.3(a) suggests that the width of the Hβ line is determined by the strengths of ionizing radiation and the lower masses of NLS1s with respect to broad-line Seyfert 1's [4]. This strength of the ionizing radiation may (as one possible explanation) determine the size of the region, which is emitting the permitted optical lines.

22.3.2 Correlation in the Hard Energy Range

Similarly to the correlation found in the soft energy range, a correlation between the slope of the 2−10 keV photon index, obtained from power-law fits to ASCA observations of broad- and narrow-line Seyfert 1 galaxies, and the width of the FWHM Hβ line, has been discovered by [10]. Figure 22.3 shows the correlation between the hard photon index and the FWHM Hβ line width. In NLS1s, a spectral flattening is seen above 2 keV. However, the 2−10 keV power-law photon indices are still steeper than previously thought, originating in a large diversity in the slopes in the hard continuum emission for NLS1s. Steep 2−10 keV X-ray continua, with values of the photon index between 1.9 and 2.6, are characteristic of NLS1s. As the power-law component is thought to arise primarily in the accretion disc corona, the distribution of data points in Fig. 22.3(b) allows in principle, statements regarding physical conditions in the accretion disc corona to be made. Detailed physical models to explain the large dispersion in the 2−10 keV slopes of Seyfert 1 galaxies are presently not available, and the origin for the different spectral continuum slopes requires further theoretical considerations. One possible explanation suggested is that NLS1s may exhibit a cooler accretion disc corona [42]. For inverse Compton scattering, the relation between the slope of the velocity distribution of the electrons in the accretion disc corona q is related to the energy index α by: $\alpha = (q+1)/2$. In this case, a cooler accretion disc results in steeper X-ray continua.

22.3.3 NLS1s with Extreme and Rapid X-ray Variability

In this Section, the discovery of extreme (amplitude variability with a factor larger than 10) and rapid (time scales of hours) X-ray variability in the two NLS1s IRAS 13224−3809 and PHL 1092 as well as their physical implications is described. The combination of extreme, rapid and persistent amplitude variability was discovered in 1996 within the first monitoring campaign on a narrow-line Seyfert 1 galaxy. Rapid and extreme X-ray variability in a second object, the X-ray luminous narrow-line quasar PHL 1092, was discovered one year later, in 1997. Although extreme amplitude X- ray variability had already been reported in a few active galaxies, RE J 1237+264 [9], E1615+061 [41], WPVS 007 [25], their time scales of variability are of the order of years.

22.3.3.1 IRAS 13224−3809, The Most Extremely Variable Seyfert Galaxy

The NLS1 IRAS 13224−3809 was observed over a time scale of 30 days, between January 11, 1996 (0:33:42 UT) and February 9, 1996 (13:23:28 UT), using the ROSAT HRI detector, the total observing time being 111,313 seconds. The number of source photons is 5602. The mean count rate is 0.05 photons per second. A careful analysis of the arrival times of the source photons shows a strong deviation from the mean count rate. This extreme X-ray variability is shown in Fig. 22.4. The minimum count rate is observed at day 5.0408 (in units of the Julian date) with $(4.7 \pm 2.5) \cdot 10^{-3}$ counts s^{-1}. The maximum count rate, observed at day 17.9861, is $(0.287 \pm 0.019)\ 10^{-1}$ counts s^{-1}. The resulting maximum amplitude variability is therefore a factor of 61. The most extreme amplitude variability on a short time scale is detected between day 16.0160 and 17.9861. The count rate increases from $(5.0 \pm 1.9) \cdot 10^{-3}$ counts s^{-1} to $(0.287\ \pm 0.019)$ counts s^{-1}. This corresponds to amplitude variability by a factor of about 57 within only two days. The resulting luminosity (the conversion between flux and luminosity was done using the relations given by [46] for a value of Hubble constant of 50, a value of q_0 of 0.5) for one HRI count per second is $2.9 \cdot 10^{45}$ erg s^{-1}. Assuming isotropic emission from IRAS 13224−3809, the resulting increase in luminosity from $1.5\ \cdot 10^{43}$ erg s^{-1} to $8.3\ \cdot 10^{44}$ erg s^{-1} within about 2 days, and the corresponding change in luminosity of $\Delta L \simeq 8.2\ \cdot 10^{44}$ erg s^{-1} is remarkable. Relativistic flux amplification in the inner accretion disc due to hot spots orbiting the central black hole provides a physical explanation. In this model, the extreme flux variations are only detected in the observers frame [5].

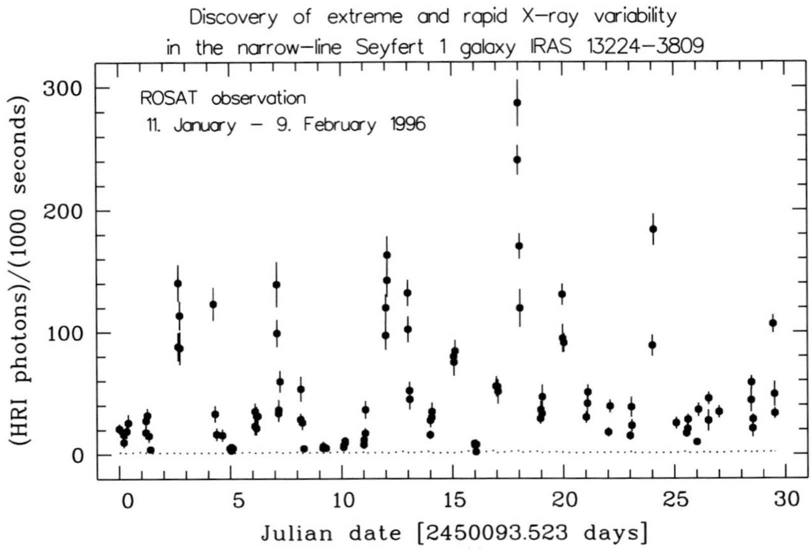

Fig. 22.4 ROSAT HRI light curve of IRAS 13224−3809, obtained during a 30 day monitoring campaign between January 11, 1996 (0:33:42 UT) and February 9, 1996 (13:23:28 UT). The x-axis gives the Julian date minus 2450093.523 days. The dashed line shows the background count rate as a function of time. At least 5 giant amplitude variations are clearly visible. IRAS 13224−3809 shows the most extreme persistent X-ray variability so far measured from active galactic nuclei

22.3.3.2 Discovery of Extreme and Rapid X-ray Variability in the Narrow-line Quasar PHL 1092

Figure 22.5 shows the X-ray light curve of PHL 1092. Strong X-ray variability is detected throughout the observation. For the most extreme combination of amplitude variability with time, at day 8.6, we obtain a change in luminosity of $\Delta L = (5.0 \pm 1.1)\ 10^{45}$ erg s^{-1} within a time interval of $\Delta t = 3580$ seconds. The observed variability per time of $\Delta L \cdot \Delta t = 1.40 \cdot 10^{42}$ erg s^{-2} is presently the largest value measured in a radio-quiet galaxy. This value exceeds the efficiency limit for accretion onto a Kerr black hole [15], and supports the model of relativistic flux boosting for the rapid and extreme variability detected in PHL 1092.

22.3.3.3 Physical Models for Rapid and Extreme X-ray Variability

In this Section, physical models for the recently discovered rapid and extreme X-ray variability are discussed. The most probable explanation is given by the effect of strong relativistic flux amplification in the inner accretion disc [24, 48]. The rapid and extreme variation, as well as the low values of absorption by neutral hydrogen along the line of sight, suggest that the innermost parts of the accretion disc in IRAS

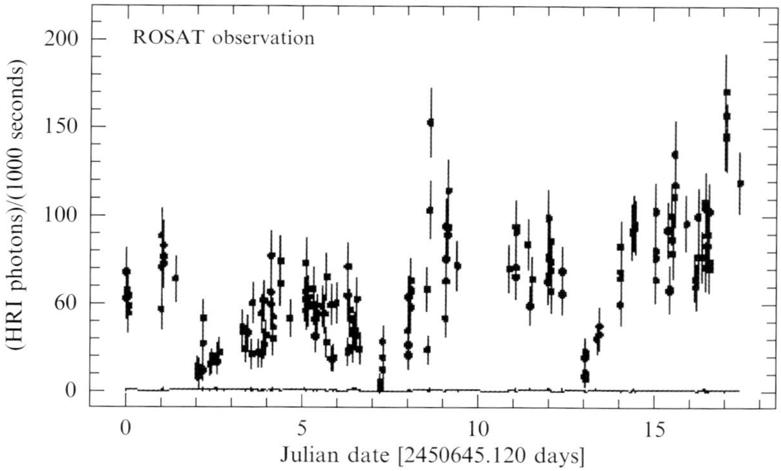

Fig. 22.5 ROSAT HRI light curve of PHL 1092, obtained within an 18-day observation between July 16, 1997 (02:47:15 UT) and August 2, 1997 (13:53:58 UT). The x-axis gives the Julian date minus 2450645.120 days). The dashed line gives the background count rate as function of time. The maximum factor of the amplitude variability is 13.9. PHL 1092 shows the most extreme change of luminosity per time interval so far found in an active galaxy

13224−3809 must be directly visible. At such distances to the central black hole, the X-ray emitting regions move with relativistic velocities and, as a consequence strong intensity variations occur due to the relativistic Doppler effect. The relativistic Doppler effect [31] gives a relation between the observed frequency ν and the frequency ν' in the system of the moving particles:

$$\frac{\nu'}{\nu} = \frac{1}{\gamma(1-\beta\, cos\Theta)} = \delta$$

The relativistic Doppler effect can result in strong apparent flux variations, if the emission region is not steady or inhomogeneous. For the system of the rotating accretion disc, this means that the X-ray emitting regions moving towards the observer are increased in their intensity in the observers frame. In the case where the temperature distribution on the accretion disc is not homogeneous, which may be caused by X-ray hot spots on the accretion disc, strong intensity variations occur in the observers frame. In the following, the resulting intensity variation in the observers frame is derived for a single hot spot orbiting the black hole. The flux value at the frequency ν' of the emission of a hot spot orbiting the black hole is called $f'_{\nu'}$ in the following. The flux value measured by the observer at the frequency ν is f_ν. From Lorentz-transformation theory [34], it follows that the ratio of the flux to the third power of the frequency is invariant for Lorentz-transformations:

$$\frac{f_\nu}{\nu^3} = \frac{f'_{\nu'}}{(\nu')^3}$$

Assuming a simple power-law model for the energy distribution[1] $f_\nu \sim \nu^{-\alpha}$ and $f'_{\nu'} \sim \nu'^{-\alpha}$ it follows that the flux ratios integrated over a limited frequency interval (ν_1, ν_2) are given by:

$$\frac{f}{f'} = \delta^{4+\alpha} = \left(\frac{1}{\gamma(1-\beta cos\Theta)}\right)^{4+\alpha}$$

The ratio of the observed flux f to the emitted flux f' is, besides the Doppler factor, a strong function of the slope of the spectral energy distribution α. Especially for narrow-line Seyfert 1 galaxies, with their extremely steep X-ray spectra, high values of the ratio $\frac{f}{f'}$ are expected.

22.4 XMM-Newton Discoveries in the High-Energy Spectra of NLS1s

22.4.1 Detection of Sharp Spectral Drops Above 7 keV

One of the most interesting spectral features detected with XMM-Newton is probably the discovery of sharp spectral power-law cut-offs at the Fe K edge energy as observed in the NLS1s 1H 0707-495 ([6, 21]) and IRAS 13224−3809 ([7]; c.f. Fig. 22.6).

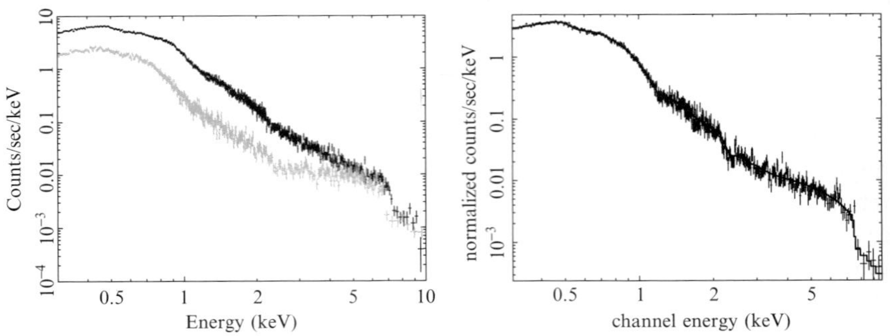

Fig. 22.6 **Left:** XMM-Newton spectrum of 1H 0707−495 obtained in October 2000. A sharp spectral drop at the energy of the neutral Fe K edge at 7.1 keV has been detected. **Right:** XMM-Newton spectrum of IRAS 13224−3809. The sharp spectral drop occurs at even higher energies at 8.2 keV

[1] For a qualitative discussion on the intensity variations caused by the relativistic Doppler effect, a approximation to a simple power law model is sufficient. The apparent intensity variations in the observers frame are mainly caused by a frequency shift of the emitted radiation in the ROSAT energy band. The steepness of the X-ray continuum slope is dominant with respect to the exact form of the spectral energy distribution.

A possible interpretation is an absorption model. The missing Fe K re-emission can be explained by a partial covering model ([6,27,50]). The partial covering model requires high densities of the absorber, located in the accretion disc environment. As they substain only a small solid angle, the Fe K re-emission remains undetected with the present generation of X-ray telescopes. In the partial covering scenario a part of the emission from the accretion disc reaches the absorber unabsorbed giving rise to the strong soft X-ray excess.

22.4.2 Neutral, Ionized Absorbers or Reflection Dominated Models

The measured edge energy and the sharpness of the edge as observed in the two objects are comparable to the energy resolution of the EPIC instruments (about 200 eV). In Fig. 22.7 is shown that in the second observation of 1H 0707−495 the energy of the sharp spectral drop has been significantly shifted to 7.5 keV. For IRAS 13224−3809 the edge energy is even higher, at 8.2 keV, corresponding to Fe IXX to Fe XXIII and a width of the feature of about 600 eV, in contrast to the observations. In addition, ionized Fe would result in a Kβ UTA feature ([39, 40]) which remains undetected in the XMM-Newton observations. Therefore, if absorption is the underlying physical process, we observe neutral Fe with significant outflow velocities of about 0.05 c (1H 0707-495; second observation) and 0.15 c for IRAS 13224−3809.

An alternative explanation is a reflection dominated model, where the sharp spectral drop is due to the blue horn of a strong relativistically broadened Fe K line [16]. The missing Fe Kβ UTA feature and the sharpness point to an absorption

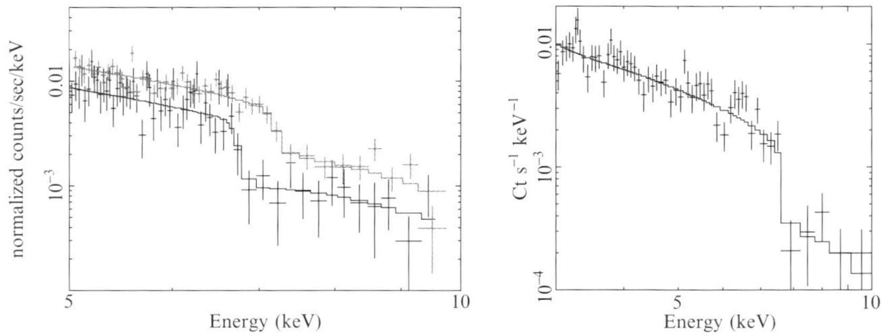

Fig. 22.7 Left: XMM-Newton spectra of 1H 0707−495 taken in October 2000 (low flux state) and in October 2002 (high flux state). The sharp spectral drop has been shifted from 7.1 keV to 7.5 keV. The sharpness of the feature suggests absorption by neutral Fe with outflow velocities of about 0.05 c. **Right:** XMM-Newton spectrum of IRAS 13224−3809. The sharp spectral drop is shifted to even higher energies of 8.2 keV, suggesting outflowing velocities of 0.15 c

process causing the sharp spectral drops in NLS1s. Finally we note that the spectrum consisting of a dominant soft X-ray component and a hard tail is strikingly similar to the spectra of the black-hole binaries at high luminosities [50]. This strongly suggests that the AGN shown in Fig. 22.6 and Fig. 22.7 are analogues of the soft-state black-hole binaries.

22.5 The Nature of the Soft X-ray Excess

As an example the new spectral complexity discovered with XMM-Newton and Chandra in the soft X-ray excess components are controversially discussed in the astrophysical community (c.f. Fig. 22.8). [11] explain the soft X-ray excess as relativistically blurred C,O and N lines. In contrast, [30] find that the high resolution soft X-ray spectral feature can be explained by absorption edges, mainly from Fe L, without requiring the line interpretation. [18] pointed out that there might be a potential problem with the line interpretation. Especially the sharpness of the spectral features, which is about 3 eV in the Chandra HETG spectra and about 10 eV in the XMM-Newton RGS data, favours the edge interpretation. This is as the O lines, assuming the line interpretation, are produced in a highly ionized medium with significant Thomson depth and the line is subject to significant line broadening considerable larger than 10 eV. Gravitational redshifts and Doppler motion effects are expected to further broaden the line. More observational and theoretical studies are necessary to solve the puzzle of the soft X-ray spectral complexity.

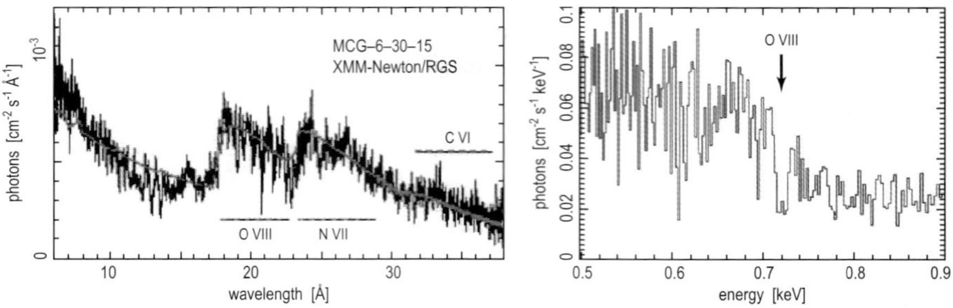

Fig. 22.8 XMM-Newton RGS (*left*) and Chandra HET (*right*) spectra of MCG-6-30-15 in the soft energy band. While [11] explain the soft X-ray excess as a superposition of accretion disc lines **left**, [30] model the spectrum by a combination of absorption edges **right**. [18] points out that there might be a problem with the line interpretation due to the sharpness of the spectral feature. More observational and theoretical work is required to achieve a better understanding of the nature of the soft X-ray excess

22.6 Matter Under Strong Gravity

22.6.1 Relativistically Blurred Fe K lines

The 6.40–6.97 keV Fe Kα lines observed from type 1 Seyfert/quasars are formed by fluorescence in optically-thick nuclear matter that is irradiated by the central X-ray source (e.g. [22]). Observations by ASCA and XMM-Newton show that the Fe Kα lines are in X-ray bright objects extremely broad with widths corresponding to motion at a substantial fraction of the speed of light (up to about 100 000 km s^{-1}; e.g. [36, 37, 49]). Such observations allow us to study matter under extreme gravity and to test the predictions from the special and relativistic theory. As predicted by the theory, the lines appear to be skewed towards lower energies, and the overall line profiles suggest an origin in the innermost parts of accretion discs where extreme Doppler/relativistic effects operate. Figure 22.9 shows the most prominent example of a relativistically blurred Fe K line, MCG-6-30-15. The left panel shows the first exciting discovery of a spectrally resolved relativistically blurred Fe K line profile obtained with the ASCA satellite [49]. The highest quality spectrum ever obtained from the Fe K line profile is shown in the right panel of Fig. 22.9 [17]. General relativity effects of gravitationally redshift and special relativity effects of relativistic Doppler boosting effects are clearly visible. The skewed profiles extend below 4 keV which is indicative for a Kerr black hole.

22.6.2 The Iron Line Background

It has been pointed out one a decade ago [33] that prominent spectral features, and especially the around 6.4 keV iron Kα emission line commonly found in Seyferts

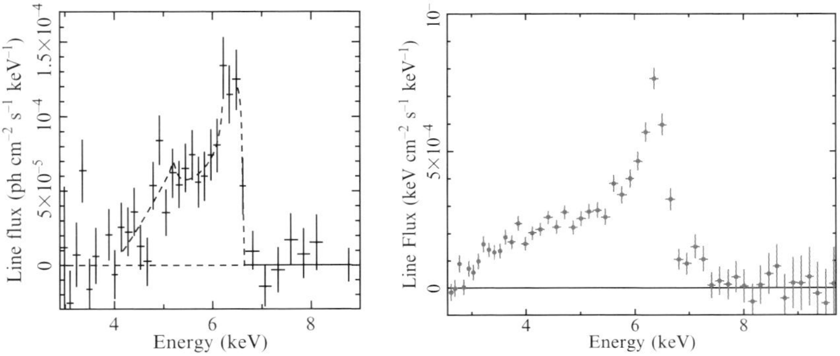

Fig. 22.9 Left: First spectrally resolved Fe K line (MCG-6-30-50), obtained with ASCA [49]. Both, relativistic redshift effects and relativistically Doppler boosting are clearly visible. **Right:** High quality spectrum obtained of the relativistic Fe K line profile obtained with XMM-Newton (MCG-6-30-50), [17]

spectra, may lead AGN synthesis models to predict a detectable signature in the spectrum of the cosmic X-ray background (XRB) around a few keV. Despite the increasing number of independent observations of the XRB spectrum below 10 keV, yielding good signal to noise ratio measurements of the extragalactic background, the accuracy reached so far is not such to detect the iron features at the level expected by the model predictions (about 3–7 per cent, [20,23]). An alternative approach, devised to avoid the line smearing due to the large redshift range over which AGN spectra are summed and the present uncertainties in the XRB spectrum, is to search for iron features over appropriate redshift bins. A careful search for iron line emission in stacked X-ray spectra has been carried out among faint X-ray sources identified in the 1Ms Chandra Deep Field South and in the 2Ms Chandra Deep Field North [12]. Individual source spectra were stacked together in seven redshift bins over the range z=0.5−4. With the exception of the highest redshift bin, a significant excess above a power-law continuum is present over the energy range $\frac{6.4}{1+z_{max}}$ to $\frac{6.4}{1+z_{min}}$ keV, where z_{min} and z_{max} are the bin boundaries (see Fig. 22.10). The measured

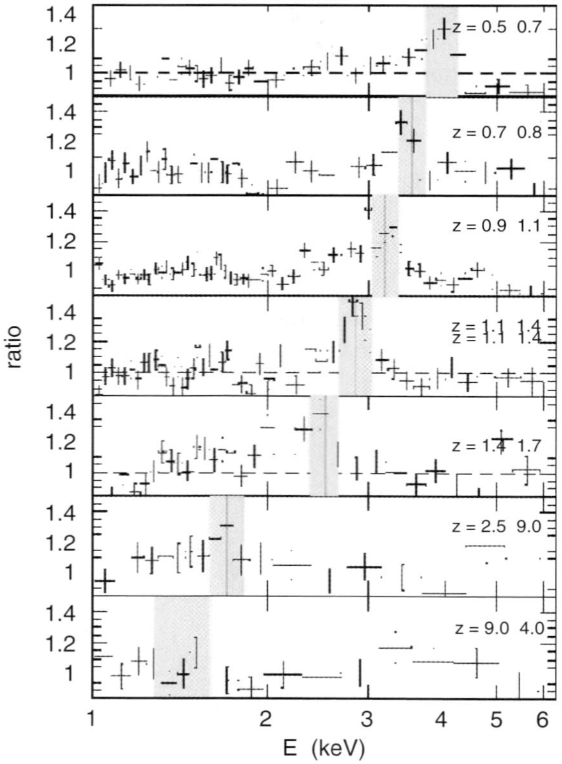

Fig. 22.10 Residuals of a simple power-law fit to the source spectra in seven different redshift bins as labelled. The vertical line in each panel is at the expected position for the redshifted 6.4 keV Fe Kα line while the shaded region encompasses the bin width defined as $\Delta E = \frac{6.4}{1+z_{max}} - \frac{6.4}{1+z_{min}}$ keV

EW are in agreement with those expected by simple pre-Chandra estimates based on X-ray background synthesis models in a scenario in which their intensity does not change significantly with redshift (and/or luminosity). Although there might be hints for the presence of gravitationally redshifted broad line components, we caution that their intensity and profile significantly depends on the modelling of the underlying continuum. The average rest frame equivalent width of the iron line does not show significant changes with redshift and the iron line emission is an ubiquitous property of X-ray sources up to $z \simeq 3$. ([12]; (c.f. Fig. 22.10)).

22.6.3 The Mean Fe K Spectrum Obtained from Stacking Analysis

The relativistic Fe K line in the mean X-ray spectra of type 1 and type 2 AGN were derived by [47]. Using a 770 ks XMM-Newton survey of the Lockman Hole, the averaged spectrum shows a strong relativistic Fe K line profile. A Laor line profile with an inner disc radius smaller than the last stable orbit of a Schwarzschild black hole is found to be consistent with the observations. This indicates, that the average black hole has a significant spin. The equivalent widths of the broad relativistic lines range between 400 and 600 eV (see Fig. 22.11).

22.6.4 Fe K Line Profile Changes

According to accretion disc theories, temperature inhomogenities in the accretion disc are expected (see Fig. 22.12). In the observers frame moving black body fields results in changes in the observed emission profiles from the multicolour disc, as

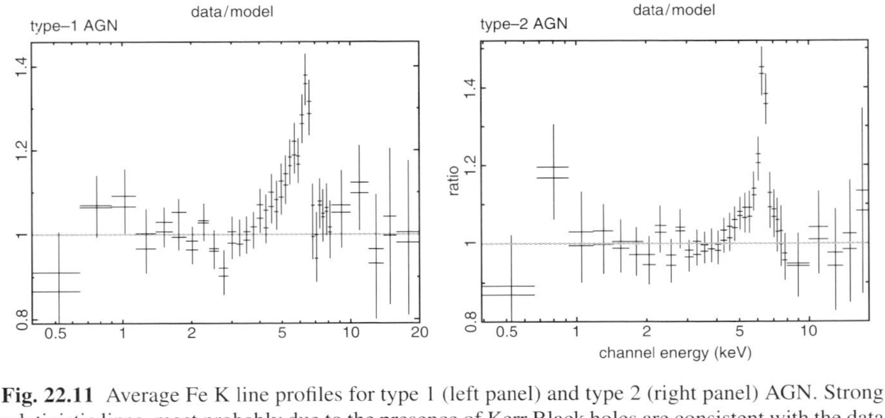

Fig. 22.11 Average Fe K line profiles for type 1 (left panel) and type 2 (right panel) AGN. Strong relativistic lines, most probably due to the presence of Kerr Black holes are consistent with the data

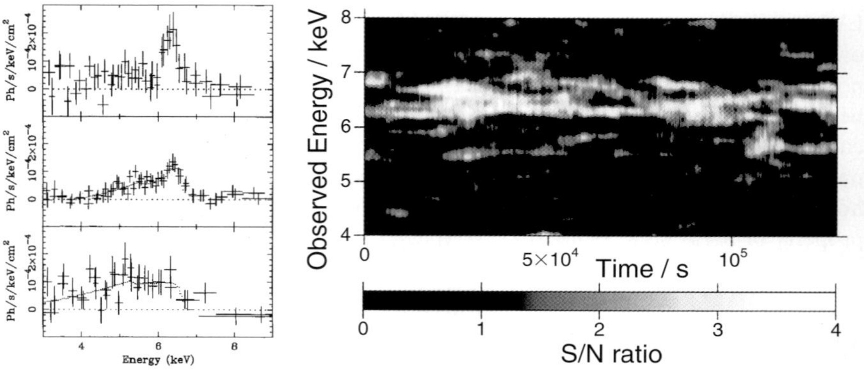

Fig. 22.12 Left panel: Changes in the Fe K line profile obtained with the ASCA satellite [28]. Changes are thought to be caused by hot spots on the accretion disc resulting in the observed changes in the Fe K line profile due to moving black body fields in the observers frame. **Right panel**: Fe K line energy as a function of time [53]. Three hot spots on the accretion disc have been postulated, resulting in the same inclination of the accretion disc

well as in Doppler boosting effects. Detailed studies of the variable Fe K line profile will be obtained with the next generation of X-ray telescopes.

References

1. Antonucci, R.R.J., 1992, in testing the AGN Paradigm, Eds. Holt S.S., Neff, S.G., Urry, C.M., Proceedings of the 2nd Annual Topical Astrophysics Conference, Univ. of Maryland, College Park, Oct. 14-16, 1991 (A93-29801 11-90), AIP Conference Proceedings, Vol. 254 p. 486-499
2. Atoyan, A.M., Dermer C.D., 2003. Astrophys. J., 586, 79
3. Baschek, B., Kegel, W., Travinv, J. 1963, Zeitschrift für Astrophysik, 56, 282
4. Boller, Th. Brandt, W.N., Fink, H. 1996, Astron. Astrophys., 305, 53
5. Boller, Th., Brandt W.N., Fabian, A.C., Fink H. 1997, MNRAS, 289, 393
6. Boller, Th., Fabian, A. C., Sunyaev, R., Trümper, J., Vaughan, S., Ballantyne et al. 2002, MNRAS, 329, 1
7. Boller, Th., Tanaka, Y., Fabian A.C., Brandt, W.N., Gallo, L. et al. 2003, MNRAS, 343, 89
8. Boroson, T.A., Green, R.F. 1992, ApJS, 80, 109
9. Brandt, W.N., Pounds, Fink, W 1995, MNRAS 273, 47
10. Brandt, W.N., Mathur, S., Elvis M. 1997, MNRAS, 285, 25
11. Branduardi G., Sako, M., Kahn, S. M., Brinkman, A. C., Kaastra J. S. et al. 2001, Astron. Astrophys., 365, 162
12. Brusa, M., Cilli, R., Com'astri A. 2005, ApJL, 621, 5
13. Byram, E.T., Chubb, T.A., Friedman, H., 1966, AJ, 71, 379
14. Elvis, M., Maccacaro, T., Wilson, A.S., Ward, M.J., Penston, M.V., et al. 1978, MNRAS, 183, 129
15. Fabian, A.C. 1979, Proc. Roy. Soc. London A, 366, 449
16. Fabian, A. C., Ballantyne, D. R., Merloni, A., Vaughan, S., Iwasawa K. et al. 2002, MNRAS, 331, 35
17. Fabian, A.C., Vaughan, S., Nandra, K., Iwasawa, K., Ballantyne, D.R. et al. 2002, MNRAS, 335, 1

18. Fabian, A.C. 2004, X-rays from AGN: relativistically broadened emission lines; In Frontiers of X-ray astronomy: Edited by A.C. Faian, K.A. Pounds, R.D. Blandford. Cambridge University Press, ISBN 0-521-53487-9, 2004, 165-173
19. Frank, J., King, A., Raine, D. 1992, Accretion Power in Astrophysics, Cambridge University Pres, Cambridge
20. Gandhi, P., Fabian, A.C. 2003, MNRAS 339, 1095
21. Gallo, L., Tanaka, Y., Boller, T., Fabian, A.C., Vaughan, S. 2004, MNRAS 353, 1064
22. George, I.M., Fabian, A.C. 1991, MNRAS, 249, 352
23. Gilli, R., Comastri A., Brunetti, G., Setti, G. 1999, New Astronomy, 4, 45
24. Guilbert, P., Fabian, A.C., Rees, M. 1983, MNRAS, 205, 593
25. Grupe, D., Beuermann, K., Mannheim, K., Thomas H.-C., Fink H. 1995, Astron. Astrophys., 300, L21
26. Harris, D.E., Birella, J.A., Junor, W., Perlman E.S., et al. 2003, Astrophys. J., 586, 41
27. Holt, S. 1981, X-Ray astronomy the 1980's; Proceedings of a Workshop held at Greenbelt, Md., 5-7 Oct. 1981
28. Iwasawa, K., Fabian, A.C., Reynolds, C.S., Nandra, K., Otani, C. et al. 1996, MNRAS 282, 1038
29. Kellermann, K.I., Sramek, R., Schmidt, M., Shaffer, D.B. et al. 1989, AJ, 98, 1195
30. Lee, J., Canizares, C.R., Fang, T., Morales R. et al. 2002, MPE Report, 279, 9
31. Lorentz, H.A. 1904, Proc. Ams. Acad. Sci., 6, 809
32. Piccinotti, G., Mushotzky, R.F., Boldt E.A., Holt, S.S., Marshall, F.E. et al. 1982, Astrophys. J., 253, 485
33. Matt, G., Fabian, A.C. 1994, MNRAS, 267, 187
34. Rybicky, G., Lightman, A. 1979, RPA, New York Weliy. Interscience
35. Mushotzky, R.F., Ferland, G.J. 1984, Astrophys. J., 278, 558
36. Mushotzky, R.F., Fabian, A.C., Iwasawa, K., Kunieda, H. 1995, MNRAS, 272, L9
37. Nandra, K., George, I.M., Mushotzky, R.F., Turner, T.J., Yaqoob, T. 1997, Astrophys. J. 477, 602
38. Netzer, H. 1990, Active Galactic Nuclei, SASS-Fee Advanced Course 20
39. Palmeri, P., Mendoza, C., Kallman, T. 2003, AAP 423, 1175
40. Palmeri, P., Mendoza, C., Kallman T. 2004, Astrophys. J., 577,
41. Piro, L., Massaro, E., Perola, G.C., Molteni, D. 1988, APJ 325, 25
42. Pounds, K. A., Done, C., Osborne, J.P. 1995, MNRAS 277, 5
43. Rees, M.J., 1984, xue..conf..138
44. Shapiro, S.L., Teukolsky, S.A. 1983, in Black Holes, White Dwarfs, and Neutron Stars, Wiley-Interscience Publication
45. Shuder, J.M., Osterbrock, D.E., 1981, Astrophys. J., 250, 55
46. Schmidt, M., Green, R.F. 1986, Astrophys. J., 305, 86
47. Streblyanska A., Hasinger G., Finoguenov A., Barcons, X., Mateos, S. et al. 2005, Astron. Astrophys., 432, 395
48. Sunyaev, R. 1973, SvA, 16, 941
49. Tanaka, Y., Nandra, K., Fabian, A.C., Inoue, H., Otani, C. et al., 1995, Nature, 375, 659
50. Tanaka, Y., Boller, T., Gallo L., Keil, R., Ueda, Y. 2004, PASJ, 56, 9
51. Tananbaum, H., in: X-ray astronomy, Proceedings of the Advanced Study institute, Erice, Italy July 1-14, 1979
52. Thorne, K.S., 1974 Astrophys. J., 191, 507
53. Turner, T.J., Miller, L., George, I.M., Reeves, J.N. 2004, AAS, 20517801
54. Veilleux, S., Osterbrock, D.E. 1987, ApJS, 63, 295
55. Walter, R., Fink, H. 1993 Astron. Astrophys., 274, 105
56. Wampler, E.J., Oke, J.B. 1967, Astrophys. J., 148, 695
57. Worrall, D.M., Boldt, E.A., Holt, S.S., Mushotzky, R.F., Serlemitsos, P.J. 1981, Astrophys. J., 243, 53

23 X-Ray Studies of Clusters of Galaxies

H. Böhringer

23.1 Introduction

Clusters of galaxies are next to quasars, the most luminous X-ray sources in the Universe with radiation powers of the order of 10^{43}–10^{46} erg s^{-1}. The first detection of a cluster source was made with M87 in 1966 by Byram et al. [30], and 5 years later also the massive nearby clusters in the constellations Coma Berenices and Perseus were detected by Gursky et al. [71] and Fritz et al. [65]. With the use of the Uhuru satellite, the extended nature of the cluster X-ray sources could be established [85]. It turns out that the diffuse X-ray emission from clusters originates in a hot intracluster plasma with temperatures of several ten Million degrees, which radiates the bulk of its thermal radiation in the soft X-ray regime. As the hot plasma is tracing the shape of the cluster, the X-ray appearance provides us with information on the cluster structure. The soft X-ray band in which clusters radiate is fortunately also the wavelength regime for which X-ray telescopes with imaging optics provide us with a detailed picture of the X-ray sky. Therefore, galaxy clusters are among the most rewarding and informative objects for X-ray imaging studies.

With the very rapid evolution of X-ray observational techniques, the X-ray studies of galaxy clusters have also experienced a breathtaking evolution. X-ray observations have provided us with a wealth of detailed knowledge on the cluster structure, composition, and formation history as well as on the statistics of the galaxy cluster population. Most of our current systematic understanding of galaxy clusters, the cluster population, and the link to the formation of large scale structure and the underlaying cosmological model is based on X-ray observations. And the importance of X-ray astronomy for cluster research is still increasing. At this moment we can provide an overview on galaxy cluster research in X-rays where the advanced X-ray observatories Chandra and XMM-Newton have unfolded and demonstrated their full capabilities. This field of research is so rich, however, that such a contribution can only provide an illustrative tour through the field rather than a comprehensive review. In particular, the references give only examples of publications as a first starting point for a literature search and cannot be complete due to space limitations. The most detailed introduction to X-ray studies of galaxy clusters is given by the review of Sarazin [127], which is now outdated in its detailed description of observational results but still provides an excellent astrophysical background. More specialized recent reviews focus on X-ray cluster appearance by Forman and

J.E. Trümper and G. Hasinger (eds.), *The Universe in X-Rays.*

Jones [63], cluster cooling flows by Fabian [53], X-ray properties of groups of galaxies by Mulchaey [104], galaxy cluster mergers by Feretti et al. [56], cluster evolution by Rosati et al. [126] and Voit [156], and the use of X-ray cluster for the test of cosmological models by Schuecker [136].

In the structural hierarchy of our Universe we can recognize three fundamental building blocks, which are with increasing size: stars, galaxies, and clusters of galaxies. The study of their physical setup is at the base of astrophysics and the assessment of their population provides us with knowledge on the structure of the next larger unit in the hierarchy. In this sequence our knowledge decreases with the size and the distance of the objects. Thus, for the largest of these building blocks, the clusters, we are just starting to gain a systematic and detailed understanding of their make-up and their role in providing useful probes to cosmology.

In contrast to their name, which characterizes them only as collections of objects, clusters of galaxies are well defined, connected structural entities. This is revealed by X-rays where the diffuse X-ray emission from the hot intracluster medium traces the whole cluster structure in a contiguous way, as shown in Fig. 23.1, for the Coma cluster of galaxies with a distance of about 100 h_{70}^{-1} Mpc. As we shall see, galaxy clusters are defined by their own proper equilibrium structure. They are the largest objects in the Universe which have such a characteristic form, that can be well assessed by observations and well described by theoretical modeling. With these properties they also form the largest astrophysical laboratories, in which the physical environmental conditions can be well observed and described. They are therefore perfect laboratory sites for studies of a wide range of astrophysical processes at large scales, as for example the investigation of galaxy evolution within a well defined environment or the evolution of the dynamical and thermal structure as well as the chemical enrichment of the intergalactic medium.

Clusters of galaxies have been formed from the densest regions in the large-scale matter distribution of the Universe at scales of the order of 10 Mpc (in comoving units) and have collapsed to form matter aggregates that have reached an approximate dynamical equilibrium giving them their proper characteristic shape. They thus form an integral part of the cosmic large-scale structure, the seeds of which have been set in the early Universe. Therefore, the evolution of the galaxy cluster population is tightly connected to the evolution of the large-scale structure and the Universe as a whole. It is for this reason that observations of galaxy clusters can be used to trace the evolution of the Universe and to test cosmological models as we will illustrate later.

Another good didactical approach to contemplate galaxy clusters is to see them as large gravitational potentials holding mostly dark matter, hot thermal plasma, and galaxies together, roughly in proportions of 87, 11, and 2%, respectively.[1] The depth of this potential is characterized by the velocity dispersion of the visible test particles that probe the potential: the galaxies with typical velocity dispersions ranging from

[1] Here and throughout the paper we adopt a Hubble constant of $H_0 = 70\,\mathrm{km\,s^{-1}\,Mpc^{-1}}$ and a concordance cosmological model with a normalized matter density parameter, $\Omega_\mathrm{m} = 0.3$, and a normalized cosmological constant parameter, $\Omega_\Lambda = 0.7$, if not stated otherwise.

Fig. 23.1 The Coma cluster of galaxies as seen in X-rays in the ROSAT All-Sky Survey [24] (underlaying red color) and the optically visible galaxy distribution in the Palomar Sky Survey Image (galaxy and stellar images from the digitized POSS plate superposed). The sky area shown is $1.42 \times 1.42\,\mathrm{deg}^2$

about $300\,\mathrm{km\,s^{-1}}$ for X-ray luminous galaxy groups up to about $1\,500\,\mathrm{km\,s^{-1}}$ for the most massive galaxy clusters. Similarly the hot X-ray luminous plasma is probing this potential in the form of an approximately hydrostatic atmosphere where the plasma temperature with values of $k_\mathrm{B}T \sim 2$–$15\,\mathrm{keV}$ gives information about the potential depth. These facts are illustrated in Fig. 23.2, which shows how the gravitational potential depth is increasing with object mass from galaxies to galaxy clusters, whereas superclusters appear just as collections of cluster potentials because they are not fully formed virialized objects. This illustrates again the point that galaxy clusters are the largest objects in our Universe, which are characterized by their own proper equilibrium structure.

As mentioned earlier, X-ray observations play currently a prime role in the observational studies of the structure and astrophysics of clusters of galaxies. This is due to the gaseous intracluster medium (ICM), a hot, highly ionized thermal plasma with a temperature of ten to hundred Million degrees that fills the whole cluster vol-

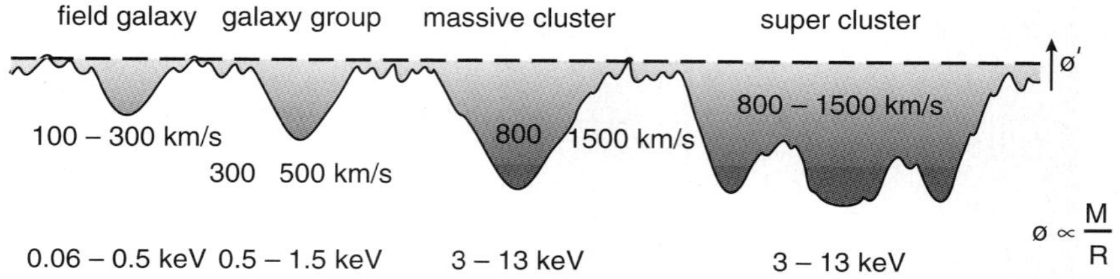

Fig. 23.2 Sketch of the gravitational potential of a part of the Universe, ϕ and $\phi\prime$, with galaxies, galaxy groups, clusters, and superclusters. (This particular potential is the potential of the matter density difference of the actual density and the mean density of the Universe and is negative for overdensities and positive in void regions.) We note the increasing depth of the potential with increasing mass of the objects, except for the last step where superclusters are not featuring as connect entities but merely as a collection of cluster potentials. The depth of the gravitational potential can be probed by the observable velocity dispersion of the stars or galaxies and the temperature of the interstellar or intracluster medium, respectively

ume and thus provides a contiguous picture of the cluster structure. The statistics of the structural information gained from X-rays thus depends on the number of photons collected, which is only limited by the observation time. This is to be compared to optical observations where the structure is traced by basically a limited number of galaxies. The X-ray imaging information is complemented by the contents of the X-ray spectra. Together they provide a wide range of insights into the galaxy cluster astrophysics as will be described in the following. In Sect. 23.2 I describe the measurements of cluster masses and the study of their composition from X-ray observations and discuss the self-similarity of the structural appearance in Sect. 23.3. In Sect. 23.4 I show observations of the Virgo cluster of galaxies and discuss the morphological variety of clusters. Sections 23.5 and 23.6 illustrate the information gained from X-ray spectroscopy on the cooling core structure in the cluster centers and the chemical composition of the ICM. The close relation of cluster masses and their X-ray luminosities make X-ray cluster surveys a good starting point for cosmological studies with clusters. These are described in Sect. 23.7 (surveys), 23.8 (measurements of the large scale structure), 23.9 (cluster evolution), and 23.10 (cosmological tests). Finally, Sect. 23.11 provides a conclusion and an outlook.

23.2 Cluster Masses and Composition

23.2.1 Mass Determination

The precise knowledge of the masses of the most massive, gravitationally confined objects is not only interesting as such, but it is also a prerequisite to many of the astrophysical and cosmological studies with clusters. The X-ray method of mass determination is based on the assumption that the ICM constitutes an atmosphere, which is approximately in hydrostatic equilibrium in the cluster potential described by the equation $\frac{1}{\rho}\nabla P = \frac{-GM(r)}{r^2}$. In most cases a spherically symmetric approximation is used and justified. Then the equation can be reformulated to yield the cluster mass profile

$$M(r) = -\frac{k_B T(r)}{G\mu m_p} r \left(\frac{d\log\rho}{d\log r} + \frac{d\log T_X}{d\log r} \right) \tag{23.1}$$

where m_p is the proton mass and μ is the mean molecular particle weight ($\sim$0.6 for a nearly fully ionized ICM plasma). The observables required for this equation are the absolute temperature profile $T_X(r)$ and the shape of the density profile $\rho(r)$, which are both obtained from X-ray observations.

Since the X-ray emission is proportional to the squared plasma density (with a weak dependence on temperature), the ICM density can be reconstructed from the observed X-ray images if we make some presumptions on the three-dimensional geometry of the clusters that allows us the deprojection of the surface brightness distribution. Also the temperature of the ICM can be determined by X-ray spectroscopy. The thermal emission (mostly bremsstrahlung with some line emission and recombination radiation) leads to optically thin radiation in which every electron–ion collision results in an emitted photon. The shape of the spectrum depends on the temperature and chemical composition of the plasma, but it is independent from the density that determines its normalization. Therefore, temperatures and abundances can be reliably determined from the spectral appearance. The only remaining ambiguities are again that the observed spectrum is a projection of thermal emission along the line of sight and the existence of a range of different temperatures can not easily and unambiguously unfolded. Thus we observe what is often called an "emission measure weighted temperature" and further assumptions play a role in the unfolding and deprojection of the measurement (see e.g. [97]). However, the best observed spectra, e.g., in the X-ray halo of M87 [94] provide good support for an uncomplicated scenario with ionization equilibrium as a valid approximation, local isothermality, and smooth temperature variations as a function of radius, which allows us a good assessment of the structure of the ICM. With the known density and temperature distribution (1) can be solved for the integral mass profile.

The new X-ray satellite observatories XMM-Newton (ESA) and Chandra (NASA) now provide advanced observational capabilities to derive spatially resolved spectroscopic information, allowing us to reconstruct the density and temperature distribution. The Chandra Observatory provides a superior angular resolution of less than 1 arcsec showing many important details in the ICM distribution like cold fronts [152], shock waves [92], and X-ray cavities blown by AGN radio lobes [7,55]. The large X-ray collecting power of XMM-Newton provides good photon statistics for the construction of detailed X-ray spectra from different regions and high resolution spectra from the reflection grating spectrometer (RGS), yielding a good overview on the temperature structure of the clusters.

Figure 23.3 gives an example of one of the best X-ray studies of a galaxy cluster observed with XMM-Newton, of Abell 1413 [119]. The figure illustrates the accuracy of the determination of the cluster surface brightness and temperature

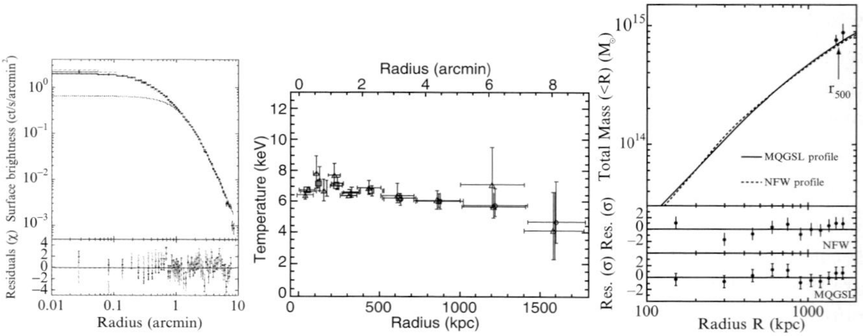

Fig. 23.3 *Left:* Surface brightness profile (the *dotted line* shows a β-model fit to the outer parts of the profile; the central excess is a signature of a cluster cooling core). *Middle:* temperature profile, and *Right:* gravitational mass profile of the cluster Abell 1413 determined from XMM-Newton observations by Pratt and Arnaud [119]. The right panel illustrates how well the profile can be fitted by the proposed, theoretically derived mass models by Navarro et al. [108] (NFW) and Moore et al. [103], (MQGSL)

profiles. The temperature profile extends to about a radius of an overdensity of 500 over the critical density of the Universe, r_{500}. This radius is often taken as a conservative measure to separate the well virialized part of the cluster from the infall and intermediate transition region (e.g. [50, 156]). The mass profile resulting from the application of (1) is shown in the right panel of Fig. 23.3 as derived by Pratt and Arnaud [119]. The profile can be reasonably well fit with a model profile by Navarro et al. (NFW) [108]. This has also been found in a series of detailed inspections of mass profiles from XMM-Newton and Chandra (e.g. [28, 116]), with the conclusion that most well-relaxed appearing clusters follow this description. Clusters showing some distortion, most probably due to recent merger activity, have cores that are often too flat to be fit by NFW profiles, however.

A possibility to test the mass measurement with an independent method is the comparison to the implications from the gravitational lensing effect of clusters. As the deepest gravitational potentials on large scale, galaxy clusters produce the largest observed angular deflections of light rays in our Universe. The more massive and compact galaxy clusters reach the critical projected mass density in the center necessary for the strong gravitational lensing effect, which gives rise to spectacular arcs (distorted images of background galaxies, e.g. [100]). It has been pointed out in some publications that the masses inferred from strong gravitational lensing tend to be larger than those determined from X-ray observations (e.g. [160]). For the comparison with the weak lensing effect which probes cluster masses on larger scales most often good agreement is found, however. Consistent measurements involving both lensing effects are in general also obtained in more regular cluster. The cluster A2390 that shows a prominent tangential and radial arc [114] provides a good example for the latter case as shown in Fig. 23.4 [1, 10].

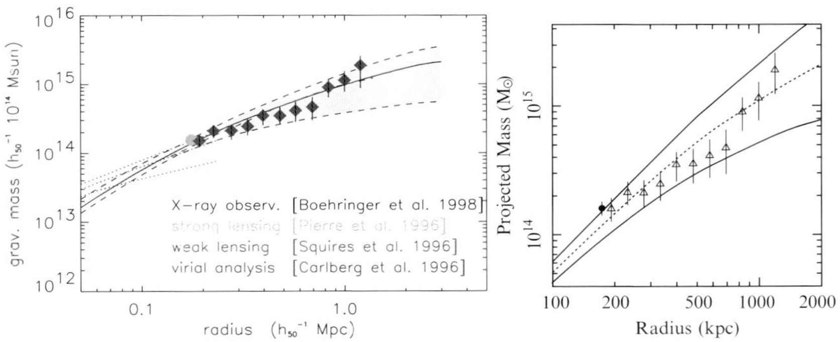

Fig. 23.4 Comparison of different methods of mass determinations for the galaxy cluster A2390. *Left:* X-ray studies with ROSAT and ASCA ([10], *light shaded region*), strong lensing mass ([114] *green dot*), weak lensing mass ([141] *diamonds*), and dynamical mass from detailed spectroscopic studies in the optical ([32], *dashed line*). *Right:* More recent mass determination based on Chandra data [1] with the same comparison to strong and weak lensing (data points)

Prominent lensing clusters for which no good agreement has been obtained comprise, for example, Abell 2218 and CL0024+17, which are famous for their very spectacular Hubble Space Telecope images. CL0024+17, for example, shows several images of the same background galaxy. The X-ray studies with XMM-Newton show smaller masses than the lensing results and indications of a disturbed cluster structure in the central regions ([87, 163] for CL0024+17, [120] for A2218). The clue to this mystery comes from the study of the galaxy velocity distributions in these clusters [39, 69], which clearly indicate that both clusters are configurations of two major subclusters merging together in the line-of-sight. Most probably due to the projection of the mass of a whole filament, in which the merging clusters are embedded along the line-of-sight, the two-dimensional mass density for lensing is greatly enhanced in these configurations, giving rise to such spectacular lensing objects. While the X-ray mass measurement provides a result on the mass of the collapsed and partly relaxed core of the structure, the lensing measurement comprises a larger fraction of the large-scale structure, which explains the observed discrepancies.

23.2.2 Matter Composition

Above, the cluster's X-ray surface brightness profile was used to determine the gas density profile. Its integration yields the integral gas mass profile of the cluster. It turns out that in rich clusters the gas mass at large radii, which makes up for about 12% of the clusters gravitational mass, is about 5–6 times larger than the mass residing in the optical galaxies as inferred from their light and the known mass-to-light ratio for different galaxy types. Thus the X-ray emitting ICM is the largest visible

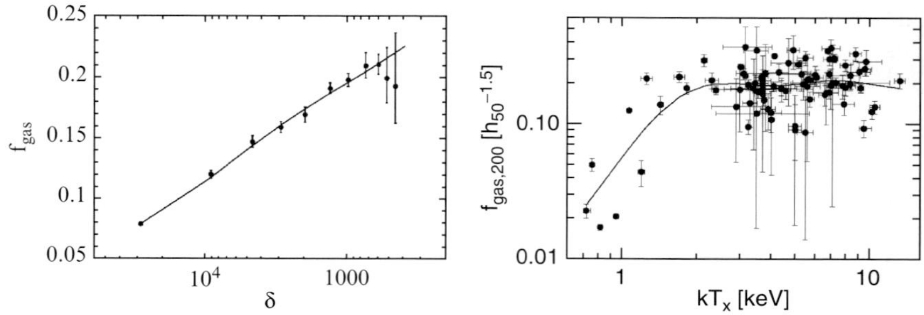

Fig. 23.5 *Left:* ICM gas mass fraction in the galaxy cluster Abell 1413 from an XMM-Newton observation as a function of the cluster mean matter density inside a given radius in units of the mean density of the Universe, δ, by Pratt and Arnaud [119]. *Right:* Gas mass fractions for the galaxy clusters in the HIFLUGCS sample of the X-ray brightest galaxy clusters in the sky studied by Reiprich [122] as a function of ther mean ICM temperature. The gas mass fraction for the massive clusters is fairly constant while a deficiency is observed for the galaxy groups with ICM temperatures below 2–3 keV. (Note that the results in the figure are given for a Hubble constant of $h_{50} = 1$)

matter component. The major part of the clusters binding mass is ascribed to an unknown form of matter usually termed "dark matter." This "missing mass" problem has been recognized as early as the thirties of the last century by Zwicky [164] from first estimates of the dynamical mass of the Coma cluster, but even though the additional matter component of the hot ICM has been subsequently discovered, the gap between inferred total mass and observed matter could not be closed. The result for the gas mass fraction profile for the example of A1413 is shown in the left panel of Fig. 23.5. In the outer region the gas mass fraction is asymptotically approaching a value of about 11–12% $h_{70}^{3/2}$ (note that the results in the figure are given for a Hubble constant of $h_{50} = H_0/(50\,\mathrm{km\,s^{-1}\,Mpc^{-1}}) = 1$). The right panel of the figure shows the results on the gas mass fraction for a whole sample of the X-ray brightest clusters in the sky [122], displaying a fairly constant gas mass fraction for massive cluster systems and smaller fractions for the groups with ICM temperatures below 2–3 keV ICM temperature. Other studies of the cluster baryon fraction based on Chandra observations, e.g., by Allen et al. [2] and Ettori et al. [49] and many other determinations yield very similar results.

Since galaxy clusters are formed essentially by gravitational collapse, which samples all forms of matter almost indiscriminently into the cluster potential, this ratio should give a rough measure of the ratio of baryons to all forms of matter including in the Universe including the dark matter.[2] This cosmic ratio has also been measured by other means, like the fluctuation spectrum of the cosmic microwave background (e.g., Spergel et al. [139]) or the primordial deuterium abundance in connection with nucleosynthesis calculations [29] with results summarized in

[2] N-body/hydrodynamical simulations suggest that the ICM has a slightly wider distribution in the final cluster, resulting in a correction of the universal versus cluster baryon fraction of the order of 10%, e.g. [48, 157].

Table 23.1 Baryon mass density fraction in the Universe determined by different methods

Method	$\Omega_b\, h^2$	Ω_b for $h = 0.7$
Cluster baryon fraction (assuming that $\Omega_m = 0.3$)	(0.3)	0.0390 (±0.004)
Nucleosynthesis and primordial deuterium	0.0205 (±0.0019)	0.0407 (±0.0038)
WMAP CMB fluct. spectrum	0.0224 (±0.0009)	0.0444 (± 0.0018)

Galaxy cluster composition, nucleosynthesis combined with the observed primordial deuterium abundance, and the relative heights of the peaks in the cosmic microwave background fluctuation power spectrum (from WAMP).

Table 23.1. There is good agreement within the quoted error limits of about 10% among all three methods. This provides good support for the reliability of the mass and gas mass determination in galaxy clusters. The baryon fraction in clusters is also used for cosmological tests (Sect. 23.10).

23.3 Exploration of Cluster Structure

23.3.1 Self-Similarity of Cluster Structure

If the structure of clusters is characterized by the approach to a dynamical equilibrium state, how much similarity should we expect between the shapes of different galaxy clusters? If cluster formation is approximately modeled by the gravitational collapse of a homogeneous spherical overdensity of noninteracting dark matter, we expect that the collapse process is self-similar as well as the produced "Dark Matter halos." That is, we expect less massive systems to be scaled down versions of the more massive clusters. The cluster dark matter central densities are, for example, expected to be the same for clusters of all masses if they have formed at the same epoch. In general the central density is then proportional to the background density of the Universe at a characteristic formation time of the cluster (e.g., the turn-around time marking the start of the collapse). The characteristic radius depends then on mass and formation time. Inhomogeneities in the initial conditions and in the collapse as well as gas dynamical processes are expected to introduce variations into this self-similar scenario. Galaxy clusters in a stage of a merging of subunits or clusters in another major phase of matter accretion are of course also expected to show deviations from the equilibrium structure. Earlier theoretical considerations of possible equilibrium structures have resulted in the popular King model [86] (based on a guess of the velocity distribution function of the mass carrying particles), which was first developed to describe the structure of globular clusters, but also found to roughly reproduce the galaxy density distribution profile in galaxy clusters. The density distribution for this model is given by the formula

$$\rho_{\rm g} = \rho_{\rm g0} \left(1 + \frac{r^2}{r_{\rm c}^2}\right)^{-3/2} , \tag{23.2}$$

where $\rho_{\rm g}$ is the gravitational mass density and $r_{\rm c}$ is called the core radius. An isothermal ICM, with a temperature scaling with the velocity dispersion of the gravitating mass particles (e.g., galaxies) as given by a parameter β, $\frac{\sigma_{\rm r}^2 \mu m_{\rm p}}{k_{\rm B} T} \equiv \beta$, has a gas density profile described by the same form as (2) but with an exponent $-3/2\beta$. The advantage of this simple description is, that it can easily be integrated along the line-of-sight to give the observed surface brightness in the case of isothermality with the surface brightness $S_{\rm X} \propto \rho_{\rm ICM}^2$:

$$S_{\rm X} = S_0 \left(1 + \frac{R^2}{r_{\rm c}^2}\right)^{-3\beta+1/2} , \tag{23.3}$$

where R is the projected radius. In fact this function, generally called the "β-model," turned out to provide a good approximate fit to the observed surface brightness profiles, except for cooling flow clusters, which may have an extra central brightness enhancement [81] and in the far cluster outskirts, where the surface brightness profiles tend to steepen [151]. Figure 23.6, shows typical surface brightness and line-of-sight integrated emission measure profiles that can be roughly fit by a β-model [4].

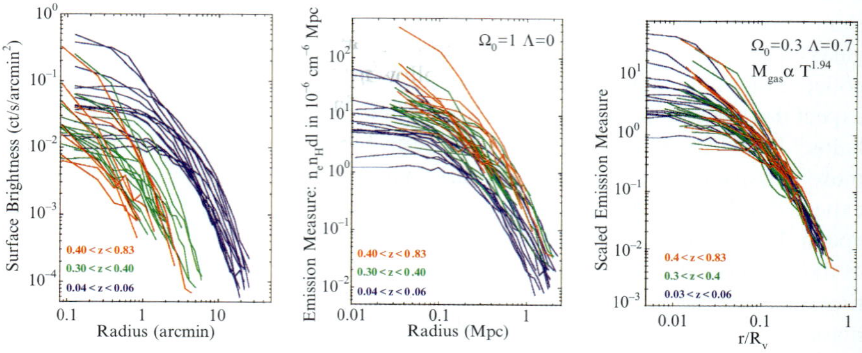

Fig. 23.6 *Left:* Surface brightness profiles of 40 relaxed appearing clusters as observed in deep ROSAT observations [4]. *Middle:* Emission measure profile determined from the surface brightness profile by accounting for the temperature variation of the emissivity function and the cosmological surface brightness dimming with increasing redshift. *Right:* Scaled emission measure (integrated along the line-of-sight) profile, $S_{\rm X}/(\Delta_z^{3/2}(1+z)^{9/2}T^{1.38})$, as a function of the scaled cluster radius. With this self-similar scaling the emission measure profiles show little dispersion at radii larger than $0.1R_{\rm v}$ while the inner profiles vary depending on the cool core structure (Arnaud et al. [4])

Recent N-body simulations with increased resolution suggest an improved description of the dark matter halo shape of the form, $\rho_g(r) \propto r^{-p}\,(r+r_s)^{p-q}$ where r_s is the scale radius. The most popular representations are $p = 1$ and $q = 3$ [108] (NFW-profile) and $p = 1.5$ and $q = 3$ [103] (Moore et al. profile). Given the mass profile the gas density and surface brightness profile can be determined with the knowledge of the temperature profile. The exact shape of nearly self-similar temperature profiles is still a matter of debate (e.g., [116, 153, 162]). As shown in Sect. 23.2 the NFW model provides a consistent description of the observed relaxed galaxy clusters.

If clusters have approximately self-similar shapes, the global parameters of clusters are also expected to follow tight relations. This was first described as an important tool for the study of the cluster population and its evolution by Kaiser [83]. If we consider clusters formed at the same epoch, it is easy to derive the basic, global relations,

$$\rho_{g0} = \text{const}; \quad M \propto R_*^3, \quad T_X \propto \sigma_r^2 \propto \frac{M}{R_*} \propto M^{2/3}, \tag{23.4}$$

where R_* is a characteristic radius and M the cluster mass. If the X-ray emission of a volume element is $\varepsilon_{\rm Xbol} \propto \rho_{\rm ICM}^2\, T_X^{1/2}$ then we can, for example, obtain a relation between bolometric X-ray luminosity, $L_{\rm Xbol}$, and ICM temperature, T_X, of the form

$$L_{\rm Xbol} \propto \rho_{\rm ICM}^2\, R_*^3\, T_X^{1/2} \propto f_b^2 \rho_g^2\, R_*^3\, T_X^{1/2} \propto f_b^2 M^2\, R_*^{-3}\, T_X^{1/2} \propto T_X^2. \tag{23.5}$$

While this is the self-similar prediction, the observed relation features an exponent more close to 3, e.g. [3]. This is the most famous relation breaking the self-similar scenario and gave rise to deeper concern.

The resolution to this discrepancy is most probably found in extra heating and cooling of the ICM, which affects the thermal structure of the cluster gas in less and more massive clusters differently [117, 118, 156]. A large effort is currently made to understand this scenario and the relevant role of heating and cooling in detail. Briefly, the general arguments are the following: while the specific thermal energy gained by the ICM from gravitational heating in the self-similar scenario is proportional to the depth of the gravitational potential and thus proportional to $M^{2/3}$ according to (4), any heating process that is tied to the galaxy population and will essentially be proportional to the galaxy light will be roughly independent of mass. In practice the energy of supernovae in early starbursts and AGN activity in the galaxies is believed to release one or a few keV per particle as specific thermal energy into the ICM (of the protocluster). While this is almost negligible for the ICM of a massive cluster in which gravitational heating has raised the gas temperatures to 5–10 keV, it will make a big effect in the ICM of a galaxy group with a temperature around 1 keV. Several theoretical works have been devoted to such preheating modeling, e.g. [23, 148]. In a similar way, the effect of radiative cooling is different for clusters within the self-similar structure picture. If the central gas density is the same but the temperature rises with mass, the cooling time is shorter in smaller clusters and cooling should have a larger effect there. In some models the predicted effect of cooling is mass condensation and the inflow of adiabatically heated gas

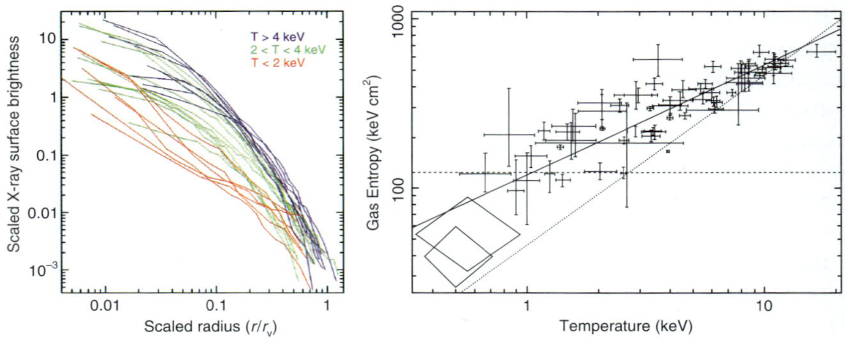

Fig. 23.7 *Left:* Surface brightness profiles scaled by $T^{1/2}$ (to account for the different extent of clusters in the line-of-sight) as a function of scaled radius, $r/r_{\rm V} \propto T^{1/2}$ [117]. *Right:* Entropy of the ICM at a radius of $r = 0.1 r_{200}$ as a function of the cluster bulk temperature [118]. The lower line shows the expectation for pure gravitational heating and the upper line the best fit of a power law function to the data (with a slope of $\sim$0.65)

from the outside. The net effect is an increase in the central entropy with a decrease in the central density and central surface brightness as shown in Fig. 23.7, e.g. [118]. Results from current refined simulations suggest that both effects play a role simultaneously. It turns out that the study of these processes are not only important for the understanding of the thermal structure of the cluster ICM but may also be the clue to understand the heating feedback regulation of galaxy formation.

It has become practice to describe the structure of the ICM in these studies by an entropy parameter, which is defined as $S = k_{\rm B} T n^{-2/3}$ expressed in units of keV cm^2. This definition differs from the classical entropy (see the review by Voit [156] for a profound explanation). In the strictly self-similar model as defined by (4) and (5) the entropy at a given scaled radius $R_* \propto M^{1/3} \propto T_{\rm X}^{1/2}$ would be given by $S(R_*) \propto T_{\rm X} \propto M^{2/3}$. Because of the effect of extra heating and cooling the observed relation as shown in Fig. 23.7, is approximately [118]

$$S(r = 0.1 r_{200}) \sim 120\,{\rm keV\,cm^2} \left(\frac{T_x}{1\,{\rm keV}}\right)^{0.65\pm0.05} . \qquad (23.6)$$

This scaling also leads to a modified scaling law for the surface brightness (emission measure) – temperature relation of the form $S_{\rm X} \propto T^{1.38}$. This scaling was used in the right panels of Fig. 23.6. Its application yields nicely self-similar emission measure profiles at radii $r \geq 0.1 R_{\rm vir}$.

There has been much recent progress in reproducing the observed ICM thermal structure in simulations (e.g. [22, 84]) but the issue is still far from being solved. In particular, the heat input not only of supernovae during star formation episodes but also AGN activity in cluster galaxies is now considered in the modeling (e.g. [140]). There is an important boundary conditions in a comprehensive modeling, including galaxy formation that has to be met. The correct amount of the stellar mass fraction of the order of 10% should be produced. For example, pure cooling models tend to overproduce the stellar mass by a large factor. This remains a very important field of research for observations with Chandra and XMM-Newton last not least because it promises to shed new light on the processes that control galaxy formation.

23.3.2 Merging Clusters of Galaxies

In the standard model of hierarchical growth of large-scale structure, where small mass aggregates collapse and form first, galaxy clusters grow by the inhomogeneous accretion of matter. Mass units of various mass fall into the cluster potential and if these mass units are large enough they are observed as cluster mergers. Major mergers of massive clusters, that is mergers with subunits of comparable mass, happen with collision velocities of the order of 2000 km s^{-1} and are with a release of gravitational energy of as much as $\geq 10^{64}$ erg, the most energetic events in the Universe next to the Big Bang [56]. At their collision velocities shocks are created in the merging ICM, which dissipate much of the gravitational energy. It is these shocks that form the major source of thermal energy due to which we observe these extremely high ICM temperatures.

An excellent overview on the astrophysics and observations of galaxy cluster mergers is given in a recent book by Feretti et al. [56]. The large amount of energy dissipated in cluster mergers make them very interesting events for further studies, as for example for the physics of shock heating. The radio halos observed in galaxy clusters are presumably created by cosmic ray acceleration in merger shocks and turbulence [67], which also boosts the strength of the intracluster magnetic fields.

Figure 23.8, shows the results of detailed XMM studies of the two most dramatic major mergers in nearby galaxy clusters. For the cluster A754 the Figure shows the image of the X-ray surface brightness [73] with contours of the radio halo observed at 20 cm [5] superposed. The system is interpreted as a merger of subclusters where core passage happened on the order of half a Gyr ago. The centers of the two subclusters are still noticeable, with the cool, dense core of the main cluster seen now in the South-East and the center of the other subcluster at the X-ray surface brightness maximum in the North-West. The radio halo is found in the region of highest pressure and probably the zone of largest turbulence. The shocks and the turbulence are most probably the sites of cosmic ray acceleration, which produces the relativistic electrons observed through their synchrotron radio radiation.

The right panel in Fig. 23.8, shows the temperature map of the merging cluster A3667 [25]. Again we see a postmerger stage with a lot of turbulence clearly displayed by the temperature map. This cluster also host the most prominent radio relics, large radio structures on opposite ends in the cluster along the merging direction which are probably created by the outgoing merger shock [67]. Observations of several of such systems, where X-ray observations indicate a cluster merger and radio observations show a radio halo inferring an intense relativistic electron population let to the expectation that radio halos should generally be expected in massive,

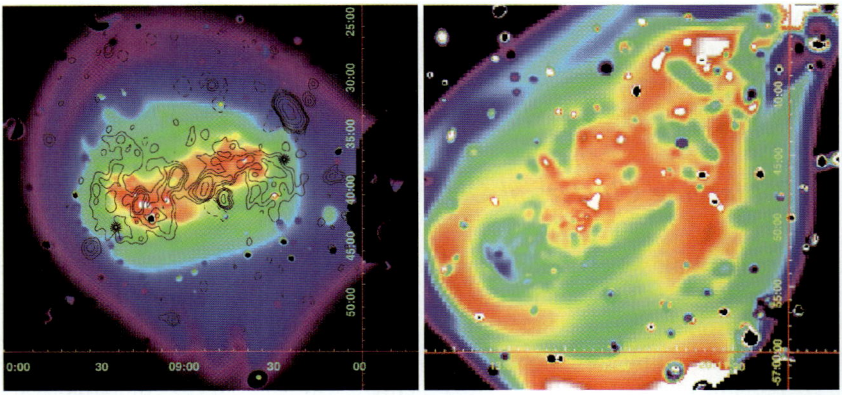

Fig. 23.8 XMM-Newton observations of two of the most dramatic nearby merging clusters. *Left:* Surface brightness distribution map of the galaxy cluster A754 (Henry et al. [73]) with radio contours overlaid (Bacchi et al. [5]). The two surface brightness maxima mark the centers of the two merging subclusters that have most probably already passed the center. *Right:* Temperature map of the cluster A3667, where red and white regions are hot and blue and green regions are cool (Briel et al. [25]). This system is also interpreted as a major cluster merger after core passage. The central region between the centers of the merging subclusters shows a high degree of complexity and turbulence

merging clusters (e.g. [57]). A statistical investigation of cluster substructure and radio halo occurrence supports this scenario [130].

To directly observe and study shock fronts in major mergers has turned out to be quite difficult and only in very few cases have shocks been observed clearly. One of the best examples is found in the cluster E0657-56 (RXCJ0568-5557) [62, 92] and is shown in Fig. 23.9. While the high resolution surface brightness image obtained with Chandra shows a bow shock type feature with a Mach cone [92], the X-ray color and spectral analysis of the cluster data reveal a region of very high entropy in front of the bow shock like feature [62]. This region of high entropy is a direct manifestation of the heating and entropy production of merger shocks.

Theoretical considerations and simulations also predict a large degree of turbulence in post-merger galaxy clusters [77, 143]. That the ICM of a postmerger cluster has the expected structure of evolved Kolmogorov-Obuchov turbulence could recently be shown with a deep study of the Coma cluster with XMM-Newton [134]. Traditionally the structure of evolved turbulence is characterized by a spatially correlated velocity field, which shows a power law power spectrum [88]. This power law behavior of the spatial correlation is also expected for the pressure distribution (with a different value for the exponent). The pressure distribution of the central part of the Coma cluster, believed to be a postmerger system [159], is shown in Fig. 23.10 [134]. The pressure values for the pixels of the image have been determined from the surface brightness and spectral hardness ratio, the two observables from which the ICM density and temperature can be deduced. An analysis of the pressure distribution shows a power law power spectrum of the kind expected for

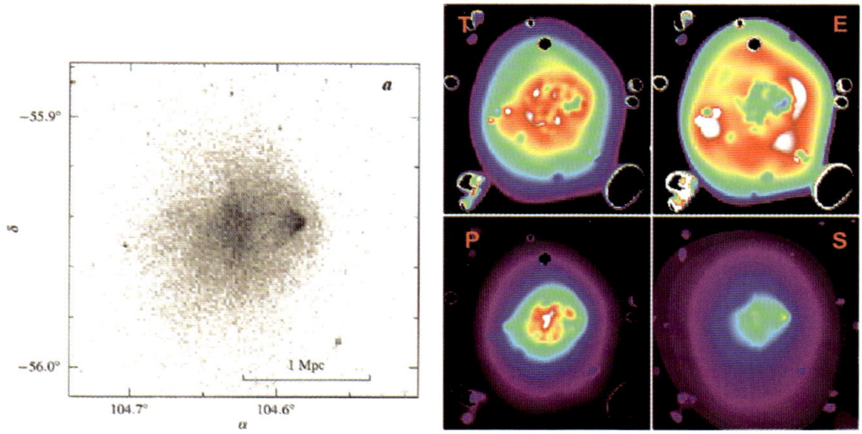

Fig. 23.9 *Left:* Chandra image of the merging cluster E0657-56 (RXCJ0568-5557) [92]. The bright feature on the upper right with a Mach cone like, sharp surface brightness structure is the core of a subcluster flying at high speed ("bullet") through the main cluster. *Right:* Projected temperature (upper left), entropy (upper right), pressure (lower left), and surface brightness map of the cluster E0657-56 (RXCJ0568-5557) constructed from XMM-Newton imaging and spectroscopic data [62] (for details of the construction of these maps see [73]). Most interesting is the high entropy feature seen in the entropy map in front of the "bullet" (which itself features a very low entropy unveiling itself as a cooling core with a cold front). This high entropy structure is most probably caused by shock heating

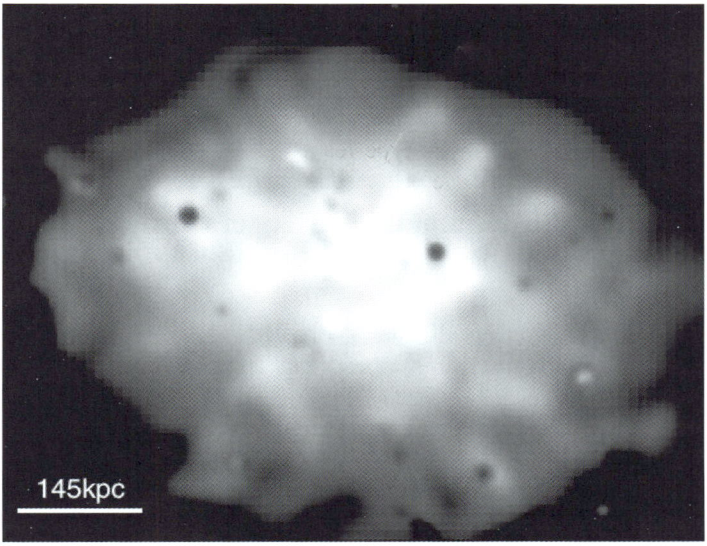

Fig. 23.10 Image of the projected pressure distribution in the COMA cluster constructed from XMM-Newton imaging and spectroscopic data by Schuecker et al. [134]. The scale of 145 kpc indicated in the figure corresponds to the largest size of turbulent eddies revealed by the turbulence power spectrum obtained from this image [134]

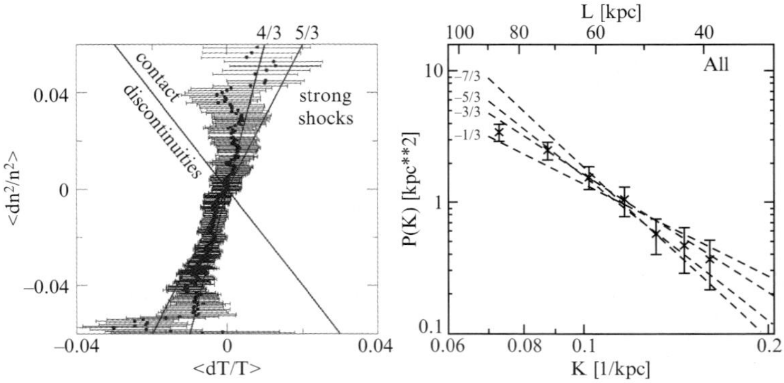

Fig. 23.11 *Left:* Correlation of the density and temperature fluctuations in the central region of the ICM of the Coma cluster of galaxies [134]. The correlation diagram is obtained by running a sliding window through the data thus that not all data points are independent. The resulting correlation slope implies that the fluctuations are close to adiabatic. *Right:* Power spectrum of the pressure fluctuations in the turbulent central region of the ICM of the Coma cluster [134]

Kologorov-Oboukhov-turbulence as shown in Fig. 23.11 (right). Further support for the interpretation of the observed features as classical turbulence are (1) a Gaussian distribution of the pressure fluctuations, and (2) the fact that the pressure fluctuations show on average roughly a correlation of temperature and density of the form $\Delta n/n \propto \Delta T/T^{4/3}$, which is close to the adiabatic value of 5/3. The latter indicates that the fluctuations also seen in the surface brightness map are neither dominated by contact discontinuities ("cold fronts") nor by strong shocks, but rather by nearly adiabatic pressure waves [134]. Power law power spectra indicating developed turbulence in the ICM have also been seen in the orientation of the ICM magnetic field as traced by Faraday rotation measurements [155].

23.4 The Virgo Cluster and the Variety of Cluster X-ray Morphology

The galaxy cluster in the constellation of Virgo is with a distance of about 17 Mpc [58] the nearest of all galaxy clusters and thus a perfect object for very detailed studies. In fact our Galaxy lays in the outskirts of the cosmic web structure surrounding the Virgo cluster, which has a recession velocity of about $1\,000\,\mathrm{km\,s^{-1}}$. The galaxy field of the cluster stretches over a region of more than $10 \times 10\,\mathrm{deg}^2$ over the sky and the galaxy distribution and population has been studied in detail by Binggeli et al. [6]. They distinguish a major norther part, A, a smaller southern part, and several galaxy sheets partly in the background. An X-ray image of the cluster

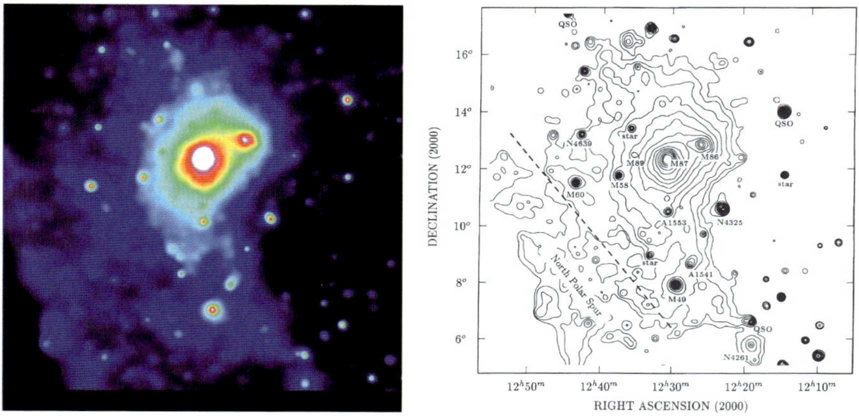

Fig. 23.12 The Virgo galaxy cluster as observed in the ROSAT All-Sky Survey in the energy band 0.5–2.4 keV [8]. The left pannel shows the surface brightness distribution as color map, while the right panel shows the same image as contour map with individual sources marked

obtained in the ROSAT All-Sky Survey is shown in Fig. 23.12, [8]. It shows a complex surface brightness distribution, which reveals Virgo as a dynamically young cluster. We note the bright nearly symmetric X-ray halo in the northern part of the cluster centered on the giant elliptical galaxy M87, a southern part of the cluster with much lower surface brightness emission with the elliptical M49 at its center, an extension in the North-East housing the massive elliptical M60 and another X-ray halo surrounding the massive elliptical M86 about 1 degree NWW of M87. Thus, apart from the central region around M87, the matter in the Virgo cluster is still collapsing and not in a virial equilibrium.

A more detailed comparison of the galaxy and X-ray emitting ICM distribution shows that the X-ray contours follow closely the distribution of the early type galaxies (ellipticals and S0s), including the early type dwarf galaxies (which are found in large numbers in Virgo), while late type galaxies (spirals and irregulars) tend to avoid the X-ray emitting regions [128]. The good match between the early type galaxies and X-rays is particularly striking in a region of a sharp contrast at the edge South-West of M87. A study of the HI content of spirals in 21 cm by Cayatte et al. [33] shows nicely that the HI deficiency is also correlated with the density of the X-ray emitting ICM. A few galaxies closer to M87 show signs of being stripped of their ICM in the 21 cm images and the modeling of this effect is well consistent with ram pressure stripping by the cluster ICM acting on these infalling galaxies [9, 33]. Overall, the Virgo cluster is considered a spiral rich cluster and the above features suggest that most of the spirals are infalling, emphasizing again the picture that the Virgo cluster consists of a smaller relaxed core and probably a much larger amount of matter on the infall, creating a cluster of the stature of the Coma cluster in the not too distant future.

One chuck of matter that is infalling with a dramatic effect at present is the galaxy M86. This galaxy is actually blueshifted as observed from Earth and thus must be falling with a high relative velocity ($\sim$1 200 km s^{-1}) from behind into the Virgo cluster. The substantial X-ray halo surrounding it and the velocity distribution of the dwarf galaxies in its vicinity [6] imply that M86 is actually the dominant galaxy of a small galaxy group falling into Virgo. The spectacular interaction effects of this event have recently been studied in detail with XMM-Newton by Finoguenov et al. [61].

Because of the large extent of the cluster in the sky it is difficult to cover Virgo in total with an X-ray study other than the sky survey. With the ASCA satellite this effort was undertaken; however, Shibata et al. [137] constructed a temperature map of the Virgo cluster ICM and found an interesting patchy structure with a correlation length of about 300 kpc, which they interpreted as the fossil traces of ICM heating by infalling galaxy groups from which the cluster was formed.

If we compare the X-ray image of Virgo with the dynamically much more evolved Coma cluster shown in Fig. 23.1, the difference is readily apparant. In general we find a broad range of cluster X-ray morphologies involving quite evolved and relaxed clusters, less frequent major cluster mergers, and also dynamically younger clusters which accrete matter in a more diffuse fashion than the major mergers. The gallery of X-ray images from our recently conducted systematical study of cluster structure with XMM-Newton (with targets statistically selected from the X-ray flux-limited REFLEX survey) gives a good impression of the typical range of cluster morphologies (Fig. 23.13).

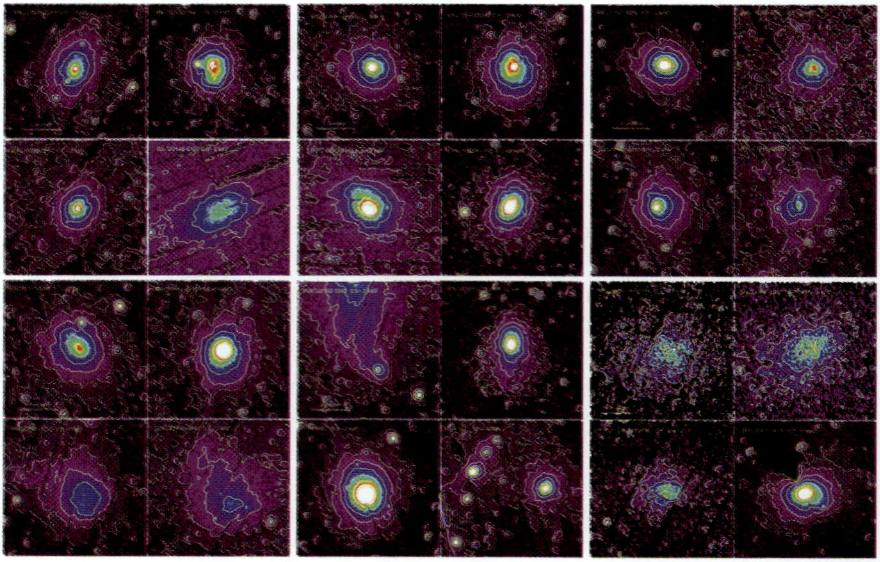

Fig. 23.13 XMM-Newton image gallery of galaxy clusters selected statistically from the REFLEX sample for a systematic study of galaxy cluster structure in the frame of an XMM Large programme [19]. The clusters cover almost homogeneously a luminosity range from $0.4 - 13. \cdot 10^{44}$ erg s^{-1} and a redshift range from $z = 0.055 - 0.183$. The sample should give a nearly unbiased impression of the range of cluster morphologies of a flux limited cluster sample

23.5 Cooling and Heating of the ICM

First detailed X-ray observations more than 25 years ago with the Uhuru and Copernicus satellites and the first rocket borne X-ray telescopes showed that the X-ray emitting hot gas in galaxy clusters reaches high enough densities in the cluster centers that the cooling time of the gas falls below the Hubble time. The consequences of these observations have first been explored in early papers by, e.g., Silk [138] and Fabian and Nulsen [51]. In the absence of a suitable fine-tuned heating source, the cooling and condensation of the gas in the central regions is a straight-forward consequence of the energy equation of the ICM. From this analysis the cooling flow scenario emerged (e.g., Fabian et al. [52, 53]). Based on the assumption of steady state cooling this model predicts spectra implying an emission distribution over a wide temperature range, a locally multi-temperature structure and as a consequence of the latter mass deposition distributed over a large fraction of the cooling flow region [89, 146]. It remained a puzzle, however, why in the regions of cooling flows with estimated mass deposition rates up to several hundred or thousand solar masses per year, and little evidence for such massive cooling of gas could be found at other wavelengths (e.g. [53]). This is a major reason why the cooling flow model did not get accepted unanimously among astronomers. Evidence for warm gas (diffuse emission line systems), cool gas, and even star formation was well detected at lower levels in the range of 1–10% of the model predictions (1 to $<100\,M_\odot\,yr^{-1}$) (e.g. [45, 53, 72] as summarized for example in the review by McNamara [98]).

23.5.1 The Observed Thermal Structure of Cool ICM Cores

New observations with XMM-Newton and Chandra lead to a revision of this scenario. XMM-Newton is now providing unprecedented detailed spectroscopic diagnostics of the central regions of clusters and new insights into the cooling flow picture. One set of observations, obtained with the XMM Reflection Grating Spectrometer (RGS), shows for several cooling core regions spectral signatures of different temperature phases ranging approximately from the hot virial temperature of the cluster to a lower limiting temperature, T_{low}, which is still significantly above the "drop out" temperature where the gas would cease to emit significant X-ray radiation. That is, the clearly observable spectroscopic features of the lower temperature gas expected for a cooling flow model are not observed. Figure 23.14 shows for example the case of the massive, cooling flow cluster A1835 with a bulk temperature of about 8.2 keV, where no lines or features for temperature phases below 2.7 keV were observed as found by Peterson et al. [112]. Systematic studies of a sample of cooling flow clusters, e.g., by Kaastra [82] support these results.

The other set of relevant XMM-Newton observations are obtained with the energy sensitive imaging devices pn and MOS. Even though the spectral resolution is less for these instruments than for the RGS, they can very well be used to

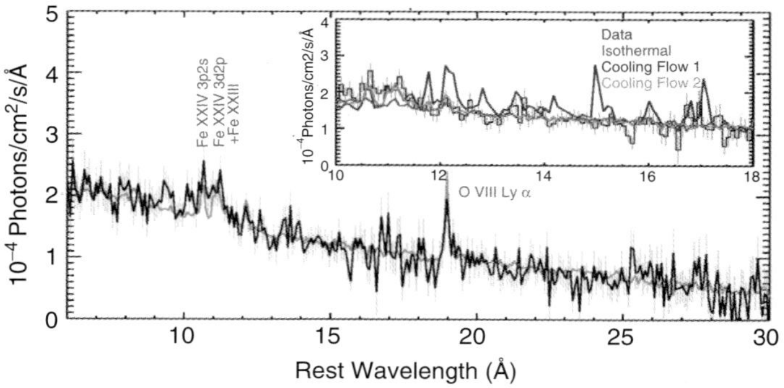

Fig. 23.14 XMM-Newton RGS spectra of the cooling core region of the massive cluster A1835 [112]. The observed spectrum is fit by several models: an isothermal model with a temperature of 8.2 keV (*red*), a classical cooling flow model with a hot temperature phase of 8.2 keV and the cooling flow component (*blue*), and a similar cooling flow model with a forced lower temperature cut-off at 2.7 keV (*green*). The inset shows clearly several lines by the blue marked cooling flow model, which are not observed in the cluster spectrum

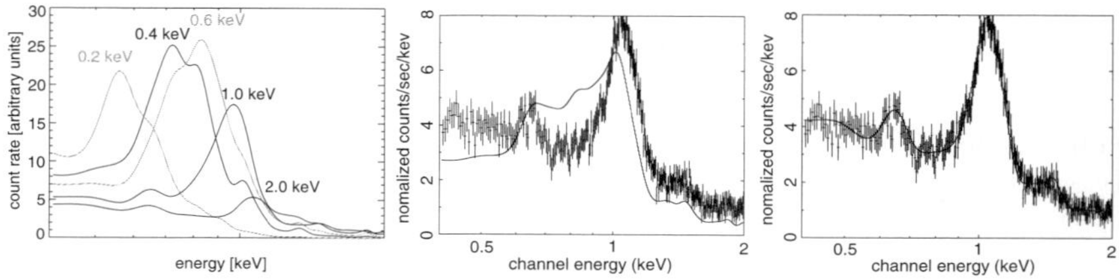

Fig. 23.15 *Left:* The Fe L-line complex in X-ray spectra as a function of the plasma temperature for a metallicity value of 0.7 solar. The simulations show the appearance of the spectra as seen with the *XMM* pn. The emission measure was kept fixed when the temperature was varied. *Middle: XMM* pn-spectrum of the central region of the M87 X-ray halo in the radial range $R = 1 - 2$ arcmin. The spectrum has been fitted with a cooling flow model with a best fitting mass deposition rate of $0.96\,M_\odot\,yr^{-1}$ and a fixed absorption column density of $1.8 \cdot 10^{20}\,cm^{-2}$, the galactic value, and a low temperature cut-off at 0.01 keV. *Right:* Same observed spectrum as in the middle panel fitted by a cooling flow spectrum artificially constraint to emission from the narrow temperature interval 1.44–2.0 keV with a mass deposition rate of $2.4\,M_\odot\,yr^{-1}$ [15]

detect temperature sensitive spectral features with high accuracy due to the good photon statistics. They provide in addition spatially resolved spectroscopic information across the entire cooling core region. Spectral fitting of the X-ray emission in the cooling flow clusters, notably M87 in the Virgo cluster, yielded two pieces of information which are incompatible with the classical cooling flow model. They showed that single temperature models provide a better representation of the data than cooling flow models [12] and they confirm the lack of low temperature components in the central ICM [15, 94]. Figure 23.15 nicely illustrates the result for the most nearby and best to study cooling core in M87. For the relevant temperature range from a few tenth of a keV to about 3 keV, the complex of iron L-shell lines provides a superb ICM thermometer as illustrated in Fig. 23.15, left. The figure shows simulated X-ray spectra for the *XMM pn* instrument in the spectral region around the Fe L-shell lines for various temperatures from 0.4 to 2.0 keV and 0.7 solar metallicity. There is a very obvious shift in the location of the peak of the blend of iron L-shell lines, caused by an increasing degree of ionization of Fe with increasing temperature.

For a cooling flow with a broad range of temperatures (as explained above) one expects a composite of several of the relatively narrow line blend features, resulting in a quite broad peak. The right panels of Fig. 23.15b shows for example the deprojected spectrum of the M87 halo plasma for the radial range 1–2 arcmin (outside the excess emission region at the inner radio lobes) and a fit of a cooling flow model with a mass deposition rate slightly less than 1 $M_\odot$ yr^{-1}, as approximately expected for this radial range of the cooling core from the analysis of the surface brightness profile, e.g., [94]. There is obviously no good agreement between the observations and the model, while a fit of a plasma with a narrow range of temperatures from 1.44 to 2 keV provides an excellent fit. Even though attempts were made to save the classical cooling flow model (e.g. [54]) it is generally accepted that the lack of evidence of cooler temperature phases in the ICM implies no massive cooling in the cool core regions of galaxy clusters.

23.5.2 Heating by a Central AGN

The prerequisite for avoiding massive cooling and condensation is a suitable heat source that is fine-tuned to just compensate the radiative losses in the cool core. This is, for example, achieved by the cycle of feeding a central AGN with cooling flow gas until the energy output of the AGN is limiting (regulating) the further cooling of the ICM, such that the gas condensation rates are reduced by factors of probably tens to hundreds [15, 34, 35]. In fact, high resolution Chandra images now show various examples of interaction effects of central AGN with the cluster ICM providing direct evidence of energy input into the radiatively cooling ICM (e.g. [7, 55, 64, 99]). The most prominent example of AGN-ICM interaction is that of the AGN in NGC1275 in the Perseus cluster of which Chandra images are shown in Fig. 23.16, [55]. Here and in now about 20 other cases we observe how the relativistic plasma ejected by the AGN in form of jets fills radio lobes, which push the X-ray emitting ICM away and forms X-ray underluminous cavities.

The observed interaction effects and the derived properties of the ambient ICM allow us to roughly estimate the power of the mechanical energy injection of the AGN and one finds values in the range of 10^{44}–10^{45} erg s^{-1}, in some cases much more than the energy that is radiated in the cooling core region. The most important question that has still to be solved is how the energy from the radio lobes is transferred to the ICM in detail. The heating has to be fine-tuned globally such that cooling is prevented effectively throughout the cooling area. How this is managed is far from clear, but recent observations of possible sound waves ("ripples" in the

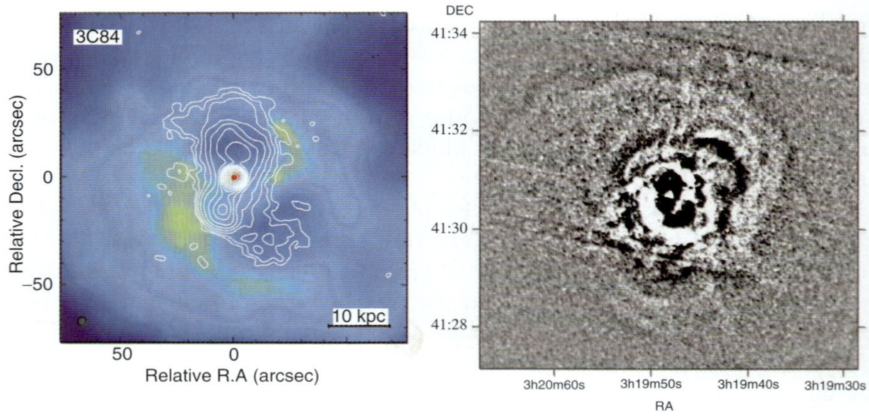

Fig. 23.16 *Left:* Chandra X-ray image of the central region of the Perseus cluster around the dominant elliptical galaxy NGC 1275 with radio contours superposed (Fabian et al. [55]). The region covered by the radio lobes show clear cavities in the X-ray surface brightness distribution where the X-ray emitting ICM has been displaced by relativistic synchrotron emitting plasma. *Right:* X-ray image of the same region produced by unsharp masking emphasizing regions of high local contrast. The concentric structures ("ripples") revealed by this picture are interpreted as sound waves or weak shock waves produced by the interaction of the expanding radio bubbles and the ambient ICM [55]

ICM [55]) observed in the Perseus cluster and shock waves observed for example in M87 (Fig. 23.17) and Hydra A [64, 90] suggest that these waves, created by the expanding radio bubbles, are the main agent to transfer the energy from the relativistic to the thermal plasma. A large effort is now also going into detailed simulations of these processes (e.g. [26]).

23.6 Heavy Element Enrichment of the Cluster ICM

At the typical cluster temperatures of several keV, the K-shell line of iron (Fe Ly-α) is the most prominent line in the X-ray spectrum. Iron was therefore the first element that was detected in the X-ray spectrum of a galaxy cluster [101]. Figure 23.18, shows modern X-ray spectra of the Centaurus and Virgo clusters observed with ASCA and XMM-Newton, respectively, displaying lines of several heavy elements. The typical iron abundances found in the early observations were in the range of 0.3–0.5 solar [107]. With the capabilities of the ASCA satellite to perform spatially resolved spectroscopy the first look at abundance profiles was possible, involving mostly the elements Fe and Si, with a few exceptions of bright sources showing also emission lines of a few other abundant elements [66]. Additional information on the spatial distribution of these elements came subsequently also from observations

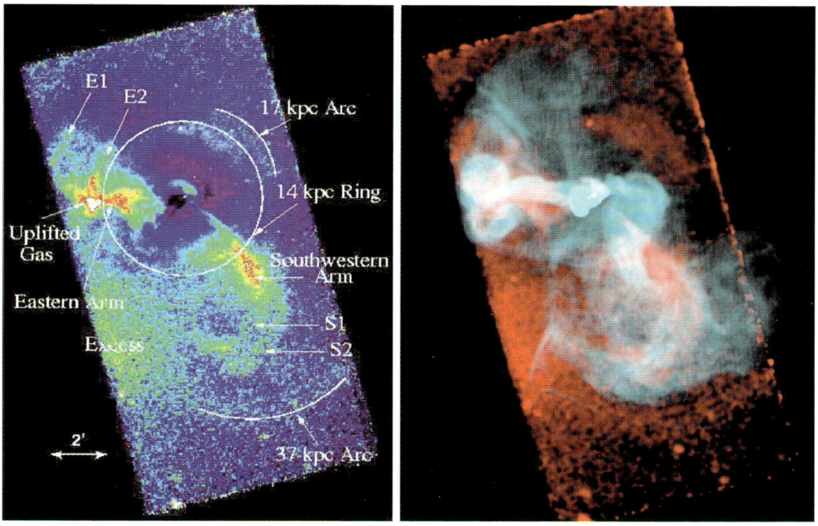

Fig. 23.17 *Left:* Chandra image of the X-ray halo around M87 showing a wealth of features due to the interaction of the jets from the central AGN in M87 and the cluster ICM. Sharp, concentric surface brightness edges with a significant brightening in the denser radio lobe regions indicate shock waves most probably created by the early supersonic expansion of the radio bubbles in the center. *Right:* Composite image of the M87 halo region with X-ray emission shown in *red* and radio emission shown in *blue* color. The spatial correlation of the radio and X-ray features (relativistic and thermal plasma) is obvious in several of the interaction zones (Forman et al. [64])

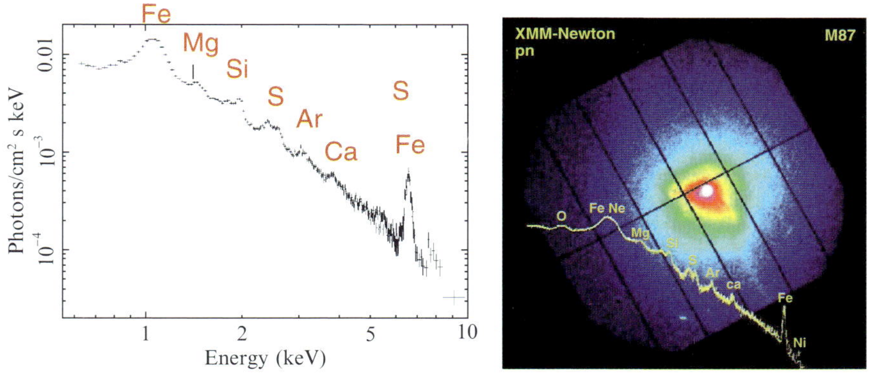

Fig. 23.18 X-ray spectra of the Centaurus cluster (*left*) observed with ASCA [76] and of the M87 X-ray halo studied with XMM-Newton (*right*) [12]

with the BeppoSAX satellite [40]. Two interesting discoveries were made with these studies. It was found that the Fe abundance profile in "cooling flow" clusters showed a pronounced central increase while the profile for "noncooling flow" clusters was essentially flat [40, 59, 60, 66, 93] as shown in Fig. 23.19. In addition the ratio of the Si to Fe abundance increases with cluster radius such that the central abundance

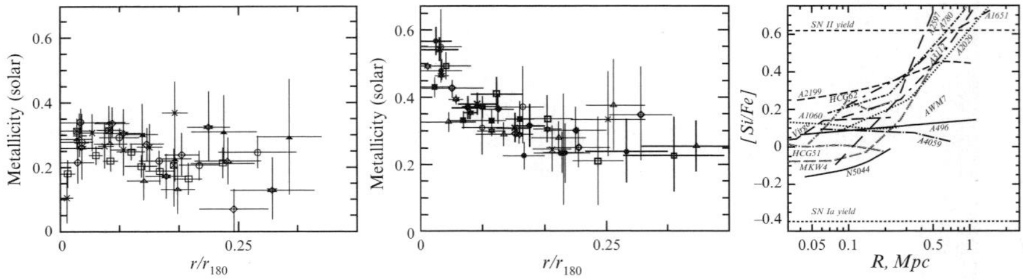

Fig. 23.19 *Left:* Abundance profiles of Fe in noncooling flow clusters (*left*) and "cooling flow" clusters (*middle*) observed with BeppoSAX by DeGrandi and Molendi [40] in solar units. *Right:* Ratio of the abundance of Si and Fe as a function of cluster radius observed with ASCA by Finoguenov et al. [59]. Also indicated are the classical values for the yields of type Ia and type II supernovae

ratio is closer to the yields of supernovae type Ia, while the global cluster ratio is closer to the yields of SN II (Fig. 23.19) [59, 60].

23.6.1 Origin of the Heavy Elements in the Central Region

With XMM-Newton and Chandra observations the information on the heavy element distribution has increased dramatically (good information on more elements, most importantly on oxygen and well resolved abundance profiles) and we can now investigate the origin of the heavy elements in more detail. The central abundance peak in cooling flow clusters with its high surface brightness and the most detailed spectra provide a special clue for this study.

We know two sources for the enrichment of the ICM with heavy elements: core collapse supernovae, type II, which produce a broad spectrum of element masses with a bias towards the lighter elements like O and Mg, and type Ia supernovae, thermonuclear explosions of white dwarfs, which dominantly yield Fe group elements and lighter elements like Si and S but very little O and Mg. Checking now which of the elements follow the central abundance peak shows that it is traced by the heavier SN Ia products but not by the lighter elements like O and Mg (Fig. 23.20, [95]. This is consistent with the picture that SN II activity happens in the early history of cluster formation when the stellar populations of the cluster galaxies are still young. These elements have time to mix well in the ICM and show a more homogeneous distribution. SN Ia are still occurring and are observed in present day cluster ellipticals and in the central cD galaxies. The more recent yields obviously lead to more local enrichments. Thus the massive stellar population of the cD galaxies dominating the centers of cooling core clusters are responsible for the central enrichment [17, 41].

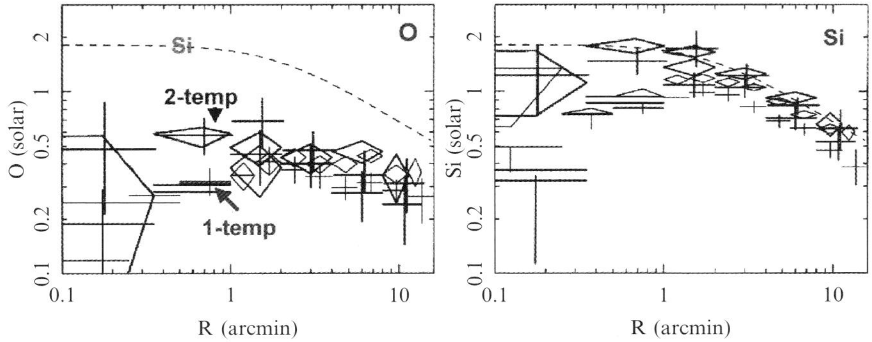

Fig. 23.20 Abundance profiles of O (*left*) and Si (*right*) in the M87 X-ray halo studied with XMM-Newton by Matsushita et al. [95]. The *thick line* symbols show the results of a more realistic two-temperature model fit (compared to a one temperature model, *thin* symbols) while the *crosses* and *diamonds* mark the data from the pn and MOS detectors, respectively. The Si abundance profile is also shown in the left panel as *dashed line* for comparison

A more careful inspection of the chemical composition of the central ICM reveals also slight increases in the relative abundances of O and Mg as found, for example, for the M87 halo and the Centaurus cluster [95, 96]. The ICM abundances of these elements are comparable to the stellar abundances which implies that the ICM in the very central region most probably reflects the stellar mass loss of the central galaxy. Consideration of the budget of the central ICM, for example, for the case of M87 shows that the inner 10 kpc region of the ICM, roughly the region of the scale radius of the cD galaxy (or approximately half light radius), contains a gas mass of about $2 \times 10^9\, M_\odot$ and an iron mass of about $6 \times 10^6\, M_\odot$. Both can be replenished by stellar mass loss from M87 if we adopt a stellar mass loss rate of $2.5 \times 10^{-11} \times L_B\, M_\odot\, yr^{-1}$ (Ciotti et al. [36]) and a Fe production with a supernova rate of 0.15 SNU (Cappelaro et al. [31]). It is important, however, for this enrichment to happen that the gas is not cooling and condensing with the classically calculated cooling flow rates, since in this case the central ICM produced by stellar mass loss would disappear faster than it can be replenished. Both, the evidence described in the previous section and the observed abundances of O and Mg support the notion that the central ICM is not condensing at the high rates inferred previously.

The central abundance peak extends much further than the central 10 kpc region inside the central galaxy. Since the central galaxy dominates the stellar population out to radii of the order of 100 kpc. It is thus responsible for most of the secular heavy elements in this zone. This implies some transport of the centrally enriched ICM to larger radii and larger enrichment times than inferred above for the 10 kpc region. Calculations for the observations of four nearby cooling core clusters (M87, Centaurs, Perseus, A1795), with constant SN Ia rate and alternatively with an increasing SN Ia rate in the past, imply very large enrichment times of 5–9 Gyrs for the region inside $\sim$50 kpc and 7–12 Gyrs for $\leq$100 kpc. If these estimates are correct the results imply that the central regions of the cool core clusters and their ICM can not have suffered major disturbances during the past 10 Gyrs [17]!

23.6.2 Supernova Yields

Adopting this scenario for the enrichment of the central heavy element abundance peak, one can also study the element abundance ratio in more detail and compare the results with predictions from supernova nucleosynthesis model calculations. The best measured abundances are those of Fe, Si, and O (the latter representing the lighter elements). Thus we will concentrate here on the use of these elements.

In Fig. 23.21 the abundance ratios of Fe/O and Fe/Si are compared for various regions in different groups and clusters of galaxies. In this diagram the Fe/O ratio is a diagnostics of the ratio of Fe yields from SN II and SNIa and the Fe/Si ratio can then be used to test various models of SN Ia yields (introduced by Matsushita et al. [95]). In the plot we have compiled a set of observational results from XMM-Newton, Chandra, and one ASCA result from the literature and compare them to various models with different deflagration speeds for the SN Ia explosion from Nomoto et al. [109] and Iwamoto et al. [79]. The results show that the bulk of the data lay below the curve of the W7 model, which fits the solar abundance and is successfully used to describe the chemical history of our galaxy. The WDD1 and WDD2 models provide yields with larger Si fractions compared to iron than supplied by the W7 model. They better fit the observational data. The data points for M87 are the most extreme in requiring a relatively large Si yield [95]. The possible reason for this is

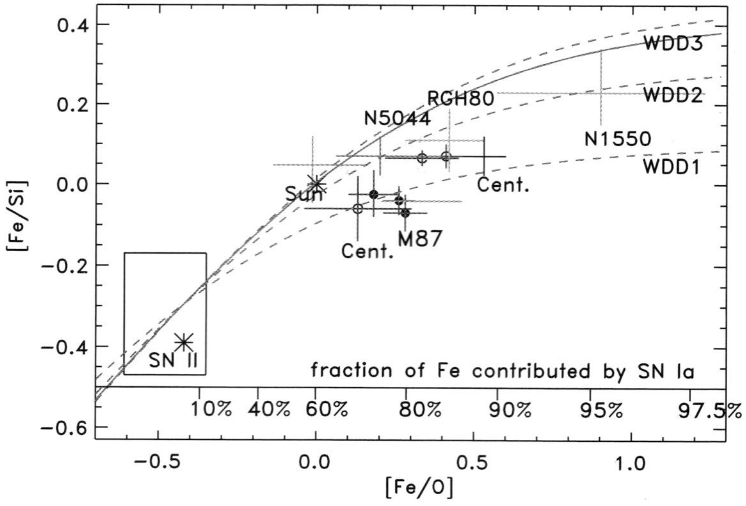

Fig. 23.21 Ratio of the Fe to Si abundance as a function of the Fe/O abundance ratio (in solar units e.g. [70]) in the ICM of groups and clusters of galaxies [27, 42, 95, 96, 142, 144, 161] as determined from Chandra, XMM-Newton, and ASCA observations. Also shown is the abundance ratio of the Sun (*asterisk*), SN II models (*asterisk*), abundance of old stars in the Milky Way (box in the lower left) [37, 47], and curves indicating the prediction for a mixture of SN II yields and varying contributions from SN Ia for various delayed-deflagration models (WDD1, WDD2, and WDD3) from Iwamoto et al. [79]. For more details see Böhringer et al. [17, 18]

a more incomplete nuclear burning in SN Ia explosions in older stellar populations [60]. This is also consistent with the observed statistically lower luminosity of SN Ia light curves observed in ellitpical galaxies as compared to spirals [78].

23.7 X-Ray Cluster Surveys

The systematic identification of galaxy clusters in X-rays across the sky has received an enormous boost by the ROSAT All-Sky Survey (RASS) [149,154], the first X-ray all-sky survey conducted with an imaging X-ray telescope and the comprehensive archive of ROSAT pointed observations. Before that event a few hundred galaxy clusters were detected in X-rays and studied mainly with the Einstein and HEAO-1 satellite observatories. The RASS Atlas contains, as estimated, 8 000–10 000 galaxy cluster X-ray sources, of which about 2 000 have been identified by now. Recent reviews by Schuecker [133] and Edge [46] provide a comprehensive account of pre-ROSAT and ROSAT-based cluster surveys, e.g. [11, 13, 16, 43, 44, 68].

The sample of the X-ray brightest $\sim$100 galaxy clusters (outside the zone-of-avoidance around the plane of the Milky Way) has presumably been detected completely and compiled by Reiprich and Böhringer in the HIFLUGCS sample [123] from which the first empirical cluster mass function has been determined. Figure 23.22 (right) shows the cumulative mass density of the clusters as a function of the lower mass limit in units of the critical density of the Universe. Only about 2% of the critical density, that is, about 6% of the matter density of the Universe (assuming a concordance cosmological model with $\Omega_m = 0.3$) is found in galaxy clusters and groups of galaxies with a mass above about $2.1 \times 10^{13} h_{70}^{-1}$ M$_\odot$ [123] . Nevertheless, this small mass fraction provides tight constraints on the large-scale structure parameters of the matter distribution in the Universe.

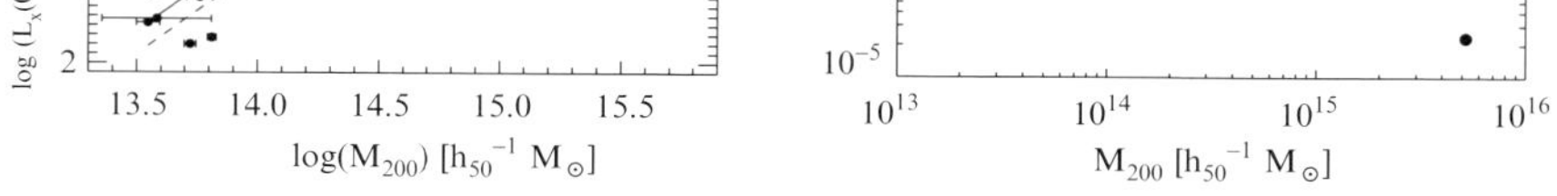

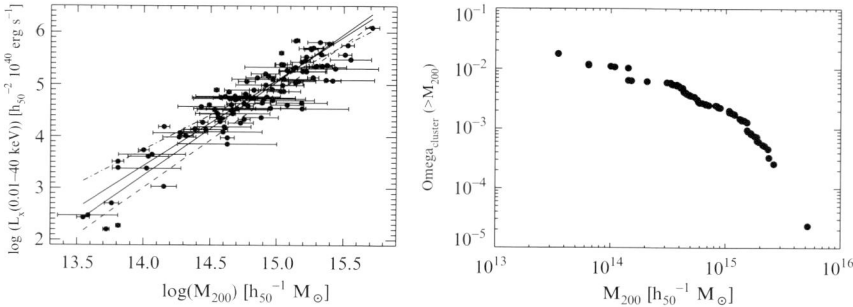

Fig. 23.22 *Left:* Mass-X-ray luminosity relation determined from the 106 brightest galaxy clusters found in the ROSAT All-Sky Survey by Reiprich and Böhringer [123]. *Right:* Mass density of galaxy clusters in the Universe as a function of the lower mass cut in units of the critical density of the Universe [123]

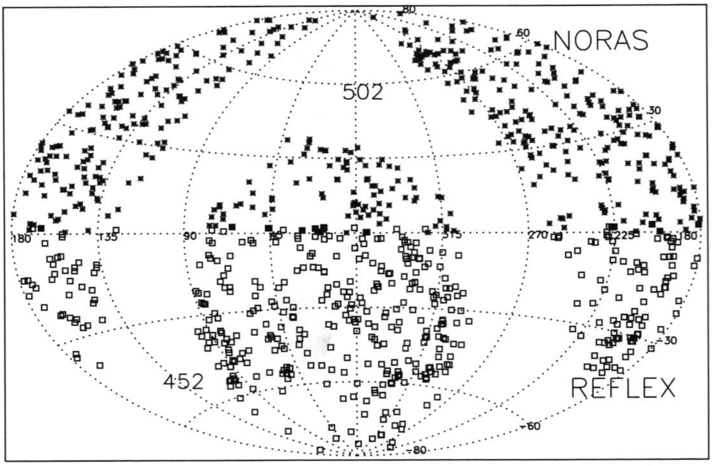

Fig. 23.23 Sky distribution of the brightest galaxy clusters found in the ROSAT All-Sky Survey investigated in the NORAS [11] and REFLEX [16] cluster redshift surveys

X-ray surveys of clusters provide several advantages for the construction of cluster samples to be used for cosmological studies. First of all the X-ray luminosity is tightly correlated to the cluster mass as shown in Fig. 23.22 (left) [123], and the presence of X-ray emission is a signature of a truly bound structure, and due to the centrally peaked surface brightness profile projection effects on the sky are minimized. For the comparison of the observed cluster abundances and spatial distributions with theoretical model predictions the masses of the clusters involved have to be known, at least approximately. This is therefore currently best provided by X-ray observations, as demonstrated in Fig. 23.22.

For the present illustration of the use of galaxy clusters as cosmological probes we use the REFLEX cluster survey, the currently largest, most homogeneous, and most complete X-ray cluster survey [13, 16]. This survey covers the southern sky up to declination $\delta = +2.5^\circ$, avoiding the band of the Milky Way ($|b_{II}| \leq 20^\circ$) and the regions of the Magellanic clouds with a total survey area of 13 924 deg^2 or 4.24 sr (Fig. 23.23). The X-ray detection of the clusters is based on the second processing of the RASS [154] with a reanalysis of the X-ray properties of the sources by means of the growth curve analysis method [11]. The current sample has a flux-limit of $F_x \geq 3 \times 10^{-12}$ erg s^{-1} cm^{-2} and comprises 447 objects. An extension of the sample termed REFLEX II to a lower flux limit is almost completed. A comprehensive optical follow-up program in the frame of an ESO key program was used to definitely identify the so far unknown X-ray cluster candidates and to measure the missing cluster redshifts. The final cluster catalogue is estimated to be better than 90% complete with a contamination from X-ray luminous AGN less than 9%. A series of tests demonstrate the high quality of the sample. Figure 23.25 (left) shows for example the Gaussianity of the cluster distribution of a study similar to counts

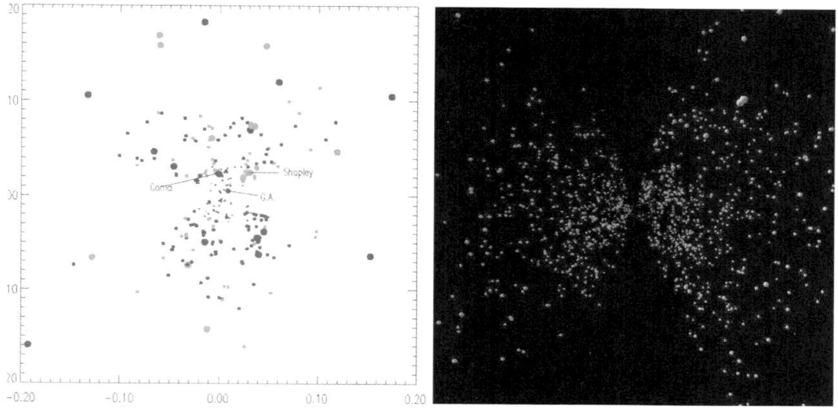

Fig. 23.24 *Left:* Slice through the three-dimensional cluster distribution in NORAS and REFLEX for a flux limit of 3×10^{-12} erg s^{-1} cm^{-2}. The slice is taken perpendicular to the galactic plane (along l_{II} = 90–270 deg). Some famous cluster concentrations in the local Universe are indicated (G.A. marks the Great Attractor region). *Right:* Distribution of the galaxy clusters in the extended REFLEX and NORAS surveys out to redshifts of about $z = 0.2$ containing about 1400 galaxy clusters. The clumpy structure of the cluster distribution is easily visible

in cells conducted in connection with the KL-structure analysis described further (see [131]).

The three-dimensional cluster distribution is sketched in Fig. 23.24, which shows a slice through the space distribution of the REFLEX and NORAS[3] clusters, which includes the region of the Milky Way. The other panel in Fig. 23.24 shows a view on a three-dimensional model of the cluster distribution in the extended REFLEX and NORAS surveys. Both pictures clearly show the clumpy structure of the cluster distribution, which results in a higher than random probability to find a cluster near another cluster than at any random point in space as reflected by the two-point correlation function [38].

23.8 Assessing the Cosmic Large-Scale Structure

The basic census of the galaxy cluster population is the density of clusters of a given mass, the mass function, which tells us much about the fluctuation amplitude of the dark matter density distribution from which the clusters form (e.g. [158]). Observationally the X-ray luminosity function is a close representation of the mass function. A binned representation of the luminosity function for the REFLEX Cluster Survey is shown in Fig. 23.25 (right). Because of the large sample size the statistical uncertainties are small. A Schechter function provides a reasonable fit to the data [14].

[3] The NORAS cluster survey is based on the northern sky region of the RASS and complements the REFLEX survey in the northern hemisphere [11].

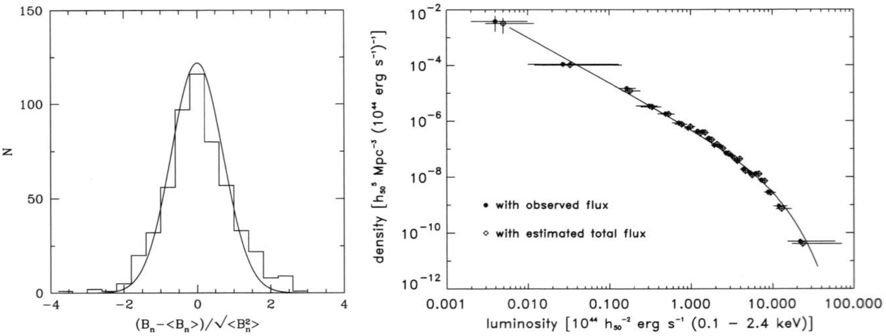

Fig. 23.25 *Left:* Histogram of the normalized KL eigenvalues used to analyze the large-scale structure statistics [132]; a fitted Gaussian function is superposed on the data. This distribution is effectively a cluster count in cells distribution, showing that the large-scale structure is approximately Gaussian on scales of $\sim 50h^{-1}$ Mpc as far as can be measured with the low number statistics of REFLEX. *Right:* X-ray luminosity function of the REFLEX cluster Survey. The filled and open data points refer to observed and corrected total luminosities, respectively. A Schechter function (*solid line*) is fitted to the data by an ML method including the flux uncertainties [14]

The spatial distribution of galaxy clusters is following the overall matter distribution in the Universe, the seeds for which were set in the early Universe. The relation of the amplitude of the cluster density and dark matter density fluctuations can be derived from statistical considerations based on the valid assumption that galaxy clusters form from the large-scale high amplitude peaks of the matter density fluctuation field, e.g. [102]. It turns out that the amplitude of the cluster density fluctuations is amplified compared to that of the matter density. This effect, called "biasing," helps us to trace the cosmic large-scale structure with increased sensitivity. Therefore, one of the very important goals of contiguous X-ray cluster surveys is the assessment of the statistics of the large-scale structure. The most fundamental statistical description of the spatial structure is based on the second moments of the distribution, characterized either by the two-point-correlation function or its Fourier transform, the density fluctuation power spectrum. The two-point correlation function of REFLEX shows a power law shaped function with a slope of 1.83, a correlation length of $18.8h_{100}^{-1}$ Mpc, and a possible zero crossing at $\sim 45h_{100}^{-1}$ Mpc [38]. The density fluctuation power spectrum (Fig. 23.26), [129] is characterized by a power law at large values of the wave vector, k, with a slope of $\propto k^{-2}$ for $k \leq 0.1h\,\mathrm{Mpc}^{-1}$ and a maximum around $k \sim 0.03h\,\mathrm{Mpc}^{-1}$ (corresponding to a wavelength of about $200h^{-1}$ Mpc). This maximum reflects the size of the horizon when the Universe featured equal energy density in radiation and matter and is a sensitive measure of the mean density of the Universe, Ω_m, [129].

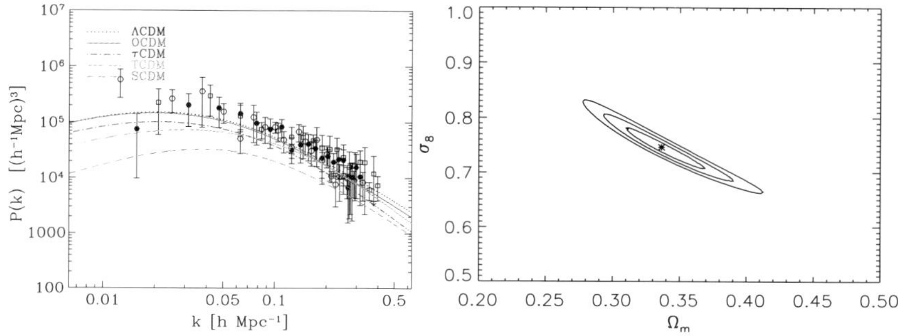

Fig. 23.26 *Left:* Power spectra of the density fluctuations in the REFLEX cluster sample together with predictions from various popular cosmological models taken from the literature. The shape of the power spectrum is best represented by the ΛCDM and OCDM models with a low matter density parameter ($\Omega_m \sim 0.3$). For details see Schuecker et al. [129]. *Right:* Constraints on the cosmological density parameter, Ω_m, and the amplitude of the matter density fluctuations on a scale of $8h^{-1}$ Mpc, σ_8, obtained from the comparison of the density fluctuation power spectrum and the cluster abundance as a function of redshift in a Karhunen-Loéve statistical analysis of the REFLEX Survey data (Schuecker et al. [131]). The likelihood contours give the 1, 2, and 3 σ limits

23.9 Cluster Evolution

The so far the best studies on cluster evolution are based on the nearby and distant cluster samples compiled from serendipitous cluster discoveries in the FoV of Einstein and ROSAT archival observations [21, 74, 105, 125, 126, 150]. For the older Einstein based sample (EMSS), detailed data were compiled in extensive follow-up observations including temperature functions for various redshifts which provide a very sensitive test of the evolution of the cluster population [74]. These data show an evolutionary trend which can approximately be explained in the frame of the concordance cosmological and current structure formation models. The ROSAT archive based distant cluster surveys reach out to larger redshifts. The largest number of high redshift clusters are contained in the ROSAT Distant Cluster Survey (RDCS) by Rosati et al. [125], while the largest number and best statistics comes from the 160 deg^2 survey lead by Vikhlinin [150]. The evolutionary effect in the luminosity function found in the RDCS is shown in Fig. 23.27 (top). There is a significant trend of a decreasing cluster abundance at the high luminosity end of the X-ray luminosity function, that is a negative evolution with redshift. This negative evolution taken together with the expected brightening of galaxy clusters of a given mass with redshift is a clear indication of the decreasing abundance of massive clusters with look-back time and lends therefore strong support to the hierarchical formation model of galaxy clusters from smaller to increasingly larger units. A similar conclusion with high significance is also obtained from the analysis of the 160 deg^2 survey [105].

The constraints on cosmological models from these results on cluster evolution are shown Fig. 23.27 (bottom) [21]. The figure shows results for various

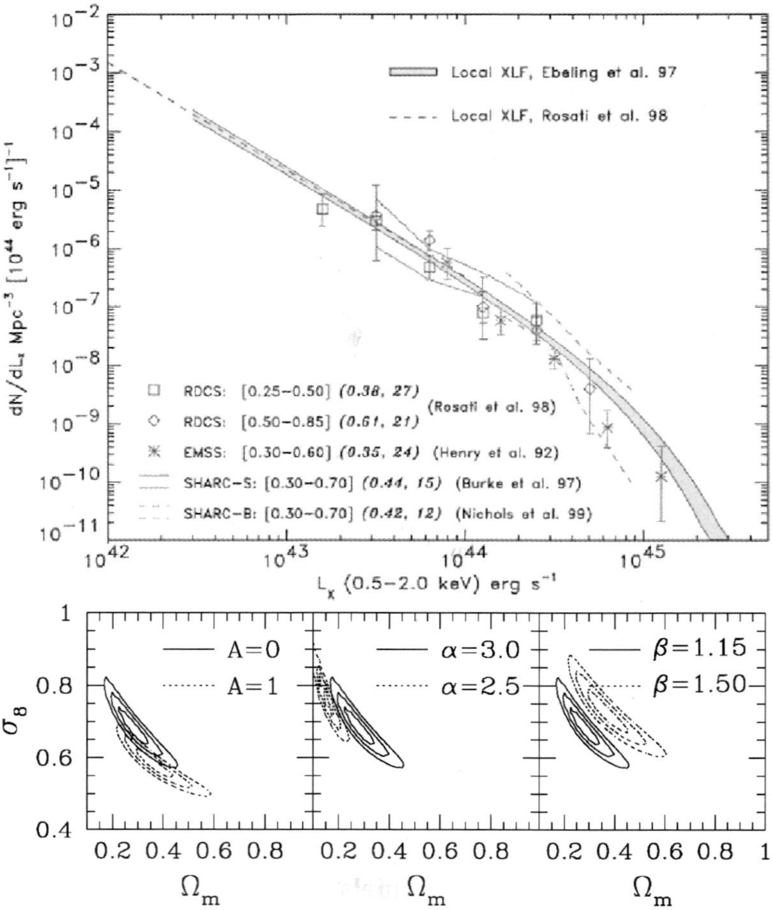

Fig. 23.27 *Top:* Evolution of the cluster X-ray luminosity function observed in the RDCS survey by Rosati et al. [126]. The abundance of the most X-ray luminous clusters is clearly lower at high redshift. *Bottom:* Constraints on the cosmological density parameter, Ω_m, and the amplitude of the matter density fluctuations on a scale of $8h^{-1}$ Mpc, σ_8, from the RDCS Survey for different assumptions on the mass-X-ray luminosity relation. The parameter β refers to the amplitude of the mass-temperature relation, while the other two parameters modify the temperature-luminosity relation, $L_X \propto T_X^{\alpha}\,(1+z)^A$. For details see Borgani et al. [21]

assumptions on how the cluster luminosity evolves for given cluster mass and temperature. They clearly favor a low density universe for the whole range of evolutionary parameters considered.

The high sensitivity of XMM-Newton is now providing a deeper look into the X-ray sky and widens the horizon to the study of even more distant galaxy clusters. So far the most distant X-ray cluster has recently been detected at a redshift of 1.39 [106]. Most surprisingly, optical and nearIR imaging shows a cluster with a

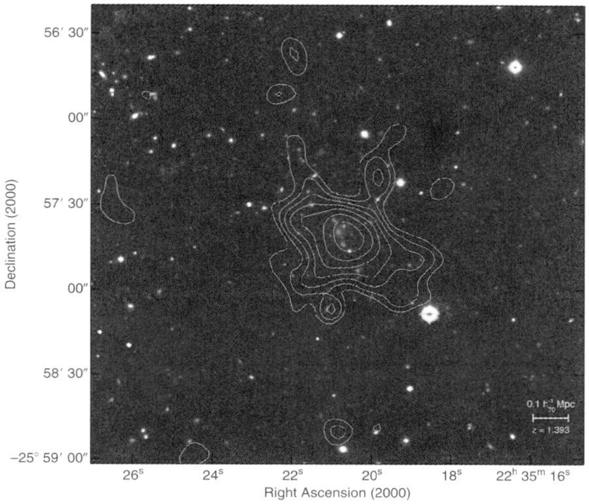

Fig. 23.28 *Left:* The most distant X-ray cluster found to date at a redshift of $z = 1.39$. The object was detected in a search for distant X-ray clusters in the fields of archival XMM-Newton observations [106]. The figure shows an optical/nearIR composite image with X-ray contours superposed. Most remarkably, even at this large look-back time of 9 Gyrs, the major cluster galaxies show very red colors, the signature of very old stellar populations

very old red population of galaxies and also the X-ray appearance of the cluster tells us that this system is already well evolved (Fig. 23.28). A large XMM-Newton survey program, the XMM-LSS, is also devoted to a distant cluster search [115].

23.10 Testing Cosmological Models

Since the characteristics and statistics of the galaxy cluster population can be predicted in the frame of cosmological and structure formation models with reasonable precision (at the state-of-the-art on the 10–20% level for clusters characterized by their mass) [80], the above results can be used for cosmological tests. The currently most critical aspect in this comparison is the uncertainty in the relation of the observable cluster parameters like X-ray luminosity and temperature and the inferred cluster mass [113]. Three different characteristics of the cluster populations provide independent constraints on cosmological models: (1) the cluster abundance at present, (2) the statistics of the spatial distribution of clusters, and (3) the evolution of the cluster abundance.

Taking only the results of (1) the most important constraints are on the matter density parameter, Ω_m, and on the amplitude of the matter density fluctuations, expressed usually in terms of the rms normalization of the fluctuations at a scale of $8h_{100}^{-1}$ Mpc, σ_8. These constraints involve a strong degeneracy of these two parameters, which can be overcome if the shape of the mass function or its observable counterpart, the X-ray luminosity function, is very precisely known. For the bright better calibrated part of the REFLEX X-ray luminosity function we obtain constraints in the range $\Omega_{\rm m} = 0.23$–0.36 and $\sigma_8 = 0.73$–0.84 [20] if we use the empirical mass-luminosity relation of Reiprich and Böhringer [123].

Including the information on the spatial distribution of the clusters greatly helps to break this degeneracy. This was exploited in a comprehensive approach to simultaneously compare the cluster abundance and spatial distribution of the REFLEX sample with cosmological predictions in a study by Schuecker et al. [131]. To fully exploit the information in the REFLEX survey the spatial distribution was not characterized by Fourier modes which are badly suited to the REFLEX survey geometry but by Karhunen-Loeve eigenmodes which by construction are ideally adapted to the survey characteristics. The constraints obtained for the parameters $\Omega_{\rm m} = 0.27 - 0.43$ and $\sigma_8 = 0.55$–0.83 (including systematic errors and uncertainties in various priors) are shown in Fig. 23.26 [131]. The parameter constraints are well consistent with the now preferred concordance cosmological model and they are also consistent with the findings from cluster evolution as shown in Fig. 23.27 [21, 113]. Similar, consistent results from cluster evolution studies have also been derived from the observations of the cluster temperature function at different redshifts [74].

To derive constraints on other very important cosmological parameters, in particular on the Λ-parameter (now often termed the dark energy density Parameter), Ω_Λ, we have to include the study of cluster evolution with higher precision. This information can in particular not be obtained from studies at a single epoch. Thus this remains one of the goals for future X-ray cluster studies [75, 91]. However, important information can be obtained on this parameter by combining the cosmological constraints from cluster studies with other observations. Figure 23.29 shows the constraints on the cosmological parameters $\Omega_{\rm m}$ and Ω_Λ for three different types of cosmological tests: (1) the study of the fluctuations in the cosmic microwave background with the WMAP satellite [139], (2) the measurements on the geometry of the Universe using supernovae type Ia as standard candles [124, 147], and the statistics of the large scale structure from X-ray cluster observations [131, 135]. The fact that all three constraints meet at the same small parameter region around the concordance cosmological model with parameters of $\Omega_{\rm m} \sim 0.3$ and $\Omega_\Lambda \sim 0.7$ is very comforting. Other measures of the large-scale structure, e.g., from the study of the large-scale galaxy distribution also provide similar constraints on the cosmological parameters as the clusters [111, 145]. The combination of several cosmological tests can also be included in one statistical parameter constraint approach. This was, for example, done for the combination of the REFLEX cluster survey and the supernova studies by Schuecker et al. [135]. The results of this test are shown in Fig. 23.29 (right), where the constraints on the parameter Ω_m and the equation of state parameter, w, of the Dark Energy are shown. For a value of $w = -1$ the Dark Energy model is equivalent to the classical Λ cosmological model. Figure 23.29 shows that the observations are fully consistent with the Λ model.

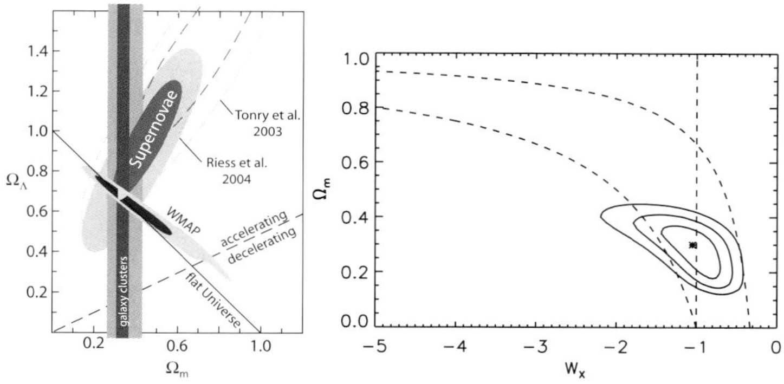

Fig. 23.29 *Left:* Constraints on the cosmological parameters Ω_m and Ω_Λ from three different cosmological tests: supernovae Ia as standard candles [124, 147], analysis of the fluctuation spectrum of the cosmic microwave background [139], and abundance and spatial clustering observations of X-ray clusters of galaxies [131, 132]. All the tests meet at the same parameter region encircling the concordance model. *Right:* Constraints on the cosmological density parameter, Ω_m, and the equation of state parameter of the dark energy, w_x, for the REFLEX cluster data combined with literature results from distant supernovae obtained by Schuecker et al. [135]. The likelihood contours give the 1,2, and 3 σ limits

Still another way of obtaining cosmological constraints from galaxy cluster studies is based on the measure of the gas mass fraction in clusters, which was described in Sect. 23.2.2., e.g. [2, 49, 121]. Using the baryon mass fraction in clusters and the results of the cosmic nucleosynthesis, as shown in Table 23.1, one can conclude on a value of Ω_m around 0.3. Taking the baryon mass fraction for massive, relaxed galaxy clusters as a universal parameter independent of redshift, as suggested by simulations, e.g. [48], one can use this quantity for standard candle cosmological tests. This is based on the fact that the mass measurement depends on the cluster distance in a linear way through the conversion of the apparent to physical size of the cluster (see (1)), while the determination of the gas mass depends on the distance to the 5/2 power. The cosmological constraints derived this way are in very good agreement with the above results [2, 49, 121].

23.11 Conclusion and Outlook

Our knowledge on the astrophysical and cosmological relevance of galaxy clusters has much increased in the last 20–30 yrs, mostly due to the spectacular results of X-ray astronomical observations. With the launch of the Chandra and XMM-Newton observatories the X-ray astrophysics of galaxy clusters has now entered a stage of precision previously only obtained in stellar or galaxy astronomy. This is due to the detailed X-ray spectroscopic capability of the observatories, Chandra's sub-arcsec angular resolution and the high photon collecting power of XMM-

Newton. Many of the results shown in this review nicely demonstrate the progress that has been made possible by these capabilities.

The tremendous impact made by the ROSAT All-Sky Survey, the so far only all-sky survey made with an imaging X-ray telescope, is spuring the dreams of a new, much improved X-ray survey effort. A less costly mission as compared to ROSAT with state-of-the art technology would provide an increased energy band, increased angular resolution, and a sensitivity increase of at least an order of magnitude. Already several proposals have been made for such a mission, but they have so far not been successful. It is interesting to note here that all the recent proposals for such a new X-ray survey had galaxy cluster cosmology as the main driver, underlining the importance that this field has currently gained. The last of these mission studies, for the planned "Dark Universe Observatory (DUO)," has made it through NASA phase A study, demonstrating the cosmological progress that could be made with such an X-ray mission [75, 91]. Thus we can be confident that such a survey will be carried out in some form in the near future.

The studies for the next generation of large X-ray observatories following the heritage of Chandra and XMM-Newton involve the NASA plans for *Constellation-X* and the ESA plans for *XEUS* [110]. The plans for the latter mission are particularly ambitious in terms of gaining a photon collecting power equivalent to an effective area of $\sim 10\,m^2$. With such a large observatory we would be able to perform structural studies and element abundance measurements of galaxy clusters at redshifts as large as $z = 2$–2.5, beyond which we should only find very poor clusters or X-ray luminous groups. Thus with some luck in the funding of such projects, the success story of X-ray astronomy with galaxy clusters will continue.

References

1. Allen, S.W., Ettori, S., Fabian, A.C., 2001, MNRAS, 324, 877
2. Allen, S.W., Schmidt, R.W., Bridle, S.L., 2003, MNRAS, 346, 593
3. Arnaud, M. & Evrard, A.E., 1999, MNRAS, 305, 631
4. Arnaud. M., Aganim, N., Neumann, D.M., 2002, A&A, 389, 1
5. Bacchi, M., Feretti, L., Giovannini, G., Govoni, F., 2003, A&A, 400, 465
6. Binggeli, B., Tammann, G.A., Sandage, A., 1987, AJ, 94, 251
7. Böhringer, H., Voges, W., Fabian, A.C., Edge, A.C., Neumann, D.M., 1993, MNRAS, 264, L25
8. Böhringer, H., Briel, U.G., Schwarz, R.A., Voges, W., Hartner, G., Trümper, J., 1994, Nature, 368, 828
9. Böhringer, H., Neumann, D.M., Schindler, S., Huchra, J.P., 1997, ApJ, 485, 439
10. Böhringer, H., Tanaka, Y., Mushotzky, R.F., Ikebe, Y., Hattori, M., 1998, A&A, 334, 789
11. Böhringer, H., Voges, W., Huchra, J.P., et al., 2000, ApJS, 129, 435
12. Böhringer, H., Belsole, E., Kennea, J., et al. 2001, A&A, 365, L181
13. Böhringer, H., Schuecker, P., Guzzo, L., et al., 2001, A& A, 369, 826
14. Böhringer, H., Collins, C.A., Guzzo, L., Schuecker, P., et al., 2002, ApJ, 566, 93
15. Böhringer, H., Matsushita, K., Churazov, E., Ikebe, Y., Chen, Y., 2002, A&A, 382, 804
16. Böhringer, H., Schuecker, P., Guzzo, L., Collins, C.A., Voges, W., et al. 2004, A&A, 425, 367

17. Böhringer, H., Matsushita, K., Churazov, E., Finoguenov, A., Ikebe, Y., 2004, A&A, 416, L21
18. Böhringer, H., Matsushita, K., Finoguenov, A., Xue, Y.-J. Churazov, E., 2005, Adv. Space Res., 36, 677
19. Böhringer, H., Schuecker, P., Pratt, G., et al., 2007, A&A 469, 363
20. Böhringer, H., et al., 2007, (in preparation)
21. Borgani, S., Rosati, P., Tozzi, P., Stanford, S.A., Eisenhardt, P.R., Lidman, C., Holden, B., Della Ceca, R., Norman, C., Squires, G., 2001, ApJ, 561, 13
22. Borgani, S., Murante, G., Springel, V., Diaferio, A., et al., 2004, MNRAS, 348, 1078
23. Bower, R.G., 1997, MNRAS, 288, 355
24. Briel, U.G., Henry, J.P., Böhringer, H., 1992, A&A, 259, L31
25. Briel, U.G., Finoguenov, A., Henry, J.P., 2004, A&A, 426, 1
26. Brüggen, M., 2003, ApJ, 592, 839; 593, 700
27. Buote, D.A., Lewis, A.D., Brighenti, F., Mathews, W.G., 2003, ApJ, 595, 151
28. Buote, D.A. & Lewis, A.D., 2004, ApJ, 604, 116
29. Burles, S., Nollett, K.M., Turner, M.S., 2001, ApJ, 552, L1
30. Byram, E.T., Chubb, T.A., Friedman, H., 1966, AJ, 71, 379
31. Cappellaro, E., Evans, R., Turatto, M., 1999, A&A, 351, 459
32. Carlberg, R.G., Yee, H.K.C., Ellingson, E., et al., 1997, ApJ, 485, L13
33. Cayatte, V., Cayatte, V., Kotanyi, C., Balkowski, C., van Gorkom, J. H., 1994, AJ, 107, 1003
34. Churazov, E., Forman, W., Jones, C., Böhringer, H., 2000, A&A, 356, 788
35. Churazov, E., Sunyaev, R., Forman, W., Böhringer, H., 2002, MNRAS, 332, 729
36. Ciotti, L., Pellegrini, S., Renzini, A., D'Ercole, A., 1991, ApJ, 376, 380
37. Clementini, G., Gratton, R.G., Carretta, E., Sneden, C., 1999, MNRAS, 302, 22
38. Collins, C.A., Guzzo, L., Böhringer, H., Schuecker, P., Voges, W., et al., 2000, MNRAS, 319, 939
39. Czoske, O., Moore, B., Kneib, J.-P., Soucail, G., 2002, A&A, 386, 31
40. De Grandi, S. & Molendi, S., 2001, ApJ, 551, 153
41. De Grandi, S., Ettori, S., Longhetti, M., Molendi, S., 2004, A&A, 419, 7
42. Dupke, R.A., White, R.E. III, 2000, ApJ, 537, 123
43. Ebeling, H., Edge, A.C., , Böhringer, H., Allen, S.W., et al., 1998, MNRAS, 301, 881
44. Ebeling, H., Edge, A.C., Henry, J.P., 2001, ApJ, 553, 668
45. Edge, A.C., 2001, MNRAS, 328, 762
46. Edge, A.C.: X-ray Surveys of Low-redshift Clusters in: *Clusters of Galaxies: Probes of Cosmological Structure and Galaxy Evolution*, ed. J.S. Mulchaey, A. Dressler, and A. Oemler, Carnegie Observatories Astrophys. Ser., 2004, p. 58
47. Edvardsson, E., Andersen, J., Gustafsson, B., et al., 1993, A&A, 275, 101
48. Eke, V.R., Navarro, J.F., Frenk, C.S., 1998, ApJ, 503, 569
49. Ettori, S., Tozzi, P., Rosati, P., 2003, A&A, 398, 879
50. Evrard, A.E., 1997, MNRAS, 292, 289
51. Fabian, A.C. & Nulsen, P.E.J., 1977, MNRAS, 180, 479
52. Fabian, A.C., Nulsen, P.E.J., Canizares, C.R., 1984, Nature, 310, 733
53. Fabian, A.C., 1994, ARA&A, 32, 277
54. Fabian, A.C., Mushotzky, R.F., Nulsen, P.E.J., Peterson, J.R., 2001, MNRAS, 321, 20
55. Fabian, A.C., Sanders, J.S., Allen, S.W., Crawford, C.S., Iwasawa, K., Johnstone, R.M., Schmidt, R.W., Taylor, G.B., 2003, MNRAS, 344, L43
56. Feretti, L., Gioia, I.M., Giovannini, G.: *Merging Processes in Galaxy Clusters*, (Kluwer Acad. Publ., Dordrecht Boston London 2002)
57. Feretti, L., Brunetti, G., Giovannini, G., Kassim, N., Orrú, E., Setti, G., 2004, JKAS, 37, 315
58. Ferrarese, L., et al., 1996, ApJ, 464, 568
59. Finoguenov, A., David, L.P., Ponman, T.J., 2000, ApJ, 544, 188
60. Finoguenov, A., Matsushita, K., Böhringer, H., Ikebe, Y., Arnaud, M., 2002, A&A, 381, 21
61. Finoguenov, A., Pietsch, W., Aschenbach, B., Miniati, F., 2004, A&A, 415, 415
62. Finoguenov, A., Böhringer, H., Y.-Y. Zhang, Y.-Y., 2005, A&A, 442, 827

63. Forman, W., Jones, C., 1982, ARA&A, 20, 547
64. Forman, W., Nulsen, P.E.J., Heinz, S., Owen, F., 2005, ApJ, 635, 894
65. Fritz, G., Davidsen, A., Meekins, J.F., Friedman, H., 1971, ApJ, 164, L81
66. Fukazawa, Y., Makishima, K., Tamura, T., Ezawa, H., Xu, H., Ikebe, Y., Kikuchi, K., Ohashi, T., 1998, PASJ, 50, 187
67. Giovannini, G. & Feretti, L.: Diffuse Radio Sources and Cluster Mergers in: *Merging Processes in Galaxy Clusters*, ed. by Feretti, L., Gioia, I.M., Giovannini, G., (Kluwer Acad. Publ., Dordrecht Boston London 2002), pp 197 – 227
68. Gioia, I.M., Henry, J.P., Mullis, C.R., Böhringer, H., Briel, U.G., Voges, W., Huchra, J.P., 2004, ApJS, 149, 29
69. Girarid, M., Fadda, D., Escalera, E., Giuricin, G., Mardirossian, F., Mezzetti, M., 1997, ApJ, 490, 56
70. Grevesse, N. & Sauval, A.J., 1998, Space Sci. Rev., 85, 161
71. Gursky, H., Kellogg, E., Murray, S., Leong, C., et al., 1971, ApJ, 167, L81
72. Heckman, T.M., Baum, S.A., van Breugel, W.J.M., McCarthy, P., 1989, ApJ, 338, 48
73. Henry, J.P., Finoguenov, A., Briel, U.G., 2004, ApJ, 615, 181
74. Henry, J.P., 2004, ApJ, 609, 603
75. Holder, G., Haiman, Z. Mohr, J.J., 2001, ApJ, 560, L111
76. Ikebe, Y., Makishima, K., Fukazawa, Y., Tamura, T., Xu, H., Ohashi, T., Matsushita, K., 1996, ApJ, 525, 581
77. Inogamov, N.A. & Sunyaev, R., 2003, AstL, 29, 791
78. Ivanov, V., Mamuy, M., Pinto, P.A., 2000, ApJ, 542, 598
79. Iwamoto, K., Brachwitz, F., Nomoto, K., et al., 1999, ApJ SS, 125, 439
80. Jenkins, A., Frenk, C.S., White, S.D.M., et al., 2001, MNRAS, 321, 372
81. Jones, C., & Forman, W., 1984, ApJ, 276, 38
82. Kaastra, J.S., Tamura, T., Peterson, J.R., Bleeker, J.A.M., et al. , 2004, A&A, 413, 415
83. Kaiser, N., 1986, MNRAS, 222, 323
84. Kay, S.T., Thomas, P.A., Jenkins, A., Pearce, F.R., 2004, MNRAS, 355, 1091
85. Kellogg, E., Gursky, H., Tananbaum, H., Giacconi, R., Pounds, K., 1972, ApJ, 174, L53
86. King, I.R., 1966, AJ, 71, 64: AJ, 71, 276
87. Kneib, J.-P., Hudelot, P., Ellis, R.S., Treu, T., et al., 2003, ApJ, 598, 804
88. Kolmogorov, A.N., 1941, Dokl. Akad. Nauk. SSSR, 30, 301
89. Nulsen, P.E.J., 1986, MNRAS, 221, 377
90. Nulsen, P.E.J., McNamara, B.R., Wise, M., David, L.P., 2005, ApJ, 628, 629
91. Majumdar, S. & Mohr, J.J., 2004, ApJ, 613, 41
92. Markevitch, M., Gonzalez, A. H., David, L., et al., 2002, ApJ, 567, L27
93. Matsushita, K., Makishima, K., Rokutanda, E., Yamasaki, N.Y., Ohashi, T., 1997, ApJ, 488, L125
94. Matsushita, K., Belsole, E., Finoguenov, A., Böhringer, H., 2002, A&A, 386, 77
95. Matsushita, K., Finoguenov, A., Böhringer, H., 2003, A&A, 401, 443
96. Matsushita, K., Böhringer, H., Takahashi, I., Ikebe, Y., 2004, A&A, in press
97. Mazzotta, P., Rasia, E., Moscardini, L., Tormen, G., 2004, MNRAS, 354, 10
98. McNamara, B., 1997, in: *Galactic and Cluster Cooling Flows* ed. by Soker, N., (Publ. Astron. Soc. Pac., San Francisco 1997), p. 109
99. McNamara, B., Wise, M., Nulsen, P.E.J., et al., 2000, ApJ, 534, L135
100. Mellier, Y., 1999, ARA&A, 37, 127
101. Mitchell, R.J., Culhane, J.L., Davison, P.J., Ives, J.C., 1976, MNRAS, 176, L29
102. Mo, H.J. & White, S.D.M., 1986, MNRAS, 282, 347
103. Moore, B., Quinn, T., Governato, F., Stadel, J., Lake, G., 1999, MNRAS, 310, 1147
104. Mulcheay, J.S., 2000, ARA&A, 38, 289
105. Mullis, C.R., Vikhlinin, A., Henry, J.P., Forman, W., Gioia, I.M., Hornstrup, A., Jones, C., McNamara, B.R., Quintana, H., 2004, ApJ, 607, 175
106. Mullis, C.R., Rosati, P., Lamer, G., Böhringer, H., Schwope, A., Schuecker, P., Fassbender, R., 2005, ApJ, 623, L85

107. Mushotzky, R.: X-Ray Observations of Clusters of Galaxies, in: *Hot Thin Plasmas in Astrophysics*, ed. R. Pallavicini NATO Advanced Science Institutes (ASI) Series C, 1988, Volume 249, p.273
108. Navarro, J.F., Frenk, C.S., White, S.D.M., 1997, ApJ, 490, 493
109. Nomoto, K., Thielemann, F.-K., Yokoi, K., 1984, ApJ, 286, 644
110. Parmar, A.N., Hasinger, G., Arnaud, M., et al., 2004, Adv. Space Res., 34, 2623
111. Peacock, J.A., Cole, S., Norberg, P., Baugh, C.M., et al. 2001, Nature, 410, 169
112. Peterson, J.R., Paerels, F.B.S., Kasstra, J.S., et al. 2001, A&A, 365, L104
113. Pierpaoli, E., Borgani, S., Scott, D., White, M., 2003, MNRAS, 342, 163
114. Pierre, M., Le Borgne, J.F., Soucail, G., Kneib, J.P., 1996, A&A, 311, 413
115. Pierre, M., et al., 2004, J. Cosmol. Astropart. Phys. 2004, Sept. No. 11, 594
116. Pointecouteau, E., Arnaud, M., Pratt, G.W., 2005, A&A, 435, 1
117. Ponman, T.J., Cannon,D.B., Navarro, J.F., 1999, Nature, 397, 135
118. Ponman, T.J., Sanderson, A.J.R., Finoguenov, A., 2003, MNRAS, 343, 331
119. Pratt, G. & Arnaud, M., 2002, A&A, 394, 375
120. Pratt, G.W., Böhringer, H., Finoguenov, A., A&A, 2005, 433, 777
121. Rapetti, D., Allen, S.W., Weller, J., 2005, MNRAS, 360, 555
122. Reiprich, T.H., 2001, Ph.D. thesis, Ludwig-Maximilians-Universität, München
123. Reiprich, T.H. & Böhringer, H., 2002, ApJ, 567, 716
124. Riess, A.G., Strolger, L.-G., Tonry, J., et al., 2004, ApJ, 607, 665
125. Rosati, P., della Ceca, R., Norman, C., Giacconi, R., 1998, ApJ, 492, L21
126. Rosati, P., Borgani, S., Colin, N., 2002, ARA&A, 40, 539
127. Sarazin, C.L., 1986, Rev. Mod. Phys., 58, 1
128. Schindler, S., Binggeli, B., Böhringer, H., 1999, A&A, 343, 420
129. Schuecker, P., Böhringer, H., Guzzo, L., Collins, C.A., Neumann, D.M., et al., 2001, A&A, 368, 86
130. Schuecker, P., Böhringer, H., Reiprich, T.H., Feretti, L., 2001 A&A, 378, 408
131. Schuecker, P., Böhringer, H., Collins, C.A., Guzzo, L., 2003, A&A, 398, 867
132. Schuecker, P., Guzzo, L., Collins, C.A., Böhringer, H., 2002, MNRAS, 335, 807
133. Schuecker, P. & Böhringer, H.: Construction of X-ray cluster surveys and their spatial analyses, in Clusters of galaxies and the high redshift universe observed in X-rays eds. D.M. Neumann & J.T.T. Van., Moriond Astrophysics Meeting, Savoie, France, March 2001, electronic publication at http://www-dapnia.cea.fr/Conferences/Morion_astro_2001/index.html
134. Schuecker, P., Finoguenov, A., Miniati, F., Böhringer, H., Briel, U. G., 2004, A&A, 426, 209
135. Schuecker, P., Caldwell, R.R., Böhringer, H., Collins, C.A., et al., 2003, A&A, 402, 53
136. Schuecker, P., 2005, Rev. Mod. Ast., 18, 76, (astro-ph/0502234)
137. Shibata, R., Matsushita, K., Yamasaki, N.Y., Ohashi, T., Ishida, M., Kikuchi, K., Böhringer, H., Matsumoto, H., 2001, ApJ, 549, 228
138. Silk, J., 1976, ApJ, 208, 646
139. Spergel, D.N., Verde, L., Peiris, H.V., Komatsu, E., Nolta, M.R., 2003, ApJ, 148, 175
140. Springel, V., Di Matteo, T., Hernquist, L., 2005, ApJ, 620, L79
141. Squires, G., Kaiser, N., Fahlman, G., Babul, A., Woods, D., 1996, ApJ, 469, 73
142. Sun, M., Forman, W., Vikhlinin, A., Hornstrup, A., Jones, C., Murray, S.S., 2003, ApJ, 598, 250
143. Sunyaev, R.A., Norman, M.L., Bryan, G.L., 2003, AstL, 29, 783
144. Tamura, T., Kaastra, J.S., den Herder, J.W.A., et al., 2004, A&A, 420, 135
145. Tegmark, M., Blanton, M.R., Strauss, M.A., Hoyle, F., et al. 2004, 606, 702
146. Thomas, P.A., Fabian, A.C., Nulsen, P.E.J., 1987, MNRAS, 228, 973
147. Tonry, J.L., Schmidt, B.P., Barris, B., et al., 2003, ApJ, 594, 1
148. Tozzi, P. & Norman, C., 2001, ApJ, 546, 63
149. Trümper, J., 1993, Science, 260, 1769
150. Vikhlinin, A., McNamara, B.R., Forman, W., Jones, C., Quintana, H., Hornstrup, A., 1998, ApJ, 502, 558
151. Vikhlinin, A., Forman, W., Jones, C., 1999, ApJ, 525, 47

152. Vikhlinin, A., Markevitch, M., Murray, S.S., 2001, ApJ, 551, 160
153. Vikhlinin, A., Markevitch, M., Murray, S.S., Jones, C., Forman, W., Van Speybroeck, L., 2005, ApJ, 628, 655
154. Voges, W., Aschenbach, B., Boller, T., et al., 1999, A&A, 349, 389
155. Vogt, C., Enßlin, T.A., 2003, A&A, 412, 373
156. Voit, M., 2005, Rev. Mod. Phys., 77, 207
157. White, S.D.M., Navarro, J.F., Evrard, A.E., Frenk, C.S., 1993, Nature, 366, 429
158. White, S.D.M., Efstathiou, G., Frenk, C.S., 1993, MNRAS, 262, 1023
159. White, S.D.M., Briel, U.G., Henry, J.P., 1993, MNRAS, 261, L8
160. Wu, X.-P., 2000, MNRAS, 316, 299
161. Xue, Y.-J., Böhringer, H., Matsushita, K, 2004, A&A, 420, 833
162. Zhang, Y.-Y., Finoguenov, A., Böhringer, H., Ikebe, Y., Matsushita, K., Schuecker, P., 2004, A&A, 413, 49
163. Zhang, Y.-Y., Böhringer, H., Mellier, Y., Soucail, G., Forman, W., 2005, A&A, 429, 85
164. Zwicky, F., 1933, Hev. Phys. Acta, 6, 110

24 Gamma-Ray Bursts

J. Greiner

24.1 The First 30 Years

24.1.1 Discovery and BATSE Era

Gamma-ray bursts (GRBs) are short bursts of high-energy radiation from an unpredictable location in the sky. The γ-ray emission can rise to maximum intensity within fractions of a millisecond. During these short times GRBs are the brightest objects in the γ-ray sky. The energy spectra are nonthermal, with most of the power radiated in the 100–500 keV range, but photons up to 18 GeV or down to a few kiloelectronvolt have also been registered. The bursts have durations of typically 0.1–100 s, with a bimodal distribution separating "short" and "long" duration GRBs at $\sim$1–2 s [57] with a relative occurrence of 1:4 [49]. In addition, short bursts are typically harder than the long bursts, supporting the conjecture that they form two classes of object (Fig. 24.1).

GRBs were first detected in 1967 with small γ-ray detectors onboard the Vela satellites [46], which were designed to verify the nuclear test ban treaty between the USA and Russia. For many years the prevailing opinion was that magnetic neutron stars (NS) in the Galactic disk were the sources of GRBs [78]. No flaring emission outside the γ-ray region could be detected, and no indisputable quiescent counterpart to a GRB could be established. Despite a distance "uncertainty" of 10 orders of magnitude, numerous theories (see [64] for a compilation) were advanced to explain the source of energy in GRBs. The measurements since 1991 of the burst and transient source experiment (BATSE) onboard the Compton gamma-ray observatory have shown unequivocally that GRBs are isotropically distributed on the sky (Fig. 24.2) even at the faintest intensities, and that there is a distinct lack of faint bursts as compared to a homogeneous distribution in Euclidian space [58]. This first indicated a cosmological origin, though also a distribution in the halo of our Galaxy was considered. The γ-ray spectra are well described by broken power laws, with mean slopes of $\alpha = -1$ at low energies (10–300 keV) and of $\beta = -2.4$ at high energies [70]. Consequently, the bulk of the emission is radiated at energies around the spectral turnover $E_{\rm break}$, more specifically at the peak energy $E_{\rm peak} = E_{\rm break}(2+\alpha)$. An unprecedented wealth of additional information on each burst as well as a number of population properties could be collected, yet the GRB origin remained a mystery.

J.E. Trümper and G. Hasinger (eds.), *The Universe in X-Rays*.

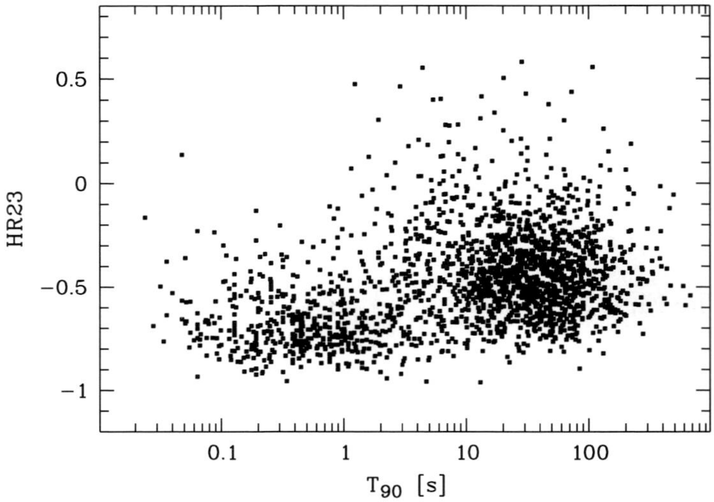

Fig. 24.1 Hardness of the prompt γ-ray emission as a function of duration T_{90} for all GRBs of the 4th BATSE catalog, showing the two classes of short/hard and long/soft GRBs as first demonstrated in 1993 [49]. T_{90} is the time interval over which a burst emits from 5% to 95% of its total measured flux, and the hardness ratio HR23 is defined as $(2-3)/(2+3)$, where "2" stands for the flux in the 50–100 keV band, and "3" for that in the 100–300 keV band, respectively. Spectrally soft GRBs have a large HR23

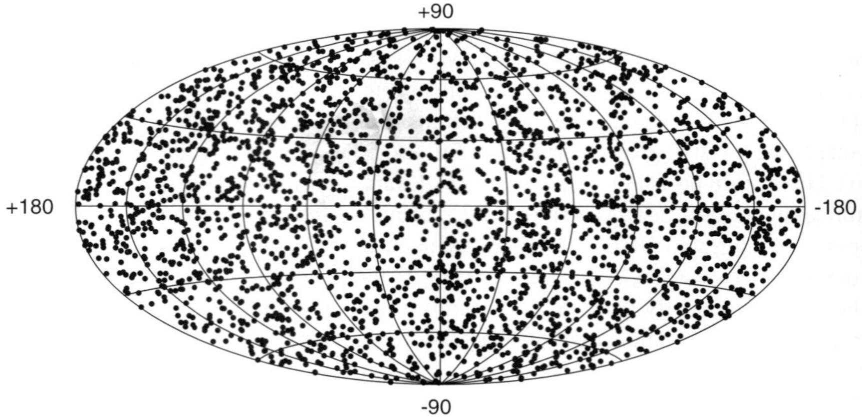

Fig. 24.2 Sky distribution of 2 704 GRBs as measured by BATSE during the full gamma-ray observatory mission lifetime from 1991 to 2000 (derived from http://f64.nsstc.nasa.gov/batse/grb/skymap.)

24.1.2 The Afterglow Era

Over the first two decades, GRB coordinates came with two mutually exclusive properties: *accurate localization*, e.g., arcmin accuracy, as provided by the Interplanetary Network [42] or *rapid localization*, e.g., short delay between GRB event and distribution of coordinates, as provided by the BATSE Coordinate Distribution Network system [6]. The launch of the Italian-Dutch "Satellite per Astronomia X," SAX (nick-named BeppoSAX after the Italian high-energy astronomy pioneer Giuseppe Occhialini), in 1996, combined a better localization accuracy of the γ-ray burst itself (of order 5 arcmin) by coded-mask imaging in the 2–35 keV range, and rapid notification within minutes of the GRB. This and the fast slewing capability of the BeppoSAX satellite with its co-aligned X-ray telescopes to the GRB position, on the timescale of few hours, allowed to discover long-lasting, fading afterglow emission of GRBs, marking a breakthrough in our understanding of GRBs. The discovery of afterglows in 1997 (see Fig. 24.3 for the first GRB afterglow [18, 94]) demonstrated their high redshifts (median of $z = 1.1$ as of end of 2005, see Fig. 24.4 for the present redshift distribution) unambiguously and proved GRBs to be the most luminous objects known. The spectral energy distribution, from radio to X-rays, suggested that the afterglow emission is predominantly synchrotron radiation.

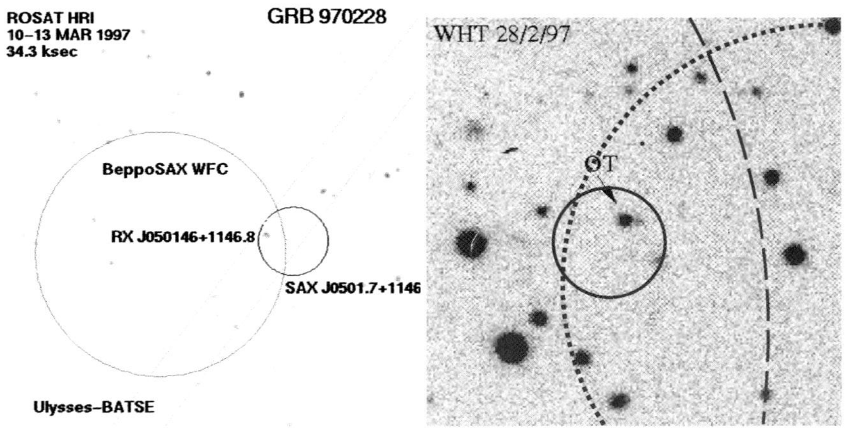

Fig. 24.3 Sequence of error circles from γ-rays to optical for GRB 970228, the first GRB for which long-wavelength afterglow emission has been identified. *Left*: The underlying image is from a 34 ks ROSAT high-resolution imager observation [24], with the large circle showing the 3σ error circle of the X-ray afterglow as determined with the BeppoSAX wide-field camera (WFC), the smaller circle being the $\approx$1 arcmin error circle of the fading source SAX J0501.7+1146 as found with the two Narrow-Field instrument (NFI) pointings, and the two straight lines marking the triangulation circle derived from the BeppoSAX and Ulysses timings [43]. *Right*: Optical image taken on Feb 28, 1997 [94] at the William Herschel Telescope (Canary Islands) with the WFC error circle marked as dashed segment, the NFI error circle with the dotted segment, and the 10$''$ ROSAT/HRI error box as full circle. The optical transient (OT) falls right into the ROSAT error box

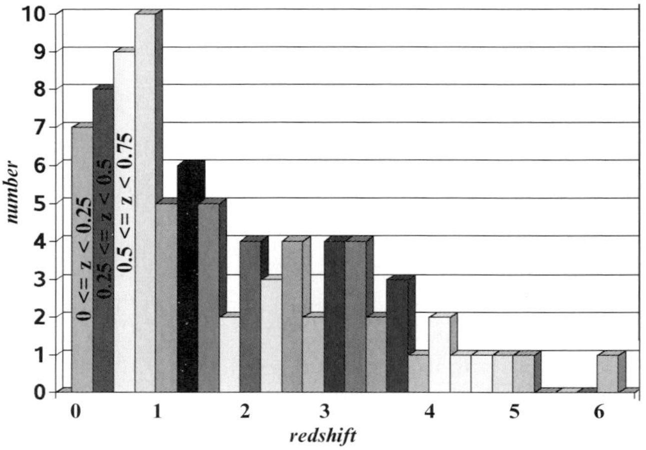

Fig. 24.4 Redshift distribution of 86 GRBs as of July 2006. (Courtesy S. Klose, priv. comm.; see www.mpe.mpg.de/~jcg/grbrsh.html for an up-to-date version) Because of, among other effects, the higher sensitivity of Swift's GRB detector, the mean redshift of Swift GRBs is above 2, higher than in the pre-Swift era

The detection of the first optical afterglow(s) sparked an international observing effort, which is unique, except perhaps for SN 1987A. All major ground-based telescopes were used at optical, infrared as well as radio wavelengths, and basically every space-born observatory since then has observed GRBs. The HETE-2 satellite [75], launched in October 2000, continued to provide rapid and arcmin sized GRB localizations at a rate of about 2 per month after BeppoSAX had been switched off in April 2003. Over the last 8 years (Feb 1997–Feb 2005) a total of 261 GRBs have been localized within a day to less than one square-degree size, and X-ray afterglows have been detected basically for each of those bursts for which X-ray observations have been done within a few days (total of 58; see http://www.mpe.mpg.de/~jcg/grb.html). Note that because of GRB detector properties only long-duration GRBs have been measured, and so no afterglow of a short GRB was known until the Swift era (see Sect. 24.5).

Interestingly, for only ~50% of the GRB (and X-ray afterglows) an optical afterglow was found, and radio afterglows for ~30%. While the low detection rate at radio wavelength can be easily explained by insufficient sensitivity, the case is more open for the optical afterglows.

Originally, those GRBs with X-ray but without optical afterglows have been coined "dark GRBs." This darkness in the optical can be due to several reasons [26]: the afterglow could (i) have an intrinsically low luminosity, e.g., due to a low-density environment or low explosion energy, (ii) be strongly absorbed by intervening material, either very local around the GRB, or along the line-of-sight through the host galaxy, or (iii) be at high redshift ($z > 6$) so that Lyα blanketing and absorption by intervening Lyman-limit systems would prohibit detection in the usually used R

band. An analysis of a subsample of GRBs, namely those with particularly accurate positions provided with the Soft X-ray Camera on HETE-2, shows that optical afterglows were found for 10 out of 11 GRBs [96]. This suggests that the majority of dark GRBs are neither at high redshift nor strongly absorbed, but just faint, i.e., the spread in afterglow brightness at a given time after the GRB is much larger than previous observations indicated.

It is probably fair to say that the progress in the GRB field since 1997 has mostly occurred in understanding the afterglow emission and the GRB surroundings. Observational X-ray astronomy has played a vital role in this progress as it allowed the identification by drastically improving the position accuracy as well as allowing to place severe constraints from measurements of the X-ray spectra and X-ray flux variability.

In addition to the nearly-default repointings of BeppoSAX for 0.5–10 keV follow-up observations with the narrow-field instruments, X-ray afterglow observations have also been done with ROSAT (8 GRBs, 5 days fastest response) [36], ASCA (7 GRBs, 1.2 days), XMM (11 GRBs until Sep. 2004 and counting, 4 h), Chandra (20 GRBs until Sep. 2004 and counting, 16 h) [30].

24.2 Major Observational Findings

24.2.1 Jets

GRBs are most likely not spherical explosions, but collimated relativistic flows (jets). Initially, the jet is ultra-relativistic and the cone ($1/\Gamma$) of emission due to relativistic beaming with Lorentz factor Γ is smaller than the geometrical opening angle (or collimation angle) of the outflow. The observer in the beam will therefore receive light only from within the relativistic cone, and the dynamical evolution is similar to the case of spherical explosion. As the jet slows down, the relativistic cone will eventually become wider than the collimation angle, and the observer will measure a reduced intensity. This "jet break" is quasi-achromatic and has been seen in many GRB afterglow light curves. For several years it has been considered the telltale of collimated outflows in GRBs.

One direct consequence of synchrotron emission is that the emission from an individual particle is polarized. Because of the probably random nature of the post-shock magnetic fields, the polarization is likely to be largely averaged out, and only a small degree of polarization left. The time at which linear polarization is expected to be detectable is thought to be around the jet-break time. Several (differing) models have been proposed, in which a collimated jet and an off-axis line-of-sight conspire to produce an asymmetry, which leads to net polarization including one or several 90° changes of the polarization angle [32,83]. This behavior could provide independent evidence for the jet structure of the relativistic outflow.

The observed polarization at optical wavelengths is less than 3% [39, 76, 100] with the exception of one report on 10% polarization [10]. Because of this low-level polarization and the rapid decline of the afterglow brightness during the first day, it has been difficult to observe changes in the polarization as predicted by theory. The by far most extensive polarization light curve with fast variability in polarization degree and angle has been obtained for the afterglow of GRB 030329 [37]. This variability pattern does not follow any of the model predictions and is also not correlated with brightness. The global behavior is consistent with the interpretation that the GRB is produced in a relativistic jet with an initial opening angle of 3°. However, in this GRB afterglow several rebrightenings relative to a power law decline have been found which likely have caused deviations from a simple single-jet model, thus making it difficult to interpret. The low level of polarization implies that the components of the magnetic field parallel and perpendicular to the shock do not differ by more than ∼10%, and suggests an entangled magnetic field, probably amplified by turbulence behind shocks, rather than a pre-existing field.

The interpretation of light curve breaks due to jets can be taken one step further. Using a sample of GRBs with known estimates of the break time (and thus jet opening angle) and redshift, a surprising clustering of the collimation-corrected energy was found around 1.3×10^{51} erg [13, 22], suggesting that GRBs are produced by a universal energy reservoir despite their large range in isotropic equivalent energies.

Considering also the additional information from the γ-ray spectra of GRBs, a correlation between the isotropic equivalent energy and the peak energy ($E_{\rm peak}$ from Sect. 24.1.1) was found [1]. Correcting for the collimation by a factor of $(1 - \cos\theta)$

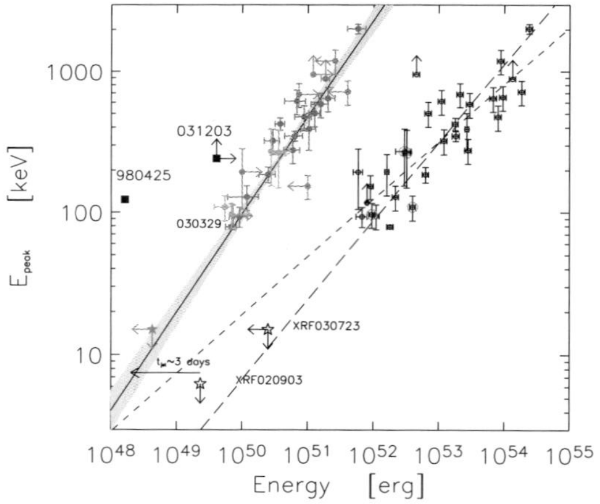

Fig. 24.5 Rest frame peak energy $E_{\rm peak}^{\rm obs}(1+z)$ vs. bolometric energy for GRBs with known redshift. The right (*black*) part is the relation as reported by [1], while the left (*color*) part is after correcting for the collimation angle (factor $1 - \cos\theta$). The solid line is the best fit with $E_{\rm peak} \sim 480 \times (E_\gamma/10^{51}\,{\rm erg})^{0.7}$ keV (from [31])

tightens this correlation even more (Fig. 24.5) [31], but also invalidates the standard candle energy of 1.3×10^{51} erg [13, 22]. While the underlying physical basis of this correlation is mysterious, its small dispersion has been considered as an indication of the robustness of the afterglow theory on which the estimate of the jet opening angle is based [31].

One problem in the above search for correlations is the assignment of a break in the light curve to the jet break. With denser and earlier sampling becoming increasingly available, GRB afterglow light curves show several breaks at different times. Also, breaks in the radio light curve typically occur substantially later than those in the optical. One prominent example is GRB 030329, where the optical break occurs 12 h after the GRB, but the radio break only 10 days later. The association of the jet break to the optical break would shift GRB 030329 (explicitly labeled in Fig. 24.5) by a factor of 100 away from the best-fit.

Another uncertainty arises from the unknown structure of the jet (Fig. 24.6) for which two alternatives have been proposed [71]: a variable jet opening angle (or uniform) model in which the emissivity is a constant independent of the angle relative to the jet axis, and a universal jet model in which the emissivity is a power-law function of the angle relative to the jet axis. Both models can explain the observed properties of GRBs reasonably well. However, if one tries to account for the properties of X-ray flashes (XRF), X-ray rich GRBs, and GRBs in a unified picture, the extra degree of freedom available in the variable jet opening angle model enables it to explain the observations reasonably well while the power-law universal jet model cannot.

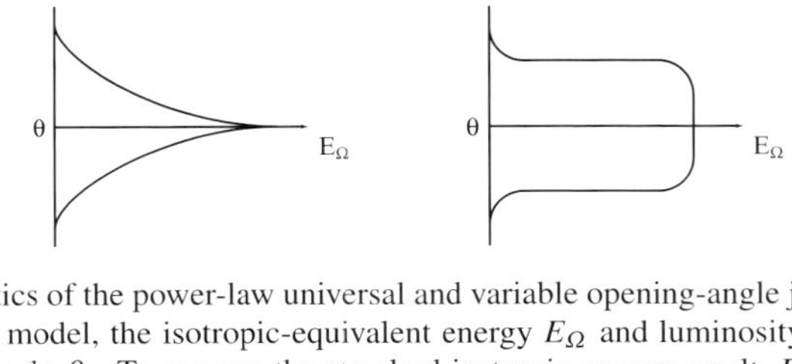

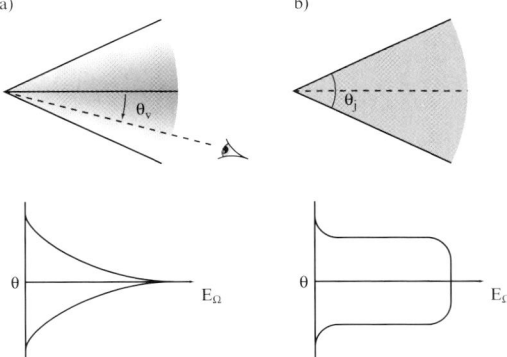

Fig. 24.6 Schematics of the power-law universal and variable opening-angle jet models of GRBs. In the universal jet model, the isotropic-equivalent energy E_Ω and luminosity decreases with increasing viewing angle θ_v. To recover the standard isotropic energy result, $E_\Omega(\theta_v) \sim \theta_v^{-2}$ is required. In the variable jet opening-angle model, the isotropic-equivalent energy and luminosity are constant across the jet (from [51])

24.2.2 Supernova Features

There is now general consensus that the long/soft GRBs are intimately connected to the death of massive stars. Massive stars undergo core-collapse, and about 70% of core-collapse supernovae are those of type II. One of the peculiar sub-classes that form part of the other 30% are type Ib/c supernovae.

While the supernova-GRB connection has been proposed some years ago [45], the unambiguous spectroscopic identification of the lowest redshift GRBs as supernovae during the last 2 yrs has provided convincing evidence for this picture [40, 87] (Fig. 24.7). The supernovae in the four spectroscopically confirmed gamma-ray bursts GRB 980425/SN 1998bw, 030329/2003dh, 031203/2003lw, and 060218/2006aj are all of type Ib/c, with unusually large kinetic energy (evidenced by very large expansion velocities of order 10–30 thousand kilometers per second after 10 days) and ejected mass of radioactive ^{56}Ni (subsequently called hypernovae [68]). In particular this last property suggests progenitors with masses >20–30 $M_{\odot}$ [65]. Theoretically, supernovae Ib/c are favored over type II because Ib/c have typically smaller envelope masses, and are thus thought to allow easier break-out of the jet, and subsequent GRB production. In fact, the lack of hydrogen lines in the spectra is consistent with model expectations that the progenitor star lost its hydrogen envelope to become a Wolf-Rayet star before collapsing.

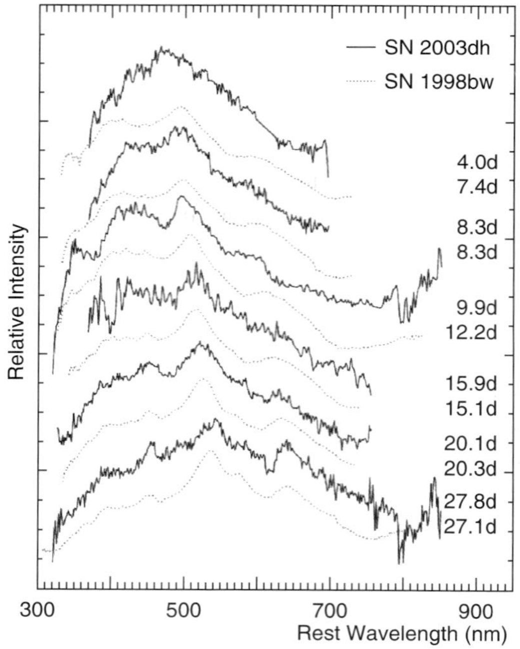

Fig. 24.7 Evolution of the optical spectrum of GRB 030329 showing the obvious signatures of a SN Ic, and compared to the spectra of SN1998bw (*dotted lines*). The age of the supernova is given on the right side (from [40])

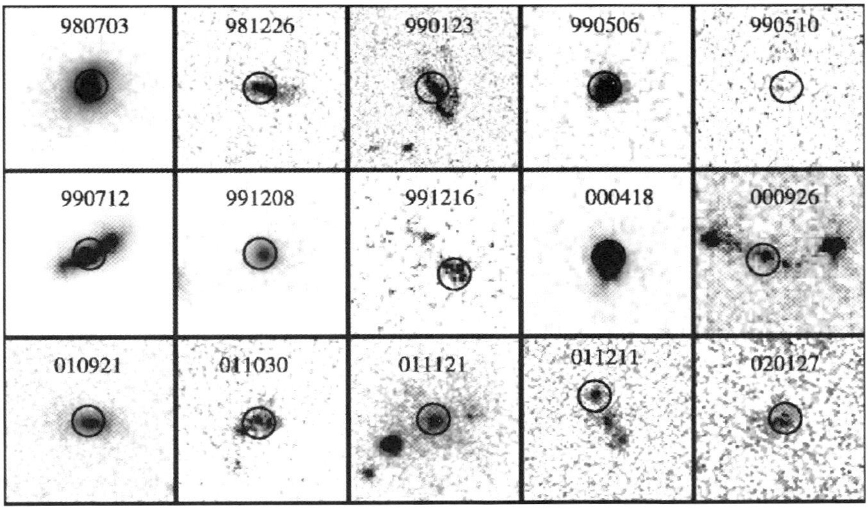

Fig. 24.8 Selection of HST images of GRB host galaxies, adapted from [17]. The *circles* mark the center of the respective galaxy and not the position of the GRB in that galaxy. GRB host galaxies are generally very compact, even for HST's high spatial resolution (image size is $\approx 4'' \times 4''$)

In contrast to these similar spectroscopic properties, the γ-ray emission properties of these events are quite different with respect to total emitted energy, temporal profile, and spectral shape, suggesting that the γ-ray properties are governed by other properties than the progenitor mass.

Rebrightenings in the light curve at $(8\text{–}20)(1+z)$ days after the GRB are highly suggestive, though not unequivocal evidence for an associated supernova also in several GRBs at higher redshift. An estimate of GRB and hypernova rates in the Universe shows that they are statistically equal [69].

24.2.3 Host Galaxies

Studies of GRB host galaxies have shown that most are faint blue galaxies [41, 52], with $R \sim 22$–30 mag, and R-K $\sim$2.5 mag. GRBs occasionally (but not always) occur within UV-bright parts of their hosts [12], consistent with their association with star forming (SF) regions.

The broad-band spectral energy distribution (SED) of practically all host galaxies are best fit with starburst galaxy templates [16, 33, 34, 72]. Also, sub-mm and radio observations have shown that at least four GRB hosts (corresponding to 20% of the observed hosts), are intensively star forming galaxies [3, 9, 23]. A recent survey [90] has led to the conclusion that the optical/infrared properties of the submillimeter-brightest GRB hosts are not typical of the galaxy population selected in sub-mm surveys. Also, there is evidence that Lyα emission is common for GRB host galaxies, while it is not for other starbursts at similar redshift [27].

The study of the SF rate in normal galaxies is a controversial topic. Though several different tracers for SF have been proposed and used, there still remain questions and problems. Light dimming by dust biases SF rate measurements, both locally [8, 89] and at high redshift [55, 88], and is the cause of uncertain cross-calibrations between different indicators. GRBs, on the other hand, provide some important advantages (though measurement of the SF in a GRB host galaxy then suffers from the same biases), namely that (i) they can be seen to high redshifts, (ii) finding the hosts by γ-ray detection is unaffected by gas/dust absorption, (iii) they can be detected in hosts which are themselves too faint to appear in any flux-limited galaxy sample.

Morphologically, GRB hosts exhibit a surprisingly broad diversity of galaxy types. Based on Hubble Space Telescope imaging, two thirds of host galaxies are found to be situated in a region of the concentration-asymmetry diagram occupied by spirals or peculiar/merging galaxies while the other are more akin to elliptical galaxies [17]. More interestingly, GRB hosts at $z > 1$ are different from the general field population at that redshift in terms of light concentration, yet have sizes similar to the general $z > 1$ population. This is the opposite of the effect at $z < 1$, where GRB hosts are smaller than average. This implies that GRB hosts trace the starburst population at high redshift, as similarly concentrated galaxies at $z > 2$ are undergoing a disproportionate amount of SF for their luminosities. Furthermore, GRBs are not only an effective tracer of SF but are perhaps ideal tracers of typical galaxies undergoing SF at any epoch, making them perhaps our best hope of locating the earliest galaxies at $z > 7$ [17].

24.2.4 X-Ray Flashes

X-ray flashes are brief transient sources of X-rays which are distributed isotropically on the sky – in fact, they are very similar to GRBs except for their softer spectra [38]. It has been shown that X-ray flashes can indeed be interpreted as a continuation of the GRB phenomenon towards lower peak energy (typically below 100 keV) [4]. Though optical/radio afterglows have been identified for only a fraction of the X-ray flashes, the afterglow properties are also consistent with those of GRBs. Interestingly, the light curve of one X-ray flash (XRF 030723) also exhibits the excess emission, suggestive of a supernova Ic at $z = 0.6$, and XRF 020903 has spectroscopic evidence for a related supernova [86].

Initially it had been speculated that XRFs are the high-redshift tail of the GRB distribution. However, the first two out of three redshift estimates of XRFs are below one, suggesting that the cause of the softer spectra is rather a different emission process and/or geometry. Different explanations have been proposed to account for the properties of XRFs: (i) they are classical GRBs but seen off-axis [74, 102],

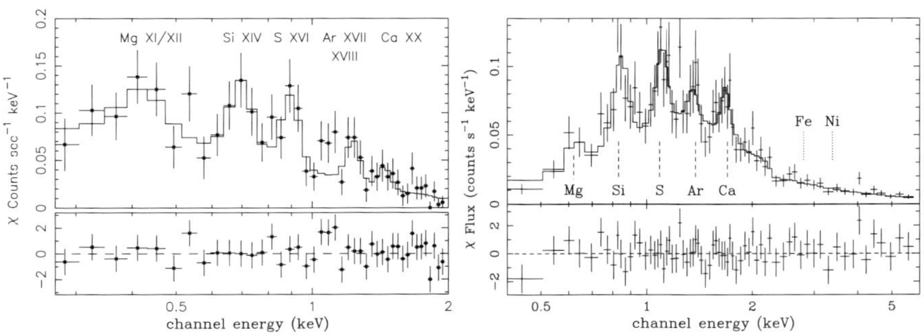

Fig. 24.9 X-ray afterglow spectra of GRB 011211 (*left*) and GRB 030227 (*right*). The detector response has not been unfolded. The Kα lines of Mg, Si, S, Ar, and Ca, redshifted by $z = 1.88$ (*left*; optical $z = 2.14$) and $z = 1.39$ (*right*) are marked. No Fe, Co, or Ni emission is detected in these two GRBs (from [73, 97])

(ii) they are "dirty" fireballs, e.g., explosions with a larger baryon load and consequently lower Γ [20]; (iii) fireballs with large Lorentz-factors and/or low baryon loading that in the case of internal shocks lead to the emission of less energetic photons [103]; (iv) low contrast between the bulk Lorentz factors of the colliding relativistic shells [5].

24.2.5 X-Ray Lines

Spectral features, in most cases time-variable, have been reported to be detected in the 0.5–10 keV X-ray spectra for several GRBs (e.g., [82] for a critical review). While in most cases only one single line has been seen, which usually has been associated to iron, there are few cases (GRB 011211, 030227) for which several emission lines were seen. There are, still, no two GRBs for which the same line(s) have been detected. In all cases, the significance of these lines is marginal, despite the improved sensitivity and reduced delay in observing them with, e.g., XMM- Newton as compared to ASCA and BeppoSAX, and therefore the issue is controversial. One way to explain the widely different lines is to assume that ^{56}Ni is mainly synthesized in the jet due to neutrino heating [63]. In this case, the ^{56}Ni production is very sensitive to the energy deposition rate, i.e., a long energy deposition rate would synthesize little ^{56}Ni, but Si and S instead. If this is true, there should be a correlation between the line features in the X-ray afterglows and the duration of the GRB.

In more general terms, the detection of X-ray lines offers a direct probe of the hot and potentially dense environment immediately surrounding the GRB. However, the interpretation of these lines or line complexes requires extreme conditions on either the parameters of the surrounding medium or its geometry. This and the diversity of emission lines are the reason why until now there is no adequate model to explain all observed features.

Beyond studying the GRB environment, X-ray lines are also of prime importance in determining the nature of their progenitors as well as remnants.

24.2.6 Time-Variable X-ray Halo

Small-angle scattering of X-rays by dust grains results in an X-ray "halo" around a distant X-ray source where the radial intensity distribution depends on the dust properties. Because of the longer path length of scattered X-rays, photons in the halo arrive later than unscattered source photons. If the X-ray source emits a time-variable signal with a unique profile which can be identified in the halo emission, then a distance estimate of the source [91] and/or the location and properties of the dust can be derived [47].

While X-ray halos have been detected from bright galactic X-ray binaries since the Einstein era, a detection of a X-ray halo due to the emission from a GRB has been achieved first for GRB 031203 using the XMM-Newton satellite [95]. GRB 031203 was located at a galactic latitude of only BII $= -4.^{\circ}6$ behind an optical extinction of $E(B-V) = 1.0$, and no optical/IR afterglow was found. The candidate host galaxy at the position of the radio afterglow is at $z = 0.105$, making this GRB the second-closest GRB. The XMM-Newton observation started about 6 h after the GRB and lasted 58 ks. Splitting the observation into contiguous time intervals revealed a X-ray halo with two distinct rings with increasing radius (Fig. 24.10). The expansion of both rings was proportional to $t^{1/2}$, consistent with small-angle scattering. A second XMM-Newton observation at 3 days after the GRB did not detect the halo anymore. The temporal profile of the evolution of the two rings implies the dust slabs to be at distances of 1388 ± 32 and 882 ± 20 pc, respectively [95]. The halo brightness implies an initial soft X-ray pulse consistent with the X-ray intensity during the GRB emission, while the afterglow contributes only about 2% to the halo emission.

24.3 The Basic Scenarios for Gamma-Ray Burst Emission

24.3.1 GRB Emission Scenarios

24.3.1.1 The Fireball

The basic scenario for the understanding of GRBs is the dissipative (shock) fireball model [59, 60]. It assumes a very large energy deposition inside a very small volume (constrained by causality and the variability timescales of GRBs to be of order

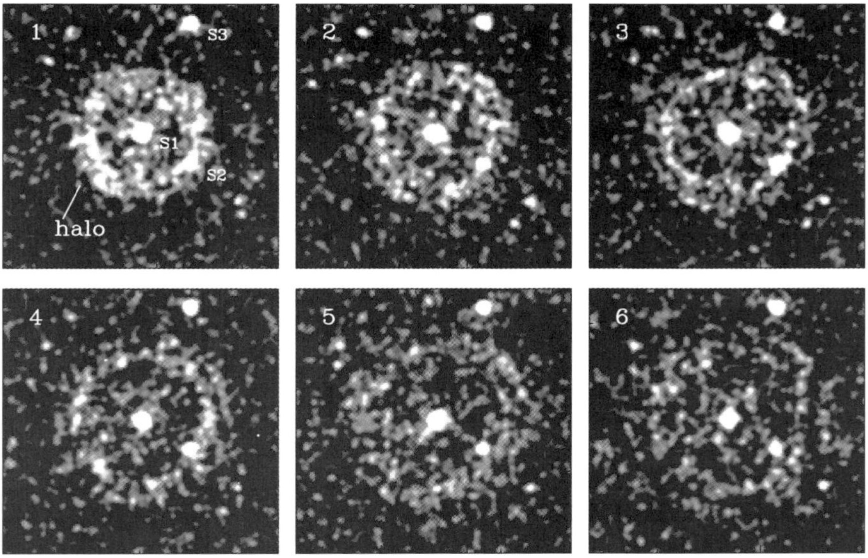

Fig. 24.10 Sequence of 5780 s slices of the X-ray halo of GRB 031203 (0.7–2.5 keV) starting at 6 h after the GRB. The images are 10′ on a side (from [95])

100 km or smaller), which leads to characteristic photon energy densities that produce an optically thick, highly super-Eddington $\gamma e^{\pm}$ fireball. The fireball initially is thermal and converts most of its radiation energy into kinetic energy, i.e., bulk motion of a relativistically expanding blast wave. Lorentz factors $\Gamma \sim 10^{2-3}$ are required by theory to avoid degradation of the GeV photons by photon–photon interactions (note that GRBs involve the fastest bulk motions known so far in the universe). The kinetic energy is tapped by shocks as the most likely dissipation mechanism, and these shocks should probably occur after the fireball became optically thin, as suggested by the observed nonthermal GRB spectra. The internal shocks are generally at least mildly relativistic due to the large Lorentz factor difference between the colliding shells.

The later emission in X-ray to optical and radio wavelengths (so-called afterglow) is dominated by synchrotron emission, i.e., emission from relativistic electrons gyrating in magnetic fields [98, 99]. This synchrotron shock model is widely accepted as the major radiation mechanism in the external shock, and the macroscopic properties of such shocks are well understood. Under the implicit assumptions that the electrons are assumed to be "Fermi" accelerated and that they have a power-law distribution with an index p upon acceleration, their dynamics can be expressed in terms of the following main parameters: (1) the total internal energy in the shocked region as released in the explosion, (2) the density n (and its radial profile) of the surrounding medium, (3) the fraction of energy carried by the electrons ε_e since only a fraction of the total electrons associated with the ISM baryons are accelerated, (4) the fraction of energy density in the magnetic field ε_B. However, there are large uncertainties in their microphysics: How are the relativistic particles accelerated? How is the magnetic field in the shocked region generated? What is its structure? Why is the afterglow polarization only few percent [37, 39], whereas synchrotron emission can be polarized up to 70%?

According to standard synchrotron emission theory, the radiation power of an electron with co-moving energy $\gamma_e mc^2$ is $P_e = 4/3\sigma_T c \gamma_e^2 (B^2/8\pi)$, so that high energy electrons cool more rapidly. For a continuous injection of electrons as is the case for ongoing plowing of the forward shock into the ISM, there is a break in the electron spectrum at $\gamma_e = \gamma_c$ above which the electron energy spectrum is steepened due to cooling. This energy is time-dependent, so that this break moves to lower energies.

Besides this cooling frequency ν_c, there are two other important frequencies, namely the frequency ν_m corresponding to the electrons accelerated in the shock to a power-law distribution with minimum Lorentz factor, and the synchrotron self-absorption frequency ν_a. The final GRB afterglow synchrotron spectrum is thus a four-segment broken power law [61, 84] separated by these frequencies ν_a, ν_m, and ν_c. The order of ν_m and ν_c defines two types of spectra, namely the "slow cooling case" with $\nu_m < \nu_c$, and the "fast cooling case" $\nu_m > \nu_c$ (Fig. 24.11).

24.3.1.2 Alternative Scenarios

Cannonball: In an alternative model to the fireball scenario, the cannonball model [19], long duration GRBs and their afterglows are produced in core-collapse supernovae by the ejection of bipolar jets of ordinary matter, hydrogenic plasma clouds ("cannonballs") with Lorentz factors of the order of 10^3. When the cannonball crosses circumburst shells with large velocity, its surface is heated up to keV temperatures and emits a single pulse, boosted by the cannonball motion towards γ-ray energies. The typically observed multi-peaked light curves in GRBs are then explained by the acceleration of multiple cannonballs. The subsequent cooling of the cannonball will produce the fading afterglow.

Electromagnetic black holes: In the electromagnetic black hole theory [80] the energy is carried away by a plasma of electron–positron pairs with a temperature of 2 MeV and total energy of a few times 10^{53} erg, created by the vacuum polarization process occurring during the gravitational collapse. Such an optically thick electron–positron plasma self-propels itself outward reaching ultrarelativistic velocities, interacts with the remnant of the progenitor star, and by further expansion becomes optically thin. The sub-structures in a long-duration GRB then originate in the collision between the accelerated ($\Gamma \sim 300$) baryonic matter pulse with inhomogeneities in the interstellar medium [81], similar to the external shock producing the afterglow in the fireball scenario.

Poynting flux: Another hypothesis is that essentially all types of ultrarelativistic outflow (AGN jets, pulsar wind nebulae, and GRBs) are electromagnetic, rather than

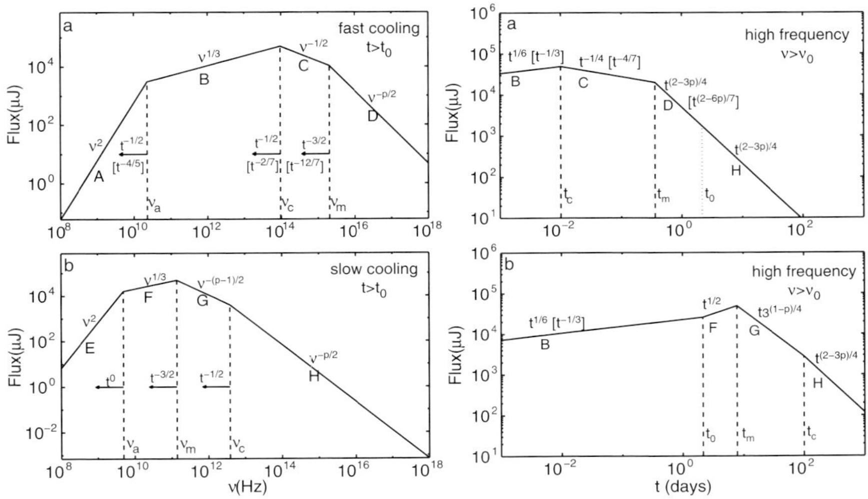

Fig. 24.11 *Left*: Synchrotron spectrum of a relativistic shock with a power-law distribution of electrons. (**a**) The case of fast cooling at early times ($t < t_0$) in a GR afterglow. Self-absorption is important below ν_a. The frequencies, ν_m, ν_c, and ν_a, decrease with time as indicated; the scalings above the arrows correspond to an adiabatic evolution, and the scalings below, in square brackets, to a fully radiative evolution. (**b**) The case of slow cooling at late times ($t > t_0$). The evolution is always adiabatic. *Right*: Light curve due to synchrotron radiation from a spherical relativistic shock, ignoring the effect of self-absorption. (**a**) The high frequency case ($\nu > \nu_0$). The four segments of the light curve are separated by the critical times, t_c, t_m, and t_0. The labels, B, C, D, H, indicate the correspondence with spectral segments in the left panel. The observed flux varies with time as indicated; the scalings within square brackets are for radiative evolution (which is restricted to $t < t_0$) and the other scalings are for adiabatic evolution. (**b**) The low frequency case ($\nu < \nu_0$) (from [84])

gas dynamical, phenomena [11,93]. Electromagnetic flows are naturally anisotropic and self-collimating so as to produce jet-like features. The generic concept is that of a source that ejects a magnetic bubble, which expands with Lorentz factor $\Gamma \sim 30000$. The magnetic field is mostly toroidal. The observed emission then traces out regions of high current density where global instabilities drive a turbulence spectrum that is ultimately responsible for the particle acceleration and the synchrotron, inverse Compton and synchro-Compton emission. The afterglow emission is created after the blast wave becomes free of its electromagnetic driver. In contrast to canonical models, an electromagnetically-driven blast wave creates an anisotropic explosion, remaining relativistic for the longest time closest to the symmetry axis.

24.3.2 *Two GRB Progenitor Models*

24.3.2.1 Collapsars

The generic nature of the fireball model is based on the fact that the details of the primary energy generation mechanism is largely undetermined, i.e., it can be reconciled with different scenarios, such as a binary compact object merger [67], a failed supernova [101], a young highly magnetic pulsar [92], or a hypernova/collapsar [54, 68].

The presently favored scenario for long-duration GRBs is the hypernova scenario, i.e., the core collapse of a single, massive star. A supernova explosion is naturally expected to be associated with this collapse and observationally proven (see Sect. 24.2.2).

A major concern with the hypernova explanation is whether baryons can be efficiently excluded from the flow, so that the expanding flow can achieve the high Lorentz factor ($\Gamma \sim 300$) needed.

24.3.2.2 Mergers

Proposed as the first cosmological scenario, the merging of a NS-NS or NS-black hole binary happens on the timescale of milliseconds, and thus is now the leading conjecture for the subclass of short/hard GRBs [25, 67]. Compact binaries provide a huge reservoir of gravitational binding energy, and due to the centrifugal force also a baryon-free fireball can easily be formed along the rotation axis of the binary system due to the compressional heating and dissipation associated with the accretion. Numerical modeling shows that due to the high densities and temperatures reached in the merger, the dominant emission is through neutrinos [77, 79].

Population analysis of double neutron star binaries suggests that the majority of mergers should happen outside the host galaxy, because they happen so late in the binary evolution that even small systemic kick velocities (induced during the collapse of one of the two constituents to a compact object) would allow most binaries to travel large distances. After the first handful of afterglows detected from short GRBs, future Swift observations are expected to verify the validity of this scenario.

24.4 Use of GRBs for Cosmology

GRB afterglows are bright enough to be used as pathfinders into the very early universe, independent of whether or not the GRB and/or afterglow phenomenon is fully understood.

In contrast to stationary sources at high redshift, GRB afterglows do not appear substantially fainter at increasing z. Relativistic time dilation implies that the observations of GRBs at the same time Δt after the GRB event in the observers frame (*on Earth*) will be observed at different times in the source frame, e.g., at *earlier* times for more distant GRB. At this *earlier* time the GRB is intrinsically brighter, thus partly compensating the larger distance.

While it seems unlikely that GRBs will soon be used to derive a Hubble-diagram and to constrain cosmological parameters, there are a few other implications of high-z GRB studies for cosmology:

- Since long-duration GRBs are related to the death of massive stars, it is likely that high-z GRBs exist. Theoretical predictions range between few up to 50% of all GRBs being at $z > 5$ [15, 50, 85]. The polarization data of the Wilkinson Microwave Anisotropy Probe (WMAP) indicate a high electron scattering optical depth, hinting at the possibility that the first stars formed as early as $z \sim 20$ [48].
- WMAP data also suggest that the onset of reionization happened at $z = 11 - 20$ [48]. Because WMAP only provides an integral constraint on the reionization history of the universe, it has led to the speculation that reionization was either an extended process or happened more than once. Since the intrinsic luminosity as well as the number density of quasars are expected to fade rapidly beyond $z \sim 6$, only GRBs are suitable to be used as bright beacons to illuminate the end of the dark age [2, 53, 62], and potentially allow to probe the reionization history of the early Universe [44].
- Extensive monitoring of afterglows would help to constrain their local environment, and could allow to tell whether GRB afterglows are decelerated by the IGM with an increasingly higher density at higher redshift, or by a stratified constant density medium in a bubble cleared by the progenitor star [35].
- Studying the distribution and absorption line properties of GRB host galaxies would shed light onto the cosmological structure formation and star forming history [56].

24.5 Outlook: First Results of the Swift Mission

The successful launch of NASA's Swift gamma-ray burst mission [28] in November 2004 and the smooth calibration/verification phase and operation since then has brought several exciting news. Swift carries three instruments: (1) the burst alert telescope (BAT) is a coded-mask telescope with a two steradian field of view, imaging the sky in the 15–150 keV band. (2) the X-ray telescope (XRT) is a Wolter-type telescope operating in the 0.2–10 keV band, and (3) the ultraviolet and optical telescope (UVOT). Within seconds of detecting a burst with BAT, Swift relays the burst's location to ground stations, allowing both ground-based and space-based telescopes around the world to observe the burst's afterglow. At the same time, Swift is autonomously slewing its XRT and UVOT instruments to the burst position, allowing to start observing the afterglow emission in the X-ray/UV/optical bands at typically 1 min after the burst onset. In the following, only few highlights are mentioned.

Because of the technical limitations of BeppoSAX to trigger on GRBs with duration shorter than 1 s, expectations were high that Swift would localize short GRBs, investigate its afterglow properties, and ultimately determine the nature of short GRBs. Swift detected the first short GRB on Feb 2, 2005, but the location was too close to the Sun to allow repointing Swift for afterglow observations. However, Swift's second short burst, GRB 050509B was better suited, and afterglow observations started within 1 min. While no optical/UV emission was detected, a fading X-ray afterglow was seen over the first 1 000 s, marking the first afterglow of a short GRB [29]. The X-ray localization near an elliptical galaxy at low redshift with no sign of star formation as well as several other constraints are consistent with the merger scenario [14]. Soon after, HETE-2 also localized a short burst, GRB 050709, and its afterglow properties [21] are surprisingly similar to that of GRB 050509B. If further observations confirm these findings, then short GRBs may indeed be powered by mergers of compact objects.

The temporal behavior of the early X-ray afterglow is certainly one of the major surprises from Swift. Starting at 50–70 s after the GRB onset, a rapidly decaying X-ray afterglow is seen which often can be described with a power law slope steeper than 3. Thereafter, the afterglow emission turns into a very shallow decay, sometimes staying even constant for 100–1 000 s. Finally, the decay slope gets steeper again (to -1.2), consistent with the standard fireball scenario. The early rapid decline, in nearly all cases, smoothly connects to the spectrally extrapolated BAT light curve, and therefore this component is now generally interpreted as the tail of the GRB. For the shallow decay, the interpretation is not yet settled. To keep the X-ray emission on for that long, the total energy in the fireball needs to increase with time, refreshing the fireball for a duration much longer than the burst duration. This could be due to different physical mechanisms, namely (1) the central engine keeps injecting energy, (2) the energy injection is short but involves a wide range of Lorentz factors, or (3) the outflow is poynting-flux dominated, so that the magnetic field energy takes a longer time to be transferred into the medium. Independent of this interpretation, the most surprising finding is that occasionally a bright X-ray flare is observed during that same time period, with a duration of few hundred seconds (Fig. 24.12). While late central engine activity is the presently favored interpretation, a clear physical mechanism still needs to be found. Most confusing in this respect is that such X-ray flares have also been seen in short bursts [7]! Thus, besides being a major problem for the merger scenario, it also seems to be independent of the rather different physical conditions of the central engine or the burst surrounding.

Finally, it is worth mentioning that despite Swift X-ray observations within less than 60 sec after the GRB trigger, no spectral lines have been found yet. It remains to be seen whether or not this is due to the generally faint X-ray afterglow intensity.

There are many more topics for which Swift is expected to significantly contribute to our understanding. With the Swift satellite and its instruments working

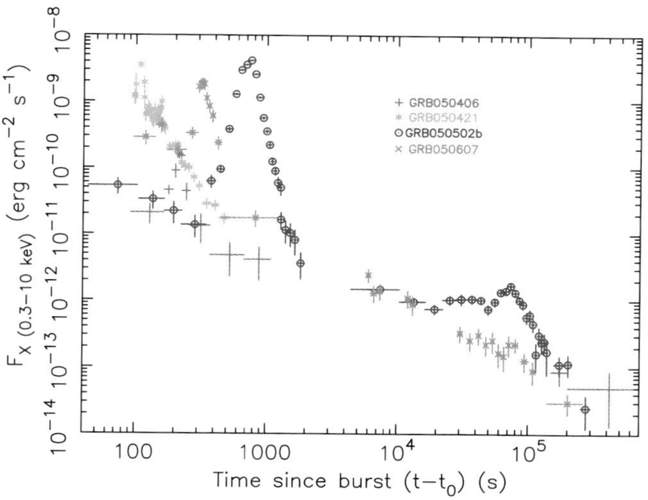

Fig. 24.12 The X-ray flux (0.3–10 keV in the observer frame) as a function of the observed time for selected GRBs showing prominent X-ray flares [66]

perfectly, the near-term future for GRB research, particularly the GRB afterglow study, is very bright.

References

1. Amati, L., Frontera, F., Tafani, M., et al. 2002, A&A, 390, 8
2. Barkana, R., Loeb, A. 2001, Phys. Rep. 349, 125
3. Barnard, V.E., Blain, A.W., Tanvir, N.R., et al. 2003, MNRAS 338, 1
4. Barraud, C., Olive, J.-F., Lestrade, J.P., et al. 2003, A&A 400, 1021
5. Barraud, C., Daigne, F., Mochkovitch, R., Atteia, J.L. 2005 A&A 440, 809
6. Barthelmy, S.D., Butterworth, P.S., Cline, T.L., et al. 1996, Gamma-Ray Bursts, Proc. 3rd Huntsville GRB workshop, Eds. Kouveliotou, C., et al. AIP 384, p. 580
7. Barthelmy, S.D., Chincarini, G., Burrows, D.N., et al. 2005, Nat 438, 994
8. Bell, E.F., Kennicutt, R.C. 2001, ApJ 548, 681
9. Berger, E. et al. 2003, ApJ 588, 99
10. Bersier, D., Cowie, L.L., Kulkarni, S.R., et al. 2003, ApJ 583, L63
11. Blandford, R.D. 2001, Lighthouses of the Universe, ESO Astroph. Symp., eds. Gilfanov, M., et al. 2002, Berlin:Springer, p. 381
12. Bloom, J.S., Kulkarni, S., Djorgovski, S.G. 2002, AJ 123, 1111
13. Bloom, J.S., Frail, D.A., Kulkarni, S.R. 2003, ApJ 594, 674
14. Bloom, J.S., Prochaska, J.X., Pooley, D. et al. 2006, ApJ 638, 354
15. Bromm, V., Loeb A. 2002, ApJ 575, 111
16. Christensen, L., et al. 2004, A&A 413, 121
17. Conselice, C.J., Vreeswijk, P.M., Fruchter, A.S., et al. 2005, ApJ 633, 29
18. Costa, E., Frontera, F., Heise, J. et al. 1997, Nat. 387, 783
19. Dado, S., Dar, A., De Rujula, A. 2002, A&A 388, 1079

25 Cosmic X-Ray Background

G. Hasinger

25.1 The Early History of the X-Ray Background (XRB)

The existence of a diffuse XRB was discovered more than 40 years ago [21]. The aim of the experiment was to measure the X-ray emission from the Moon, but the data showed a strong X-ray source about 30° away from the Moon (Sco X-1) and a diffuse emission approximately constant from all the directions observed during the flight. In 2002, Riccardo Giacconi received the Nobel prize for this discovery, among others.

Improvements in our understanding of the XRB have been achieved with the first all-sky surveys (Uhuru and Ariel V) at the beginning of the seventies. The high degree of isotropy revealed by these surveys immediately led to realize that the origin of the XRB has to be mainly extragalactic. Active galactic nuclei (AGNs) were already considered the most likely candidates for the production of the XRB [76].

In the same years a number of experiments were set up to measure the spectrum of the XRB over a large energy range. At the beginning of the eighties, the HEAO-1 data showed that in the energy range 3–50 keV the shape of the XRB is very well fit by an isothermal bremsstrahlung model corresponding to an optically thin, hot plasma with kT of the order of 40 keV [52]. It was also shown that essentially all Seyfert 1 galaxies with reliable 2–20 keV spectra were well-fit by a single power law with an average spectral index of the order of 0.65, significantly different from the slope of the XRB in the same energy range [61], the so called "spectral paradox," which was originally taken as additional support of the hypothesis, that the XRB may be of diffuse thermal origin.

Observations with the Einstein Observatory were painting a different picture: pointed observations of previously known objects very soon showed that AGNs, as a class, are luminous X-ray emitters [86]. Deep Einstein surveys showed that about 20% of the soft XRB (1–3 keV) is resolved into discrete sources [22, 26] at fluxes of the order of a few $\times 10^{-14}$ erg cm^{-2} s^{-1}. A large fraction of these faint X-ray sources were identified with AGNs. Subtraction of the flux already resolved into discrete objects from the total XRB spectrum would destroy the apparent thermal nature of the X-ray background spectrum [23].

The debate between the supporters of the discrete source vs. thermal plasma hypothesis continued until the final resolution of the controversy came from the very convincing results obtained with the FIRAS instrument on board COBE: the absence

J.E. Trümper and G. Hasinger (eds.), *The Universe in X-Rays.*

to the detector coordinates. The wobble mode, while introducing artificial periodic power into the measured light curves of X-ray sources, turned out to be very valuable for the study of the X-ray background or other diffuse sources. Figure 25.1 shows the PSPC flat-field expected after the wobble motion has been applied to it. The rms variation of the wobbled PSPC flat field is about 3% [31]. The idea of the wobble mode was later picked up by the Chandra Observatory, which uses a Lissajous figure pointing drift as standard observing mode.

25.2.2 The Lockman Hole

The Lockman Hole is actually several degrees across. At first we chose a $\sim$0.3 deg^2 sky region close to the absolute minimum of N_{H}, but away from bright stars. Well in advance of the actual X-ray observations we performed optical UBVRI and radio observations in this field.

Later many other groups have picked up this field as a deep study area, so that the Lockman Hole became one of the best studied extragalactic survey fields. It has been covered by deep VLA [10, 13, 39], SCUBA [75], ISO [16, 42], and Spitzer [37] imaging studies, was targeted in deep NIR surveys from Calar Alto, the Hawaii Institute for Astronomy wide field weak lensing studies with CFHT and Subaru, as well in high-z supernova searches. In X-rays, the Lockman Hole has been covered with deep ASCA [38], BeppoSAX [25], Chandra [45], and XMM-Newton [35] observations.

Deep X-ray survey observations in the Lockman Hole commenced in the ROSAT AO-1 (spring of 1991) with a 100 ks PSPC pointing exposure. The PSPC exposure was aimed from the beginning to reach the ultimate instrumental limits, while an HRI raster scan was planned to improve the PSPC positions. Since the spatial homogeneity of ROSAT observations is very good (see above), their ultimate sensitivity limit is set by confusion. To fight confusion, we had to obtain a very good understanding of the PSPC point-spread function, developed a completely new algorithm for X-ray crowded field analysis, and were running large numbers of simulations in order to understand the systematic subtleties and the limitations of the observations. A substantial fraction of this work was already available before the actual observations. The knowledge about instrumental limitations has led us to the final choice of the exposure time and to a conservative selection of flux limits and off-axis ranges for the complete samples to be analyzed with PSPC data.

Because of the expected confusion in the PSPC, it was also clear from the beginning that HRI data would be necessary to augment the PSPC identification process. Because of the HRI smaller field-of-view, lower quantum efficiency, and higher intrinsic background, we figured that it was necessary to invest about a factor four more HRI time than PSPC time to cover the same field with the same sensitivity, based on prelaunch knowledge. We therefore started a raster scan in AO-1, with $\sim$100 pointings of 2 ks each across the survey region. The remaining 200 ks of HRI raster observations were approved in AO-2. When, however, the first HRI in-flight performance figures became available, we realized that the anticipated sensitivity would not be reached with the HRI raster scan. Because of the increased quantum efficiency, compared to the Einstein HRI, the ROSAT HRI is also more susceptible to background induced by particles in orbit. An increased halo of the HRI point-spread function as well as irreproducible attitude errors of about 5$''$ are responsible for a further loss of sensitivity. Knowing this, we were able to trade the 200 ks HRI time remaining for AO-2 into an extra PSPC observation of 100 ks, which was performed in spring of 1992.

Two years later, after the PSPC had run out of gas, we started to apply for an HRI ultradeep survey aimed for a total observing time of 1 Ms in a single pointing direction. This survey was planned to push the unconfused sensitivity limit deeper than the PSPC exposure in a substantial fraction of the PSPC field. To allow an X-ray "shift and add" procedure, correcting for the erratic ROSAT pointing errors, we selected a pointing direction for the ultradeep HRI exposure which is inside the PSPC field of view, but shifted about 8 arcmin to the North-East of the PSPC center, this way covering a region containing about 10 relatively bright X-ray sources known from the PSPC and the HRI raster scan.

The observations in the *Lockman Hole* represented the deepest X-ray survey ever performed at that time. The total observing time invested (about 1.4 Ms) is quite comparable to that of other major astronomical projects, like, e.g., the Hubble Deep Field, and set the stage for future deep X-ray surveys. Figure 25.2 shows a color composite of the 200 ks ROSAT PSPC, and the 1.2 Ms HRI image of the Lockman Hole. About 70–80% of the X-ray background has been resolved into discrete sources at a flux limit of $\sim 10^{-15}$ erg cm^{-2} s^{-1} in the 0.5–2.0 keV energy band [33].

Fig. 25.2 *Left*: ROSAT PSPC/HRI false color image of the Lockman Hole region. The HRI sources are shown in green. Red and blue colors indicate PSPC sources in the 0.1–0.5 keV and 0.5–2.0 keV energy bands, respectively. The field size is $\sim$30 arcmin. North is up, east to the left (from [44]). *Right*: Color composite image of the $\sim$800 ks XMM-Newton image of the Lockman Hole. The image was obtained combining three energy bands: 0.5–2 keV, 2–4.5 keV, 4.5–10 keV (respectively, red, green, and blue). The image has a size of 43 $\times$ 30 arcmin^2 (from [35])

25.2.3 Optical Identifications of ROSAT Surveys

As one of the first ROSAT surveys, Shanks et al. [78] have carried out a program of optical spectroscopy of sources detected in a 30 ks PSPC verification phase observation in one of the AAT deep optical QSO fields and could quickly identify an impressive fraction of faint X-ray sources as classical broad-line AGNs (mainly QSOs). Using data from medium-deep ROSAT fields combined with the Einstein medium sensitivity survey, Boyle et al. [6] could significantly improve on the derivation of the AGN XLF and its cosmological evolution. Their data was consistent with pure luminosity evolution proportional to $(1+z)^{2.7}$ up to a redshift $z_{\max} \approx 1.5$, similar to what was found previously in the optical range. This result has been confirmed and improved later on by more extensive or deeper studies of the AGN XLF, e.g., the RIXOS project [64] or the UK deep survey project [40].

All these studies agree that at most half of the faint X-ray source counts and, correspondingly, half of the soft X-ray background can be explained by classical broad-line AGN based on the luminosity evolution models. There was, indeed, mounting evidence that a new class of sources might start to contribute to the XRB at faint X-ray fluxes. The faintest X-ray sources in ROSAT deep surveys on average show a harder spectrum than the identified QSOs [2, 31]. In medium-deep pointings a number of optically "innocent" narrow-emission line galaxies (NELGs) at moderate redshifts ($z < 0.4$) were identified as X-ray sources, which was in excess of those expected from spurious identifications with field galaxies [20]. Roche et al. [70] have found a significant correlation of X-ray fluctuations with optically faint galaxies. Finally, in an attempt to push optical identifications to the so far faintest X-ray fluxes, McHardy et al. [54] claimed that broad-line AGN practically cease to exist at fluxes below 5×10^{-15} erg cm^{-2} s^{-1}, while the NELG number counts still keep increasing, so that they would dominate below a flux of 10^{-15} erg cm^{-2} s^{-1}.

While this is obviously an interesting possibility, it is useful to remind that these results were only statistical in nature. All these findings were based on identifications near the limit of deep PSPC surveys, at fluxes where our simulations suggested that the PSPC data start to be severely confused. Because moderate-redshift field galaxies almost all show emission lines [44], there is the possibility to either misidentify a field galaxy as the counterpart of an X-ray source, which in reality is associated to a different optical object (e.g., a fainter AGN) or to misclassify an intrinsically faint AGN hidden in a NELG-type spectrum as a new class of X-ray sources. (Later Chandra observations confirmed, that both problems indeed existed.) On the contrary, the X-ray positions in the ultradeep Lockman Hole survey were largely determined by the HRI raster scan and ultradeep pointing. Instead of confused PSPC error boxes of (realistically) 15–20 arcsec radius, they therefore had error box radii of 2–5 arcsec.

Using optical spectroscopy from the Keck telescopes, the optical identifications in the Lockman Hole could be completed down to a flux limit of 1.2×10^{-15} erg cm^{-2} s^{-1} in 0.5–2.0 keV energy band [44, 73]. Nearly complete spectroscopic identifications (90%) of the sample of 94 X-ray sources were presented based on low-resolution Keck spectra. In this survey, which has one of the highest rates of optical identifications among existing deep X-ray surveys and has a high degree of reliability, no evidence for the emergence of a new non-AGN source population at low X-ray fluxes was found. However, the identified type-II AGN provided the first glimpse of the missing source population to explain the hard X-ray background spectrum. Using a set of ROSAT surveys with a large range of flux limit and solid angle coverage, Miyaji et al. [56] have determined a state of the art X-ray luminosity function and its cosmological evolution.

25.3 AGN Spectra and Fits to the XRB Spectrum

The significant advances in X-ray spectroscopy made possible with Ginga, ASCA, and BeppoSAX had changed substantially our views on the spectral characteristics of AGNs and have shown that the X-ray spectra of AGNs are much more complex than what was thought originally. Detailed observations of an increasing number of relatively bright AGNs have detected a number of spectral features in addition to the power law continuum, such as the Compton reflection "hump" due to reprocessed emission, absorption from cold material probably in the torus, the warm absorbers, the Fe lines, the soft excess at low energies, and optically thin bound–bound transitions from the ICM.

Already Ginga data had shown convincingly [67] that the typical spectrum of Seyfert 1 galaxies is flattening at ~10 keV, due to the reflected component, with respect to the observed power law slope in the range 2–10 keV.

The Ginga data have immediately led a number of groups to construct models for fitting the XRB spectrum with various combinations of AGN spectra and evolution (see, for example, [15,59]). Most of these models require an AGN population whose hard X-ray spectrum is dominated by the reflected component. Actually, the main difficulty of these models is the extremely large contribution required for such a component.

Already in 1989, Setti and Woltjer [77] had concluded that unified theories of AGN predict the existence of numerous AGN with a low energy X-ray cut off in the 5–10 keV range and that the resulting spectrum of the X-ray background produced by all AGN may easily reproduce the observed hard spectrum. Alternative studies [11, 24, 47] have thus explored in detail the possibility that the dominant contribution to the XRB is due to the combination of the emission from a population of unabsorbed and absorbed AGNs, folded over the corresponding luminosity function and cosmic evolution. This class of models appears to be highly successful not only in fitting the spectrum of the XRB but also in reproducing a number of other observational constraints.

As an example, Fig. 25.3 (left panel) shows the results of fits to the XRB spectrum obtained by Comastri et al. [11]. As shown in the figure, the model predicts that in the hard X-ray band most of the contribution to the XRB comes from significantly

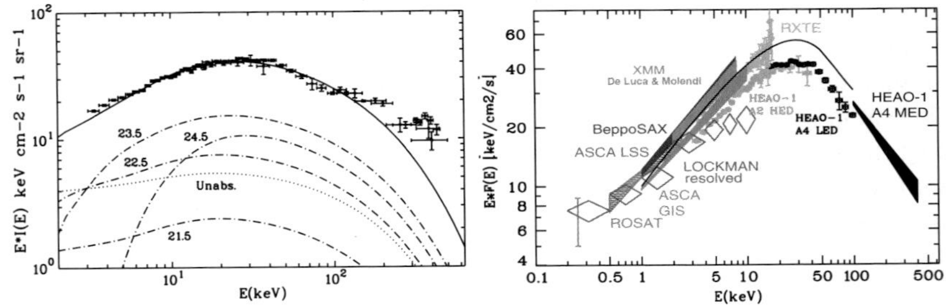

Fig. 25.3 *Left*: The XRB spectrum above 3 keV: comparison between model (continuous line) and data. The curves, labeled with values of log N_{H}, show the expected contribution from AGNs with different amount of intrinsic absorption (from [11]). *Right*: Compilation of the X-ray background spectrum in the 0.1–100 keV range, as measured from a number of different satellites over the past 30 yrs: In order of increasing energy band, the data are from ROSAT [20], BeppoSAX [89], XMM-Newton [12], RXTE [69], HEAO-1 A2 HED + A4 LED [29], and HEAO-1 A4 MED [43]. The solid black line is the analytic model given in [28], scaled to fit the XMM-Newton data in the 5–10 keV range

absorbed objects, which are almost absent in the soft band, even at the faintest ROSAT limit. The modeling of the background spectrum is complicated by the fact that its absolute normalization is still a matter of debate. The HEAO-1 background spectrum [52], used as a reference over many years, has a ~30% lower normalization than several earlier and later background measurements. Recent determinations of the background spectrum with XMM-Newton [12] and RXTE [69] strengthened the consensus for a 30% higher normalization, which has been assumed in the most recent population synthesis model by Ueda et al. [88], but a recent determination of the 5–100 keV background with INTEGRAL [9] is more consistent with the HEAO-1 data.

25.4 Deep Surveys with Chandra and XMM-Newton

Further progress was only possible with deep Chandra and XMM-Newton surveys selected in the 0.5-10 keV band. A comprehensive account of deep surveys with Chandra and XMM-Newton and their source content is given in a recent review by Brandt and Hasinger [7]. The Chandra and XMM-Newton observatories, the current workhorses for X-ray astronomy provide powerful capabilities to pursue the X-ray background studies. Those capabilities are largely complementary between the two observatories. Chandrs's excellent sub-arcsecond point spread function allows very long, highly sensitive exposures in the ≈0.5–8 keV band. The faintest Chandra sources detected have count rates of about one count every 3 days. The Chandra images are far enough away from the background and confusion limits, that even stacking analyses with effective observing times of more than 10 Ms are feasible [7]. XMM-Newton, on the other hand, has a substantially larger telescope collecting area and CCD chips with higher quantum efficiency both in the soft and in the hard band, albeit with a significantly larger point spread function than Chandra. The field of view of XMM-Newton is also more than twice as large as that of the Chandra CCDs. XMM-Newton is thus ideal to cover large area surveys and to obtain high quality spectra of moderately bright X-ray sources.

The Chandra X-ray Observatory has performed deep X-ray surveys in a number of fields with ever increasing exposure times [62, 63, 82] and has completed a 1 Ms exposure in the Chandra Deep Field South (CDF-S, [71]) and a 2 Ms exposure in the Hubble Deep Field North (HDF-N, [1]). A similar set of deep fields has also been observed with XMM-Newton [34, 35, 57, 65, 83]. In Fig. 25.4 (left), the color composite Chandra image of the CDF-S is shown. This was constructed by combining images smoothed with a Gaussian with $\sigma = 1''$ in three bands (0.3–1 keV, 1–3 keV, 3–7 keV), which contain approximately equal numbers of photons from detected sources. Blue sources are those undetected in the soft (0.5–2 keV) band, most likely due to intrinsic absorption from neutral hydrogen with column densities $N_H > 10^{22}\,\mathrm{cm}^{-2}$. Figure 25.1 (right) shows a similar image for the 2 Ms observation of the HDF-N [1]. Very soft sources appear red. A few extended low surface brightness sources are also readily visible in the image.

The deepest XMM-Newton survey is shown in Fig. 25.2 (right panel). The different colors of X-ray sources indicate different spectral hardness and thus predominantly different amounts of intrinsic absorption. The fact that there are so many faint green and blue sources immediately confirmed one of the major predictions of the population synthesis models, i.e., that there is a significant fraction of absorbed sources, required to produce the hard spectrum of the X-ray background.

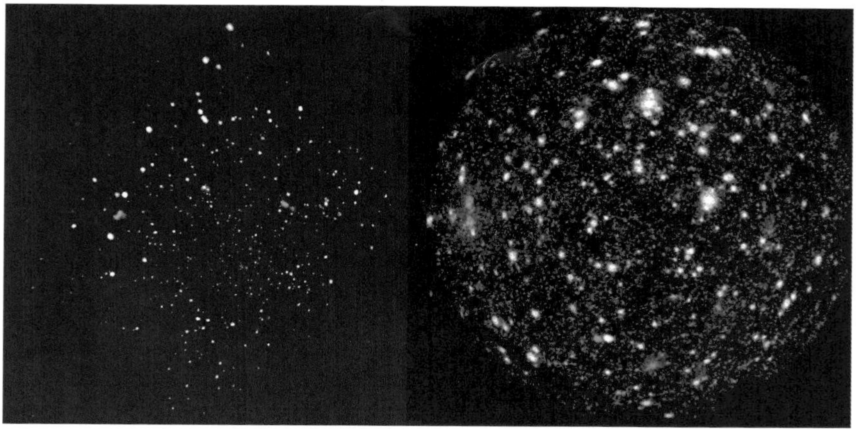

Fig. 25.4 *left*: Color composite image of the Chandra Deep Field North of 2 Ms. The image was obtained combining three energy bands: 0.3–1 keV, 1–3 keV, 3–7 keV (respectively, red, green, and blue), from [1]. *right*: Deep survey image obtained with XMM-Newton (370 ks) of the Chandra Deep Field South using the same energy bands as in Fig. 25.2 right [84]

25.5 A Multi-cone Survey AGN-1 Sample

Extensive optical and near-infrared identification programmes using many tens of nights on the largest telescopes of the world were performed on the deep Chandra and XMM-Newton survey fields. Because of the average faintness of the optical counterparts, typical exposure times had to be in excess of 4–5 h, but still the faintest sources escape spectroscopic diagnostics and have to be identified using photometric redshift techniques. For cosmological studies well-defined flux-limited samples of active galactic nuclei have been chosen in [36], with flux limits and survey solid angles ranging over five and six orders of magnitude, respectively. To be able to utilize the massive amount of optical identification work performed previously on a large number of shallow to deep ROSAT surveys, the analysis was restricted to samples selected in the 0.5–2 keV band. In addition to the ROSAT surveys already used in [56], data from the recently published ROSAT North Ecliptic Pole Survey (NEPS) [60], from an XMM-Newton observation of the Lockman Hole [49], as well as the Chandra Deep Fields South (CDF-S) [50,85,98] and North (CDF-N) [3] were included.

Based on deep surveys with Chandra and XMM-Newton, the X-ray $\log(N) - \log(S)$ relation has now been determined down to fluxes of 2.4×10^{-17}, 2.1×10^{-16}, and 1.2×10^{-15} erg cm^{-2} s^{-1} in the 0.5–2, 2–10, and 5–10 keV band, respectively [5]. Figure 25.5 shows the normalized cumulative source counts $N(> S_{\mathrm{X}14})S_{\mathrm{X}14}^{1.5}$

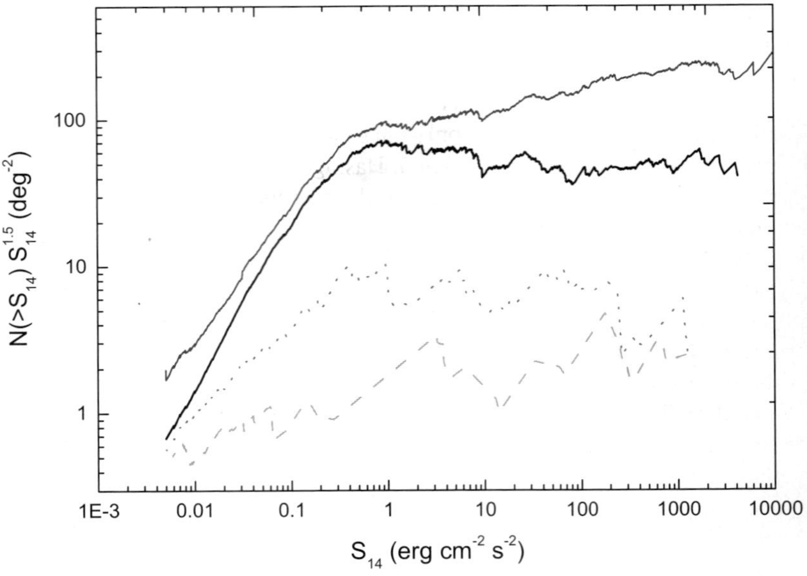

Fig. 25.5 Cumulative number counts $N(> S)$ for the total sample (*upper blue thin line*), the AGN-1 subsample (*lower black thick line*), the AGN-2 subsample (*red dotted line*), and the galaxy subsample (*green dashed line*). From [36]

for the total soft X-ray selected sample, as well as for the subsamples of AGN-1, AGN-2, and normal galaxies. For simplicity we define $S_{X14} = 10^{-14}$ erg cm^{-2} s-1. Euclidean source counts would correspond to horizontal lines in these graphs. A broken power law fitted to the differential source counts yields power law indices of $\alpha_b = 2.34 \pm 0.01$ and $\alpha_f = 1.55 \pm 0.04$ for the bright and faint end, respectively, a break flux of $S_{X14} = 0.65 \pm 0.10$, and a normalisation of $dN/dS_{X14} = 103.5 \pm 5.3$ deg^{-2} at $S_{X14} = 1.0$ with a reduced χ^2=1.51. We see that the total source counts at bright fluxes, as determined by the ROSAT All-Sky Survey data, are significantly flatter than Euclidean, consistent with the discussion in [31]. Moretti et al. [58], on the other hand, have derived a significantly steeper bright flux slope ($\alpha_b \approx 2.8$) from ROSAT HRI pointed observations. This discrepancy can probably be attributed to the selection bias against bright sources, when using pointed observations where the target area has to be excised.

The ROSAT HRI Ultradeep Survey had already resolved 70–80% of the extragalactic 0.5–2 keV XRB into discrete sources, the major uncertainty being in the absolute flux level of the XRB. The deep Chandra and XMM-Newton surveys have now increased the resolved fraction to 85–100% [58,92]. Above 2 keV the situation is complicated by the uncertain normalization of the background measurements. The 30% higher normalization discussed above would require that the resolved fractions above 2 keV have to be scaled down correspondingly. On the other hand, the 2–10 keV band has a large sensitivity gradient across the band. A more detailed investigation, dividing the recent 770 ks XMM-Newton observation of the Lockman Hole and the Chandra Megasecond surveys into finer energy bins, comes to the conclusion that the resolved fraction decreases substantially with energy, from over 90% below 2 keV to less than 50% above 5 keV [92]. The flux resolved into discrete sources is indicated in Fig. 25.3b.

To avoid systematic uncertainties introduced by the varying and a priori unknown AGN absorption column densities, only unabsorbed (type-1) AGN classified by optical and/or X-ray methods were selected. Hasinger et al. [36] are using a definition of type-1 AGN, which is largely based on the presence of broad Balmer emission lines and small Balmer decrement in the optical spectrum of the source (optical type-1 AGN, e.g., the ID classes a, b, and partly c in [73], which largely overlaps the class of X-ray type-1 AGN defined by their X-ray luminosity and unabsorbed X-ray spectrum [85]. However, as Szokoly et al. show, at low X-ray luminosities and intermediate redshifts the optical AGN classification often breaks down because of the dilution of the AGN excess light by the stars in the host galaxy, so that only an X-ray classification scheme can be utilized. Schmidt et al. [73] have already introduced the X-ray luminosity in their classification. For the deep XMM-Newton and Chandra surveys in addition the X-ray hardness ratio was used to discriminate between X-ray type-1 and type-2 AGN.

Type-1 AGN are the most abundant population of soft X-ray sources. For the determination of the AGN-1 number counts we include a small fraction of unidentified sources (6%), which have hardness ratios consistent with AGN-1. Figure 25.5 shows that the break in the total source counts at intermediate fluxes is produced by type-1 AGN, which are the dominant population there. Both at bright fluxes and at the faintest fluxes, type-1 AGN contribute about 30% of the X-ray source population. At bright fluxes, they have to share with clusters, stars, and BL-Lac objects, and at faint fluxes they compete with type-2 AGN and normal galaxies (see Fig. 25.5 and [5]). A broken power law fitted to the differential type-1 AGN source counts yields power law indices of $\alpha_{\mathrm{b}} = 2.55 \pm 0.02$ and $\alpha_{\mathrm{f}} = 1.15 \pm 0.05$ for the bright and faint end, respectively, a break flux of $S_{\mathrm{X14}} = 0.53 \pm 0.05$, consistent with that of the total source counts within errors, and a normalization of of $\mathrm{d}N/\mathrm{d}S_{\mathrm{X14}} = 83.2 \pm 5.5\,\mathrm{deg}^{-2}$ at $S_{\mathrm{X14}} = 1.0$ with a reduced $\chi^2 = 1.26$. The AGN-1 differential source counts normalized to a Euclidean behavior ($\mathrm{d}N/\mathrm{d}S_{\mathrm{X14}} \times S_{\mathrm{X14}}^{2.5}$) is shown with filled symbols in Fig. 25.5.

25.6 The Soft X-Ray Luminosity Function and Space Density Evolution

Using the sample of $\sim$1 000 type-1 AGN, Hasinger, Miyaji, and Schmidt [36] have employed two different methods to derive the AGN-1 X-ray luminosity function and its evolution. The first method uses a variant of the $1/V_{\mathrm{a}}$ method, which was developed in [56]. The binned luminosity function in a given red-shift bin z_i is derived by dividing the observed number $N_{\mathrm{obs}}(L_x, z_i)$ by the volume appropriate to the red-shift range and the survey X-ray flux limits and solid angles. To evaluate the bias in this value caused by a gradient of the luminosity function across the bin, each of the luminosity functions is fitted by an analytical function. This function is then used to predict $N_{\mathrm{mdl}}(L_x, z_i)$. Correcting the luminosity function by the ratio $N_{\mathrm{obs}}/N_{\mathrm{mdl}}$ takes care of the bias to first order. Figure 25.6 shows the luminosity function derived this way in different red-shift shells. A change of shape of the luminosity function with red-shift is clearly seen and can thus rule out simple density or luminosity evolution models. In a second step, instead of binning into red-shift shells, the sample has been cut into different luminosity classes and the evolution of the space density with red-shift was computed.

The second method uses a variant of the $1/V_{\mathrm{max}}$ method. Individual V_{max} of the ROSAT Bright Survey (RBS) sources [74] are used to evaluate the zero-redshift luminosity function. This is free of the bias described above: using this luminosity function to derive the number of expected RBS sources matches the observed numbers precisely. In the subsequent derivation of the evolution, i.e., the space density as a function of redshift, binning in luminosity and red-shift is introduced to allow evaluation of the results. Bias at this stage is avoided by iterating the parameters of an analytical representation of the space density function. Together with the zero-redshift luminsity function this is used to predict $N_{\mathrm{mod}}(L_x, z_i)$ for the surveys. The observed densities in the bins are derived by multiplying the space density value by the ratio $N_{\mathrm{obs}}(L_x, z_i)/N_{\mathrm{mod}}(L_x, z_i)$.

The other difference between the two methods is in the treatment of missing red-shifts for optically faint objects. In the binned method, all AGN without

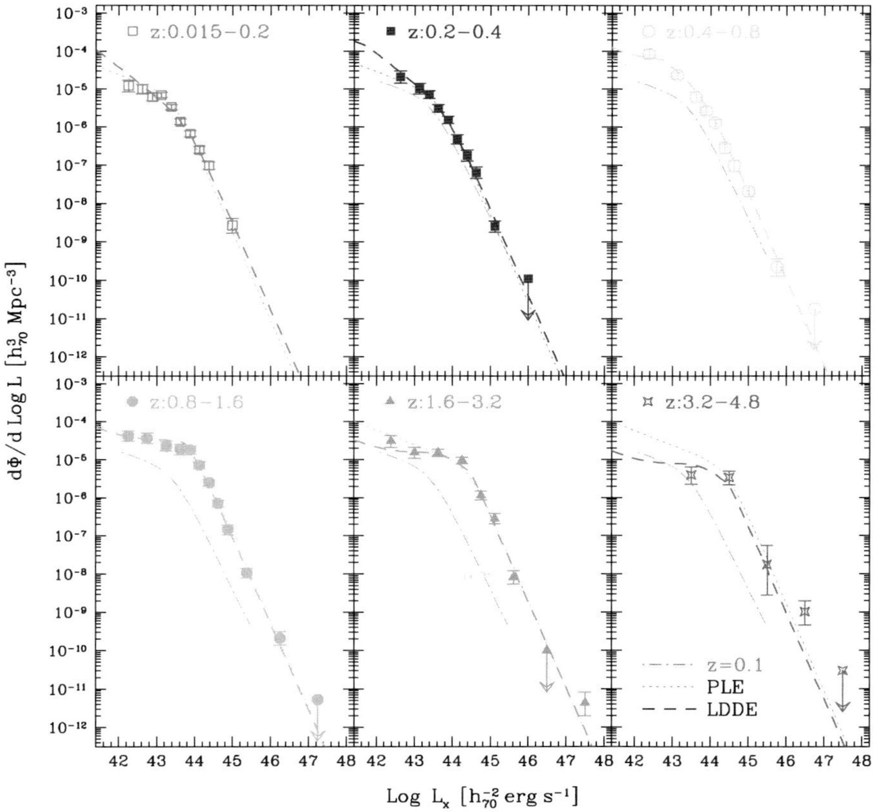

Fig. 25.6 The soft X-ray luminosity function of the type-1 AGN sample in different red-shift shells for the nominal case as labeled. The error bars correspond to 68% Poisson errors of the number of AGNs in the bin. The best-fit two power-law model for the $0.015 < z < 0.2$ shell are overplotted in the higher red-shift panels for reference. The dotted and dashed lines give the best-fit PLE and LDDE models (from [36])

red-shift with $R > 24.0$ were in turn assigned the central red-shift of each red-shift bin to derive an upper boundary to the luminosity function. In the unbinned method, the optical magnitudes of the RBS sources were used to derive the optical red-shift limit corresponding to $R = 24.0$. The $V_{\max}$ values for surveys spectroscopically or photometrically incomplete beyond $R = 24.0$ (such as CDF-N) were based on the smaller of the X-ray and optical red-shift limits.

Figure 25.7 shows a direct comparison between the binned and unbinned determinations of the space density, which agree very well within statistical errors. The fundamental result is that the space density of lower-luminosity AGN- peaks at significantly lower red-shift than that of the higher-luminosity (QSO-type) AGN. Also, the amount of evolution from red-shift zero to the peak is much less for lower-luminosity AGN. The result is consistent with previous determinations based on

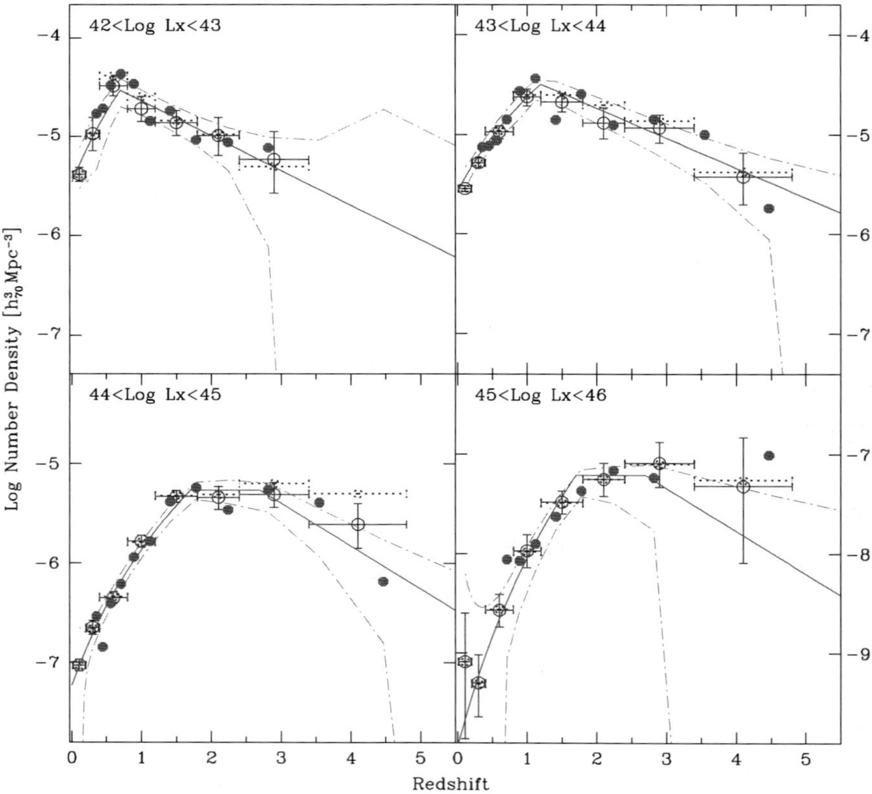

Fig. 25.7 Comparison between the space densities derived with two different methods. The data points with error bars refer to the binned treatment using the N_{obs}/N_{mdl} method, the dashed error bars corresponding to the maximum contribution of unidentified sources. The solid lines, the filled circles and the dashed lines refer to the unbinned method. (from [36])

less sensitive and/or complete data, but for the first time our analysis shows a high-red-shift decline for all luminosities $L_X < 10^{45}\,\mathrm{erg\,s^{-1}}$ (at higher luminosities the statistics is still inconclusive). Albeit the different approaches and the still existing uncertainties, it is very reassuring that the general properties and absolute values of the space density are very similar in the two different derivations in.

A luminosity-dependent density evolution (LDDE) model has been fit to the data. Even though the sample is limited to soft X-ray-selected type-1 AGN, the redshift dependence of the space density is similar to that obtained by Ueda et al. (2003) for the intrinsic (deabsorbed) luminosity function of hard X-ray selected obscured and unobscured AGN, except for the normalization, where Ueda et al. reported a value about five times higher. However, with ~250 AGN selected at considerably brighter fluxes than the 1 000 AGN-1 sample of [36], the statistical quality and parameter range of the Ueda et al. sample is not sufficient to, e.g., constrain the decline of the

AGN space density at high red-shift. Very recently, Barger et al., [4] have presented X-ray luminosity function analyses both in the hard and soft X-ray bands, based on the CDF-N, CDF-S, CLASXS, and ASCA surveys. Again, their results are in good agreement with the soft XLF discussed above and the hard XLF presented by Ueda et al.; however, they still suffer from substantial identification incompleteness. Also, their results on broad-line AGN are not directly comparable to the type-1 AGN sample discussed here, because they only include the optically classified type-1 AGN and thus miss most of the low-luminosity unabsorbed AGN-1. The space density of soft X-ray selected high-luminosity QSOs has been compared to the one of luminous optically- and radio-selected QSOs by [90], who concluded that a substantial high-red-shift decline is observed in all wavebands.

These new results paint a dramatically different evolutionary picture for low-luminosity AGN compared to the high-luminosity QSOs. While the rare, high-luminosity objects can form and feed very efficiently rather early in the Universe, with their space density declining more than two orders of magnitude at red-shifts below $z = 2$, the bulk of the AGN has to wait much longer to grow with a decline of space density by less than a factor of 10 below a red-shift of one. The late evolution of the low-luminosity Seyfert population is very similar to that which is required to fit the Mid-infrared source counts and background [18] and also the bulk of the star formation in the Universe [48], while the rapid evolution of powerful QSOs traces more the merging history of spheroid formation [17].

This kind of antihierarchical Black Hole growth scenario is not predicted in most semi-analytic models based on Cold Dark Matter structure formation models (e.g. [41, 94]). This could indicate two modes of accretion and black hole growth with radically different accretion efficiency. A self-consistent model of the black hole growth which can simultaneously explain the antihierarchical X-ray space density evolution and the local black hole mass function derived from the $M_{\rm BH} - \sigma$ relation assuming two radically different modes of accretion has recently been presented in [55].

25.7 X-ray Constraints on the Growth of SMBH

The AGN luminosity function can be used to determine the masses of remnant black holes in galactic centers, using Sołtan's continuity equation argument [80] and assuming a mass-to-energy conversion efficiency ε. For a nonrotating Schwarzschild BH, ε is expected to be 0.057, while for a maximally rotating Kerr BH, ε can be as high as 0.29. The AGN demography predicted, that most normal galaxies contain supermassive black holes (BH) in their centers, which is now widely accepted. Recent determinations of the accreted mass from the optical QSO luminosity function are around $2\varepsilon_{0.1}^{-1} \cdot 10^5\,\mathrm{M}_\odot\,\mathrm{Mpc}^{-3}$ [8, 96]. Probably the most reliable recent determination comes from an integration of the X-ray luminosity function. Using the Ueda et al. [88] hard X-ray luminosity function including a correction for

Compton-thick AGN normalized to the X-ray background, as well as an updated bolometric correction ignoring the IR dust emission, Marconi et al [51] derived $\rho_{accr} \sim 3.5\varepsilon_{0.1}^{-1} \cdot 10^5\,\mathrm{M}_\odot\,\mathrm{Mpc}^{-3}$.

The BH masses measured in local galaxies are tightly correlated to the galactic velocity dispersion [19] and less tightly to the luminosity of the host galaxy bulge. Using these correlations and galaxy luminosity (or velocity) functions, the total remnant black hole mass density in galactic bulges can be estimated. Scaled to the same assumption for the Hubble constant ($\mathrm{H}_0 = 70\,\mathrm{km\,s^{-1}\,Mpc^{-1}}$), recent papers arrive at different values, mainly depending on assumptions about the intrinsic scatter in the BH-galaxy correlations. The most recent estimate is $\rho_{\mathrm{BH}} = (4.6^{+1.9}_{-1.4})\ h_{70}^2 \cdot 10^5\,\mathrm{M}_\odot\,\mathrm{Mpc}^{-3}$ [51]. The local dark remnant mass function is thus fully consistent with the above accreted mass function, if black holes accrete with an average energy conversion efficiency of $\varepsilon = 0.1$ [51], which is the classically assumed value and lies between the Schwarzschild and the extreme Kerr solution. However, taking also into account the widespread evidence for a significant kinetic AGN luminosity in the form of jets and winds, it is predicted that the average supermassive black hole should be rapidly spinning fast (see also [14, 96]).

25.8 Conclusions

The quest of the X-ray background, detected by Giacconi and colleagues in 1962, is now largely resolved. The interpretation in terms of discrete emission from active galactic nuclei put forward by Setti and Woltjer in 1973 and 1985 is basically correct. Almost all of the background has been resolved below 2 keV, while around 10 keV only 50% and above 10 keV a very small fraction has been resolved. There is thus sufficient discovery space for future high-resolution imaging telescope missions with hard X-ray response. In particular a better determination of the background flux and spectral shape is required above 10 keV. Population synthesis models can in principle explain the shape and intensity of the X-ray background by a mixture of absorbed and unabsorbed AGN and, in particular including a correction for the Compton-thick objects undetectable at X-rays, are roughly consistent with the mass distribution of dormant black holes in the local galaxies. The detailed evolutionary behavior of the AGN population is still a matter of intense study. The AGN evolution seems to go hand in hand with the evolution of galaxies, but one of the big surprises was the finding of an "anti-hierarchical" growth of black holes: while massive black holes (QSOs) are formed very early in the universe, the lower-luminosity, smaller mass black holes are formed considerably later. A theoretical understanding of this phenomenon is still lacking.

In the future, X-ray surveys that are both wide and deep are necessary in order to provide enough volume for a better measurement of the space density function of the rare high-luminosity AGN at large red-shifts. Several new surveys towards this goal are already underway, e.g., the *Chandra Multiwavelength Project (Champ)* [79]), the *Chandra Large Area Synoptic X-Ray Survey (CLASXS)* [95], the *Extended Chandra*

Deep Field South (PI: W.N. Brandt), or the XMM-Newton *COSMOS Field* (PI: G. Hasinger), which together should enrich the sample of $z > 4$ objects by about an order of magnitude. Ultimately, a new X-ray survey mission, like ROSITA [68] and the *Dark Universe Observatory* [27], which aim to survey large solid angles on the sky to considerable depth, could provide a factor of 100 increase in the AGN sample size.

References

1. Alexander, D.M., Bauer, F.E., Brandt, W.N., et al. 2003, Astron. J., 126, 539
2. Almaini, O., Shanks, T., Boyle, B.J., et al. 1996, Mon. Not. Roy. Astron. Soc., 282, 295
3. Barger, A.J., Cowie, L.L., Capak, P., et al. 2003, Astrophys. J., 126, 632
4. Barger, A.J., Cowie, L.L., Mushotzky, R.F., et al. 2005, Astrophys. J., 129, 578
5. Bauer, F.E., Alexander, D.M., Brandt, W.N., et al. 2004, Astrophys. J., 128, 2048
6. Boyle, B.J., Shanks, T., Georgantopoulos, I., et al. 1994, Mon. Not. Roy. Astron. Soc., 260, 49
7. Brandt, W.N., Hasinger, G. 2005, Annu. Rev. Astron. Astrophys., 43, 827
8. Chokshi, A., Turner, E.L. 1992, Mon. Not. Roy. Astron. Soc., 259, 421
9. Churazov, E., Sunyayev, R., Revnivtsev, M., et al., 2007, A&A 467, 529
10. Ciliegi, P., Zamorani, G., Hasinger, G., et al. 2003, Astron. Astrophys., 398, 901
11. Comastri, A., Setti, G., Zamorani, G., Hasinger, G. 1995, Astron. Astrophys., 296, 1
12. De Luca, A., Molendi, S., 2004 Astron. Astrophys., 419, 837
13. de Ruiter, H.R., Zamorani, G., Parma, P., et al. 1997, Astron. Astrophys., 319, 7
14. Elvis, M., Risaliti, G., Zamorani, G. 2002, Astrophys. J., 565, L75
15. Fabian, A.C., George, I.M., Miyoshi, S., Rees, M.J. 1990, Mon. Not. Roy. Astron. Soc., 242, L14
16. Fadda, D., Flores, H., Hasinger, G., et al. 2002, Astron. Astrophys., 383, 838
17. Franceschini, A., Hasinger, G., Miyaji, T., Malquori, D. 1999, Mon. Not. Roy. Astron. Soc., 310, L15 (1999)
18. Franceschini, A., Braito, V., Fadda D. 2002, Mon. Not. Roy. Astron. Soc., 335, L51
19. Gebhardt, K., Bender, R., Bower, G., et al. 2002, Astrophys. J., 539, L13
20. Georgantopoulos, I., Stewart, G.C., Shanks, T., et al. 1996, Mon. Not. Roy. Astron. Soc., 280, 276
21. Giacconi, R., Gursky, H., Paolini, F.R., Rossi, B.B. 1962, Phys. Rev. Lett., 9, 439
22. Giacconi, R., Bechtold, J., Branduardi, G., et al. 1979, Astrophys. J., 234, L1
23. Giacconi, R., Zamorani, G. 1987, Astrophys. J., 313, 20
24. Gilli, R., Salvati, M., Hasinger, G. 2001, Astron. Astrophys., 366, 407
25. Giommi, P., Perri, M., Fiore, F. 2000, Astron. Astrophys., 362, 799
26. Griffiths, R.E., Murray, S.S., Giacconi, R., et al. 1983, Astrophys. J., 269, 375
27. Griffiths, R.E., Petre, R., Hasinger, G., et al. 2004, SPIE 5488, 837
28. Gruber, D.E. 1992, in Barcons, X., Fabian, A.C. (eds.), The X-ray Background, Cambridge University Press, Cambridge, 44
29. Gruber, D.E., Matteson, J.L., Peterson, L.E., Jung, G.V. 1999, Astrophys. J., 520, 124
30. Hamilton, T.T. 1992, in Barcons X., Fabian A.C. (eds.), The X-ray Background, Cambridge University Press, Cambridge, 138
31. Hasinger, G., Burg, R., Giacconi, R., et al. 1993, Astron. Astrophys., 275, 1
32. Hasinger, G., Zamorani, G., in Gursky, H., Ruffini, R., Stella, L. (eds.), Exploring the Universe: A Festschrift in Honor of Riccardo Giacconi, World Scientific, Singapore, p. 119 (astro-ph/9712341) (2000)
33. Hasinger, G., Burg, R., Giacconi, R., et al. 1998, Astron. Astrophys., 329, 482
34. Hasinger, G., Altieri, B., Arnaud, M., et al. 2001, Astron. Astrophys., 365, 45

35. Hasinger, G. 2004, Nucl. Phys. B (Proc. Supp.), 132, 86
36. Hasinger, G., Miyaji, T., Schmidt, M. 2005, Astron. Astrophys., 441, 417
37. Huang, J.-S., Barmby, P., Fazio, G.G., et al. 2004, Astrophys. J. Suppl., 154, 44
38. Ishisaki, Y., Ueda, Y., Yamashita, A., et al. 2001, Pub. Astron. Soc. Jpn., 53, 445
39. Ivison, R.J., Greve, T.R., Smail, I., et al. 2002, Mon. Not. Roy. Astron. Soc., 337, 1
40. Jones, L.R., McHardy, I.M., Merrifield, M.R., et al. 1997, Mon. Not. Roy. Astron. Soc., 285, 547
41. Kauffmann, G., Haehnelt, M. 2000, Mon. Not. Roy. Astron. Soc., 311, 576
42. Kawara, K., Matsuhara, H., Okuda, H., et al. 2004, Astron. Astrophys., 413, 843
43. Kinzer, R.L., Jung, G.V., Gruber, D.E., et al. 1997, ApJ 475, 361
44. Lehmann, I., Hasinger, G., Schmidt, M., et al. 2001, Astron. Astrophys., 371, 833
45. Lehmann, I., Hasinger, G., Murray, S.S., Schmidt, M. 2002, Astron. Soc. Pac., 262, 105
46. Lockman, F.J., Jahoda, K., McCammon, D. 1986, Astrophys. J., 302, 432 (1986)
47. Madau, P., Ghisellini, G., Fabian, A.C. 1994, Mon. Not. Roy. Astron. Soc., 270, 117
48. Madau, P., Pozzetti, L., Dickinson, M. 1998, Astrophys. J., 498, 106
49. Mainieri, V., Bergeron, J., Rosati, P., et al. 2002, Astron. Astrophys., 393, 425
50. Mainieri, V., Rosati, P., Tozzi, P., et al., 2005, Astron. Astrophys., 437, 805
51. Marconi, A., Risaliti, G., Gilli, R., et al. 2004, Mon. Not. Roy. Astron. Soc., 351, 169
52. Marshall, F.E., Boldt, E.A., Holt, S.S., et al. 1980, Astrophys. J., 235, 4
53. Mather, J.C., Cheng, E.S., Cottingham, D.A., et al. 1994, Astrophys. J., 420, 439
54. McHardy, I.M., Jones, L.R., Merrifield, M.R., et al. 1998, Mon. Not. Roy. Astron. Soc. 295, 641
55. Merloni, A. 2004, Mon. Not. Roy. Astron. Soc., 353, 1035
56. Miyaji, T., Hasinger, G., Schmidt, M. 2001, Astron. Astrophys., 369, 49
57. Miyaji, T., Griffiths, R.E., Lumb, D., et al. 2003, Astron. Nachr., 324, 24
58. Moretti, A., Campana, S., Lazzati, D., et al. 2003, Astrophys. J., 588, 696
59. Morisawa, K., Matsuoka, M., Takahara, F., Piro, L. 1990, Astron. Astrophys., 263, 299
60. Mullis, C.R., Henry, J.P., Gioia, I.M., et al. 2004, Astrophys. J., 617, 192
61. Mushotzky, R.F. 1984, Adv. Space Res., 3, 157
62. Mushotzky, R.F., Cowie, L.L., Barger, A.J., Arnaud, K.A. 2000, Nature, 404, 459
63. Nandra, K., Laird, E.S., Adelberger, K., et al. 2005, Mon. Not. Roy. Astron. Soc., 356, 568
64. Page, M.J., Carrera, F.J., Hasinger, G., et al. 1996, Mon. Not. Roy. Astron. Soc., 281, 579
65. Page, M.J., McHardy, I.M., Gunn, K.F., et al. 2003, Astron. Nachr., 324, 101
66. Pfeffermann, E., Briel, U.G. 1986, SPIE 733, 519
67. Pounds, K.A., Nandra, K., Stewart, G.C., et al. 1990, Nature, 344, 132
68. Predehl, P., Friedrich, P., Hasinger, G., et al. 2003, Astron. Nachr., 324, 128
69. Revnivtsev, M., Gilfanov, M., Sunyaev, R., et al. 2003, Astron. Astrophys., 411, 329
70. Roche, N., Shanks, T., Almaini, O., et al. 1995, Mon. Not. Roy. Astron. Soc., 273, L15
71. Rosati, P., Tozzi, P., Giacconi, R., et al. 2002, Astrophys. J., 566, 667
72. Schmidt, M., Green, R.P. 1983, Astrophys. J., 269, 352
73. Schmidt, M., Hasinger, G., Gunn J.E., et al. 1998, Astron. Astrophys., 329, 495
74. Schwope, A., Hasinger, G., Lehmann, I., et al. 2000, Astron. Nachr., 321, 1
75. Scott, S.E., Fox, M.J., Dunlop, J.S., et al. 2002, Mon. Not. Roy. Astron. Soc., 331, 817
76. Setti, G., Woltjer, L. 1973, in Bradt, H., Giacconi, R. (eds.), X- and Gamma-Ray Astronomy, IAU Symposium No. 55, Reidel, Dordrecht, p. 187
77. Setti, G., Woltjer, L. 1989, Astron. Astrophys., 224, L21
78. Shanks, T., Georgantopoulos, I., Stewart, G.C., et al. 1991, Nature, 353, 315
79. Silverman, J.D., Green, P.J., Barkhouse, W.A., et al. 2005, Astrophys. J., 624, 630
80. Sołtan, A.M. 1982, Mon. Not. Roy. Astron. Soc., 200, 115
81. Sołtan, A.M. 1991, Mon. Not. Roy. Astron. Soc., 250, 241
82. Stern, D., Moran, E.C., Coil, A.L., et al. 2002, Astrophys. J., 568, 71
83. Streblyanska, A., Bergeron, J., Brunner, H., et al. 2004, Nucl. Phys. B (Proc. Supp.), 132, 232
84. Streblyanskaya, A., Hasinger, G., Finoguenov, A., et al., 2007, A&A (submitted)
85. Szokoly, G., Bergeron, J., Hasinger, G., et al. 2004, Astrophys. J. Suppl., 155, 271

86. Tananbaum, H., Avni, Y., Branduardi, G., et al. 1979, Astrophys. J., 234, L9
87. Trümper, J. 1984, Physica Scripta T7, 2
88. Ueda, Y., Akiyama, M., Ohta, K., Miyaji, T. 2003, Astrophys. J., 598, 886
89. Vecchi, A., Molendi, S., Guainazzi, M., et al. 1999, Astron. Astrophys., 349, L73
90. Wall, J.V., Jackson, C.A., Shaver, P.A., Hook, I.M. & Kellermann, K.I. 2005, Astron. Astrophys., 434, 133
91. Warwick, R.S., Roberts, T.P 1998, Astron. Nachr., 319, 59
92. Worsley, M., Fabian, A.C., Bauer, F.E., et al. 2005, Mon. Not. Roy. Astron. Soc., 357, 1281
93. Wright, E.L., Mather, J.C., Fixsen, D.J., et al. 1994, Astrophys. J., 420, 450
94. Wyithe, J.S.B., Loeb, A. 2003, Astrophys. J., 595, 614
95. Yang, Y., Mushotzky, R.F., Barger, A.J., et al. 2003, Astrophys. J., 585, L85
96. Yu, Q., Tremaine, S. 2002, Mon. Not. Roy. Astron. Soc., 335, 965
97. Zamorani, G., Mignoli, M., Hasinger, G., et al. 1999, Astron. Astrophys., 346, 731
98. Zheng, W., Mikles, V.J., Mainieri, V., et al. 2004, Astrophys. J. Suppl., 155, 73

26 The Future

G. Hasinger

26.1 Introduction

We are currently in a"golden age" of astrophysics. Several fundamental discoveries have led to paradigm shifts in recent years. Among the most important are the establishment of a high precision cosmological "Standard Model," with dark energy, dark matter, and hot baryons as the dominating forms of the matter/energy density, and the confirmation of an inflationary phase in the early Universe, the discovery of extrasolar planets, the understanding that Gamma ray bursts mark the birth of black holes in distant galaxies, and the realization that supermassive black holes are an essential element of galaxy formation and evolution. Astrophysics has become an integral component of modern physics and leads the way to concepts beyond the standard models of elementary particle physics and relativity. X-ray astronomy is a crucial element of modern astrophysics and is particularly well suited to study matter under extreme conditions, black holes, and warm/hot baryons in the dark matter potential wells of the cosmic web.

The current observational capabilities for X-ray astronomy are excellent. The two X-ray observatory work horses, NASA's Chandra and ESA's XMM-Newton, are still performing extremely well after more than 7 years in orbit and we just can support every effort and hope that they will continue operation for the next several years. NASA's RXTE and ESA's INTEGRAL satellites have complementary capabilities in the hard X-ray energy range. The NASA satellite Swift has been launched in November 2004 with important European instrument contributions and is now routinely chasing gamma ray bursts and other violent phenomena "on the fly." The Japanese satellite Suzaku with important US instrument contributions been launched in July 2005 and offers an unprecedented sensitivity over the widest X-ray band pass. With four working satellites equipped with imaging X-ray telescopes and several nonimaging detectors in orbit, research possibilities in the fields covered by this book are very bright in the next few years. The goal of high-resolution nondispersive X-ray spectroscopy is still a very high priority for the future.

Both NASA and ESA are discussing next generation X-ray observatories Constellation-X and XEUS, respectively, which either separately or joined to a next generation global X-ray observatory, should surpass the sensitivity of the instruments on the current major observatories as well as the intermediate dedicated missions planned for the next decade by more than an order of magnitude.

J.E. Trümper and G. Hasinger (eds.), *The Universe in X-Rays.*

This sensitivity is well matched to future observatories in other wavebands, e.g., the extremely large groundbased telescope(s) (ELT) in the optical, the James Webb space telescope (JWST) in the near-infrared, the Atacama large millimeter array (ALMA) in the submillimeter or the square kilometer array (SKA) in the radio band. The time interval between large space missions in any wave band and in any space agency are very long (15–25 yrs) and significant technology development is required from one step to the next. In order to maintain the scientific competitiveness and the technological expertise of scientific groups worldwide, as well as the excitement and career possibilities of young scientists, it is of utmost importance to develop intermediate, smaller projects like those planned in Japan, France, Italy, Germany, UK, Russia, USA, Brasil, India, and China. Already approved, under development, or close to launch are the two gamma-ray satellites AGILE (ASI) and GLAST (NASA), the Indian mission astrosat, the Brasilian mission MIRAX, the Japanese all-sky monitor MAXI and their next X-ray telescope mission (NeXT), the Chinese hard X-ray mission HXMT, the French/Italian/German formation flying mission SIMBOL-X, and the revived Russian Spektrum–Roentgen–Gamma mission SRG, with significant instrument contributions from Germany (eROSITA) and other partners. These missions, addressing exciting scientific topics like dark energy, dark matter, and the cosmic Web, the complete census of obscured black holes and time variable phenomena, will hopefully be launched before 2012.

26.2 Space Agency Strategic Planning

26.2.1 NASA "Beyond Einstein" Roadmap

The NASA Office for Space Science has currently two strategic planning roadmaps, "Origins" and "Beyond Einstein." For the subjects discussed in this book, the latter one, developed by NASA's "Exploration of the Universe" division, is most important. The "Beyond Einstein" Roadmap published in 2003 [13] identifies three main science objectives: (1) Find out what powered the Big Bang; (2) Observe how black holes manipulate space, time, and matter; and (3) Identify the mysterious dark energy pulling the universe apart. The major elements are two new large observatories, the gravitational wave Laser interferometer space antenna (LISA), and the X-ray mission Constellation-X [14], as well as moderate sized, PI-led missions, so called Einstein probes, which focus on the three scientific themes. For the scope of this article, the large X-ray observatory (Con-X) and the black hole finder probe (EXIST) [5] are most relevant. The current NASA budget unfortunately provides a very pessimistic outlook for these missions. Funding for LISA and Con-X, together with several other missions from the Origins roadmap, has been stalled or significantly reduced for the mid-term future. Candidate missions for NASA's "Beyond Einstein" program are being reviewed by the "Beyond Einstein Programme Assessment Committee" (BEPAC) of the National Academy of Sciences with the aim to recommend to NASA which of the five beyond Einstein missions should be the first mission to be implemented, hopefully with a new start in the 2010 time frame. This development will also have important consequences for the joint NASA/ESA missions in planning.

26.2.2 ESA Cosmic Vision 2015–2025

ESA is performing a long-term strategic planning exercise typically every 10–15 yrs. This was the case for the "Horizon 2000" (1984) and "Horizon 2000 Plus" (1994–1995) programs, which laid out the long-term plan according to which the major ESA missions have been developed and scheduled for launch up to the 2014 time frame. The planning exercise for the following decade is called "Cosmic Vision 2015–2025" and has been published in 2005 [15]. The planning exercises cover all fields of space science. The high priority questions addressed in cosmic vision (CV) 2015–2025 are (1) What are the conditions for planet formation and the emergence of life? (2) How does the Solar System work? (3) What are the fundamental physical laws of the Universe? (4) How did the Universe originate and what is it made of? A large aperture X-ray observatory is identified as one of the high priority tools to be developed, because it will address two of these four fundamental questions. A call for new missions within CV 2015–2025 has been issued in spring 2007. The X-ray evolving universe spectroscopy mission (XEUS) [6], for which substantial development for mirror and detector technologies has been going on for the last 10 yrs, is proposed as an answer to this quest. Planning and priority setting on the ESA side is supported by white papers in the ESA member status and by an Europe-wide planning exercise "Astronet."

26.3 Spektrum–Roentgen–Gamma

Among the smaller projects mentioned in the introduction, the Spektrum–Roentgen–Gamma (SRG) project is of special relevance. It addresses the nature of the mysterious dark energy, which is one of the most exciting questions facing astronomy and physics today. It may be the vacuum energy providing the cosmological constant in Einstein's theory of General Relativity, or it may be a time-varying energy field. The solution could require a fundamental revolution in physics. The discovery of Dark Energy has come from three complementary techniques: observations of distant supernovae, the microwave background, and clusters of galaxies.

In 2005, a new concept was defined for a revived Spektrum–Roentgen–Gamma (SRG) mission [10]. A medium size satellite will be launched by the Russian space agency Roskosmos in 2011 into a 600 km low-Earth orbit, which provides a significantly lower particle background than those of Chandra and XMM-Newton, and will allow detailed studies of low-surface-brightness diffuse objects. The payload includes eROSITA (extended Roentgen survey with an imaging telescope array) provided by a German consortium led by MPE [12], the wide field X-ray monitor Lobster provided by a consortium led by Leicester University, UK, and a hard X-ray telescope (ART) provided by IKI, Russia. The eROSITA X-ray telescope consists of seven mirror modules (Wolter-I optics) each having a CCD-detector in the focus. By using frame store pn-CCDs with a field of view of about 1 square degree, newly developed by our dedicated semiconductor laboratory [10] and based on the technology of the successful pn-CCD detector aboard XMM-Newton, eROSITA achieves a grasp (solid angle area product) which is 3–4 times larger than that of XMM-Newton. Recently the addition of a calorimeter instrument has been discussed.

The mission will conduct the first all-sky survey with an imaging telescope in the medium energy band up to 12 keV, with an unprecedented spectral and angular resolution and a factor of 30 and 100 more sensitive than the previous ROSAT and HEAO-1 surveys in their respective energy bands. This survey will detect several Million new X-ray sources, most importantly about one hundred thousand clusters of galaxies which can be used to investigate the nature of dark matter and dark energy, and several hundred thousand obscured supermassive black holes. With Lobster the first all-sky imaging X-ray time variability survey will be performed. In addition to the all-sky surveys it is foreseen to do pointed observations of selected sources. The new SRG mission would thus be a highly significant scientific and technological step beyond Chandra and XMM-Newton and would provide important and timely inputs for the next generation of giant X-ray observatories like XEUS or Con-X.

26.4 The Next Generation Large X-ray Observatory

26.4.1 Evolution of Large Scale Structure and Nucleosynthesis

About 96% of the energy density of the Universe exists in the form of dark matter and dark energy, which govern the structure and evolution of the Universe on the largest possible scales. Clusters of galaxies are the largest collapsed objects in the Universe. Their formation and evolution is dominated by gravity, i.e., dark matter, while their large scale distribution and number density depends on the geometry of the Universe, i.e., dark energy. They are filled with hot baryonic gas, which is enriched with elements by star formation and stellar explosions, and is preferably detected by its high energy radiation. X-ray observations of clusters provide information on the dark matter and dark energy content of the Universe, on the amplitude of primordial density, on the complex physics governing the formation and evolution of structures in the Universe and on the history of metal synthesis. Whilst nearby clusters of galaxies have been studied at great detail with existing X-ray satellites, nothing is known about their formation and evolution in the early Universe. In addition, the fate of almost 50% of the baryons in the Universe, believed to reside in warm/hot filamentary structures observable with X-ray absorption spectroscopy, is still a mystery. To study the genesis of groups and clusters of galaxies and the cosmic web at up to $z \sim 2$, and the evolution of the physical state and chemical abundances of the intergalactic medium, an X-ray telescope combining a very large collecting power with excellent energy resolution and good spatial resolution is necessary.

26.4.2 Coeval Evolution of Galaxies and their Supermassive Black Holes

The first stars and galaxies formed when gravity overpowered the pressure of the ambient baryons. Ultimately, gravity dominated and the first stellar mass black holes were formed, very likely in gamma-ray burst explosions. Supermassive black holes can grow in cataclysmic feeding events. The highest redshift accreting black holes known are around $z = 6.5$. The WMAP studies of the microwave background show that the first light must have ionized the universe already as early as $z =$ 10–20. The fact that practically all galaxy bulges in the local Universe contain supermassive black holes, with a tight relation between black hole mass and the stellar velocity dispersion, indicates a coexistence and coevolution of stars and central black holes early in the universe. Supermassive black holes must thus be an important constituent of the evolving universe. Only recently has the importance of feedback of stellar explosions and accreting black holes into the intergalactic and interstellar medium and thus their role for star and galaxy formation been realized. The study of the birth and growth of supermassive black holes at $z \sim 10$ requires an unprecedented combination of large spectral throughput, high angular resolving power, and large field of view in the X-ray regime, matching those of future optical and radio telescopes.

The XEUS requirement is to detect and to study X-ray emitting black holes out to $z = 10$ and to investigate their nature. These objects could be either continuously accreting black holes in their quasar growth phase, or, as discovered recently, mature black holes which are tidally capturing, disrupting and consuming individual stars. In either case, their X-ray luminosity should be $>10^{42.5}$ erg s^{-1} in order to discriminate them from star forming emitters.

Assuming that 10% of the bolometric luminosity is emitted in X-rays and that these objects accrete at the Eddington limit, this corresponds to a black hole mass of $3 \times 10^6\,M_\odot$. At $z = 8$, such a black hole would have an X-ray flux of $4 \cdot 10^{-18}$ erg cm^{-2} s^{-1} (assuming standard WMAP cosmology: $H_0 = 70$ km s^{-1} Mpc^{-1}, $\lambda_o = 0.73$, and a flat Universe). To study the overall properties of such objects the spectral shape, the amount of absorption, and the properties of any Fe lines need to be measured.

26.4.3 Matter Under Extreme Conditions

Black holes play a key role in the evolution of galaxies and ultimately in the star formation history, but they also distort the shape of space in their vicinity. X-rays resulting from accretion of matter onto a compact object (a supermassive or stellar mass black hole, or neutron star) probe the motion of matter in strongly curved space time, in which general relativity is no longer a small correction to the classical laws of motion. This is a regime in which fundamental predictions of general relativity are still to be tested, such as the existence of an event horizon for black holes, or the dragging of inertial frames. These predictions cannot be confirmed using weak field measurements from Earth orbiting gravity probes. High throughput timing, hard X-ray spectroscopic and polarimetric measurements, e.g., the fast time variability of a relativistic Fe X-ray line or the energy and time dependence of the polarization angle of the accretion disk emission probe the strong field region and constrain the physical parameters of the compact object (mass and spin). Deviations from general relativity in the strong field limit can be studied in a way complementary to measurements by gravitational wave detectors such as LISA.

In addition, black holes and neutron stars provide a unique laboratory for studying matter under strong gravity, while neutron stars also allow the study of matter in the presence of extreme magnetic fields and at supra-nuclear densities. The structure of a neutron star is set by the nuclear equation of state, whose determination is one of the priorities of physics today. The composition of a neutron star can vary from neutrons and protons to hyperons – particles that contain strange quarks – and possibly even free quarks. X-rays emerging from the strongly curved space time of neutron stars encode information on the mass and radius of the compact object, hence the equation of state, EOS, of matter at supra-nuclear density. A large area, high spectral, and high time resolution X-ray telescope is required to constrain physics in the strong field and high density limit.

26.4.4 The Current XEUS Concept

In order to meet the scientific requirements discussed above, an X-ray telescope is required providing more than ten times the collecting area of XMM-Newton and an angular resolution of better than 5 arcsec, with a goal of 2 arcsec. This will be achieved with a X-ray telescope of 35 m focal length, consisting of a mirror satellite and a detector satellite in formation flight.

For the large mirrors a novel design is under development, which is based on light weight silicon pore optics [2]. The detector spacecraft will carry a Wide-Field Imager, utilizing modern silicon pixel sensor technology, which will provide a limiting sensitivity in the 0.1–10 keV band around hundred times deeper than XMM-Newton. Photons above 10 keV will be picked up by a hard X-ray detector mounted behind the silicon device in a hybrid configuration. A cryogenically cooled narrow-field-imager, most likely based on a Transition-Edge-type microcalorimeter will provide an energy resolution of 2–3 eV. In addition to these two work-horse instruments smaller, more specialized instruments are planned, in particular a high-time-resolution spectrometer (HTRS) based on silicon drift detectors providing very high time resolution and count rate capability to study the brightest X-ray sources on the sky on the dynamical time scale of their compact objects, and an X-ray polarimeter (XPOL) utilizing the huge effective area of XEUS for unprecedented physical emission diagnostics.

26.5 Conclusions

Although still rather uncertain in its individual facets and albeit hampered by severe funding shortages in the major international space agencies, the future potential of high-energy astrophysics appears rather bright. Two gamma-ray missions will be launched rather soon. Several smaller and medium-size X-ray missions are planned, prepared, or already approved in international space agencies. They provide tremendous discovery space in certain, specialized areas and keep the international high-energy community and the technological know-how alive. The realization of the next generation of large X-ray observatories is probably still more than a decade away, but requires substantial technology development and international support to become reality.

References

1. Bavdaz, M. et al. 2006, SPIE, 6266, p62661S
2. Grindlay, J.E. 2005, NewAR, 49, 436
3. Hasinger, G., et al. 2006, SPIE, 6266, 62661N
4. Meidinger, N., et al. 2006, SPIE, 6276, 627618
5. Pavlinsky, M., et al. 2006, SPIE, 6266, 62660O
6. Predehl, P., et al. 2006, SPIE, 6266, 62660P
7. http://universe.nasa.gov/be/
8. http://constellation.gsfc.nasa.gov/
9. http://esa.int/esapub/br247/br247.pdf

Appendix: More Information About X-Ray Missions

During the last 40 years almost three dozen X-ray satellite missions have been flown. Twenty-four of them have been mentioned in the various chapters of this book (see index). The most comprehensive collection of mission and instrument characteristics can be found on the web pages of NASA's High Energy Astrophysics Archive (`http://heasarc.gsfc.nasa.gov/docs/`). This database not only comprises all NASA led satellite projects, but also collaborative projects of NASA led by other institutions like ESA, ISAS (Japan), MPE (Germany), SERC (UK), NIVR (Netherlands) etc. More information on ESA satellites (EXOSAT, XMM-Newton, INTEGRAL, XEUS) are found on the web pages of ESA's space science department (`http://www.rssd.esa.int`). Japanese satellites (Hakucho, Tenma, Ginga, ASCA, Suzaku, Yohkoh) are presented on the web pages of ISAS (`www.isas. ac.jp`). The Netherlands (ANS): `www.nivr.nl`, United Kingdom (UK-6): `www. serc.ac.uk`, Germany (ROSAT, MIR HEXE): `www.mpe. mpg.de`.

Table 1 Past and current space missions mentioned in this book: names and acronyms

Missions		Period	Lead country/ institution
Ariel V	The fifth Ariel Satellite	1974–1980	United Kingdom
ANS	Astronomische Nederlandse Satelliet	1974–1976	The Netherlands
ASCA	The Advanced Satellite for Cosmology and Astrophysics	1993–2001	Japan
BeppoSAX	Satellite per Astronomia X (SAX)	1996–2002	Italy
CGRO	Compton Gamma Ray Observatory	1991–2000	NASA
Chandra	Short for Chandrasekhar (AXAF, CXO)	1999–	NASA
COBE	Cosmic Background Explorer	1989–1993	NASA
Copernicus	The Copernicus Satellite (OAO-3)	1972–1991	NASA
Einstein	Einstein Observatory (HEAO-2)	1978–1981	NASA
EUVE	Extreme Ultraviolett Explorer	1992–2001	NASA
EXOSAT	European X-ray Observatory Satellite	1983–1986	ESA
Ginga	Japanese for Galaxy	1987–1991	Japan
Granat		1988–1998	USSR/Russia
HEAO-1	High Energy Astrophysics Observatory 1	1977–1979	NASA
HETE-2	High Energy Transient Explorer	2000–	NASA
INTEGRAL	International Gamma-Ray Astrophysics Laboratory	2002–	ESA
ROSAT	Roentgen Satellite	1990–1999	Germany
RXTE	Rossi X-Ray Timing Explorer	1995–	NASA
SAS-3	The Third Small Astronomy Satellite	1975–1979	NASA
Suzaku	Japanese for Zhũ Què (Astro-E2)	2005–	Japan
Swift		2004–	NASA
Tenma	Japanese for Pegasus	1983–1985	Japan
Uhuru	Swahili for freedom	1970–1973	NASA
XMM-Newton	(=XMM X-ray Multi-Mirror Mission)	1999–	ESA
Yohkoh	Japanese for sunbeam	1991–2001	Japan

Acknowledgements

The authors of this book are indebted to many colleagues in a wide scientific community for their excellent cooperation over many years. L. Strüder and N. Meidinger (Chap. 7) thank their friends and colleagues at the Halbleiterlabor, PNSensor, MPE, University Tübingen, Max-Planck-Institut für Physik, Politechnico di Milano, Brokhaven National Laboratory for many stimulating discussions. K. Werner (Chap. 11) thanks Martin Barstow, Stefan Jordan, Thomas Rauch, Sonja Schuh and Burkhard Wolff for useful discussions and comments. Y. Tanaka (Chap. 16) thanks Bernd Aschenbach, K. Makashima, S. Mineshige, and H. Spruit for valuable comments and discussions. R. Petre is indebted to U. Hwang and S. Immler for insightful comments on the manuscript. Finally, G. Hasinger and J.E. Trümper acknowledge the work of many people of MPE's High Energy Astrophysics groups and of our technical staff who contributed to all the high energy astrophysics projects we have been involved in - from the early balloon and rocket experiments to the past and current satellite missions EXOSAT, Mir-HEXE, ROSAT, BeppoSAX, Compton Gamma Ray Observatory, Chandra, XMM-Newton, INTEGRAL, Swift and the future projects GLAST, eROSITA and XEUS.

Index

ASTRONOMY AND ASTROPHYSICS LIBRARY

A&A LIBRARY

Series Editors: G. Börner · A. Burkert · W. B. Burton · M. A. Dopita
A. Eckart · T. Encrenaz · E. K. Grebel · B. Leibundgut
J. Lequeux · A. Maeder · V. Trimble

The Stars By E. L. Schatzman and F. Praderie

Modern Astrometry 2nd Edition
By J. Kovalevsky

The Physics and Dynamics of Planetary Nebulae By G. A. Gurzadyan

Galaxies and Cosmology By F. Combes, P. Boissé, A. Mazure and A. Blanchard

Observational Astrophysics 2nd Edition
By P. Léna, F. Lebrun and F. Mignard

Physics of Planetary Rings Celestial Mechanics of Continuous Media
By A. M. Fridman and N. N. Gorkavyi

Tools of Radio Astronomy 4th Edition, Corr. 2nd printing
By K. Rohlfs and T. L. Wilson

Tools of Radio Astronomy Problems and Solutions 1st Edition, Corr. 2nd printing
By T. L. Wilson and S. Hüttemeister

Astrophysical Formulae 3rd Edition (2 volumes)
Volume I: Radiation, Gas Processes and High Energy Astrophysics
Volume II: Space, Time, Matter and Cosmology
By K. R. Lang

Galaxy Formation By M. S. Longair

Astrophysical Concepts 4th Edition
By M. Harwit

Astrometry of Fundamental Catalogues The Evolution from Optical to Radio Reference Frames
By H. G. Walter and O. J. Sovers

Compact Stars. Nuclear Physics, Particle Physics and General Relativity 2nd Edition
By N. K. Glendenning

The Sun from Space By K. R. Lang

Stellar Physics (2 volumes)
Volume 1: Fundamental Concepts and Stellar Equilibrium
By G. S. Bisnovatyi-Kogan

Stellar Physics (2 volumes)
Volume 2: Stellar Evolution and Stability
By G. S. Bisnovatyi-Kogan

Theory of Orbits (2 volumes)
Volume 1: Integrable Systems and Non-perturbative Methods
Volume 2: Perturbative and Geometrical Methods
By D. Boccaletti and G. Pucacco

Black Hole Gravitohydromagnetics
By B. Punsly

Stellar Structure and Evolution
By R. Kippenhahn and A. Weigert

Gravitational Lenses By P. Schneider, J. Ehlers and E. E. Falco

Reflecting Telescope Optics (2 volumes)
Volume I: Basic Design Theory and its Historical Development. 2nd Edition
Volume II: Manufacture, Testing, Alignment, Modern Techniques
By R. N. Wilson

Interplanetary Dust
By E. Grün, B. Å. S. Gustafson, S. Dermott and H. Fechtig (Eds.)

The Universe in Gamma Rays
By V. Schönfelder

Astrophysics. A New Approach 2nd Edition
By W. Kundt

Cosmic Ray Astrophysics
By R. Schlickeiser

Astrophysics of the Diffuse Universe
By M. A. Dopita and R. S. Sutherland

The Sun An Introduction. 2nd Edition
By M. Stix

Order and Chaos in Dynamical Astronomy
By G. J. Contopoulos

Astronomical Image and Data Analysis
2nd Edition By J.-L. Starck and F. Murtagh

The Early Universe Facts and Fiction
4th Edition By G. Börner

ASTRONOMY AND ASTROPHYSICS LIBRARY

The Design and Construction of Large Optical Telescopes By P. Y. Bely

The Solar System 4th Edition
By T. Encrenaz, J.-P. Bibring, M. Blanc, M. A. Barucci, F. Roques, Ph. Zarka

General Relativity, Astrophysics, and Cosmology By A. K. Raychaudhuri, S. Banerji, and A. Banerjee

Stellar Interiors Physical Principles, Structure, and Evolution 2nd Edition
By C. J. Hansen, S. D. Kawaler, and V. Trimble

Asymptotic Giant Branch Stars
By H. J. Habing and H. Olofsson

The Interstellar Medium
By J. Lequeux

Methods of Celestial Mechanics (2 volumes)
Volume I: Physical, Mathematical, and Numerical Principles
Volume II: Application to Planetary System, Geodynamics and Satellite Geodesy
By G. Beutler

Solar-Type Activity in Main-Sequence Stars
By R. E. Gershberg

Relativistic Astrophysics and Cosmology
A Primer By P. Hoyng

Magneto-Fluid Dynamics
Fundamentals and Case Studies
By P. Lorrain

Compact Objects in Astrophysics
White Dwarfs, Neutron Stars and Black Holes
By M. Camenzind

Special and General Relativity
With Applications to White Dwarfs, Neutron Stars and Black Holes
By N. K. Glendenning

The Universe in X-Rays
By J. E. Trümper and G. Hasinger (eds.)

Galaxy Formation
By M. S. Longair

Printing: Krips bv, Meppel, The Netherlands
Binding: Stürtz, Würzburg, Germany